Tunnel Boring Fundamentals

Richard Skiba

AFTER MIDNIGHT
PUBLISHING

Skiba, Richard (Author)

Tunnel Boring Fundamentals

ISBN 978-1-7644453-1-3 (Paperback) 978-1-7644453-2-0 (eBook)

Non-fiction

Contents

Introduction

Underground construction has become one of the defining engineering disciplines of the modern era. As cities expand, infrastructure demands intensify, and surface space becomes increasingly constrained, the need to build safely and efficiently below ground has never been greater. Tunnel boring, and in particular the use of Tunnel Boring Machines (TBMs), has emerged as a cornerstone technology enabling the delivery of transport systems, water infrastructure, energy networks, and utilities with minimal disruption to the surface environment.

This book, *Tunnel Boring Fundamentals*, provides a comprehensive and structured exploration of the principles, technologies, and practices that underpin modern mechanised tunnelling. As outlined in the foundational material of the book, , tunnel boring is not simply an excavation process, it is a complex interaction between ground behaviour, machine systems, structural response, and operational control. Understanding this interaction is essential for designing, constructing, and managing tunnels that are safe, durable, and economically viable.

The purpose of this book is to bridge the gap between theory and practice. It introduces the fundamental concepts of tunnelling engineering, such as ground conditions, stress redistribution, tunnel geometry, and support systems, before progressing into the detailed mechanics of TBM design, operation, and performance. The text examines both the engineering science and the practical realities of tunnelling, recognising that successful outcomes depend not only on technical knowledge but also on informed decision-making in uncertain and variable ground conditions.

A key theme throughout the book is the evolution of tunnelling from a labour-intensive, empirical craft to a highly mechanised, data-driven engineering discipline. Modern TBMs operate as integrated systems, combining excavation, support installation, monitoring, and control within a single continuous process. Advances in automation, digital monitoring, predictive maintenance, and digital twins are transforming how tunnels are planned, constructed, and managed, enabling greater precision, safety, and efficiency.

The book also emphasises the importance of adaptability. Despite advances in technology and modelling, the ground remains inherently uncertain. No two tunnelling projects are identical, and even the most detailed investigations cannot fully predict subsurface conditions. As a result, successful tunnelling requires a balance between planning and real-time response—an ability to interpret data, adjust parameters, and manage risk as conditions evolve.

Designed for students, engineers, and industry professionals, this text provides both a foundational understanding and a practical reference for tunnel boring operations. It integrates concepts across geotechnical engineering, structural design, mechanical systems, and construction management, reflecting the inherently multidisciplinary nature of tunnelling.

Ultimately, *Tunnel Boring Fundamentals* presents tunnelling as a dynamic and evolving field that sits at the intersection of engineering science, technological innovation, and real-world problem solving. As global demand for underground infrastructure continues to grow, the knowledge and principles presented in this book will play a critical role in shaping the next generation of tunnelling projects.

Chapter 1

Introduction to Tunnel Boring

Underground construction has become an essential component of modern infrastructure development, enabling the delivery of transport, water, energy, and utility systems in environments where surface construction is increasingly constrained. As urbanisation intensifies, populations grow, and climate-related pressures increase, the demand for reliable, resilient, and low-impact infrastructure has expanded beneath cities, waterways, and sensitive landscapes. Tunnel boring has emerged as a critical engineering response to these challenges, allowing long-distance underground passages to be constructed safely, accurately, and with minimal surface disruption.

This chapter introduces the fundamental concepts of tunnel boring and situates the technology within its broader historical, technical, and infrastructural context. It begins by outlining the purpose and applications of tunnel boring, highlighting why mechanised tunnelling—particularly through Tunnel Boring Machines (TBMs)—has become the preferred method for many contemporary projects. The chapter then traces the historical evolution of tunnelling practices, demonstrating how centuries of empirical knowledge, industrial innovation, and engineering science have shaped modern underground construction methods.

The chapter further examines the key differences between tunnel boring and conventional drill-and-blast excavation, providing a foundation for understanding method selection, risk management, and project optimisation. It also introduces the cyclical excavation method as a traditional yet still widely applied approach in variable ground conditions. Finally, the chapter explores modern infrastructure drivers, the advantages and limitations of tunnel boring technology, and the complete lifecycle of a tunnel boring project—from planning and design through construction, commissioning, and handover.

Together, these sections establish the conceptual and practical framework required to understand tunnel boring as both an engineering discipline and a critical enabler of modern infrastructure. This foundation prepares the reader for subsequent chapters, which examine tunnel geometry, ground behaviour, machine systems, operations, calculations, and construction methods in greater technical detail.

Purpose and Applications of Tunnel Boring

Tunnel boring is the process of excavating underground passages in a controlled, continuous, and engineered manner to create safe and durable tunnels while minimising impacts on the surface environment. Its primary purpose is to allow infrastructure to pass through soil and rock in situations where surface construction would be impractical, unsafe, environmentally disruptive, or technically impossible. By operating below ground, tunnel boring enables essential infrastructure to be delivered without extensive disturbance to existing land use.

At its core, tunnel boring is intended to provide reliable underground access for transport systems, utilities, water supply, energy infrastructure, and other essential services. It is designed to maintain ground stability throughout excavation, manage groundwater and in-situ soil or rock pressures, and minimise surface settlement, particularly in sensitive urban areas. Tunnel boring also allows for the construction of long, precise, and repeatable tunnel alignments, resulting in predictable performance and long-term structural reliability.

Modern tunnel boring using Tunnel Boring Machines (TBMs) enables excavation and ground support to occur simultaneously. This integrated approach significantly improves safety, productivity, and construction quality when compared with traditional excavation methods, as the tunnel is continuously supported as it advances.

Figure 1: Tunnel boring machine in Sofia - construction of the subway, Line 2, Vitosha Boulevard, Sofia, Bulgaria. Brejnev, CC0, via Wikimedia Commons.

Tunnel boring, notably through Tunnel Boring Machines (TBMs), is increasingly selected over surface excavation methods in urban and sensitive ecological environments. This preference is attributed to numerous factors, including minimizing disruption, managing geological challenges, and meeting stringent construction requirements.

One of the primary advantages of tunnel boring is its capacity to mitigate the impacts on surface activities. Surface excavation can cause severe disruptions to urban infrastructure, traffic, and natural ecosystems, a concern well-documented in tunnelling research. For example, TBMs allow for the construction of tunnels beneath cities without necessitating the demolition of buildings or prolonged road closures, significantly reducing urban disruption compared to cut-and-cover methods [1]. Additionally, the construction of tunnels in proximity to existing transport corridors minimizes the need for disruptive surface alterations [2].

Another significant consideration for the use of TBMs is their efficiency in unfavourable geological conditions. In environments characterized by deep alignments, weak ground, or high groundwater pressures, conventional open-cut methods may pose safety and economic challenges. Advancements in TBM technology enable successful tunnelling through mixed face conditions, where two or more geological formations with differing mechanical properties may coexist [3, 4]. Such technology reduces the risks associated with tunnel stability and unforeseen geological anomalies, thus promoting a safer excavation environment [5].

Tunnel boring is particularly advantageous when projects necessitate consistent depth and alignment over lengthy distances. TBMs are engineered to operate efficiently, ensuring tight construction tolerances essential for long civil engineering projects [6]. The reliance on mechanized tunnelling as opposed to manual excavation enhances precision, critical for the quality of infrastructure like railways and roads [7]. Additionally, TBMs substantially reduce construction timelines due to their high operational efficiency, which is critical in densely populated areas where time constraints are a constant consideration in project planning [1, 8].

When considering environmental impacts, tunnel boring is often the method of choice due to its reduced surface disruption and lower levels of noise and vibration during construction. Studies indicate that TBM operations can lead to less environmental degradation compared to traditional methods, such as cut-and-cover tunnelling, which generates substantial surface noise and vibration that may affect residential and commercial areas nearby [9, 10]. Traffic-conscious tunnelling practices, alongside fellow stakeholders, underline the importance of minimizing disturbances in densely populated urban environments [8, 11].

Figure 2: The southbound Central Subway bore viewed north from Union Square/Market Street station in November 2022. Evan0512, CC BY-SA 4.0, via Wikimedia Commons.

Major Applications of Tunnel Boring

In transport infrastructure, tunnel boring is widely used for metro and subway systems, rail tunnels for both freight and passenger services, and road tunnels or underground bypasses. These tunnels help reduce surface congestion, improve safety, and allow transport routes to pass beneath cities, mountains, and waterways. TBMs are particularly effective for long, circular tunnels where consistent geometry and high-quality linings are essential.

Figure 3: Nanhu Road Tunnel (West Entrance), the entrance is located at Xiaoxiangnan Road (Northbound lane), Changsha, China. Ddayzzz, CC BY 4.0, via Wikimedia Commons.

Tunnel boring also plays a vital role in water supply and sewerage infrastructure. It is commonly used for drinking water conveyance tunnels, sewer interceptors, wastewater tunnels, stormwater drainage systems, and flood mitigation or storage tunnels. These applications typically require large tunnel diameters, precise gradients, and long service lives, making TBMs well suited to installing smooth, durable linings with low maintenance demands.

In the energy and hydropower sector, tunnel boring is used to construct headrace and tailrace tunnels, pressure tunnels, penstocks, and underground power stations, as well as corridors for oil, gas, and hydrogen pipelines. Hydropower tunnels often carry water under high pressure over long distances, requiring excellent alignment control and structural integrity—key strengths of TBM excavation.

Utilities and service infrastructure also rely heavily on tunnel boring and microtunnelling techniques. These methods are used to install electrical cable tunnels, telecommunications corridors, district heating and cooling systems, and multi-utility service tunnels. By placing services underground, these tunnels enable future maintenance and upgrades without repeated surface excavation, supporting resilient and expandable infrastructure networks.

In mining and resource infrastructure, although drill-and-blast methods remain dominant, tunnel boring is applied in specific situations such as access and haulage tunnels, conveyor and service tunnels, and declines or ventilation tunnels in competent rock. TBMs produce smoother tunnel walls, predictable profiles, and reduced ground disturbance, which can improve long-term stability and ventilation efficiency.

Tunnel boring is also increasingly used in environmental and climate-resilience projects. Applications include flood diversion tunnels, underground stormwater storage systems, wastewater overflow control tunnels, and large-scale water transfer schemes in drought-prone regions. These projects help cities and regions adapt to climate extremes while preserving valuable surface land and urban functionality.

Across all applications, tunnel boring offers significant strategic advantages. Mechanised and shielded excavation enhances worker safety, while consistent excavation methods deliver high precision and uniform tunnel geometry. Tunnel boring reduces the environmental footprint of construction, lowers long-term maintenance costs through high-quality linings, and provides predictable construction performance, making it a cornerstone technology for modern underground infrastructure development.

History and Evolution of Tunnelling Methods

Whenever a technical subject such as tunnelling is examined in depth, it is both instructive and revealing to begin with a historical perspective. The evolution of tunnelling reflects broader advances in engineering knowledge, materials, tools, and scientific understanding. For clarity, the history of tunnelling can be divided into three principal eras: the Ancient Era, extending from biblical times to approximately 1880; the Industrial Era, spanning roughly from 1880 to 1970; and the Modern Era, from about 1970 to the present day. Across these eras, tunnelling methods progressed from manual excavation guided by experience and intuition to highly mechanised, instrumented, and data-driven construction processes [12]. This evolution was driven primarily by the need to manage water, provide defence, and support transportation and urban infrastructure [12].

The Ancient Era, spanning from approximately 4500 BC to the late eighteenth century, was characterised by manual labour, rudimentary tools, and practical engineering insight developed through observation and experience. Early tunnels were excavated using stone hammers, chisels, and later simple metal tools, with progress dependent entirely on human effort. In strong rock, ancient builders sometimes employed fire-quenching techniques, heating the rock face with fire and then rapidly cooling it with water to induce cracking and spalling. One of the most remarkable achievements of this period was the development of qanats by ancient Persian civilisations around 2000 BC. These underground water conveyance systems consisted of gently sloping tunnels connected to the surface by vertical shafts and enabled irrigation and water supply over long distances, a technique that remains in use in some regions today.

Surveying and alignment techniques during the Ancient Era were far more sophisticated than often assumed. A notable example is the tunnel of Eupalinos in sixth-century BC Greece, where engineers used string lines, plumb bobs, and geometric principles to excavate simultaneously from both ends of a mountain, meeting in the middle with astonishing accuracy [12]. Ground support during this era relied primarily on timber for temporary support, followed by permanent linings constructed from brick or stone masonry where required. Early tunnels were often constructed in competent rock, allowing builders to avoid extensive support systems, but as tunnelling applications expanded to drainage, water supply, sewers, and transportation, tunnel alignments increasingly passed through difficult ground conditions that demanded more complex support solutions.

A comprehensive account of ancient tunnelling practices was provided by Henry S. Drinker in his 1878 treatise *Tunneling, Explosive Compounds, and Rock Drills*. Drinker documented nearly 2,000 tunnel case histories from across Europe and the United States, with additional examples from more than fifteen other countries [12]. His work describes tunnels of extraordinary scale, including openings comparable to two-lane roadways extending for thousands of feet, constructed with immense labour forces. In one instance, Drinker records a project employing approximately 30,000 workers over eleven years. These achievements were accomplished without analytical design methods, relying entirely on experience, judgment, and empirical observation [12].

As tunnelling expanded into softer ground and water-bearing strata, timbering became the primary method of support whenever unstable conditions were encountered. Timber was used both at the face and along the tunnel length, often followed by a permanent lining of brick, stone, or cement installed inside the timber framework. This approach required large excavation profiles to accommodate both the temporary and final linings, resulting in significant material consumption and labour demands. Entire forests were often cleared to supply the necessary timber. Drinker emphasised that effective timbering aimed to prevent ground movement rather than resist it after deformation occurred, and he highlighted the importance of early drainage in soft ground to control instability and swelling.

Figure 4: Portland & Southwestern Railroad Tunnel, Willamette Meridian, Chapman, Columbia County, OR. Library of Congress, CC0, via Picryl.

Despite the absence of formal analytical methods, ancient tunnel builders demonstrated remarkable skill in managing ground behaviour. Decisions regarding initial support, lining thickness, and structural form—including the use of inverted arches in heavy ground—were based entirely on observational judgment. The lack of predictive engineering tools makes the accomplishments of this era particularly striking. Near the end of the Ancient Era, a transformative concept emerged with the invention of the tunnelling shield. Patented by Marc Isambard Brunel in 1828, the shield was designed to support soft ground during excavation beneath the Thames River. Brunel's device was mechanically complex and labour-intensive, requiring numerous screw jacks to be reset repeatedly. Although the Thames Tunnel was eventually completed, the process was slow and prohibitively expensive, and shield tunnelling was largely abandoned for several decades [12].

Significant progress in shield technology occurred in the late nineteenth century with innovations introduced by engineers such as Barlow, Greathead, and Beach. These developments included cylindrical shields, cast-iron segmental linings, external grouting, and the use of screw jacks or hydraulic rams to propel the shield forward as a single unit. These innovations laid the foundation for modern shield tunnelling and marked the transition into the Industrial Era. During this period, compressed air was also introduced to counteract groundwater pressure in soft ground, although it posed serious health risks to workers [12].

The Industrial Era, extending from approximately 1880 to 1970, was marked less by revolutionary ideas than by steady, incremental refinement of existing methods. Rock tunnelling benefited from improved drilling equipment, more effective explosives, enhanced ventilation, and better surveying techniques. Soft-ground tunnelling continued to rely on shields, timbering, and cast-iron linings, with gradual improvements in mechanisation and construction efficiency. Steel ribs and liner plates began to replace timber in the early twentieth century, offering greater strength, faster installation, and reduced tunnel cross-sections. Cast-in-place concrete became the dominant final lining material, while pressurised air tunnelling gradually gave way to dewatering techniques as groundwater control methods improved [12].

The Modern Era of tunnelling emerged in the mid-twentieth century and accelerated rapidly from the 1960s onward. A critical underlying development was the formalisation of geotechnical engineering and engineering geology, driven largely by the work of Karl Terzaghi. Terzaghi's emphasis on soil and rock behaviour, pore water pressure, and the importance of field observation transformed tunnel design philosophy. His 1946 publication on rock defects and tunnel support loads was seminal in linking ground conditions to support requirements, establishing principles that remain influential today [12].

Technological advances during the Modern Era fundamentally reshaped tunnelling practice. Rock bolts enabled engineers to reinforce the ground itself, creating stable arches and distributing loads more uniformly. Shotcrete evolved into a versatile support system capable of being applied close to the face and adapted as excavation progressed. Tunnel boring machines extended shield concepts into hard rock, producing uniform circular openings with minimal disturbance and enabling efficient installation of initial and permanent support systems. Pressurised face shields expanded the range of soil conditions that could be safely excavated, while ground improvement techniques such as dewatering, grouting, and freezing allowed engineers to modify ground properties to suit construction needs [12].

Figure 5: Shotcreting Over Steel Ribs in Preperation for Waterproofing in Access Tunnel 4 (3-23-11). MTA Construction & Development Mega Projects, CC BY 2.0, via Flickr.

By the late twentieth century, these innovations had reduced tunnelling costs, improved safety, and expanded the range of feasible applications. Tunnels became increasingly attractive for transportation, water supply, and environmental infrastructure, particularly in urban settings where surface disruption was unacceptable. As Drinker observed more than a century earlier, the art of tunnelling has consistently advanced in parallel with civilisation itself. Modern tunnelling, with its integration of engineering science, technology, and skilled human judgment, represents the culmination of thousands of years of accumulated knowledge applied to the challenge of constructing safely and reliably in the hidden depths below the surface [12].

Timeline of Tunnel Boring Development

c. 4500 BC – Early Civilisations (Ancient Era)

The earliest tunnels were excavated entirely by hand using stone hammers, chisels, and basic metal tools. Construction relied on human labour and empirical knowledge of ground behaviour. In competent rock, builders exploited natural strength to create stable openings with minimal support.

c. 2000 BC – Qanat Systems (Ancient Persia)
Persian engineers developed **qanats**—underground water conveyance tunnels constructed using vertical access shafts connected by gently sloping tunnels. These systems enabled reliable irrigation and water supply across arid regions and remain in use in some parts of the world today.
6th Century BC – Advanced Surveying Techniques
Greek engineers, notably Eupalinos, demonstrated sophisticated surveying methods using string lines and plumb bobs. Tunnels were driven from both ends of a mountain and met with remarkable accuracy, highlighting early mastery of geometry and alignment.
Classical to Medieval Periods (Ancient Era)
Tunnelling expanded from tombs and religious structures to drainage, sewers, canals, and early transportation works. Timber was commonly used for temporary support, followed by permanent linings of brick or stone masonry where ground conditions required additional stability.

Late 17th Century – Introduction of Explosives (Early Industrial Transition)
Gunpowder began to be used in public-works tunnelling, notably for France's Canal du Midi. This marked the first major departure from purely manual excavation and significantly improved progress in hard rock.
1818–1843 – The Tunnelling Shield
Marc Isambard Brunel patented the tunnelling shield in 1818, applying it during construction of the Thames Tunnel (completed 1843). The iron shield supported soft ground during excavation beneath the river, preventing collapse while workers excavated manually at the face.
Mid–Late 19th Century – Compressed Air and Mechanisation
Engineers such as James Henry Greathead improved shield technology and introduced compressed air to counteract groundwater pressure in subaqueous tunnels. At the same time, steam-powered and later compressed-air drilling equipment was adopted in major projects such as the Hoosac Tunnel (USA) and the Mont Cenis Tunnel (Alps), dramatically increasing excavation rates in rock.
Late 19th to Early 20th Century – Steel Ground Support
Steel ribs and liner plates progressively replaced timber supports. These materials offered greater strength, durability, and reduced tunnel cross-sections, enabling more efficient construction and improved long-term performance.

1950s – Birth of Modern TBMs (Modern Era)
The first successful hard-rock tunnel boring machines were developed in the 1950s by James Robbins, using disc cutters to fracture rock mechanically. This innovation allowed continuous, mechanised excavation with reduced ground disturbance and highly uniform tunnel profiles.
1960s – New Austrian Tunnelling Method (NATM)
NATM, also known as the Sequential Excavation Method, emerged in the 1960s. It introduced a flexible "design-as-you-go" philosophy that uses the inherent strength of the surrounding ground, combined with shotcrete, rock bolts, and mesh, to form a load-bearing support ring.
1950s–Present – Immersed Tube Tunnels

For underwater crossings, the immersed tube method became a widely used alternative to shield tunnelling. Prefabricated tunnel elements are floated into position, sunk into a dredged trench, and joined underwater, enabling rapid construction beneath rivers and harbours.

Late 20th Century to Present – Data-Driven and Automated Tunnelling

Modern tunnelling integrates durable precast concrete linings, advanced numerical modelling, real-time sensors, and digital monitoring systems. Contemporary TBMs function as highly automated underground factories, capable of operating in a wide range of ground conditions using Earth Pressure Balance and slurry shield technologies, with increasing use of data analytics and artificial intelligence to optimise safety, performance, and cost.

TBM vs Conventional Drill-and-Blast Tunnelling

Tunnel Boring Machine (TBM) tunnelling and conventional drill-and-blast tunnelling represent two fundamentally different approaches to underground excavation. While both aim to create safe, stable tunnels through soil or rock, they differ markedly in their excavation mechanics, construction sequence, risk profile, cost structure, and suitability for specific ground conditions and project requirements. Understanding these differences is essential for selecting the most appropriate method for a given tunnel project.

TBM tunnelling is a highly mechanised, continuous process in which excavation, spoil removal, and initial ground support occur simultaneously. A TBM advances through the ground using a rotating cutterhead equipped with disc cutters or cutting tools that fracture rock or condition soil at the tunnel face. As the machine advances, it either installs a permanent segmental lining or provides space for immediate ground support behind the cutterhead. This integrated process results in a uniform, circular tunnel profile with tight geometric tolerances and predictable performance. TBMs are particularly well suited to long tunnels with consistent alignment, diameter, and ground conditions, where high upfront capital investment can be offset by productivity and reduced operational risk.

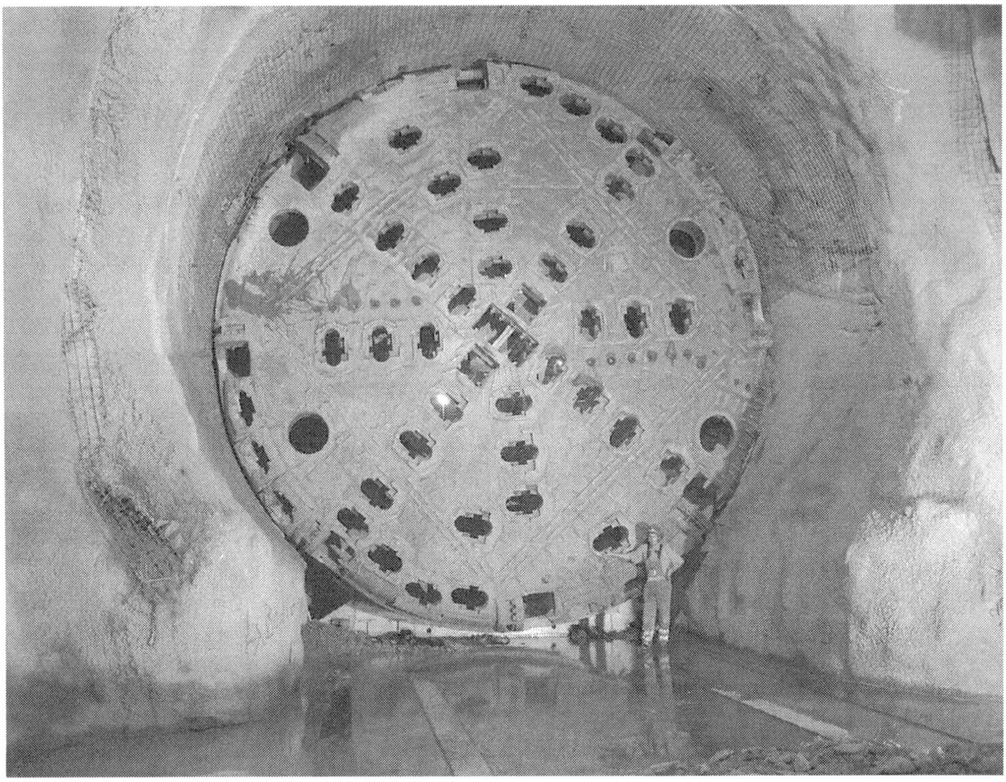

Figure 6: Head of the S-210 Tunnel boring machine used to excavate the eastern tube of the Gotthard Base tunnel between Bodio and Faido, Canton Ticino, Switzerland. Cooper.ch 16:58, 19 September 2006 (UTC), CC BY-SA 2.5, via Wikimedia Commons.

In contrast, drill-and-blast tunnelling is a cyclic excavation method that relies on drilling a pattern of holes into the rock face, charging them with explosives, detonating the blast, and then removing the fragmented rock. After each blast, the tunnel must be ventilated to remove fumes, inspected for loose material, and supported using rock bolts, shotcrete, steel ribs, or other support systems before the next cycle can begin. This method is inherently discontinuous and depends heavily on skilled judgment at the tunnel face to adapt excavation and support to changing geological conditions. Drill-and-blast remains the dominant method in hard rock tunnelling, particularly where tunnel length is shorter, geometry is complex, or ground conditions vary significantly.

Figure 7: Koralmtunnel Construction Lot KAT1 – Cyclical Excavation Method. Pfundner Michael, CC BY-SA 3.0 DE, via Wikimedia Commons.

Cyclical Excavation Method (CEM)

The Cyclical Excavation Method is a traditional tunnelling approach in which excavation advances in repeating, discrete cycles rather than continuously. Each cycle consists of a defined sequence of operations—excavation, spoil removal, ground support, and preparation for the next advance. The method is most commonly associated with conventional drill-and-blast tunnelling in rock, but the term "cyclical" refers to the process structure, not solely to the use of explosives.

At its core, the cyclical method divides tunnel advance into short increments, typically ranging from 1 to 5 metres per cycle, depending on ground conditions, tunnel size, and support requirements. Progress only continues once each step of the cycle is completed and the ground is stabilised. This staged approach allows close observation of geological conditions and flexible adaptation of excavation and support measures.

A standard cyclical excavation cycle begins with face preparation and drilling. In hard rock, a drilling pattern is designed and drilled into the tunnel face using pneumatic or hydraulic rock drills. In softer or mixed ground, mechanical excavation tools may be used instead. The drilling pattern is carefully planned to control fragmentation, minimise overbreak, and protect the surrounding rock mass.

Once drilling is complete, the excavation phase follows. In drill-and-blast tunnelling, explosives are charged into the drilled holes and detonated in a controlled sequence to fragment the rock. After blasting, the tunnel is ventilated to remove fumes and dust, and the face is inspected for stability. In non-blasting cyclical methods, excavation may be carried out using roadheaders, excavators, or hydraulic breakers, followed by similar inspection and ventilation steps.

The next stage is mucking, where the excavated material (spoil) is removed from the tunnel using loaders, conveyors, rail cars, or trucks. Efficient mucking is critical, as it directly affects the duration of the cycle and overall advance rate. Once the face is cleared, scaling is performed to remove any loose or unstable rock from the crown and sidewalls.

After excavation and muck removal, ground support is installed. This may include rock bolts, shotcrete (sprayed concrete), steel ribs, lattice girders, wire mesh, or a combination of these systems. The type and extent of support are determined based on observed ground behaviour and predefined support classes. Installing support promptly is essential to control deformation and ensure safety before the next cycle begins.

The cycle concludes with surveying, quality checks, and preparation for the next advance. Alignment is verified, instrumentation readings may be reviewed, and the next drilling pattern is set out. Only once the ground is stable and the tunnel meets design tolerances does the next excavation cycle commence.

A defining feature of the cyclical excavation method is its flexibility. Because excavation proceeds in short increments, engineers and supervisors can adjust excavation length, support type, and support density in response to changing geology. This makes the method particularly well suited to variable or complex ground conditions, fault zones, and mixed rock types.

The method is also highly observational, relying on continuous assessment of ground behaviour rather than purely predictive models. This aligns well with observational design philosophies such as the New Austrian Tunnelling Method (NATM), where support is adapted based on measured performance rather than fixed assumptions.

However, the cyclical nature of the method means that advance rates are generally slower and less uniform than those achieved with continuous systems such as Tunnel Boring Machines. Downtime between cycles for ventilation, inspection, and support installation is inherent to the process, making labour efficiency and cycle optimisation critical to project performance.

From a ground disturbance perspective, TBMs generally induce less damage to the surrounding rock mass. Tunnel excavation methods significantly influence the surrounding rock mass's integrity and the overall effectiveness of construction projects. Among these methods, Tunnel Boring Machines (TBMs) are known for their controlled cutting action and reduced ground disturbance compared to traditional drill-and-blast techniques. This distinction leads to smoother tunnel walls, more predictable load transfers, and ultimately, lower long-term maintenance requirements and enhanced structural performance. In contrast, drill-and-blast methods introduce dynamic loading that contributes to overbreak, fracture propagation, and increased variability in tunnel geometry, necessitating thicker linings or additional ground support to compensate for disturbed zones.

TBMs operate with a shielded mechanism that allows for controlled excavation, minimizing the disturbance of surrounding rock. Research indicates that compared to drill-and-blast methods, TBMs generally exhibit less deformation and damage to adjacent rock due to their consistent cutting action, resulting in a more stable excavation surface [13, 14]. TBMs contribute to less ground disturbance, ultimately leading to smoother tunnel walls and a reduced need for extensive maintenance interventions [13]. This controlled environment enhances the overall performance of tunnel lining systems and contributes to safer and more reliable tunnelling operations [13].

Moreover, the effectiveness of TBMs, particularly in challenging geological conditions, where minimizing environmental impact is crucial [15]. The mechanism by which TBMs exert thrust allows for precise control during excavation, offsetting potential rock collapses and thereby preserving the integrity of the surrounding mass [13].

In contrast, the drill-and-blast method is associated with notable disruptions in the surrounding rock mass due to explosive loading. This conventional technique can induce significant overbreak and fracturing beyond the intended excavation limits, as described by Yang et al., where dynamic loading and stress redistribution lead to complex deformation patterns and unstable rock conditions [16]. The presence of an excavation damage zone (EDZ) is commonly observed in drill-blast projects, exacerbating the need for additional structural support to counterbalance the resultant instability [16, 17].

According to Rana et al., the inefficiencies associated with drill-blast methods often require extensive remedial works, such as thicker linings or reinforcement measures, to address these disturbances effectively [18]. The cumulative effect of repeated blasting operations can significantly escalate construction costs and timelines due to the need for frequent adjustments in support structures to accommodate varying rock conditions [16, 18].

Due to the controlled nature of TBMs, long-term maintenance associated with tunnels excavated using this method tends to be lower. The precise and smooth finish provided by TBMs enhances the load transfer characteristics of the tunnel lining, which can lead to improved structural performance over time [19, 20]. Conversely, the unpredictable geometrical outcomes from drill-and-blast operations require additional ongoing maintenance, as the variability in the tunnel's architecture can lead to increased operational costs over the lifetime of the structure [16, 18].

Initial damage from drill-and-blast activities can deepen over time, necessitating updates in the support systems to maintain operational integrity [17]. The habit of employing reinforced linings increases upfront costs and complicates future repairs and upgrades due to the added complexity in accessing potentially compromised areas of the tunnel [16].

The choice between Tunnel Boring Machines (TBM) and the drill-and-blast method for tunnelling significantly impacts the safety and working conditions of construction personnel. TBM tunnelling is widely regarded as offering a more controlled working environment, primarily because the excavation face is shielded. This shielding minimizes worker exposure to unsupported ground, explosives, and harmful emissions, thus contributing to a safer operational framework. Specifically, TBM operations employ a higher degree of mechanization and automation, which can lead to reduced risks associated with manual labour and human error during excavation [21].

TBM technology not only promotes safety through shielding but also centralizes operations, which reduces the potential for human exposure to hazardous elements such as blast fumes. A study demonstrated that TBMs are preferred due to their higher efficiency and safety compared to conventional blasting methods [21]. The mechanized operations of TBMs allow for better dust and gas control, further enhancing worksite safety and reducing health risks associated with exposure to harmful substances [22].

Conversely, the drill-and-blast method presents notable safety concerns. Regular handling of explosives and repeated exposure to unsupported faces create significant operational hazards for workers. Studies have shown that tunnel construction workers face increased risks for respiratory issues due to exposure to blasting fumes, including acute irritations and chronic conditions such as chronic obstructive pulmonary disease [23]. The reliance on post-blast ventilation strategies is essential but often insufficient to entirely eliminate exposure to toxic gases like nitrogen dioxide and carbon monoxide, which are released during blasting operations [24]. The unshielded nature of blast sites contributes to hazardous conditions, where personnel are subject to injuries due to rockfalls and other incidents stemming from sudden geological changes [24].

Table 1: Summary of Safety Considerations in TBM and Drill-and-Blast Methods.

Aspect	TBM Tunnelling	Drill-and-Blast Tunnelling
Working Environment	Controlled and shielded	Open and prone to hazards
Mechanization	High	Low
Worker Exposure to Hazards	Low to moderate	High
Toxic Fumes	Controlled emissions	Significant exposure post-blasting
Flexibility	Limited adaptability to sudden changes	High, with immediate response to conditions

Tunnel Boring Fundamentals

Despite the increased risks associated with drill-and-blast tunnelling, this method provides certain advantages, particularly in terms of flexibility in adapting to unexpected geological conditions. Drill-and-blast crews often have better visibility of ground conditions and can make immediate adaptations when encountering unforeseen geological challenges [25]. This flexibility is crucial for minimizing delays and optimizing excavation timelines, particularly in complex geology, where TBM operations may face limitations.

Cost and project economics further distinguish the two approaches. The selection of tunnelling methodologies can have profound implications on the overall cost and effectiveness of construction projects. Among the various approaches, Tunnel Boring Machines (TBMs) and drill-and-blast techniques stand out due to their unique characteristics and economic viability.

TBMs necessitate substantial upfront investments. Costs include the procurement of the machine itself, transportation, assembly, and launch infrastructure. This high initial capital requirement can be justified economically when applied to long tunnels with consistent cross-sections and minimal interruptions, as they afford a higher penetration rate over time, significantly enhancing project efficiency [26, 27]. For instance, research indicates that the maximum monthly advance using TBMs can exceed 1,800 meters under optimal conditions, which reflects the considerable potential of TBMs in suitable geological environments [27].

Conversely, drill-and-blast methods generally incur lower initial capital costs. Their flexible nature allows for varied tunnel sizes and complex layouts. This flexibility makes drill-and-blast techniques particularly appealing for shorter tunnels and situations where geological conditions vary widely [28]. However, it is crucial to note that while the initial investment is lower, the unit costs associated with drill-and-blast techniques can escalate in challenging ground conditions due to slower advance rates and heightened labour intensity, along with additional support requirements [29, 30].

Research indicates that TBMs deliver numerous advantages in terms of excavation speed and safety. They generally perform faster than conventional drilling and blasting techniques, yielding higher overall work efficiency [21]. For instance, TBM operations are typically less hazardous, reducing labour risk factors associated with blasting operations, which can lead to accidents involving flying debris and unexpected geological conditions [19].

Drill-and-blast techniques, though initially less expensive, often face higher operational risks in difficult terrains. These include slower advancement rates and increased costs associated with stabilization efforts, which can complicate project timelines and budgets [31]. Studies reveal that granular variations in rock types can substantially impact the efficiency of these methods, leading to exacerbated costs if the geological formations are not well understood or pre-evaluated [8].

The applicability of each method heavily influences its economic feasibility. TBMs are best suited for extensive projects where uniformity in tunnel profile and reduced environmental impact are priorities. They excel in urban areas where ground disturbance must be minimized and structural integrity preserved [30]. Conversely, drill-and-blast is favoured in scenarios requiring flexibility in design and adaptability to variable geological formations, making it suitable for shorter tunnels or caverns where conditions can change abruptly [32].

Furthermore, the literature illustrates that projects incorporating both methods, such as the TBM for initial tunnelling followed by drill-and-blast for expansions, have seen successful implementations, showcasing a hybrid approach that optimizes both methods' strengths [28, 33].

TBM tunnelling is generally characterized by its reduced surface vibration and noise levels compared to drill-and-blast methods, making it preferable in urban settings where sensitive structures are prevalent. Research indicates that TBM operations, particularly those employing pressurized-face TBMs, can effectively control ground settlement, a critical factor in minimizing impacts on existing infrastructure and nearby communities. According to Hishammuddin et al. [34], TBMs, when properly operated, can avoid excessive settlement issues that often accompany conventional tunnelling methods.

Moreover, the use of the right face support pressures is highlighted as crucial for mitigating ground deformations during TBM operations. A study found that maintaining appropriate support pressures significantly reduces ground settlement, reinforcing the notion that TBM tunnelling, particularly with careful operational management, minimizes adverse impacts on surrounding structures [35].

Further research emphasizes that the careful management of TBM operations can lead to less disruption in both urban environments and sensitive ecological zones compared to other tunnelling methods [36]. This aligns with findings regarding the lower noise induced by TBMs during construction processes, highlighting the need for incorporating TBM methodologies in urban planning to enhance community acceptance and environmental sustainability [37].

In contrast, drill-and-blast tunnelling methods are often associated with increased vibration and air overpressure, posing risks to nearby structures and communities. These vibrations can lead to physical damage to buildings and infrastructure, making drill-and-blast techniques less suitable for densely populated areas unless extensive mitigation measures are employed. Studies have elaborated on how construction-induced noise and vibrations adversely affect the quality of life in urban settings, particularly for vulnerable populations [38].

The environmental impact created through blasting can lead not only to community annoyance but also to more severe structural impacts when detonations are unregulated. Rapid urban development can exacerbate subsidence related to drill-and-blast tunnelling, which may lead to significant ecological and infrastructural disruptions [39]. This necessitates adopting comprehensive mitigation strategies, such as real-time monitoring of ground vibrations and implementing buffer zones during construction to safeguard immediate communities.

The variations in ground settlement associated with these tunnelling methods can significantly affect urban infrastructure. Research indicates that even minor ground movements resulting from tunnelling can lead to serviceability issues, necessitating precise control measures in urban settings [40]. The findings from another study concerning surface settlements occurring during shield construction further reinforce the critical need for continuous monitoring and adaptive management strategies to mitigate adverse environmental impacts [41].

Moreover, integrating community feedback into tunnelling operations can foster a collaborative approach to managing the social impacts of tunnelling. The health effects of noise and vibration exposure should be considered in community engagement strategies for tunnelling projects [42].

In practice, the choice between TBM and drill-and-blast tunnelling is rarely absolute. Many large projects use a combination of both methods, deploying TBMs for long, straight tunnel sections while relying on drill-and-blast for portals, cross-passages, caverns, and areas of highly variable geology. Ultimately, the selection depends on a careful balance of geological conditions, tunnel length and geometry, environmental constraints, risk tolerance, schedule requirements, and overall project economics.

Modern Infrastructure Drivers (Transport, Water, Energy, Utilities)

The landscape of modern infrastructure development is intricately tied to various global challenges, including population growth, urbanization, and climate change. This paper outlines the impacts on four critical areas: transport, water, energy, and utilities, with a focus on how underground construction and tunnelling have become pivotal in addressing the needs of urbanized and rapidly evolving environments.

Transport Infrastructure Drivers

Transport systems are foundational to modern infrastructure, primarily driven by rapid urbanization and population growth. Demand for high-capacity, reliable transport systems has surged as cities face constraints on surface space. Tunnelling has emerged as a viable solution, with underground metro systems and rail corridors significantly minimizing surface congestion, reducing travel times, and improving safety without displacing existing infrastructure or communities [43, 44].

Governments are increasingly pressured to decarbonize their transport networks by transitioning from private vehicle dependency to sustainable public transport alternatives. Underground rail and metro systems, in particular, facilitate this transition by promoting high-frequency, low-emission travel options within densely populated urban centres [45, 46] . Additionally, long-distance tunnels enhance regional connectivity, which is crucial for economic integration, as they can traverse challenging terrains while maintaining reliability across various weather conditions [44, 45].

Water Infrastructure Drivers

Water infrastructure faces pressing challenges driven by population growth and climate variability. As cities evolve, the need for secure drinking water and effective wastewater systems has become critical. Tunnelling enables the creation of large-diameter water conveyance tunnels and sewer interceptors, essential for managing urban water resources with minimal surface disruption [47]. This technology is particularly vital for flood mitigation and stormwater management, which are increasingly necessary due to climate-induced extreme weather events [45, 47].

Moreover, climate change has amplified the urgency for advanced water management solutions, such as underground water storage and flood diversion tunnels. Such infrastructure is designed to withstand high pressure and fluctuating flow conditions, ensuring long-term viability while protecting urban waterways from pollution during significant rainfall [48, 49].

Energy Infrastructure Drivers

The energy sector is undergoing a substantial transition aimed at sustainability and resilience, which prominently influences underground construction strategies. Hydropower facilities, for instance, rely on extensive headrace and tailrace tunnels that efficiently convey water through complex terrains to turbines, necessitating advanced tunnelling methods for structural integrity [46, 50].

Beyond conventional energy sources, there is a growing trend toward incorporating underground transmission and distribution systems. Underground cable tunnels not only facilitate the transport of electricity and gas but also improve safety and reduce visual and land-use impacts in urban areas [51, 52]. As energy infrastructure evolves to be more interconnected and decentralized, these underground systems are vital for enhancing reliability and supporting the transition to cleaner energy solutions [48, 49].

Utilities and Digital Infrastructure Drivers

The rapid growth of digital technologies necessitates an expansion of utilities infrastructure, including telecommunications and data networks. As cities embrace smart technology initiatives, the construction of underground utility tunnels becomes increasingly important [45, 52]. These tunnels serve multiple purposes, hosting utilities such as water, sewage, electricity, and communication systems, thus minimizing the disruption associated with surface excavation for maintenance and upgrades [46, 53].

Moreover, enhanced resilience against environmental extremes prompts urban authorities to manage infrastructure strategically. Multi-utility tunnels centralize services, leading to improved asset management while reducing lifecycle costs and enhancing community accessibility [51, 54]. This coordination is critical for ensuring the continual delivery of essential services, particularly during operational disruptions caused by extreme weather or urban stresses [55].

Integrated and Strategic Drivers

Modern infrastructure planning increasingly prioritizes sustainability and resilience, and tunnelling offers a solution that aligns with such goals. By constructing underground systems, cities can minimize surface impacts and protect critical assets from environmental threats, facilitating urban growth without compromising safety or accessibility [48, 54]. Advancements in tunnelling technologies, such as tunnel boring machines (TBMs) and real-time monitoring systems, have made underground projects more economically viable and reliable [52, 56].

The drivers of modern infrastructure in transport, water, energy, and utilities are inextricably linked to global challenges like urban growth, decarbonization, and climate change. Tunnelling and underground construction provide effective solutions to these pressing needs, allowing for the expansion of essential services beneath urban landscapes where surface options are limited. As these pressures further intensify, the role of underground infrastructure will become even more critical in supporting sustainable and resilient urban societies.

Advantages and Limitations of Tunnel Boring Technology

Tunnel Boring Machine (TBM) technology represents one of the most advanced methods of underground construction, offering significant benefits in productivity, safety, and environmental performance. At the same time, it has inherent limitations that must be carefully considered during project planning and method selection. Understanding both aspects is essential for determining whether tunnel boring is the most appropriate solution for a given project.

Advantages of Tunnel Boring Technology

One of the principal advantages of using Tunnel Boring Machines (TBMs) is their high level of mechanization, which greatly enhances operational safety. TBMs create a shielded environment at the tunnel face, thereby significantly reducing workers' exposure to hazards such as unstable ground conditions, blasting operations, and hazardous fumes associated with conventional tunnelling techniques. The automation of many processes, including remote operation capabilities, allows for reduced personnel presence at the tunnel face. This mechanization leads to more consistent safety outcomes, as highlighted in studies that emphasize the safety and excavation speed of pressurized-face TBMs, particularly in urban scenarios where minimizing environmental impacts is crucial [7, 57].

Furthermore, TBMs can effectively control face pressure, contributing to the stability of the tunnel face and minimizing ground deformations, which is critical for urban construction [7, 58]. The advancements in TBM technology ensure that workers operate in safer conditions compared to traditional methods that rely heavily on manual intervention.

In terms of geometric accuracy, TBMs are designed to excavate a uniform circular profile with minimal overbreak, which is crucial for applications requiring stringent tolerances, such as transport and utility tunnels. The ability to achieve smooth tunnel walls allows for reliable load transfer to the lining, ensuring lasting structural integrity [59, 60]. The precision offered by TBM operations is critical for installations that demand exact placements, such as drainage systems and ventilation shafts.

Research indicates that the segments produced and installed under controlled conditions can meet high standards of strength and watertightness, enhancing the overall durability of tunnelling projects [59, 61]. The factory-produced precast concrete segments benefit from a consistent manufacturing process, thereby reducing variability often seen in in-situ construction methods.

Another significant advantage of TBM operations is their reduced surface impact, particularly in dense urban settings. TBM excavation methods result in lower vibrations and noise levels compared to traditional drill-and-blast approaches, making them suitable for projects beneath sensitive infrastructure and historical buildings [57, 62]. Pressurized-face TBMs are particularly adept at controlling groundwater, minimizing settlement, and reducing surface damage risks, as noted in several studies [40, 57, 62].

The accurate control of face pressure not only stabilizes the tunnel face but also helps in preventing subsidence, which is crucial in urban environments where surface stability is vital [7, 58, 63]. From a productivity standpoint, TBMs demonstrate high and predictable advance rates in appropriate ground conditions. Once operational, TBMs can maintain a continuous excavation process with minimal interruptions [57, 64]. This predictability allows for advanced project scheduling and cost certainty over extensive tunnel drives, particularly as TBMs can simultaneously excavate and install permanent lining segments, enhancing overall efficiency [61, 65].

Moreover, empirical studies show that the operational parameters of TBMs, including face pressures, directly influence the performance outcomes during tunnelling operations, resulting in further time and cost savings [62]. This transactional efficiency makes TBMs a preferred choice for modern tunnelling projects.

The design and operational standards of TBMs contribute to long-term durability and lower maintenance requirements for the tunnels constructed. The durability of the precast concrete segments installed during the TBM operation ensures that the structures have prolonged service life with diminished lifecycle costs [59, 61]. Reducing the need for maintenance not only results in cost savings but also extends the operational viability of the tunnels, making TBMs a sustainable choice for infrastructure projects [61, 62].

Limitations of Tunnel Boring Technology

Tunnel Boring Machines (TBMs) represent a significant technological advancement in the construction of underground tunnels. They offer various advantages, including reduced disturbance to the surrounding environment and improved safety measures during excavation. However, despite these benefits, TBMs come with considerable upfront capital costs and engineering challenges that can limit their applicability.

The initial investment for TBMs is high due to several factors. TBMs are complex machines requiring sophisticated design, engineering, and manufacturing [66]. They involve substantial costs in transportation and assembly, as well as the construction of launch and retrieval shafts that are essential for their operation [67]. Given their high cost, these machines are typically justified economically only for long tunnels with uniform diameters and alignments, where the initial investment can be amortized across the length of the project [68].

Once a TBM begins excavation, its design imposes limitations on modifications. Changes to the diameter, shape, or alignment of the tunnel mid-excavation can be prohibitively expensive or outright impossible [69]. This rigidity makes TBM technology less suitable for projects with complex geometries, such as those requiring frequent size adjustments or intricate junctions. The fixed design of TBMs means that a pre-defined path must be adhered to closely, often resulting in suboptimal performance in less predictable or variable geological conditions [70].

Ground conditions are another critical factor that influences the efficiency and effectiveness of TBMs. While modern machines can accommodate various soil types and rock conditions, heterogeneous or inconsistent geological environments pose significant challenges [71]. For instance, encountering abrupt transitions from soft to hard rock or dealing with fault zones can significantly slow advance rates and increase wear on the cutting tools, leading to increased operational costs and potential project delays [72].

Maintenance operations for TBMs are complex and can become even more challenging under adverse underground conditions. For example, replacing or repairing the cutter head often requires high-risk interventions, such as conducting work under compressed air or hyperbaric conditions, which can delay construction timelines and lead to substantial recovery costs if a TBM becomes stuck or damaged [73]. Such interventions necessitate specialized knowledge and safety protocols to ensure the well-being of personnel involved [74].

Successful TBM projects depend heavily on strategic planning and expertise in project execution. Comprehensive geotechnical investigations are crucial for selecting the appropriate TBM type and operational strategy [70]. If there is misalignment between geological conditions, machine capabilities, and project needs, the inherent advantages of

TBM technology can be diminished [10]. The intricacies involved in TBM deployment and operation necessitate skilled teams capable of navigating the complexities of tunnelling projects [75].

Overview of the Tunnel Boring Process Lifecycle

The tunnel boring process lifecycle describes the full sequence of activities involved in delivering a bored tunnel, from early planning through to commissioning and operation. Unlike surface construction, tunnel boring requires long-term, integrated decision-making because many critical choices—such as tunnel alignment, diameter, and machine type—cannot be easily changed once excavation begins. The lifecycle can be broadly divided into planning and design, construction and excavation, and completion and handover, with each phase influencing the safety, cost, and performance of the final asset.

The lifecycle begins with strategic planning and feasibility assessment. During this phase, project objectives are defined, such as transport capacity, water conveyance, or utility routing. Preliminary route selection considers geology, topography, environmental constraints, land use, and connections to existing infrastructure. Feasibility studies evaluate whether tunnel boring is technically and economically viable compared to alternatives such as cut-and-cover or drill-and-blast. Early risk identification is critical, as ground conditions and groundwater behaviour are the dominant uncertainties in tunnelling projects.

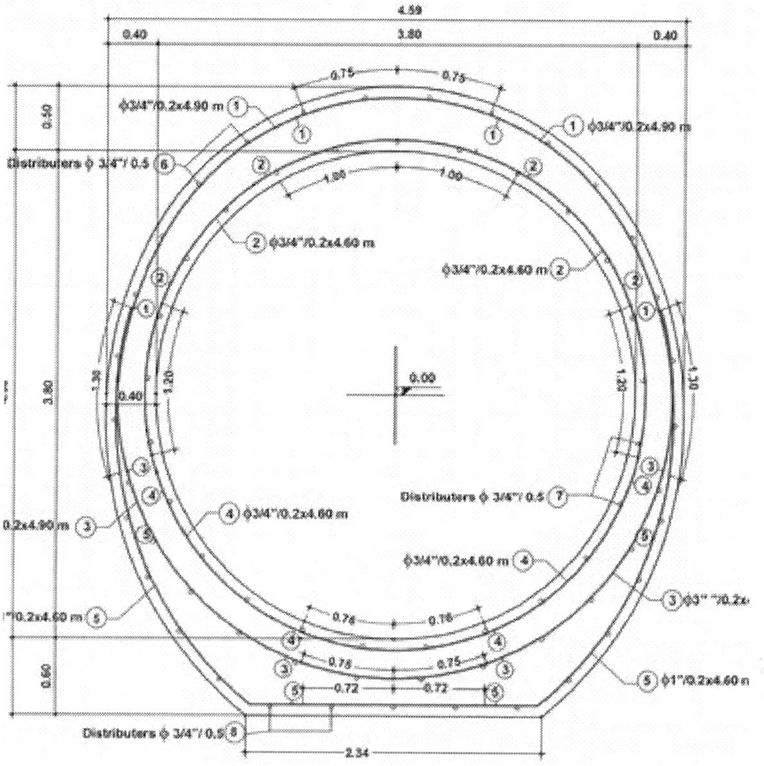

Figure 8: Tunnel design.

Following feasibility, the project enters detailed investigation and design. Comprehensive geotechnical and hydrogeological investigations are undertaken, including boreholes, in-situ testing, laboratory analysis, and geological modelling. This information informs tunnel alignment refinement, lining design, ground support philosophy, and—critically—the selection of the appropriate tunnel boring machine type. During this phase, designers also develop construction methodologies, settlement control strategies, environmental management plans, and safety systems. Decisions made here largely determine the success or failure of the construction phase.

Figure 9: Canada Line TBM Installation. Whatever2009, CC BY-SA 3.0, via Wikimedia Commons.

The next stage is TBM procurement, fabrication, and site preparation. Once the tunnel design and ground conditions are sufficiently understood, a TBM is specified and manufactured to suit the project. In parallel, construction sites are prepared, including launch shafts, retrieval shafts, portals, spoil handling facilities, segment factories, power supply, and logistics systems. The TBM is delivered in components, assembled underground or at the portal, and thoroughly tested before excavation begins. This stage represents a major investment and marks the transition from planning to physical construction.

Figure 10: Borehole – TBM Ulrikke, assembly in Arna in September 2015. Svein Harkestad, CC BY-SA 4.0, via Wikimedia Commons.

The core of the lifecycle is the excavation and tunnelling phase. During this stage, the TBM advances through the ground in a continuous or semi-continuous process, excavating material at the face, transporting spoil to the surface, and installing initial or permanent support. In most modern projects, precast concrete lining segments are installed immediately behind the cutterhead, forming a finished tunnel as excavation progresses. Throughout this phase, performance data such as thrust, torque, penetration rate, face pressure, and ground movement are continuously monitored. Operations are adjusted in response to changing ground conditions to maintain safety, control settlement, and optimise advance rates.

Figure 11: CQ031 - "A" TBM Assembly - Concrete segment feeder. MTA Construction & Development Mega Projects, CC BY 2.0, via Flickr.

As excavation progresses, maintenance and intervention activities occur as part of the lifecycle. Cutter wear, equipment servicing, and occasional repairs are unavoidable and must be planned into the construction program. In difficult ground, interventions such as ground conditioning, grouting, or hyperbaric access may be required. Effective maintenance management is essential to avoid unplanned downtime and protect the long-term integrity of the TBM and tunnel lining.

The excavation phase concludes with breakthrough and TBM retrieval. When the TBM reaches its destination shaft or portal, breakthrough is carefully managed to maintain ground stability and safety. The machine is then dismantled and removed, either for refurbishment and reuse or for demobilisation. At this point, the structural tunnel is complete, but it is not yet operational.

The final stages of the lifecycle involve fit-out, commissioning, and handover. Fit-out works depend on the tunnel's purpose and may include track installation, road pavement, drainage, waterproofing, ventilation systems, power supply, communications, and safety systems. Once installation is complete, the tunnel undergoes testing and commissioning to verify structural performance, system functionality, and compliance with design and regulatory requirements. After successful commissioning, the tunnel is handed over for operation and long-term maintenance.

Chapter 2

Fundamentals of Tunnelling Engineering

This chapter introduces the fundamental principles that govern tunnelling engineering, providing a structured framework for understanding how underground openings are designed, constructed, and stabilised within the ground. It establishes the essential links between tunnel geometry, alignment, ground conditions, stress redistribution, and support interaction, emphasising that tunnelling is not merely an excavation process but a carefully managed ground–structure system. By progressing from basic geometric and alignment considerations through to stress redistribution, load paths, confinement, and modern construction philosophies such as the New Austrian Tunnelling Method, the chapter equips readers with the conceptual tools needed to interpret tunnel behaviour and engineering decisions. These fundamentals form the basis for all subsequent analysis, design, and construction practices discussed throughout the remainder of the text, ensuring that technical choices are grounded in a clear understanding of ground response and stability mechanisms.

Basic Tunnel Geometry and Alignment

Basic tunnel geometry and alignment define the shape, size, and spatial positioning of a tunnel and form the foundation for all subsequent design, construction, and operational decisions. Geometry determines whether a tunnel can safely accommodate its intended function, while alignment governs how the tunnel fits within the ground, connects to surface infrastructure, and responds to geological, environmental, and operational constraints.

Tunnel Geometry

Tunnel Boring Fundamentals

Tunnel geometry significantly impacts the structural efficiency, operational functionality, and economic viability of underground constructions. The cross-sectional shape and dimensions of tunnels are crucial in determining how loads are distributed and how effectively tunnels can be excavated and supported.

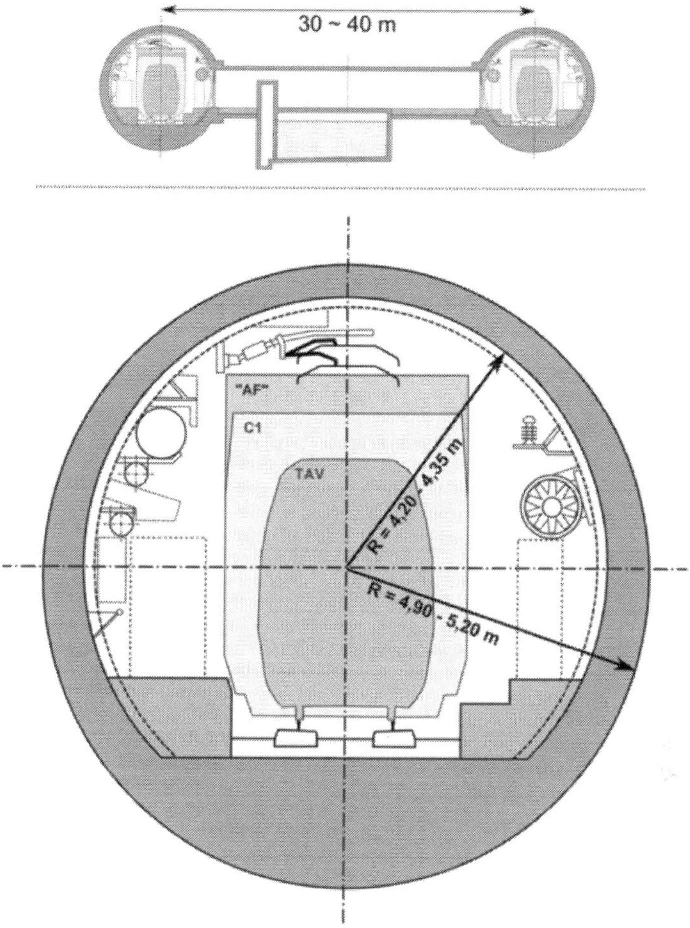

Figure 12: Typical tunnel cross-section of the Turin–Lyon Line. Cheminvento at Italian Wikipedia, CC BY-SA 3.0, via Wikimedia Commons. The Turin–Lyon Line is a major cross-border high-speed and freight railway project linking Turin in Italy with Lyon in France through a new base tunnel beneath the Alps to improve capacity, reduce travel times, and shift transport from road to rail.

Dominance of Circular Cross-Section: The circular cross-section is the predominant geometry employed in modern tunnelling, particularly for Tunnel Boring Machine (TBM)-driven tunnels. This geometry is considered structurally efficient as it uniformly distributes ground and water pressures around the tunnel lining, thereby minimizing bending stresses and reducing the requisite lining thickness. Circular cross-sections are most commonly used due to their inherent strength and stability under variable pressure conditions [76]. Furthermore, circular tunnels generally perform well in seismic conditions, providing reliable performance under dynamic loads [77]. Such attributes make circular tunnels favourable for rapid and safe excavation utilizing TBMs, which are adept at handling the specific challenges posed by circular formations [78].

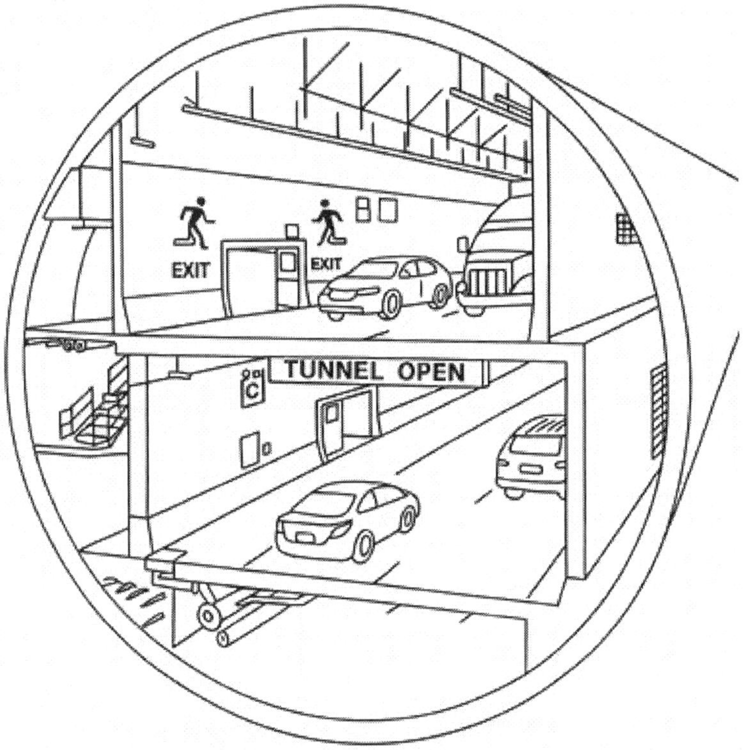

Figure 13: Circular cross-section.

Alternative Geometries and Their Applications: While circular profiles dominate, alternative geometric forms such as horseshoe, D-shaped, and rectangular cross-sections are utilized, particularly in drill-and-blast or sequential excavation methods. Each of these shapes serves specific functional needs—whether maximizing usable interior space, accommodating road or rail clearances, or integrating essential services like drainage. Noncircular profiles can enhance construction efficiency and adaptability to specific site conditions, despite their added complexities in support and lining designs due to uneven stress distributions [79]. Additionally, studies illustrate how different tunnel shapes, including semi-elliptical and horseshoe forms, could be optimized for specific excavation conditions and ground stability [80].

Key Geometric Elements: Key geometric elements of tunnel cross-sections include the crown, sidewalls, and invert. The crown signifies the highest point of the tunnel and plays a critical role in structural stability, while the sidewalls provide lateral support. The invert, or base of the tunnel, is pivotal for resisting uplift pressures, ensuring structural integrity, and facilitating drainage systems [79]. Different studies emphasize that the optimal dimensions of a tunnel, influenced by functional requirements such as vehicle envelopes, track spacing, and maintenance access, are foundational for successful tunnelling projects. Factors like allowances for linings and construction tolerances further complicate the design process, making precise geometric planning essential [77].

Tunnel geometry is a multifaceted aspect of civil engineering that directly impacts tunnelling efficiency, safety, and design. The circular cross-section remains the preferred choice for TBM-driven tunnels due to its structural advantages, while alternative shapes offer flexibility for varied operational needs. Understanding the geometric features of tunnel design is fundamental to optimizing excavation methods and ensuring the feasibility of underground infrastructures.

Horizontal Alignment

Horizontal alignment defines the tunnel's plan-view path. It consists of straight sections connected by horizontal curves with defined radii and transition curves where required. Alignment design must satisfy operational requirements—such as train speed, vehicle steering limits, and driver comfort—while also avoiding geological hazards, existing foundations, utilities, and property boundaries.

Tunnel Plan View Path

The tunnel plan view path describes the tunnel's alignment as seen from above, showing how it defines a horizontal route through the ground between the entry and exit points. It is essentially the tunnel's *map-view trajectory* and is a fundamental element of tunnel design, influencing constructability, operational performance, safety, and interaction with surface and underground features.

In plan view, a tunnel alignment is typically composed of straight sections (tangents) connected by horizontal curves. Straight sections are used wherever possible because they simplify construction, surveying, and operation. Curves are introduced to navigate around geological hazards, existing buildings, underground utilities, property boundaries, waterways, or environmental constraints. The curvature must be carefully designed to meet functional requirements such as vehicle speed, comfort, and safety, particularly for rail and road tunnels.

The radius of horizontal curves is a critical parameter in the plan view path. Rail tunnels generally require large curve radii to allow higher speeds, reduce wheel and rail wear, and improve passenger comfort, while road tunnels can tolerate tighter curves but must satisfy sight-distance and lane geometry standards. In water and utility tunnels, curvature is often minimised to reduce hydraulic losses and simplify maintenance, but limited curvature may still be necessary to fit within available corridors.

Plan view alignment must also consider interaction with the ground and surrounding infrastructure. Designers aim to keep tunnels within competent ground, avoid fault zones or weak strata where possible, and maintain adequate separation from nearby tunnels, foundations, and buried services to prevent ground interaction or instability. In urban areas, the plan view path is often constrained by dense development, requiring careful optimisation to balance engineering feasibility with planning and property considerations.

From a construction perspective, the plan view path directly affects tunnelling accuracy and control. Long straight alignments are easier to survey and maintain within tolerance, while curved alignments require more sophisticated guidance systems and tighter operational control, especially for tunnel boring machines. Poorly designed curvature can lead to increased wear on equipment, reduced advance rates, and difficulties during breakthrough or connection to other underground structures.

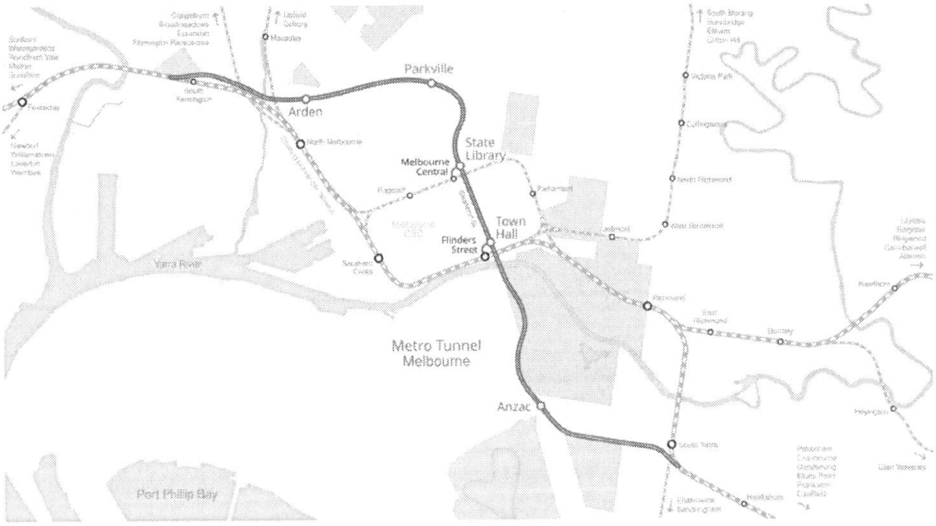

Figure 14: Map of the Metro Tunnel project in Melbourne (Australia), new 9km twin rail tunnels running through the CBD with five underground stations. Gracchus250, CC BY-SA 4.0, via Wikimedia Commons.

For rail tunnels, large curve radii are preferred to allow higher speeds and reduce wear, while road tunnels can tolerate tighter curves but must meet sight-distance and safety standards. In utility and water tunnels, alignment is often governed by available corridors and hydraulic efficiency. Horizontal alignment must also allow sufficient clearance from adjacent tunnels or underground structures to prevent interaction or ground instability.

Vertical Alignment

Vertical alignment describes the elevation profile of the tunnel along its length. It is defined by gradients (slopes) and vertical curves. Gradients are controlled by operational constraints, such as traction limits for rail vehicles, vehicle performance in road tunnels, and flow requirements in water and sewer tunnels.

Tunnel Elevation Profile

The tunnel elevation profile, also known as the vertical alignment, describes how the tunnel rises, falls, or remains level along its length when viewed from the side. It defines the tunnel's depth below ground, gradients, and vertical curves, and is a critical component of tunnel design because it directly affects constructability, drainage, operational performance, and long-term safety.

The elevation profile is made up of gradients (slopes) connected by vertical curves. Gradients are expressed as a percentage or ratio and indicate the rate of ascent or descent along the tunnel. The selection of gradients is governed by the tunnel's function: rail tunnels require gentle gradients to accommodate traction and braking limits, road tunnels must consider vehicle performance and driver safety, and water or sewer tunnels are often designed with continuous slopes to control flow and prevent sediment accumulation. In pressurised water tunnels, the elevation profile also influences internal pressure and structural loading.

Drainage is one of the most important considerations in vertical alignment. Tunnels are typically designed so that water flows toward low points where sumps, pumps, or outlets are located, preventing water ponding and reducing maintenance requirements. Poorly designed elevation profiles can lead to standing water, corrosion, ice formation, or operational disruptions. For transport tunnels, drainage gradients must be coordinated with track or pavement design and emergency access requirements.

Vertical curves are introduced where gradients change to provide smooth transitions. In transport tunnels, these curves improve ride comfort, reduce dynamic loads on vehicles, and maintain adequate sight distance. In construction terms, vertical curves also help control changes in overburden thickness and ground stress, reducing the risk of abrupt loading changes on the tunnel lining.

The tunnel elevation profile must also respond to geological and environmental constraints. Designers aim to maintain sufficient ground cover above the tunnel to ensure stability and limit surface settlement, while avoiding excessive depth that would increase construction cost and risk. The profile may be adjusted to pass beneath rivers, foundations, or environmentally sensitive areas, or to avoid weak strata and high groundwater pressures.

During construction, accurate control of the elevation profile is essential. Survey systems, laser guidance, and real-time monitoring are used to keep excavation within tight vertical tolerances. Deviations can lead to drainage problems, reduced clearances, or difficulties connecting to portals, shafts, or adjacent tunnel sections.

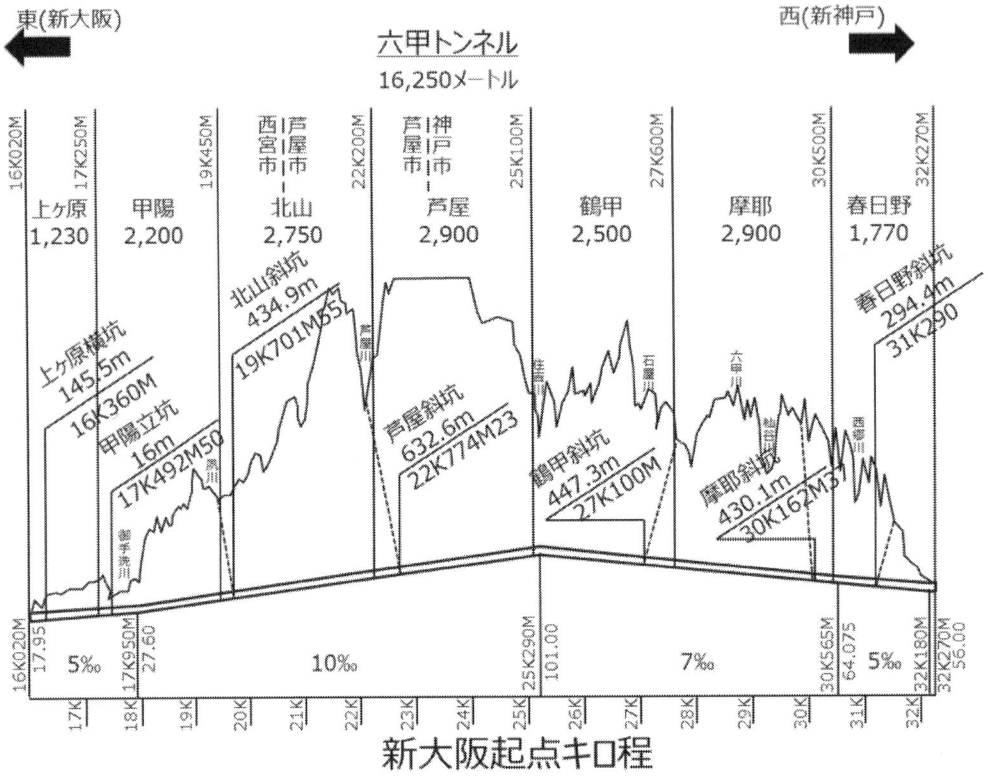

Figure 15: Vertical alignment of Rokko Rail Tunnel, Sanyo Shinkansen. Tam0031, CC BY-SA 4.0, via Wikimedia Commons.

Drainage is a critical consideration in vertical alignment. Tunnels are often designed with continuous gradients to allow water to flow toward sumps or outlets, preventing ponding and reducing maintenance issues. In pressurised water tunnels, vertical alignment must also consider internal pressure changes and structural loading. Vertical curves are introduced to provide smooth transitions between gradients and to maintain comfort and safety for users.

Alignment Control and Accuracy

The construction of long tunnels, particularly those driven from multiple headings, requires meticulous control over geometry and alignment to ensure structural integrity and operational functionality. The utilization of survey control networks, laser guidance systems, and advanced navigation technology plays a crucial role in maintaining these tunnels within specified positional tolerances. These technologies facilitate real-time monitoring and adjustments during the tunnelling process, effectively addressing small alignment errors that, if uncorrected, tend to accumulate over long distances. Such misalignment could lead to costly correction measures or inadequate fit at breakthroughs, underscoring the importance of precise positional control in tunnel construction [81, 82].

Modern tunnelling techniques incorporate advanced survey control networks that utilize laser scanning technology for high-accuracy measurements. The use of 3D laser scanning allows for accurate spatial mapping and documentation of tunnel geometries during construction. This technology not only aids in maintaining alignment but also enables the identification of any deviations from the intended path early in the construction process. Such proactive measures are critical in preventing misalignment that could occur due to accumulating errors over the lengthy spans of a tunnel [81].

Laser guidance systems, integral to contemporary tunnelling machinery, further enhance the precision of alignment during the excavation process. These systems provide feedback that allows for real-time adjustments of the tunnelling shield, keeping it aligned according to the predetermined design specifications. Gyroscopic navigation complements these technologies by providing orientation stability and accuracy, particularly in the curved segments of tunnels where conventional methods may falter [82].

When designing tunnel alignment, it is vital to account for various tolerances, including construction tolerances, expected ground deformation, and the accuracy of lining installation. Ground movement due to geological activities and the stresses of the construction process can result in significant deviations unless well-planned allowances are integrated into the design. This ensures that even after potential settlement or shifts in the surrounding ground, the tunnel can meet functional and regulatory standards [83, 84].

Research highlights the necessity for integrating substantial tolerances into the alignment process, which accounts for variances in construction techniques and materials. For instance, the standardization of geometric tolerances, as outlined in various International Organization for Standardization (ISO) specifications, is fundamental in ensuring that all facets of the tunnelling project adhere to established guidelines, thus maintaining structural integrity post-construction [84, 85].

Integration with Ground Conditions

In the field of civil engineering, particularly in tunnel design, it is essential that both geometry and alignment are adapted to the prevailing geological and hydrogeological conditions. Designers must ensure that tunnels are aligned to traverse the most competent ground available while avoiding fault zones or weak strata. Such considerations are crucial for the overall stability of the tunnel structure and to mitigate risks associated with groundwater pressures and settlement beneath sensitive structures.

Geological conditions play a pivotal role in the design and construction of tunnels. For instance, a comprehensive understanding of geological behaviour through geotechnical surveys and laboratory tests informs engineers in their design processes, allowing for the effective accommodation of complex geological environments [86]. Mixed-face ground conditions, characterized by the simultaneous presence of geological formations with different properties, significantly impact tunnel design strategies [4]. Misalignment with geological structures, for instance, could lead to increased construction risks and unstable conditions.

Moreover, the analysis from Blasco et al. [87] indicates a direct correlation between ground displacement, variations in groundwater levels, and geological characteristics, illustrating the substantial influence of geology on urban subsidence phenomena and the stability of tunnel structures. This interplay necessitates that tunnel designers factor in geological conditions to avoid undesirable outcomes such as excessive settlement or structural failure.

Groundwater management is integral to the planning and execution of tunnel projects. Groundwater pressures need to be carefully evaluated as they can severely compromise tunnel stability. The complexity of groundwater interactions and pressure dynamics has been referenced in several studies. For example, Dematteis et al. [88] emphasize utilizing advanced techniques, such as borehole temperature monitoring, to discern groundwater flow patterns that influence hydrogeological behaviour in tunnel environments. The integration of such data into the design process enables designers to adapt tunnel alignments historically aligned with field observations, optimizing groundwater management and structural stability.

Additionally, the methodology for managing groundwater pressures is further supported by García-Gil et al. [89], who discuss innovative approaches to regulate groundwater flow and reduce hydraulic connectivity, thus safeguarding tunnel structures from varying seasonal water levels. These adaptations are critical in environments with fluctuating hydrological conditions.

Adjustments in tunnel alignment may also be necessary to reduce construction risk associated with geological anomalies and groundwater dynamics. A study by Gruslin and Hennebaut [90] illustrates how unforeseen geological complexities and inadequate groundwater management can lead to catastrophic failures. Here, reconnaissance drilling and geophysical surveys revealed significant geological challenges, underscoring the necessity for thorough pre-construction analysis and flexibility in design options to adapt to encountered conditions.

Furthermore, Bogaard and Greco [91] have demonstrated the importance of understanding pore water pressures in preventing geological disasters during tunnel construction. Designers must be prepared to implement strategies to manage groundwater inflow that could threaten the integrity of both the tunnel and adjacent structures.

Basic tunnel geometry and alignment establish the structural efficiency, functionality, and constructability of a tunnel. Geometry governs the tunnel's cross-sectional form and load-carrying behaviour, while alignment controls its horizontal and vertical position through the ground. Together, they ensure the tunnel can be built safely, perform its intended function, and integrate effectively with surrounding infrastructure and geological conditions.

Stress Redistribution Around Underground Openings

When an underground opening such as a tunnel is excavated, the natural stress state that existed in the ground before excavation is disturbed. This process, known as stress redistribution, is fundamental to tunnelling engineering because it governs ground deformation, stability, and the loads transferred to tunnel support and lining systems.

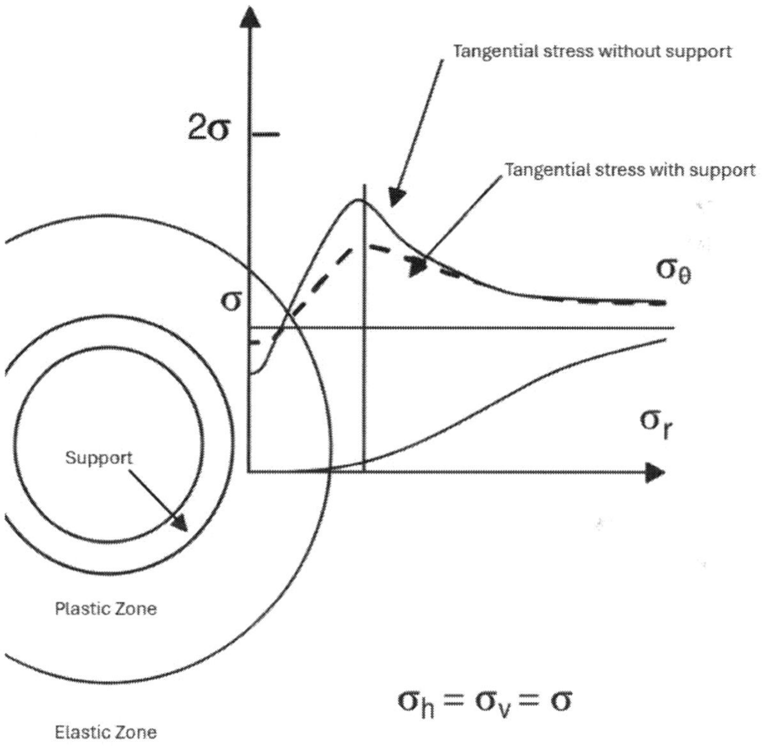

Figure 16: Stress distribution around a tunnel in elastic rock masses.

Figure 16 illustrates how stresses redistribute around a circular underground opening (such as a tunnel) after excavation, and how the installation of support modifies those stresses to improve stability. It combines a cross-sectional view of the tunnel and surrounding ground with stress profiles plotted as a function of distance from the tunnel wall.

In-situ Stress Condition (Before Excavation)

The notation $\sigma_h = \sigma_v = \sigma$ indicates an idealised case where the ground is subjected to uniform in-situ stress—the horizontal and vertical stresses are equal. Before excavation, the ground is in equilibrium and stresses are evenly distributed.

Excavation and Stress Redistribution

When the tunnel is excavated, the material inside the opening is removed, and the stresses that were previously carried by that material must be redistributed into the surrounding ground.

- Radial stress (σ_r) is the stress acting perpendicular to the tunnel wall.
- Tangential stress ($\sigma\theta$) is the stress acting circumferentially around the tunnel.

At the tunnel wall:

- Radial stress drops sharply because the surface is exposed.
- Tangential stress increases as the surrounding ground carries additional load.

Elastic Zone and Plastic Zone

The diagram shows two distinct regions around the tunnel:

- Plastic zone (shaded region around the tunnel):
  This is the zone where the redistributed stresses exceed the strength of the ground, causing yielding or irreversible deformation. Rock or soil in this zone may crack, shear, or squeeze toward the opening.
- Elastic zone (outside the dashed boundary):
  Beyond the plastic zone, the ground behaves elastically. Deformations are small and reversible, and stresses gradually return to the original in-situ value σ with increasing distance from the tunnel.

The size of the plastic zone depends on ground strength, in-situ stress level, tunnel size, and the timing and stiffness of support.

Effect of Tunnel Support

The inner dark ring labelled Support represents a lining or support system (such as shotcrete, rock bolts, steel ribs, or a segmental lining).

The stress curves on the right show the critical effect of support:

- Tangential stress without support (solid curve) rises sharply near the tunnel wall, reaching a high peak. This high stress concentration can lead to failure, excessive deformation, or collapse.
- Tangential stress with support (dashed curve) has a lower peak and a smoother distribution. The support shares load with the surrounding ground, reducing stress concentration and limiting the extent of the plastic zone.

Support also increases the radial stress at the tunnel boundary, which helps confine the ground and improve its strength.

Engineering Significance

This diagram captures several key tunnelling principles:

- Excavation causes unavoidable stress redistribution.
- Excessive tangential stress leads to yielding and plastic deformation.
- Early and adequate support reduces peak stresses and controls deformation.
- Stability is achieved when the ground–support system reaches a new equilibrium.

Before excavation, soil or rock is in a state of in-situ stress equilibrium, generated primarily by the weight of the overlying ground (vertical stress) and horizontal tectonic or geological stresses. Once material is removed to create a tunnel, the stresses that were previously carried by that material must be redistributed into the surrounding ground. This causes a complex reorganisation of stresses around the opening.

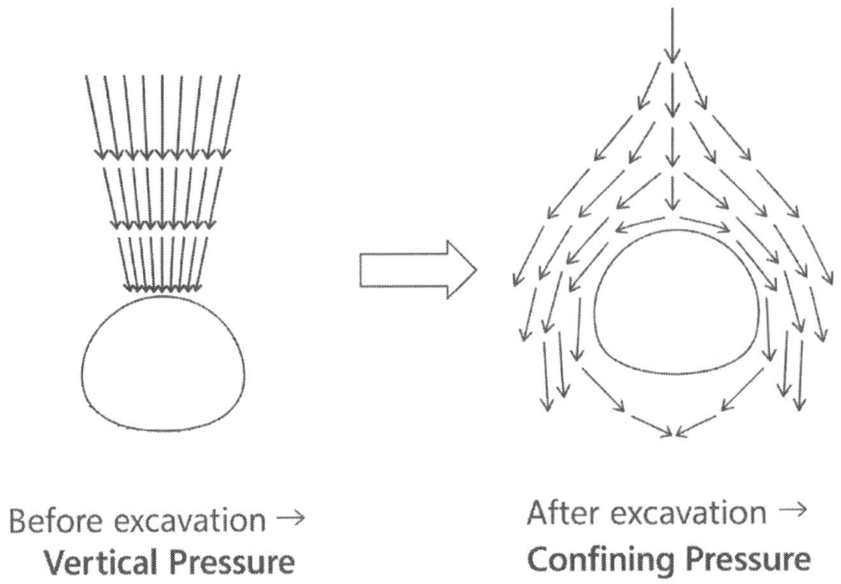

Figure 17: Redistribution of stress caused by tunnel excavation.

Figure 17 illustrates a fundamental concept in tunnelling engineering: how excavation transforms vertical ground pressure into confining (arching) pressure around an underground opening. It compares the stress condition before excavation with the stress condition after excavation, showing how the ground adapts to the creation of a tunnel.

Before Excavation – Vertical Pressure

In the left-hand image, the ground is intact and continuous. The arrows pointing downward represent vertical in-situ stress, which is primarily caused by the weight of the overlying soil or rock (overburden). At this stage, stresses are transmitted almost vertically through the ground, and the material is in a natural state of equilibrium. There is no void, so the stresses are distributed uniformly through the ground mass.

After Excavation – Confining Pressure

In the right-hand image, an underground opening has been created. The excavation removes material that previously carried vertical load, so the stress field must reorganise. Instead of continuing straight downward, the stress paths curve around the opening, forming an arch-like pattern. This redirection of stress is known as ground arching.

As a result, the dominant loading around the tunnel becomes confining pressure, acting radially and tangentially around the opening rather than purely vertically. The ground above and beside the tunnel carries the load by transferring stresses around the void, reducing the direct load acting on the tunnel crown.

Engineering Meaning of the Arching Effect

This stress redirection is highly beneficial in tunnelling. When arching develops effectively, the surrounding ground becomes self-supporting to a degree, and the tunnel lining or support system is required mainly to control deformation and maintain confinement, rather than to carry the full overburden load. Strong, stiff ground promotes efficient arching, while weak or heavily fractured ground may only develop limited arching and therefore require more substantial support.

Role of Tunnel Support

Tunnel support systems—such as shotcrete, rock bolts, steel ribs, or segmental linings—help preserve and enhance this confining pressure. By providing radial support at the tunnel boundary, they prevent excessive deformation that would otherwise disrupt the arching mechanism. Early support installation is therefore critical to maintaining confinement and preventing collapse.

Immediately after excavation, stresses at the tunnel boundary are reduced—often to near zero at the exposed surface—while stresses in the surrounding ground increase. This creates a stress concentration zone around the tunnel perimeter. In rock, the highest stress concentrations typically occur at the sidewalls for deep tunnels where horizontal stress dominates, and at the crown and invert where vertical stress is higher. In soil, stress redistribution is often accompanied by deformation and consolidation rather than brittle fracture.

As stresses redistribute, the surrounding ground attempts to deform toward the opening, a process known as convergence. The magnitude and rate of this deformation depend on ground strength, stiffness, in-situ stress levels, groundwater pressure, and excavation method. If the ground is strong and stiff, it may arch naturally around the opening, transferring loads away from the tunnel void. This phenomenon, known as ground arching, is beneficial because it reduces the load acting directly on the tunnel lining.

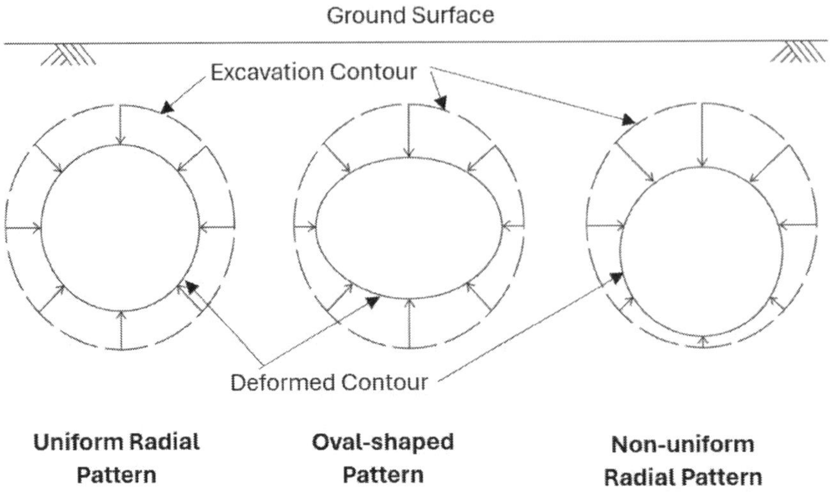

Figure 18: Convergence patterns at the tunnel opening.

Figure 18 illustrates how the ground surrounding a tunnel deforms following excavation and highlights three characteristic deformation patterns. These patterns develop in response to the interaction between in-situ stress conditions, ground properties, tunnel depth, and the effectiveness and timing of installed support. In the diagram, the dashed circle represents the original excavation contour, while the solid inner shape shows the deformed contour that results from stress redistribution and ground movement toward the opening.

The horizontal line at the top of the diagram represents the ground surface and indicates the influence of surface loading and overburden on the stress regime acting around the tunnel. The dashed outlines surrounding each tunnel depict the intended excavation shape, which is typically circular for mechanised tunnelling. Once excavation takes place, the surrounding ground relaxes and moves inward toward the void, producing deformation that alters the original excavation profile.

In the uniform radial deformation pattern, ground movement toward the tunnel occurs evenly around the entire circumference. This behaviour is typically observed when in-situ stresses are approximately equal in all directions, the tunnel is relatively deep compared to its diameter, the ground is homogeneous and isotropic, and support is installed symmetrically and without delay. Under these conditions, the tunnel retains a near-circular shape, and the lining is subjected to relatively uniform loading. This represents the most structurally efficient and predictable deformation response.

The oval-shaped deformation pattern develops when the stress field is anisotropic, most commonly in shallow tunnels where vertical stress exceeds horizontal stress due to gravity loading. In this situation, inward movement at the tunnel crown and invert is greater than at the sidewalls, causing the tunnel cross-section to deform into a vertically compressed oval. This pattern is common in soft ground and shallow cover conditions and leads to increased bending moments in the lining. As a result, careful design and reinforcement of the crown and invert are required to control deformation and maintain stability.

The non-uniform radial deformation pattern reflects highly irregular ground movement around the tunnel. This behaviour occurs when ground properties vary significantly around the excavation, faults, joints, or weak zones are present, excavation or support installation is uneven, or external influences such as nearby structures or adjacent excavations affect the stress field. In these cases, deformation becomes concentrated in specific areas, increasing the risk of cracking, instability, or localised failure. Tunnel support systems must therefore be adapted to address these asymmetric loads, often requiring additional reinforcement in critical zones.

Overall, the diagram emphasises that tunnel behaviour is governed by the combined effects of ground conditions, the in-situ stress regime, and the timing and configuration of support. Recognising the likely deformation pattern enables engineers to select suitable excavation methods, design effective support and lining systems, anticipate areas of high stress or deformation, and manage surface settlement and structural risk throughout construction and operation.

In weaker ground or highly fractured rock, stress redistribution may exceed the ground's strength, leading to yielding, cracking, or squeezing. In such cases, the ground cannot sustain the redistributed stresses without support, and excessive deformation or collapse may occur. The role of tunnel support—such as rock bolts, shotcrete, steel ribs, or segmental linings—is to control deformation, stabilise the ground, and share load with the surrounding material until a new equilibrium state is reached.

The timing of support installation in tunnelling is crucial for managing stress redistribution, especially in methods such as the New Austrian Tunnelling Method (NATM). Early support installation effectively limits deformation and reduces the extent of plastic or damaged zones around the tunnel. This is supported by studies indicating that prompt support installation helps mitigate the stresses that develop in the surrounding rock, thus stabilizing the tunnel excavation and minimizing ground movements [92-94]. Conversely, delayed installation allows the ground to relax and deform more significantly, which can increase support loads and elevate long-term lining demands [95, 96]. Therefore, the strategic timing of support is fundamental to addressing the engineering challenges associated with tunnelling.

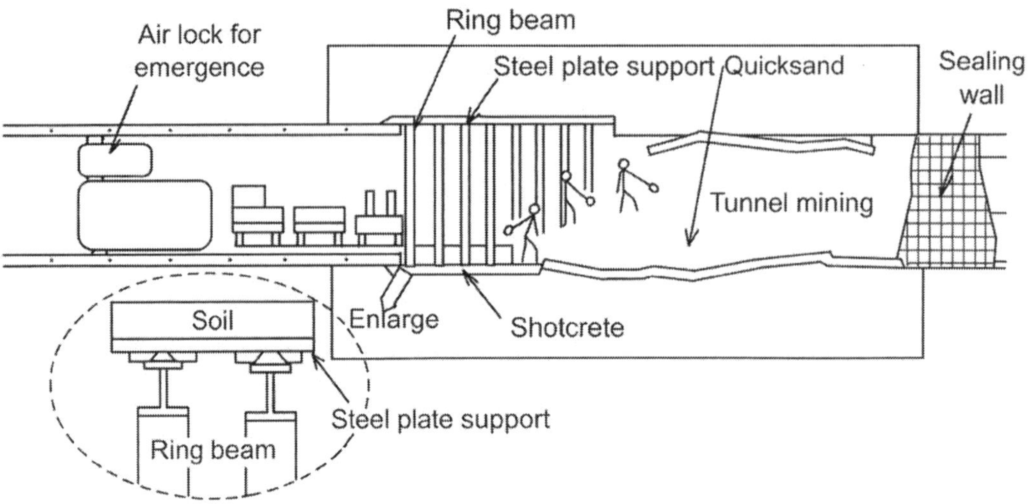

Figure 19: New Austrian Tunnelling Method (NATM), based on Terzi & Manolis (2023) [97].

New Austrian Tunnelling Method (NATM)

The New Austrian Tunnelling Method (NATM), also known as the Sequential Excavation Method (SEM), is a notable advancement in tunnelling techniques that emphasizes the collaborative interaction between the ground and the support structures. NATM emerged in Austria during the mid-20th century and marks a paradigm shift from traditional rigid lining approaches to a system that values the inherent strengths of surrounding geological formations.

The fundamental principle of NATM is to harness the natural load-bearing capacity of the surrounding rock, allowing controlled deformation during tunnel construction. Instead of employing a rigid lining immediately after excavation, NATM supports the excavated opening by allowing certain movements within the ground. This induces natural phenomena such as arching and confinement, which contribute to tunnel stability [98, 99]. The method emphasizes a responsive approach to tunnelling, enabling adjustments to be made in real-time based on observed ground behaviours, particularly when deformations surpass predictive models [100].

NATM advocates for a sequential excavation strategy where the cross-section of the tunnel is divided into multiple stages, including top heading, bench, and invert [101]. This approach minimizes ground displacement and enhances overall tunnel stability, especially in varying geological conditions, which is a notable advantage over conventional tunnelling techniques that typically employ a uniform approach to support design [102, 103].

NATM involves the use of various temporary support mechanisms immediately after the excavation of each round. These supports often include shotcrete, rock bolts, and steel meshes [104, 105]. The rapid installation of shotcrete provides immediate confinement, while rock bolts ensure long-term stability by securing the rock mass [106]. The design of these supports is adaptive; support systems are optimized based on continuous monitoring of ground movements and forces acting on the tunnel [107, 108].

Instrumentation plays a crucial role in NATM, with the implementation of devices such as extensometers and load cells to track behaviour during the construction process [100, 109]. This observational approach allows engineers to alter excavation methods or enhance support measures as necessary. For example, excessive ground deformation might prompt the addition of more rock bolts or the adjustment of the excavation sequence to better accommodate the geological realities at hand [107, 109].

NATM is versatile and has found applicability across various geological settings, including soft ground and complex ground conditions where traditional methods may not perform well [102, 110]. Its success has been documented globally, particularly in areas with diverse geological challenges, such as loess in China and seismic zones like Lombok Island [108, 111]. However, the method's reliance on skilled engineering judgment and meticulous workmanship is critical, as deficiencies in execution or timing can lead to severe ground instability [105, 110].

Once the tunnel stabilizes and further deformations cease, a cast-in-place concrete lining may be applied. This lining primarily serves long-term durability and serviceability, as NATM's interim supports typically manage a significant portion of the ground loads during the excavation phase [101, 108].

Moreover, observational approaches integrated within methods like NATM involve controlled deformation while maintaining close monitoring of ground conditions. Studies emphasize the importance of dynamic monitoring processes in evaluating support effectiveness and adjusting excavation procedures accordingly [92, 93, 112]. Such adaptability is essential as it allows engineers to respond effectively to changes in ground behaviour while managing stresses in the rock mass surrounding the tunnel.

In addition to mechanical stresses, groundwater critically influences stress redistribution in tunnel environments. The presence of pore water pressure reduces the effective stress in both soil and rock, potentially leading to increased deformation [94]. Groundwater alters effective stress and affects excavation stability by modifying groundwater flow paths, thus influencing pore pressures that dictate the ground's mechanical behaviour. Research highlights the necessity for excavation activities to account for hydrological changes to maintain stability and ensure safety [113, 114].

Designing and constructing tunnels must incorporate effective drainage or pressure control mechanisms to mitigate groundwater's adverse effects on excavation [92, 93, 115]. This strategy is vital, particularly in areas with variable groundwater conditions, where the potential for inflow can exacerbate challenges related to tunnel stability and deformation. Moreover, empirical data support the need for comprehensive monitoring systems that can account for both geotechnical and hydrological changes in real-time to promote informed decision-making during construction [107, 115].

Ground–Structure Interaction Concepts

Ground-structure interaction (GSI) is an important concept in understanding how the surrounding soil or rock and the tunnel support system work together to maintain stability after excavation. This interaction is critical specifically regarding load distribution, deformation control, and the establishment of a new equilibrium state following a tunnelling operation. The complexities of GSI are underscored by the need for a shared load between the ground and the structural support system, as excess rigidity in support can restrict necessary stress redistribution and potentially lead to structural failure.

When a tunnel is excavated, surrounding material is removed, resulting in a loss of confinement and a subsequent redistribution of in-situ stresses. The soil or rock surrounding the tunnel tends to deform inward toward the void, while the support system is designed to resist this inward movement, indicating a complex interplay between support and surrounding ground conditions.

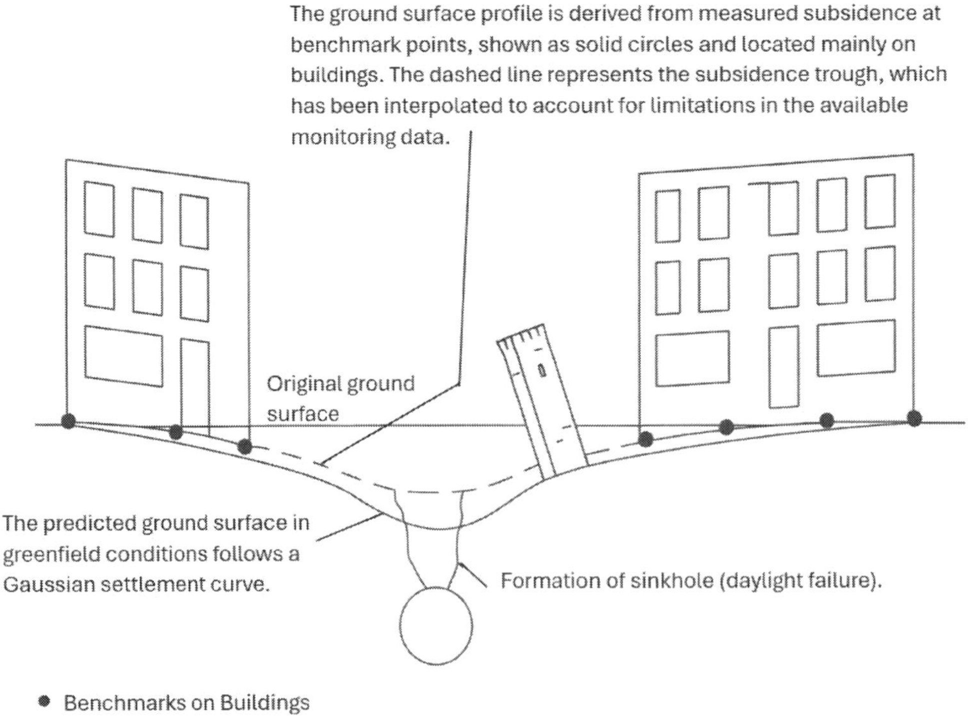

Figure 20: A conceptual model to explain ground loss, adapted from Kontogianni & Stiros (2020) [116].

Figure 20 illustrates ground surface settlement caused by underground tunnelling, showing how excavation-induced ground movement propagates upward and affects surface structures. It combines measured data, predicted behaviour, and a worst-case failure mechanism to explain different surface response scenarios.

At the surface, solid black circles represent benchmark points, which are monitoring markers installed mainly on buildings. These benchmarks record actual vertical movements of the ground surface during and after tunnelling. The solid ground line passing through these points shows the measured subsidence profile based on available monitoring data. Because benchmark locations are discrete and unevenly distributed, the full settlement shape cannot be directly measured everywhere.

The dashed line represents the interpreted subsidence trough, which has been derived by interpolating between benchmark readings. This interpolation accounts for limitations in monitoring coverage and provides an estimated continuous settlement profile across the affected area. The label "original ground surface" indicates the pre-construction ground level, allowing the magnitude and shape of settlement to be clearly visualised.

The diagram also shows the predicted settlement profile under greenfield conditions, where no buildings or structures influence ground behaviour. In such conditions, tunnelling-induced settlement typically follows a Gaussian curve, characterised by a smooth, symmetrical trough with maximum settlement above the tunnel centreline and decreasing settlement with horizontal distance. This curve is widely used in tunnelling design to predict surface movements and assess potential risk to structures.

Beneath the surface, the tunnel is shown as the source of ground loss. In the central area above the tunnel, the image illustrates the formation of a sinkhole (daylight failure). This occurs when ground loss is excessive and the soil or rock arch above the tunnel collapses upward to the surface, creating a sudden and localised failure rather than a smooth settlement trough. The leaning structure in the image highlights the severe structural damage that can result from such a failure.

Overall, the image demonstrates three key concepts:

1. Measured settlement using benchmarks on buildings,
2. Predicted settlement behaviour using an idealised Gaussian curve in greenfield conditions, and
3. Extreme failure conditions, where insufficient ground support or excessive ground loss leads to sinkhole formation.

Together, these elements illustrate how tunnelling can affect the ground surface, why monitoring and prediction are essential, and how controlled deformation differs fundamentally from catastrophic ground collapse.

Research indicates that these interactions can significantly influence both the magnitude and distribution of loads on the support system, emphasizing the need for careful monitoring and modelling of these effects [117]. For instance, utilizing the Convergence-Confinement Method (CCM) allows for a structured approach to predict the loads acting on the support system while also controlling the displacement at the tunnel face to maintain stability [117].

The Convergence–Confinement Method (CCM) is a conceptual and analytical framework used in tunnelling to describe and predict how the ground around a tunnel deforms (converges) and how tunnel support provides confinement to control that deformation. Rather than treating the ground and the support as separate systems, CCM views them as an interacting pair that evolves toward a new equilibrium after excavation.

At the heart of CCM is the recognition that excavation causes a reduction in radial stress at the tunnel boundary, which leads to inward ground movement. This convergence continues until either the ground develops sufficient self-supporting capacity through arching or the installed support provides enough confinement to arrest further deformation. The final tunnel shape and support load are therefore determined by the balance between ground behaviour and support response.

CCM is commonly represented using two curves plotted on the same axes. The ground reaction curve describes how the ground pressure acting on the tunnel boundary decreases as tunnel convergence increases. It reflects the stiffness and strength of the ground and whether behaviour is elastic, elasto-plastic, or fully plastic. The support characteristic curve shows how the support pressure increases as convergence occurs, depending on the stiffness, thickness, and type of support installed. The point where these two curves intersect defines the system's equilibrium state, indicating the final deformation and the load carried by the support.

A critical aspect of CCM is the timing of support installation. Support installed immediately after excavation limits convergence but attracts higher loads, shifting the equilibrium point toward smaller deformation and higher support pressure. Delayed support allows more ground relaxation and stress redistribution before confinement is applied, resulting in larger deformation but lower support loads. CCM helps engineers assess this trade-off and select an installation timing that balances safety, constructability, and economy.

The method is particularly valuable in observational tunnelling approaches, such as the New Austrian Tunnelling Method, where support measures are adapted based on measured ground behaviour. By monitoring convergence and comparing it with predicted curves, engineers can adjust support stiffness, thickness, or installation sequence to keep deformation within acceptable limits while avoiding unnecessary overdesign.

CCM also provides insight into the extent of plastic zones around the tunnel. In weaker ground, significant plastic deformation may occur before confinement is mobilised, requiring flexible and staged support systems such as shotcrete and rock bolts. In stronger ground, convergence may be limited, and lighter support may be sufficient.

Groundwater conditions, tunnel depth, and excavation method all influence the shape of the reaction curves and therefore the final equilibrium.

A significant aspect of GSI is that support does not necessarily bear the full ground load from the outset. In certain tunnelling methods, notably the New Austrian Tunnelling Method (NATM), allowing for some ground deformation is beneficial as it enables natural arching and the mobilization of ground strength, effectively reducing loads on the support system [118]. Rigid supports installed prematurely may inhibit beneficial deformations, thereby increasing loads on structural elements and compromising tunnel integrity [119]. Therefore, an optimized balance between ground support stiffness and ground response becomes essential for maintaining stability.

The timing of support installation plays a critical role in GSI. Early installation can limit deformation, ensuring a more stable tunnel profile; however, it concurrently raises the load on the structural support. Conversely, delayed installation may permit more ground relaxation, facilitating better stress redistribution, but introduces risks of excessive deformation or instability should the surrounding ground conditions be poor [118]. Thus, defining an optimal installation timeline is vital for effective GSI, integrating factors such as the excavation method and geological conditions [4].

Ground Reaction Curves and Support Characteristic Curves

Engineers typically utilize graphical tools such as ground reaction curves (GRC) and support characteristic curves (SCC) to model and assess GSI effectively. These curves illustrate the relationship between ground pressure and tunnel convergence, allowing for a clear understanding of the interacting forces at play. The intersection of these curves represents the equilibrium state where both ground and support conditions are optimized, providing insight into necessary adjustments in support systems or excavation sequence [120]. This approach assists in foreseeing how changes in tunnel design or construction methodology will affect both performance and stability.

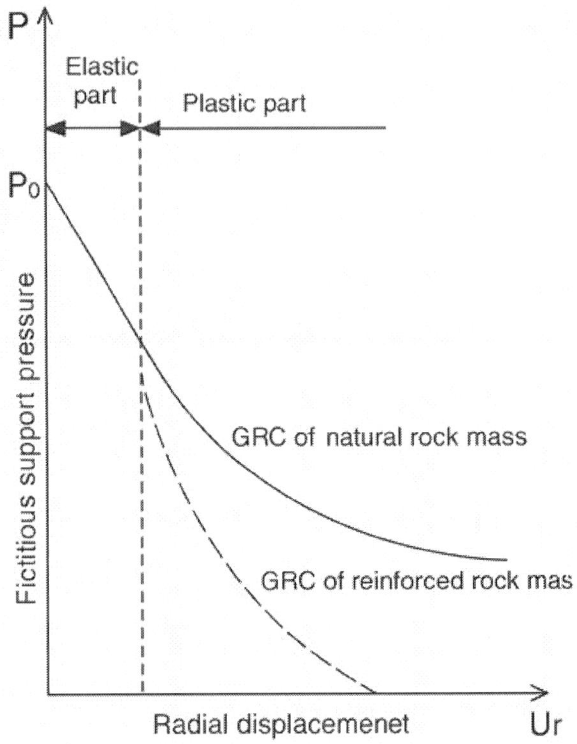

Figure 21: Typical ground reaction curve (GRC) in the presence of rock reinforcement, where P represents the fictitious support pressure and u_r denotes the radial displacement of the tunnel wall. Adapted from Osgoui & Ünal (2009) [121].

Figure 21 illustrates a key concept of the Convergence–Confinement Method (CCM) by showing how support pressure and radial displacement of the tunnel boundary are related through the behaviour of the surrounding rock mass. It compares the response of a natural (unreinforced) rock mass with that of a reinforced rock mass, and distinguishes between elastic and plastic ground behaviour.

The horizontal axis (Ur) represents radial displacement, or convergence, of the tunnel wall as the ground moves inward after excavation. The vertical axis (P) represents a fictitious support pressure, which is the pressure that would need to be applied at the tunnel boundary to restrain this displacement. The point P_0 corresponds to the in-situ stress acting on the tunnel boundary before excavation.

The solid curve labelled "GRC of natural rock mass" is the ground reaction curve for unreinforced rock. Immediately after excavation, support pressure drops from P_0, causing inward displacement. Initially, the rock mass behaves elastically, meaning deformation is small and reversible. This region is shown to the left of the vertical dashed line and labelled as the elastic part. As displacement increases beyond this point, the rock strength is exceeded and the ground enters the plastic regime, where irreversible deformation occurs and additional convergence develops with relatively little increase in confinement.

The dashed curve labelled "GRC of reinforced rock mass" shows how ground reinforcement—such as rock bolts, shotcrete, or grouting—modifies the ground response. Reinforcement increases the stiffness and strength of the rock mass, shifting the ground reaction curve downward and to the left. This means that, for a given radial displacement, the reinforced ground requires less support pressure and exhibits smaller convergence than the natural rock mass.

The transition from elastic to plastic behaviour is highlighted at the dashed vertical line. Reinforcement delays the onset of plasticity and reduces the extent of the plastic zone around the tunnel. As a result, the reinforced ground reaches equilibrium at a lower displacement and with more stable load-sharing between the ground and the support system.

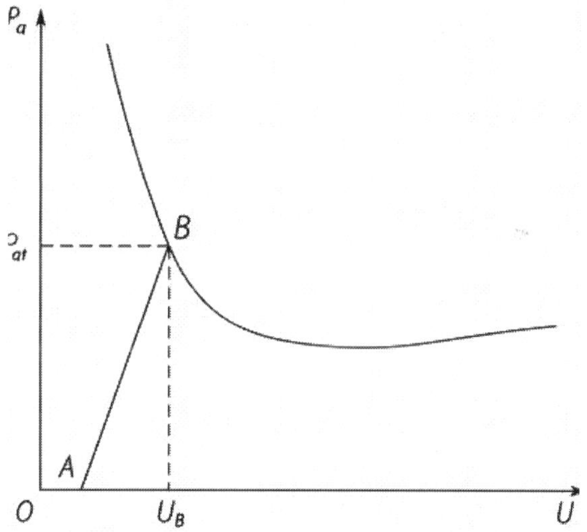

Figure 22: Support characteristic curves showing: (a) a single primary support system, and (b) a double primary support system. Adapted from Liu (2020) [122].

Figure 22 illustrates a support characteristic curve, which describes how a tunnel support system develops confining pressure as the tunnel wall undergoes radial displacement (convergence). It is a core component of the Convergence–Confinement Method, used to understand how support interacts with ground deformation.

The horizontal axis (U) represents the radial displacement of the tunnel wall, while the vertical axis (P_a) represents the active support pressure mobilised by the support system. Point O is the origin, indicating zero displacement and zero support pressure. At this stage, the support has not yet engaged with the ground.

The straight line from A to B represents the behaviour of the support system itself. This line reflects the stiffness of the support: a steeper slope indicates a stiffer support that mobilises pressure rapidly with small deformation, while a flatter slope would indicate a more flexible support. Point A marks the installation of the support, which typically occurs after some initial ground convergence has already taken place.

Point B is the working equilibrium point of the support. At this point, the support has mobilised a pressure P_{at} corresponding to a tunnel wall displacement U^B. The dashed horizontal and vertical lines highlight these equilibrium values. This point is significant because it defines the deformation at which the support becomes fully effective in resisting further convergence.

The curved line to the right of point B represents the ground reaction curve, which shows how the surrounding ground responds to increasing deformation. The intersection at B indicates the balance between ground behaviour and support response: beyond this point, the support pressure is sufficient to restrain further deformation, and the tunnel system reaches a stable equilibrium.

Engineering-wise, this graph shows that support does not act instantaneously but is mobilised progressively as deformation occurs. The location of point B depends on when the support is installed, its stiffness, and its load-bearing capacity. Installing support earlier or using a stiffer system shifts the equilibrium point to lower displacements but higher support pressures, while later or more flexible support allows greater deformation with lower support loads.

The support characteristic curve explains how tunnel support pressure develops as a function of tunnel convergence and how equilibrium is achieved through interaction with the surrounding ground. It provides engineers with a powerful tool for selecting support type, stiffness, and installation timing to control deformation safely and efficiently.

Influence of Groundwater Conditions

Another pivotal factor in GSI is groundwater management, as varying pore water pressures can weaken the surrounding soil and impact its ability to interact favourably with structural support [123]. Draining or depressurizing the tunnel site is often required to restore effective stress and enhance cooperation between ground conditions and tunnel linings. This aspect of tunnel design underlines the necessity of considering hydrogeologic factors to maintain stability in saturated conditions.

Rock Mass Vs Soil Behaviour

Rock mass and soil behave very differently when subjected to excavation and loading, and understanding these differences is fundamental to tunnel design, support selection, and construction method choice. While both are geological materials, their mechanical behaviour, failure mechanisms, and response to stress redistribution are distinct.

Rock Mass Behaviour

A rock mass is characterized by intact rock blocks that are separated by discontinuities such as joints, bedding planes, faults, and fractures. The behaviour of such rock masses is influenced not only by the inherent strength of the intact rock but, more significantly, by the characteristics of the discontinuities, which include their orientation, spacing, persistence, and condition. For example, studies have indicated that different classes of jointed rock masses can exhibit anisotropic properties, significantly influencing how stress is distributed and managed within the rock mass during tunnelling operations [124].

When excavation occurs, the in-situ stresses within the rock mass are redistributed, which can lead to complex deformation and instability. This deformation is often a result of the release of built-up stress due to the disturbance caused by tunnelling activities. Deformation may be expressed as a relaxation of the rock mass surrounding the excavated area, primarily defined as changes in shape or size of the tunnel [125]. Such relaxation can lead to the development of various failure mechanisms, such as block detachment, joint sliding, or stress-induced cracking, rather than a uniform compressive failure of the rock mass [126].

In addition to elastic behaviour, tunnelling in competent rock masses may also exhibit elasto-plastic characteristics, particularly when influenced by joints and other discontinuities in the rock. The effectiveness of stress arching, which is a natural mechanism through which the surrounding rock redistributes loads following an excavation, becomes central in enhancing stability under such conditions [126]. Effective use of support systems like rock bolts and shotcrete serves not only to reinforce the rock mass but also to facilitate the function of such arching, allowing the surrounding rock to carry a major portion of the load instead of relying solely on the support systems [120, 127, 128].

The anisotropic nature of rock mass behaviour plays a critical role in the design and assessment of tunnelling operations. The strength and deformation response of rock masses tend to vary significantly depending on the orientation relative to pre-existing joint sets. This is substantiated by research indicating that a tunnel aligned unfavourably to these joint directions is vulnerable to wedge failures or asymmetric deformations, which can complicate the excavation and increase the likelihood of structural failure [129]. Furthermore, multiple studies have confirmed that the success of tunnelling operations often hinges on the intricate relationship between joint orientation and the resultant morphological response of the surrounding rock, intensifying stress concentrations at the joints [130, 131].

To mitigate the risks associated with tunnelling through jointed rock masses, effective support systems must be employed. Such systems, which may include rock bolts, shotcrete, and advanced engineering techniques, are designed to bind rock blocks together and to promote arching mechanisms within the rock mass. Support measures that use an integrated approach, such as combining steel arches with shotcrete, have been shown to enhance the stability of the excavated sections by closing fractures and harnessing the self-bearing strengths of the rock mass [120, 132]. As articulated by multiple studies, the choice of support system relies heavily on an accurate classification of the rock mass, which considers factors such as the spacing and orientation of joints as well as the overall geological conditions present [133-135].

The behaviour of rock masses in tunnelling is complex and largely determined by the interaction between the rock's inherent properties and the characteristics of inherent discontinuities. The understanding of stress redistribution, anisotropy, and the deployment of effective reinforcement techniques is paramount in ensuring successful tunnelling operations. To optimize stability and minimize risks, continuous research and technological advancements in rock mass classification and support systems are essential.

Soil Behaviour

Soil is fundamentally a particulate material characterized by grains that possess limited or negligible tensile strength. Its behaviour is influenced predominantly by factors such as particle arrangement, density, cohesion, friction, and pore water pressure. Distinctively, soil exhibits a ductile and continuous response, as opposed to the relatively rigid behaviour observed in rocks. This is exemplified in its deformation processes, including compression, shear, and consolidation, rendering it vital for tunnelling operations.

In tunnelling operations, the deformation of soil around an opening tends to be more uniform compared to other mediums, leading to phenomena such as stress redistribution that can result in ground loss, settlement, and volume changes. These factors are especially prevalent in loose or saturated soils, where different failure mechanisms, including shear failure, squeezing, flowing, and collapse, may occur. For example, the study by Zhang et al. [136] underscores how excess pore water pressure and soil consolidation can significantly affect ground behaviour during tunnelling, particularly in clayey soils, where such pressure often leads to substantial deformation and settlement impacts.

The interaction between tunnelling and surrounding soil dynamics necessitates that professionals account for the effects of excavation on soil fabric. Work by Lin et al. [137] indicates that volumetric deformation increases with variations in depth above a tunnel, with soil contraction being a significant contributor to ground volume loss. Furthermore, findings highlight that soft soils, characterized by high water content and compressibility, are especially prone to larger volume changes, necessitating careful intervention and monitoring during construction phases [138].

Groundwater presents a complex challenge in tunnelling, as elevated pore water pressure can markedly reduce effective stress in soil, thereby weakening its structural integrity and increasing susceptibility to deformation. A pivotal study elucidates how scenarios involving tunnelling in under-consolidated soils, such as those observed in Jakarta, can be exceedingly precarious due to interactions between construction activities and groundwater levels [139]. The dissipation of excess pore water pressure must be managed to mitigate the risks associated with instability and surface settlement [140].

Maintaining tunnel stability requires that engineers effectively manage the confinement around the excavation face and throughout the tunnel perimeter. This management can involve techniques like using pressurized Tunnel Boring Machines, implementing ground improvement strategies, or immediate installation of linings, as emphasized by several studies [141, 142]. The importance of recognizing external factors, such as soil heterogeneity and pore water pressure distribution, is underscored in research indicating that these parameters substantially influence the support pressures required for safe excavation under varying conditions [142].

The failure modes of soil during tunnelling include shear failure, which can be exacerbated under high pore water pressure conditions. For example, a comprehensive evaluation of base excavations and their impacts on tunnel stability reveals that not addressing consolidation and pore water issues can lead to significant long-term deformation as the soil continues to settle and consolidate over time [143]. Additionally, research focusing on excess pore water pressure confirms that it acts as a pivotal factor in soil stability during tunnelling operations, necessitating proactive management strategies to mitigate failure risks [144, 145].

The behaviour of soil as a particulate material is essential in understanding its implications for tunnelling processes. Soil's lack of tensile strength, influenced by its physical properties and groundwater interactions, results in complex behaviour during tunnelling. The successful management of these variables is crucial to ensuring the stability of tunnels and minimizing associated risks. Thus, ongoing research and methodical approaches are pivotal in improving tunnelling safety and efficiency in variable soil conditions.

Deformation Characteristics

Rock masses typically exhibit limited deformation prior to failure, especially in strong, brittle rock, whereas soils may undergo large deformations before reaching failure. In rock, deformation is often localised along discontinuities, while in soil it is more distributed throughout the ground mass.

These differences explain why tunnels in rock can often tolerate some unsupported span for short periods, whereas tunnels in soil usually require continuous support or face pressure to prevent collapse.

Implications for Tunnel Design and Construction

Because of these fundamental behavioural differences, tunnelling approaches diverge significantly between rock and soil environments. In rock tunnelling, the primary objective is often to reinforce the existing ground so that its inherent strength can be mobilised. This is achieved through systems such as rock bolts and shotcrete, which bind individual rock blocks together, control movement along discontinuities, and promote the formation of a stable load-bearing arch around the tunnel.

In contrast, soil tunnelling depends much more on maintaining confinement and controlling deformation from the outset. Because soils have little tensile strength and are highly sensitive to stress changes and groundwater pressure, stability is achieved through continuous face support, pressurised excavation methods, and relatively rigid linings installed promptly to prevent collapse, excessive settlement, or groundwater inflow.

These differences also influence construction philosophy. Observational and adaptive methods are particularly well suited to tunnelling in variable rock masses, where ground behaviour can be monitored and support adjusted as conditions change. Soil tunnels, however, generally allow far less deformation and uncertainty, requiring tighter control of excavation, support installation, and groundwater management to ensure safety and protect surface structures.

Load Paths and Confinement

When a tunnel is excavated, the original in-situ stresses existing in the material must be redistributed through the surrounding earth. This phenomenon is important for understanding how tunnels can remain stable post-excavation. As stresses that were previously supported by the removed material find new routes, they develop what are known as load paths. These load paths usually redirect stresses around the tunnel opening, creating curved trajectories resembling an arch, a process termed ground arching. Such ground arching is significant as it contributes to stabilizing the tunnel walls by producing radial and tangential stress that act towards the tunnel wall and circumferentially along the opening, respectively [146, 147].

In competent rock and solid ground, load paths can develop quickly, stabilizing the construction rapidly through self-supporting mechanisms [148]. The quality of the rock and the methodology of excavation greatly influence this process. If the surrounding material is strong enough, it can support the redistributed stresses effectively, allowing for engineering designs that rely on minimal additional support. Conversely, in weaker or water-bearing materials, the load paths may be less defined, necessitating the implementation of robust support systems to maintain stability and mitigate the risk of collapse [149, 150].

Competent Rock and Solid Ground

Competent rock and solid ground are terms used in geotechnical and tunnelling engineering to describe ground that is strong, stable, and capable of supporting loads with limited deformation—but they are used in slightly different contexts.

Competent rock refers specifically to rock masses with sufficient strength, stiffness, and continuity to stand with minimal support after excavation. It typically has intact rock of good quality, limited fracturing or jointing, low weathering, and adequate confinement at depth. In competent rock, stresses can redistribute efficiently around a tunnel, allowing natural arching to develop. This means that deformation is small, failure is unlikely, and support systems such as rock bolts or thin shotcrete layers are often sufficient. Examples include massive limestone, granite, basalt, and strong sandstone with favourable joint conditions.

Solid ground is a broader, more informal term that can apply to both rock and soil. It describes ground that is dense, stiff, and stable, with good load-bearing capacity and low compressibility. In soil mechanics, solid ground may include dense sands and gravels, stiff clays, or cemented soils that behave predictably under load. While solid ground may not be rock, it can still provide adequate support for excavations if groundwater conditions are controlled and deformation is managed.

The key difference lies in material behaviour. Competent rock derives its stability primarily from its inherent strength and structural integrity, while solid ground may rely more on density, cohesion, or confinement to remain stable. In tunnelling, competent rock is ideal because it promotes arching and limits support demands, whereas solid ground may still require more systematic support and groundwater control to prevent settlement or collapse.

Confinement in tunnelling refers to the radial pressure exerted by the surrounding ground at the tunnel boundaries, which prevents excessive deformation or failure. This radial pressure is crucial as it enhances the effective strength of the soil or rock by limiting mechanisms such as dilation, cracking, and shear movement [151, 152]. The effectiveness of this confinement is pivotal to successful tunnelling; it not only relies on the physical properties of the surrounding material but is also greatly enhanced by various support systems employed within the tunnel structure.

Support systems, including shotcrete, rock bolts, steel ribs, and segmental linings, are integral in maintaining effective confinement. By applying radial pressure to the tunnel walls, these supports prevent the loosening of surrounding ground and thus preserve the integrity of the load paths. Early installation of such support systems is particularly vital. Delayed reinforcement can lead to uncontrolled deformation, which can disrupt load paths and exacerbate ground instability, further increasing support requirements [150, 153] .

Modern tunnelling methodologies, such as the New Austrian Tunnelling Method (NATM) and the use of Tunnel Boring Machines with immediate lining installation, have been specifically developed to manage load paths and confinement effectively. These techniques recognize that the tunnel lining is not solely responsible for supporting the overburden load; instead, it acts in synergy with the natural ground properties to create a stable structural system [147, 154]. The convergence-confinement method illustrates this well, emphasizing the interaction between rock and lining systems under stress [153, 154].

Understanding load paths and confinement is essential for designing stable tunnelling systems. The management of stresses through careful engineering allows tunnels to maintain structural integrity despite the removal of material that had previously supported the overburden. Successful tunnelling not only requires appropriate excavation methods and support system implementation but also relies on the integration of these factors to maintain stability over time.

Tunnel Stability Fundamentals

Tunnel stability is a fundamental aspect of underground construction that concerns the integrity and geometric configuration of the ground surrounding tunnel openings. This stability is primarily influenced by a combination of ground conditions, in-situ stress, methods of excavation, and the capacity and timing of support installations. Maintaining stability involves understanding how the surrounding soil or rock responds when excavation disrupts existing stress equilibrium, leading to a redistribution of forces around the excavation site.

Before excavation begins, the ground is in a state of stress equilibrium characterized by vertical stresses arising from overburden and horizontal stresses due to geological history. The excavation process disrupts this equilibrium, as material loss forces the surrounding ground to accommodate new stress distributions. The effectiveness of this accommodation is critical; excessive deformation can lead to structural failure [155, 156].

Gong et al. [157] describe the challenges posed by high-pressure water conditions, illustrating how pore pressure alterations can affect ground stability by modifying effective stress levels. Similarly, Zhao et al. [158] conducted a study on significant ground displacements observed during tunnel excavation, which ties back to these principles of stress redistribution and effective stress management.

A significant concept in understanding tunnel stability is ground arching, where the redistribution of stresses forms load paths that allow overlying ground to support much of the load above the tunnel [159]. This arching is highly effective in competent rock, leading to minimal deformation. However, in weaker ground conditions, the inefficiency of arching coupled with reliance on artificial support systems can result in significant instability [160].

Monitoring the deformation of tunnel walls is paramount, as some ground movement—known as convergence—is often beneficial. This movement aids in mobilizing the ground's inherent strength, but if it exceeds the ground's capacity, risks such as spalling and beam collapse arise [161]. Hence, deformation control is essential in ensuring that structural integrity is maintained through monitoring systems that provide real-time feedback [160].

Support systems play a vital role in managing the stresses acting on a tunnel. Various support mechanisms, such as shotcrete, rock bolts, and steel ribs, are employed to enhance confinement and improve effective ground strength [161]. Early installation of these supports is crucial, as delaying this process can lead to uncontrolled stress anomalies and increased demands for support due to the expanding plastic zone around tunnels [162].

Research by Fahimifar and Zareifard [162] demonstrated how support methods could mitigate the impact of pore-water pressure on tunnel linings, emphasizing that the design must accommodate these pressures to maintain structural integrity. The coordination between excavation timing and support placement is, therefore, a key factor in ensuring stability throughout a tunnel's service life.

Groundwater significantly impacts tunnel stability by affecting pore water pressure, thus reducing effective stress in surrounding ground materials, which can increase the potential for deformation and other instabilities [157]. Excavation activities can alter the natural flow of groundwater, necessitating effective drainage and pressure control systems following excavation to ensure stability [156, 163].

Watershed management is paramount during excavation, as seen in case studies where groundwater control through advanced drainage techniques has shown to mitigate positive pore pressures and bolster ground strength significantly [164]. Without such management, tunnels are susceptible to phenomena like heaving and piping, which can destabilize the structure [156].

The method of excavation chosen also dictates the stability of a tunnel. For example, Tunnel Boring Machines provide more consistent openings and facilitate immediate support, resulting in more predictable ground behaviour. In contrast, drilling and blasting techniques necessitate vigilant monitoring of excavation stages and support adjustments due to the variability of conditions encountered during these methods [159].

Furthermore, Aygar [161] highlights how excavation methods can affect the collapse mechanisms within tunnel systems, particularly in soft grounds, emphasizing the need for methodical approaches tailored to specific soil conditions.

Tunnel stability is governed by a complex interaction of ground conditions, pore-water pressures, and support mechanisms, influenced by excavation methods and timing. Engineers must strike a balance between these factors to design and implement effective excavation and support systems that ensure a stable and safe tunnelling process. Continuous monitoring and adherence to adaptive management principles remain integral to realizing and maintaining this balance throughout the tunnel's lifecycle.

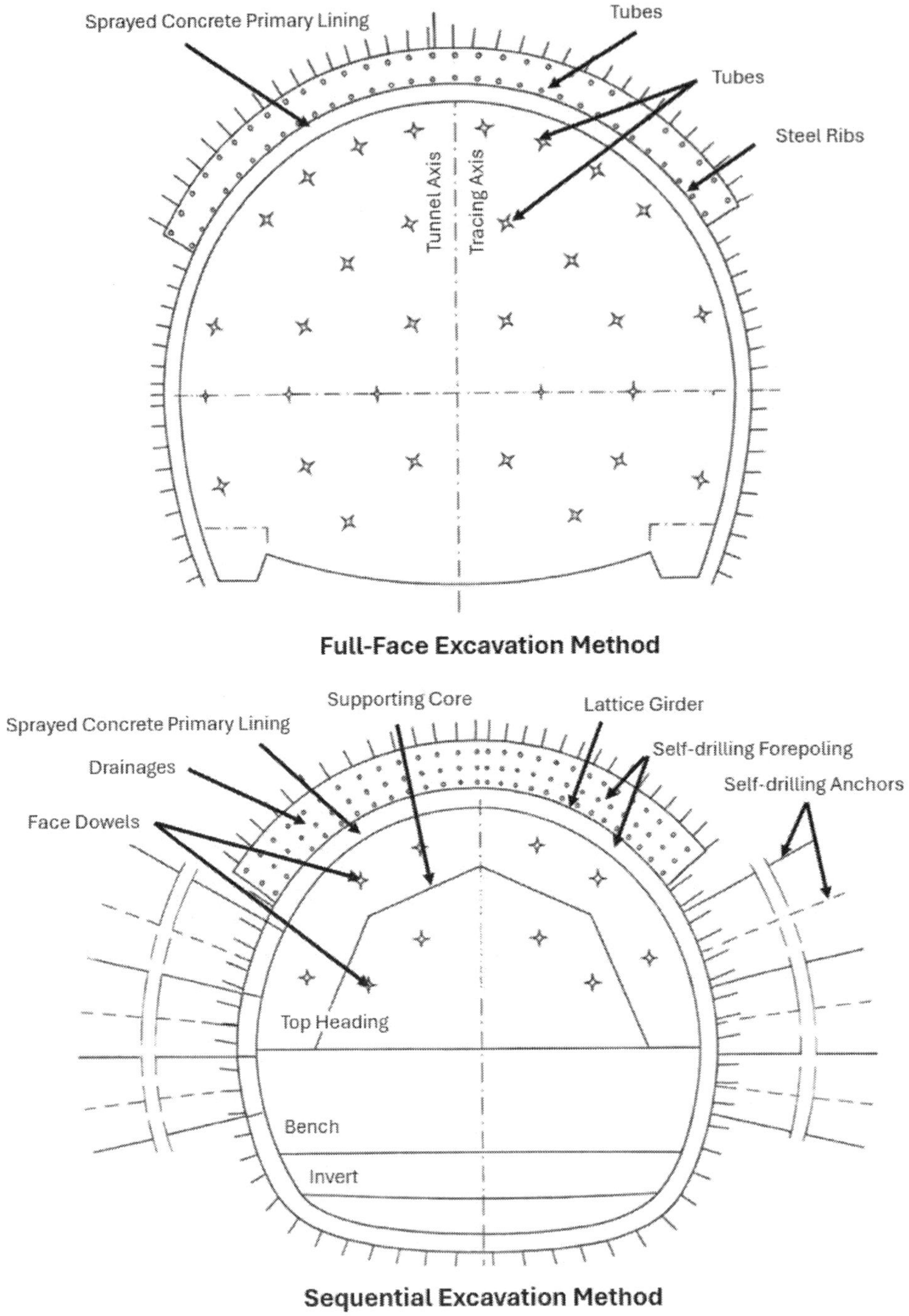

Figure 23: Comparison of Core-Face Stabilisation Design for Full-Face and Sequential Excavation Methods. Adapted from Asadollahi (2016) [165].

The diagram shown as Figure 23 compares core-face stabilisation concepts for two tunnelling approaches: the Full-Face Excavation Method and the Sequential Excavation Method (SEM / NATM). In both approaches, the fundamental objective is to maintain stability at the tunnel face and within the surrounding ground by controlling deformation, preserving confinement, and managing groundwater conditions throughout excavation.

In the full-face excavation method, the entire tunnel cross-section is excavated in a single operation. Because the full opening is exposed simultaneously, stability depends on the rapid installation of uniform and continuous support around the excavation perimeter and at the face. Sprayed concrete primary lining is applied immediately behind the face to provide circumferential confinement and prevent ravelling of the ground. Steel ribs embedded within the sprayed concrete increase stiffness and load-carrying capacity, particularly in weak or fractured ground. Ahead of the face, tubes such as forepoling or umbrella pipes are installed to reinforce the ground and reduce deformation before excavation advances. This method relies on an even distribution of support around the full perimeter to control deformation uniformly. While efficient in relatively consistent ground conditions, full-face excavation offers limited flexibility when ground conditions change abruptly.

The sequential excavation method divides the tunnel cross-section into stages, allowing stability to be managed progressively as excavation advances. The diagram illustrates a typical sequence consisting of the top heading, bench, and invert. A central supporting core of ground is intentionally left in place during the early stages to provide temporary face and structural support. Face dowels are installed into the excavation face to reinforce weak ground and limit face extrusion or collapse. Self-drilling forepoling and anchors are placed ahead of the face to pre-support the ground, particularly in poor or mixed geological conditions. Lattice girders and sprayed concrete primary lining are applied incrementally to provide flexible confinement as each stage progresses, while drainage elements are used to reduce groundwater pressure and improve effective stress conditions.

By advancing in stages, the sequential excavation method allows controlled deformation, enabling the ground to mobilise its inherent strength while maintaining safety. Support systems can be adjusted in response to observed ground behaviour, making SEM highly adaptable to variable and uncertain ground conditions.

The diagram highlights several key differences between the two methods. Full-face excavation removes the entire cross-section at once, whereas sequential excavation progresses in controlled stages. Face stability in full-face excavation relies on rapid, uniform lining and pre-support, while sequential excavation relies on a supporting core, staged excavation, and targeted reinforcement. As a result, sequential excavation offers greater flexibility in changing ground conditions, whereas full-face excavation requires higher immediate support capacity and concentrates risk at the face.

Overall, Figure 23 underscores the engineering significance of core-face stabilisation in ensuring tunnel safety and performance. Full-face excavation prioritises speed and uniform confinement, while sequential excavation emphasises adaptability and observational control. The choice between these methods depends on ground quality and variability, groundwater conditions, construction risk, and project-specific performance requirements.

Chapter 3

Ground and Geological Conditions

Building on the fundamental principles of tunnel construction and ground behaviour introduced in Chapters 1 and 2, this chapter focuses on how subsurface conditions are identified, characterised, and interpreted prior to and during tunnelling works. While an understanding of excavation methods and ground response is essential, successful tunnelling ultimately depends on the quality of information available about the ground before excavation begins and the ability to refine that understanding as construction progresses.

Chapter 1 established the purpose, methods, and evolution of tunnelling, highlighting how modern techniques seek to control deformation, manage risk, and optimise performance. Chapter 2 then examined the mechanical behaviour of rock and soil, showing how strength, deformability, discontinuities, and groundwater govern tunnel stability. Together, these chapters demonstrate that tunnel behaviour is not defined by excavation alone, but by the interaction between construction methods and the ground through which the tunnel passes.

This chapter addresses the practical challenge of predicting that interaction. It introduces the investigation methods used to define geological and geotechnical conditions, including boreholes, in-situ testing, geophysical surveys, and sampling techniques. Emphasis is placed on how investigation data are integrated into geological and geotechnical models, how uncertainties are identified, and how risks associated with variability, groundwater, and adverse ground conditions are interpreted and managed.

The chapter also highlights the limitations of investigation methods and the importance of progressive investigation and the observational approach during construction. By understanding how ground investigation informs excavation method selection, support design, groundwater control, and contractual baselines, this chapter provides the critical link between theoretical ground behaviour and practical tunnel engineering decision-making.

Engineering Geology for Tunnelling

Engineering geology for tunnelling is the discipline that applies geological knowledge to the planning, design, construction, and operation of tunnels. Its purpose is to understand how the geological environment will behave when excavated and how that behaviour influences tunnel stability, construction method selection, support design, risk, and long-term performance.

Unlike pure geology, which focuses on describing earth materials and processes, engineering geology focuses on how geological conditions interact with engineering works. In tunnelling, this interaction is critical because excavation disturbs the natural stress state of the ground and exposes materials that may behave very differently once confined support is removed.

A central task of engineering geology is the characterisation of the ground mass, not just the intact material. Tunnels are rarely governed by the strength of intact rock or soil alone; instead, behaviour is controlled by features such as joints, bedding planes, faults, shear zones, weathering, and fabric orientation. Engineering geologists assess these features to determine block size, potential failure mechanisms, permeability, and deformability.

Geological structure and anisotropy are especially important. The orientation of bedding, foliation, joints, and faults relative to the tunnel axis can strongly influence stability. For example, unfavourably oriented joints may daylight into the excavation, causing wedge failures, while folded or faulted zones may introduce highly variable ground conditions over short distances. Engineering geology helps anticipate these changes so excavation sequences and support systems can be adapted.

The Importance of Geological Structure and Anisotropy in Tunnelling

Understanding the geological structure and anisotropy of rock and soil is critical for successful tunnelling. Discontinuities such as bedding planes, foliation, joints, and faults play crucial roles in how the ground interacts with engineered structures. Depending on their orientation concerning the tunnel axis, these geological features can significantly influence the stability of a tunnel once it is excavated.

Ground masses generally contain various planes of weakness that dictate both the deformation characteristics and the load-bearing capacity of the tunnelling environment. When these discontinuities are favourably oriented, they contribute positively to the stress redistribution around the tunnel, promoting stability and enhancing the natural arching effect of the surrounding rock [166, 167]. Conversely, unfavourably oriented discontinuities can lead to failures known as "daylighting," where these planes intersect the tunnel excavation, resulting in potential rock blocks or wedges that can become unstable [168]. A keen understanding of these dynamics is essential for predicting potential failure mechanisms in tunnelling projects.

Faulted and folded zones introduce additional complexities in tunnelling, as properties of ground materials can vary markedly over short distances. Transitioning from strong, competent rock to crushed or weathered materials often leads to reduced strength and increased deformability, complicating tunnel support strategies [169]. Such geological heterogeneity can disrupt load-bearing pathways, amplifying the risk of deformation and instability during and post-excavation, often resulting in excess water inflows or squeezing conditions [170].

Engineering geology plays a pivotal role in mitigating these risks through careful identification and analysis of geological structures prior to excavation. Techniques such as borehole investigations, surface mapping, and tunnel face mapping allow engineers to predict where unfavourable structures might be encountered, facilitating proactive measures [171]. Knowledge gained from these analyses can lead to adaptations in excavation strategies, such as changing round lengths, utilizing staged excavation methods, or enhancing support systems with targeted interventions like rock bolts or shotcrete applications in critical areas [172].

Adapting to geological variations not only enhances stability but also shifts tunnelling from a reactive strategy to a more predictive and adaptive framework, ensuring longevity and security in the tunnel alignment [173].

Groundwater conditions are another critical focus. Engineering geologists assess aquifers, permeability, pore pressures, and potential inflow paths such as faults or karst systems. Groundwater reduces effective stress, weakens ground, increases deformation, and can destabilise the tunnel face. Understanding groundwater behaviour allows engineers to plan drainage, depressurisation, grouting, or pressurised excavation methods.

Engineering geology also underpins the use of rock mass classification systems, such as RMR, Q-system, and GSI. These systems combine geological observations with engineering judgement to provide guidance on expected ground behaviour and suitable support types. While not design tools on their own, they provide a common language between geologists and engineers and support rapid decision-making during construction.

Rock Mass Classification Systems: RMR, Q-system, and GSI

Rock mass classification systems are used in tunnelling to describe the quality and behaviour of rock masses and to guide excavation and support decisions. Because rock masses are controlled not just by intact rock strength but by joints, faults, groundwater, and stress conditions, these systems provide a structured way to translate geological observations into engineering judgement. The three most widely used systems are RMR, the Q-system, and GSI.

Rock Mass Rating (RMR)

The Rock Mass Rating (RMR) system provides an overall numerical score that represents rock mass quality on a scale from 0 (very poor) to 100 (excellent). It is based on six parameters:

1. Intact rock strength

2. Rock Quality Designation (RQD)
3. Spacing of discontinuities
4. Condition of discontinuities (roughness, weathering, infilling)
5. Groundwater conditions
6. Orientation of discontinuities relative to the excavation

Each parameter is assigned a rating, and the sum gives the RMR value. The resulting class (very poor to very good rock) is then linked to stand-up time, excavation method, and typical tunnel support measures, such as rock bolts, shotcrete thickness, or steel ribs. RMR is widely used because it is intuitive, systematic, and easy to apply in the field. However, it provides general guidance rather than detailed design, and it is less sensitive to stress conditions and tunnel size.

Q-system (Tunnelling Quality Index)

The Q-system was developed specifically for tunnelling and places strong emphasis on block size, stress effects, and groundwater, which are critical to tunnel stability. It expresses rock mass quality as:

$$Q = \frac{RQD}{J_n} \times \frac{J_r}{J_a} \times \frac{J_w}{SRF}$$

Where:

- **RQD** = Rock Quality Designation
- **Jn** = Joint set number
- **Jr** = Joint roughness number
- **Ja** = Joint alteration number
- **Jw** = Joint water reduction factor
- **SRF** = Stress reduction factor

Q-values typically range from 0.001 (extremely poor rock) to 1000 (exceptionally good rock). The system links Q-values to tunnel span, excavation method, and detailed support recommendations, including bolt length, bolt spacing, and shotcrete thickness. The Q-system is particularly strong in addressing stress-related problems, such as squeezing ground or rock bursting, and is well suited to deep tunnels and variable rock conditions. Its main limitation is that it can appear complex and requires experienced judgement to select appropriate parameter values.

Geological Strength Index (GSI)

The Geological Strength Index (GSI) is a descriptive system used primarily to estimate the mechanical properties of rock masses for numerical modelling and design analysis. Rather than producing direct support recommendations, GSI characterises rock masses based on:

- Rock mass structure (blocky, massive, disintegrated, laminated, etc.)
- Surface condition of discontinuities (roughness, weathering, infilling)

GSI is typically estimated visually using charts and ranges from about 10 (very poor, heavily fractured rock) to 90 (intact or massive rock). The GSI value is then used with empirical strength criteria (such as Hoek–Brown) to derive rock mass strength and deformation parameters.

GSI is especially valuable in analytical and numerical modelling, where realistic rock mass properties are required. It is less prescriptive than RMR or Q-system and relies heavily on geological judgement and experience.

How the Systems Are Used Together
In practice, these systems are often used in combination rather than in isolation. RMR provides a quick, broad assessment of rock quality, the Q-system offers detailed tunnelling-focused support guidance, and GSI supplies the parameters needed for advanced analysis and modelling. Together, they allow engineers and engineering geologists to understand rock behaviour, compare conditions consistently, and design excavation and support systems that are appropriate for the ground encountered.

A defining feature of engineering geology in tunnelling is the emphasis on observation during excavation. Ground conditions encountered underground often differ from predictions based on boreholes and surface mapping. Engineering geologists map tunnel faces, record discontinuities, assess groundwater inflows, and update ground models in real time. This observational role is fundamental to adaptive approaches such as the New Austrian Tunnelling Method, where support design evolves in response to actual ground behaviour.

Engineering geology also plays a key role in risk management. By identifying geological hazards—such as squeezing ground, swelling clays, fault crossings, gas, karst cavities, or abrasive rock—engineering geologists help engineers anticipate construction challenges, select appropriate excavation methods, and implement mitigation measures before problems escalate.

Rock Types and Mechanical Properties

Understanding rock types and their mechanical properties is fundamental to tunnelling and underground engineering because excavation behaviour, stability, support requirements, and construction method selection are all strongly controlled by how different rocks respond to stress redistribution and deformation.

Major Rock Types and Their Engineering Significance

Igneous rocks, including granite, basalt, diorite, and gabbro, form from the solidification of magma and are generally strong, dense, and durable. From a tunnelling perspective, these rocks typically exhibit high uniaxial compressive strength, high stiffness, low porosity and permeability, and brittle failure behaviour. Intact igneous rock is often highly competent, allowing tunnels to remain stable with minimal support where jointing is limited. However, tunnel stability in igneous formations is frequently governed more by the presence and orientation of joints and fractures than by intact rock strength. In deep tunnels subjected to high in-situ stresses, the brittle nature of igneous rock can lead to spalling or rockbursting, requiring careful stress management and robust support systems.

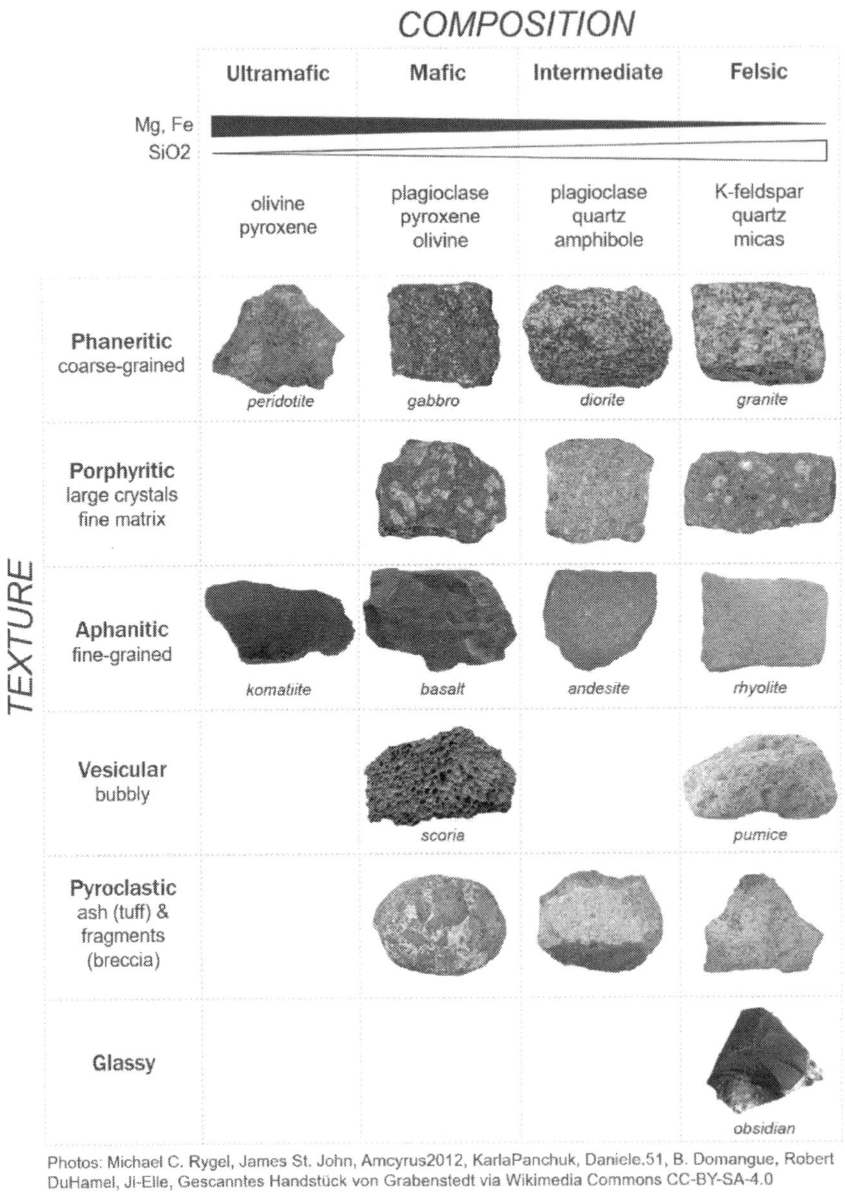

Figure 24: Table of very general types of igneous rocks, categorized by composition and texture. Carie Frantz (Carie027), CC BY-SA 4.0, via Wikimedia Commons.

Spalling and rockbursting are stress-induced failure mechanisms that occur mainly in strong, brittle rock when a tunnel is excavated and the in-situ stress field is suddenly redistributed [174].

Spalling is a progressive form of brittle failure in which thin slabs or flakes of rock detach from the tunnel wall or crown. It occurs when high tangential stresses develop near the excavation boundary and exceed the rock's tensile strength, causing cracking parallel to the tunnel surface. Spalling is most common in deep tunnels in hard rock (such as granite or quartzite), where confinement is reduced rapidly by excavation. While spalling may begin as shallow peeling, it can deepen over time if stresses remain high or support is delayed, leading to overbreak and increased support demand [174].

Rockbursting is a more violent and sudden failure mechanism. It occurs when large amounts of strain energy stored in highly stressed, brittle rock are released abruptly during excavation. This release can eject rock fragments explosively into the tunnel, sometimes accompanied by loud noises and seismic events. Rockbursts are typically associated with very deep tunnels, high horizontal stresses, and stiff, low-porosity rock masses that store elastic energy efficiently. Unlike spalling, rockbursting poses a significant safety hazard due to its sudden and unpredictable nature [174, 175].

The key distinction between the two lies in failure rate and energy release. Spalling is generally gradual and localised, allowing time for observation and intervention, whereas rockbursting is instantaneous and energetic. Both mechanisms are strongly influenced by tunnel depth, stress magnitude and orientation, excavation rate, and the brittleness of the rock mass.

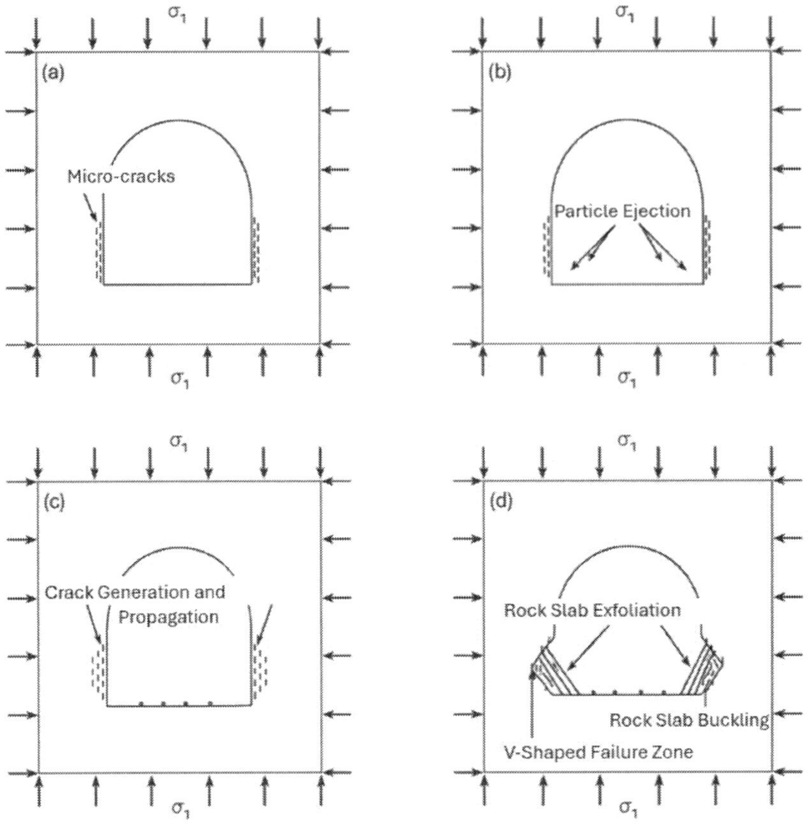

Figure 25: Schematic of the evolution of spalling failure in a D-shaped tunnel with increasing vertical stress. Adapted from Si et al. (2023) [176].

Mitigation strategies for spalling and rockbursting focus on stress management and confinement. These include reducing excavation advance length, installing support early to restore confinement, using yielding support systems (such as energy-absorbing rock bolts), altering tunnel shape to reduce stress concentrations, and, in extreme cases, preconditioning the rock mass through destress blasting or pilot tunnels.

Sedimentary rocks include sandstone, limestone, shale, siltstone, and conglomerate, and their engineering behaviour varies widely depending on factors such as cementation, bedding, and degree of weathering. Sandstone may range from weak to very strong and generally performs well in tunnelling when it is well cemented. Limestone can also be strong, but it often presents additional challenges due to the potential presence of karst features, cavities, and solution channels, which can significantly affect stability and groundwater inflow. Shale and mudstone are typically weaker, finely laminated, and highly anisotropic, with strength and stiffness strongly influenced by moisture content. Pronounced bedding planes are a defining characteristic of many sedimentary rocks and can introduce directional weakness that dominates tunnel behaviour when unfavourably oriented relative to the excavation.

Figure 26: Instances of sedimentary rocks. Adapted from ManuRoquette, CC BY 4.0, via Wikimedia Commons.

Metamorphic rocks, such as gneiss, schist, slate, and quartzite, have been altered by heat and pressure, resulting in foliated or banded structures that strongly influence their mechanical behaviour. Quartzite is generally very strong and highly abrasive, behaving in many respects like hard igneous rock. Gneiss and schist may have high intact strength but exhibit strong anisotropy due to well-developed foliation, which can control deformation and failure mechanisms. Slate commonly splits along cleavage planes, making it susceptible to planar instability. In tunnelling, metamorphic rocks often perform well when foliation is favourably oriented, but when foliation planes daylight into the excavation, they can lead to planar or wedge failures that require targeted support and careful excavation sequencing.

Figure 27: Gneiss. Siim, CC BY-SA 3.0, via Wikimedia Commons.

Key Mechanical Properties Relevant to Tunnelling

Uniaxial Compressive Strength (UCS) defines the maximum axial stress that intact rock can withstand before failure. It is a fundamental indicator of excavatability, potential failure mechanisms, and overall support demand. However, UCS alone does not determine tunnel behaviour and must always be interpreted in the context of rock mass structure, jointing, and in-situ stress conditions.

A tunnel driven through massive granite with a UCS of 200–300 MPa may appear highly favourable based on laboratory strength alone. However, if the rock mass is heavily jointed or intersected by faults, the tunnel may still experience block falls and instability despite the high UCS. Conversely, a tunnel excavated in moderately strong sandstone (UCS 40–80 MPa) with wide joint spacing and good cementation may remain stable with minimal support. These examples show that UCS provides an indication of excavatability and stress capacity, but tunnel behaviour is ultimately controlled by rock mass structure and stress conditions rather than intact strength alone.

Tensile strength of rock is typically only about 5–15% of its compressive strength, making rock far more vulnerable to failure in tension than in compression. Tensile cracking governs many common tunnelling problems, including spalling, overbreak, and face instability, particularly in strong but brittle rock types where stress concentrations develop rapidly after excavation.

In deep hard-rock tunnels excavated in quartzite or granite, high tangential stresses develop around the excavation boundary. Because the tensile strength of these rocks is only a small fraction of their compressive strength, tensile cracking can occur parallel to the tunnel wall, leading to spalling. This is commonly observed as thin slabs peeling from the crown or sidewalls shortly after excavation. In extreme cases, such tensile failure can progress to violent rockbursting, even though the intact rock is very strong in compression.

Elastic modulus and deformability describe how much a rock mass deforms under load. Stiff rocks tend to deform very little before failure, while softer or more fractured rocks can undergo significant deformation. Deformability has a strong influence on tunnel convergence, lining loads, and the interaction between ground and support systems. Highly deformable ground often requires flexible support arrangements and staged excavation to accommodate controlled movement.

A tunnel excavated in stiff crystalline rock may show minimal convergence, with ground movements of only a few millimetres, resulting in relatively low lining deformation but potentially high stress concentrations. In contrast, a tunnel driven through weak mudstone or highly fractured shale may experience significant convergence as the ground deforms plastically around the opening. In such deformable ground, rigid linings can crack or fail, requiring flexible support systems such as yielding bolts, thick shotcrete, and staged excavation to accommodate controlled movement.

Shear strength and discontinuities are often more critical to tunnel stability than intact rock strength. Shear behaviour along joints, bedding planes, faults, and shear zones controls whether rock blocks remain stable or move. Factors such as joint roughness, infilling material, degree of weathering, and water pressure determine the likelihood of sliding, rotation, or collapse around the excavation.

In a tunnel excavated through bedded limestone, stability may be governed by the shear strength along bedding planes rather than by intact rock strength. If these bedding planes dip toward the excavation, blocks can slide into the tunnel, forming wedge failures. Similarly, in faulted ground with clay-filled joints, even moderate stress changes can trigger sliding or rotation of rock blocks. Groundwater pressure acting along discontinuities can further reduce shear resistance, increasing the likelihood of collapse unless drainage or reinforcement is provided.

The contrast between brittleness and ductility is another key consideration. Brittle rocks tend to fail suddenly and violently, leading to spalling or rockbursting under high stress conditions. In contrast, ductile or weak rocks deform progressively, which can result in squeezing behaviour or time-dependent tunnel closure. Recognising this distinction is essential when selecting excavation methods and designing appropriate support systems.

A deep tunnel excavated in hard, brittle granite may experience sudden spalling or rockbursting shortly after excavation due to rapid stress release and limited deformation capacity. By contrast, a tunnel driven through ductile rock such as weak shale or evaporitic formations may not fail suddenly but instead undergo gradual inward movement of the walls and invert, a phenomenon known as squeezing ground. This progressive deformation can continue long after excavation, requiring long-term support measures and careful control of excavation advance.

Permeability and porosity govern groundwater inflow, pore pressure development, and long-term tunnel durability. Highly permeable rocks can lead to significant water inflows that complicate excavation and reduce effective stress, while low-permeability rocks may trap pore pressures that weaken the ground and compromise stability if not properly managed.

In highly permeable sandstone, excavation may trigger large groundwater inflows that reduce effective stress and destabilise the tunnel face, necessitating dewatering or grouting. Conversely, a tunnel in low-permeability claystone or mudstone may experience little initial inflow, but high pore pressures trapped in the rock can reduce strength and lead to delayed instability or swelling if not properly drained. Both scenarios demonstrate how permeability and porosity directly affect tunnel stability and construction strategy.

Finally, a critical distinction in tunnelling engineering is between intact rock properties and rock mass behaviour. Laboratory tests describe the strength of intact rock specimens, but tunnel performance is controlled by the condition of the rock mass as a whole. Even very strong intact rock can behave poorly if it is heavily fractured, weathered, or faulted. As a result, engineering decisions are based on rock mass quality and behaviour rather than on intact strength values alone.

Laboratory testing may show very high strength for intact basalt cores, suggesting excellent tunnelling conditions. However, if the basalt is intensely fractured due to cooling joints or tectonic activity, the rock mass may behave poorly, with frequent block falls and high support demand. Conversely, a weaker intact rock, such as moderately cemented sandstone, may form a stable tunnel if jointing is limited and well interlocked. These examples reinforce that tunnelling design must be based on rock mass behaviour, not just intact material properties.

Engineering Implications of Rock Type and Mechanical Properties

The engineering implications of rock type and its mechanical properties are critical in the field of underground excavations. These properties dictate the choice of excavation method and influence support types, face stability, and long-term maintenance strategies for tunnels. This paper synthesizes recent studies to explore various factors impacted by rock types and mechanical properties, such as excavation methods, support systems, face stability, groundwater management, and tunnel durability.

The selection of the excavation method, Tunnel Boring Machine (TBM) versus drill-and-blast, is heavily influenced by rock type. TBMs are favoured in conditions with hard rock as they provide greater efficiency and less surface disturbance [177, 178]. Research indicates that different rock masses require distinct operational characteristics; thus, the performance of

TBMs can be predicted using geomechanical parameters derived from actual performance data, including cutterhead torque and thrust force [14, 179]. Conversely, in rock types that are less homogeneous or brittle, drill-and-blast methods may still be necessary despite their potential for increased ground disturbance [14, 180].

The mechanical properties of surrounding rock directly affect the choice of support systems and their installation timing. For instance, shotcrete is often employed due to its ability to enhance bearing capacity and facilitate significant stress redistribution within tunnel support systems, thereby improving overall stability [181-183]. The interaction dynamics between shotcrete and the surrounding rock mass must be evaluated to determine optimal reinforcement techniques [183, 184]. Moreover, support structures such as rock bolts and segments are designed based on stability analysis that considers factors like the excavation-damaged zone [14, 185].

Face stability is primarily governed by the mechanical characteristics of the excavation environment. Understanding the rock mass's stress state is essential, as it significantly influences deformation behaviour during excavation [14, 181]. The use of robust optimization methods for bolt-shotcrete support structures can significantly mitigate uncertainties related to rock and soil parameters, ultimately enhancing face stability [184]. Effective design and timely implementation of support systems can considerably control deformations that may arise during and after excavation [182].

Groundwater management strategies are intricately linked to the geological and geotechnical properties of the rock mass through which tunnels are excavated. The hydraulic conductivity and porosity of the rock influence groundwater behaviour, necessitating tailored strategies based on actual excavation conditions [186, 187]. The utilization of TBMs, which tend to introduce less disturbance to the ground, can facilitate better groundwater management in sensitive geological settings [177, 178]. Continuous monitoring of geotechnical conditions, as gathered from TBM parameters, can provide valuable early warnings for potential groundwater-related issues affecting tunnel stability [178].

The long-term durability of tunnels is contingent upon the performance of tunnel linings, which are influenced by the properties of the rock mass. Studies have shown that the performance of shotcrete linings, affected by their application timing and mechanical properties, is foundational to structural longevity [182, 183, 188, 189]. The maintenance of tunnel infrastructure must consider the degradation of rock and lining materials over time, which is exacerbated by geotechnical challenges such as groundwater infiltration and seismic activity [190, 191]. A comprehensive understanding of the material properties is thus pivotal for developing effective maintenance protocols to ensure the longevity of tunnel systems.

Soil Classification and Behaviour

Soil classification and behaviour are fundamental to tunnelling and underground construction because soils respond very differently to excavation compared with rock. Unlike rock, soil has no tensile strength and relies on particle friction, cohesion, and confinement to maintain stability. Understanding how soils are classified and how they behave under changing stress and groundwater conditions is essential for selecting excavation methods, support systems, and groundwater control strategies.

Particle Friction, Cohesion, and Confinement

Particle friction, cohesion, and confinement are the three fundamental mechanisms that give soils and weak rocks their strength and stability, particularly during excavation. Together, they explain why ground can stand temporarily, why it may suddenly collapse, and why timely support is so effective in tunnelling.

Particle friction arises from the resistance to sliding between individual soil grains. In sands and gravels, strength comes almost entirely from friction and interlocking between particles. The rougher and more angular the grains, and the denser the soil, the greater the frictional resistance. When confinement is reduced by excavation, frictional resistance drops sharply, which is why loose sands can flow into an unsupported tunnel face even though they may appear stable before excavation.

Cohesion is the apparent “stickiness” that holds particles together independently of friction. In clays, cohesion is produced by electrochemical bonding between fine particles and by suction effects associated with pore water. This cohesion allows cohesive soils to stand vertically for short periods, which is why cuts in clay may appear stable immediately after excavation. However, cohesion is highly sensitive to water content; increases in moisture or pore pressure can rapidly reduce cohesive strength, leading to delayed collapse or squeezing behaviour.

Confinement refers to the pressure acting on the soil or rock that restrains particle movement and increases strength. Confinement enhances both friction and cohesion by preventing particles from dilating, sliding, or separating. In tunnelling, excavation removes confinement at the tunnel boundary, causing a sudden reduction in strength. Support systems such as shotcrete, segmental linings, or pressurised TBM faces restore confinement, allowing the ground to mobilise its shear strength safely.

The interaction between these three mechanisms governs tunnel behaviour. For example, in dense sand, friction provides high strength as long as confinement is maintained; once confinement is lost, collapse can be rapid. In clay, cohesion may provide short-term stability, but long-term behaviour depends on drainage and confinement. In all cases, restoring confinement early is the most effective way to stabilise the ground.

Soil Classification Systems

The most widely used engineering classification system is the Unified Soil Classification System (USCS), which groups soils based on grain size distribution and plasticity. Soils are broadly divided into coarse-grained and fine-grained categories.

Coarse-grained soils include sands and gravels. Their behaviour is dominated by particle size, density, and interlocking rather than cohesion. Gravels and dense sands typically have high shear strength and good drainage, but they can be unstable during excavation if not confined. Loose sands are particularly problematic, as they can collapse or flow into an opening unless face pressure or immediate support is provided.

Fine-grained soils include silts and clays. Their behaviour is governed by plasticity, water content, and mineralogy. Clays exhibit cohesive behaviour and can stand temporarily in unsupported excavations, but they are highly sensitive to moisture changes. Silts often behave unpredictably, combining low cohesion with poor drainage, which makes them prone to running or flowing under disturbance.

Organic soils and peat are generally unsuitable for tunnelling without extensive ground improvement due to their very low strength and high compressibility.

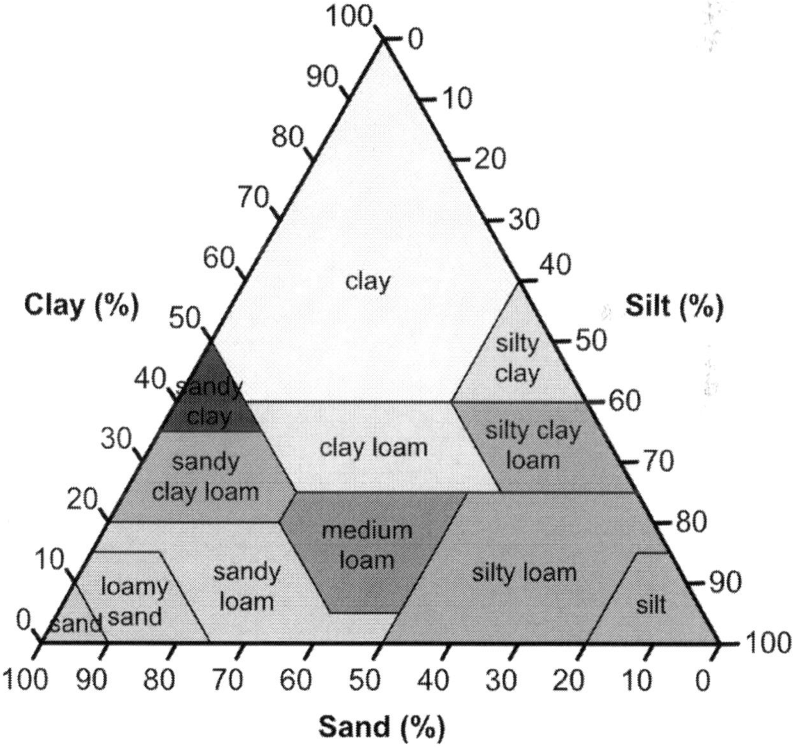

Figure 28: Soil types by clay, silt and sand composition. Zephyris at English Wikipedia, CC BY-SA 3.0, via Wikimedia Commons.

Unified Soil Classification System (USCS)

The Unified Soil Classification System (USCS) is a widely used engineering framework for classifying soils based on their grain size distribution and plasticity characteristics. It was developed to provide a consistent and practical way for engineers and geologists to describe soil behaviour and to anticipate how soils will perform under loading, excavation, and groundwater conditions. This makes USCS particularly important in tunnelling, foundation design, and earthworks, where soil behaviour directly influences stability, deformation, and construction risk.

USCS is more than a naming convention; it links soil description to expected engineering behaviour. By grouping soils with similar mechanical and hydraulic characteristics, the system helps engineers predict strength, compressibility, permeability, and response to excavation. This information supports informed decisions on excavation methods, support systems, and groundwater control measures, especially in complex underground environments.

The system divides soils into three primary groups—coarse-grained soils, fine-grained soils, and organic soils—based mainly on particle size. Coarse-grained soils are those in which more than 50 percent of the particles are larger than 0.075 mm, meaning they are retained on the No. 200 sieve. This group includes gravels, with particle sizes greater than 4.75 mm, and sands, with particle sizes between 4.75 mm and 0.075 mm.

Coarse-grained soils are further classified according to their grain size distribution and the amount of fines they contain. Well-graded soils contain a wide range of particle sizes and generally exhibit good strength and stability, while poorly graded soils have a narrow range of particle sizes and can be more susceptible to instability. The presence of fines is also critical, leading to classifications such as well-graded gravel (GW), poorly graded gravel (GP), well-graded sand (SW), poorly graded sand (SP), and gravels or sands with silt (M) or clay (C), such as GM, GC, SM, and SC. In tunnelling, coarse-grained soils often have good drainage and frictional strength but can become unstable if they are loose or water-bearing.

Fine-grained soils contain more than 50 percent particles smaller than 0.075 mm and are classified primarily by their plasticity rather than particle size. These soils are divided into silts (M), which have low plasticity, and clays (C), which exhibit moderate to high plasticity. Plasticity is determined using the Atterberg limits—the liquid limit and plastic limit—and the results are plotted on a plasticity chart. Fine-grained soils are further subdivided into low-plasticity (L) and high-plasticity (H) groups, resulting in classifications such as ML, CL, MH, and CH. These soils are particularly significant in tunnelling because they are often compressible, sensitive to changes in water content, and prone to time-dependent deformation and settlement.

Soils with significant organic content are classified separately because of their poor engineering performance. Organic silts and clays (OL and OH) and peat (PT) typically have very low strength and high compressibility. Such soils are generally unsuitable for tunnelling without extensive ground improvement, such as replacement, stabilisation, or ground treatment.

USCS uses two-letter symbols to describe soil types clearly and consistently. In cases where soil behaviour lies between two categories, dual symbols are used, such as SW–SM or CL–ML. These dual classifications indicate transitional soils with mixed characteristics that may lead to variable behaviour during excavation and support installation.

In tunnelling and underground works, USCS classification assists engineers in anticipating face stability and collapse risk, assessing groundwater inflow potential, estimating compressibility and settlement, selecting appropriate excavation methods, and designing suitable support and ground improvement measures. For example, a dense SW sand may remain stable if confinement is maintained, while a loose SP sand below the groundwater table may require pressurised excavation techniques. Similarly, a CH clay may stand temporarily but experience long-term squeezing and settlement.

While USCS is a powerful and widely used system, it has limitations. It does not directly account for in-situ stress conditions, cementation or bonding, soil structure and fabric, or time-dependent behaviour. For this reason, USCS should always be used in conjunction with site investigation data, in-situ and laboratory testing, and sound engineering judgement to achieve safe and effective tunnelling outcomes.

Fundamental Soil Behaviour Relevant to Tunnelling

Soil strength is defined by cohesion and friction angle, rather than compressive strength as in rock. Excavation reduces confinement, which can rapidly reduce shear strength and lead to collapse if support is delayed. In sands, strength is almost entirely frictional, while in clays, apparent cohesion plays a major role.

A tunnel excavated in dense, dry sand may remain stable for a short time because particle friction provides adequate shear strength. However, once excavation removes confinement at the face, even dense sand can rapidly lose strength and collapse if support or face pressure is not applied promptly. In contrast, a tunnel driven through stiff clay may stand temporarily with a near-vertical face due to apparent cohesion. If excavation is paused or groundwater conditions change, that cohesion can diminish, leading to sudden face collapse or slabbing.

Soils deform much more than rock under load. Some soils behave elastically at small strains but exhibit plastic deformation at relatively low stress levels. This means tunnel excavation in soil often leads to significant convergence and surface settlement, especially in soft clays and loose sands.

In soft clay deposits, tunnel excavation often results in significant inward movement of the tunnel walls and crown as the soil yields plastically under reduced confinement. This convergence can continue well after excavation is completed. By comparison, loose sands may experience immediate deformation during excavation, with ground loss occurring rapidly at the face. In both cases, soil deformation translates into surface settlement, which can be substantial in urban environments.

Soil behaviour is strongly influenced by drainage conditions. In sands and gravels, pore pressures dissipate quickly, leading to immediate strength mobilisation. In clays, pore pressures dissipate slowly, meaning that short-term behaviour may differ significantly from long-term behaviour. This time dependency is critical in tunnel design, as tunnels may appear stable initially but deform or fail later as consolidation occurs.

A tunnel driven through coarse sand or gravel typically experiences rapid dissipation of pore water pressures, allowing soil strength to be mobilised almost immediately after excavation. Conversely, in low-permeability clay, excess pore pressures generated by excavation dissipate slowly. A tunnel may appear stable during construction but gradually deform or settle over weeks or months as consolidation occurs. This delayed response is a common cause of long-term lining distortion and surface settlement in clayey soils.

Groundwater plays a dominant role in soil behaviour. Elevated pore water pressures reduce effective stress and shear strength, increasing the risk of face instability, base heave, and inflow. Many tunnelling methods in soil—such as Earth Pressure Balance (EPB) and slurry TBMs—are specifically designed to control groundwater pressure at the face to maintain stability.

In water-bearing sands below the groundwater table, uncontrolled excavation can lead to running ground, face instability, or sudden inflow of soil and water into the tunnel. To prevent this, Earth Pressure Balance or slurry TBMs are used to apply controlled pressure at the face, counteracting groundwater pressure and maintaining effective stress. In contrast, a tunnel excavated in partially saturated clay may show little initial inflow but can experience instability if rising groundwater pressures reduce shear strength around the opening.

Soils are generally compressible, especially fine-grained soils and organic deposits. Excavation-induced stress redistribution can lead to volume loss around the tunnel, causing surface settlement that may damage buildings and utilities. Predicting and controlling settlement is one of the most critical challenges in urban tunnelling.

Urban tunnels excavated in soft alluvial clays commonly induce ground volume loss, leading to surface settlement troughs that can damage buildings, roads, and buried services. Even small tunnel deformations can translate into noticeable surface settlement. In soils with high organic content, such as peat, compressibility is extreme, and excavation without extensive ground improvement can result in unacceptable settlement or structural failure of nearby infrastructure.

Typical Soil Behaviour Types in Tunnelling

Understanding the behaviour of soil in tunnelling is crucial for the successful execution of projects, as various soil types exhibit distinct mechanical properties and responses during excavation.

1. Dense Sands and Gravels:

Dense sands and gravels are characterized by high strength and stability under confining conditions, making them suitable for tunnelling. However, these materials can become unstable when subjected to groundwater, as water can undermine their structural integrity by introducing hydraulic forces that lead to 'running' conditions in saturated layers [136]. Consequently, adequate confinement must be maintained during excavation to prevent collapses [192]. Furthermore, techniques such as soil conditioning can be employed to improve the flowability of denser materials, further mitigating risks [193].

2. Loose Sands

Loose sands present significant challenges due to their instability; these soils tend to collapse under their own weight when excavated. They can be highly prone to face instability, necessitating the application of face pressure or ground improvement strategies to maintain structural integrity [194]. Predictive models have been developed to address these aspects, with an emphasis on time and cost estimation for tunnelling projects in loose sandy conditions [194]. Moreover, soil conditioning through methods such as foam injection has proven effective in reducing friction forces and enhancing excavation performance in loose sand [193, 195].

3. Stiff Clays

Stiff clays can temporarily support tunnels under specific conditions but are known to exhibit long-term deformations due to their plasticity and time-dependent behaviour [30]. When excavation exposes them to changes in stress regimes, these clays may undergo consolidation, leading to significant settlements that must be anticipated and modelled [196]. The interaction between stiff clays and tunnelling infrastructure is complex; thus, numerical simulations using finite element analysis are often employed to predict behaviour under varying loading conditions [197].

4. Soft Clays

Soft clays are notorious for their low strength and high compressibility, making them particularly problematic during tunnelling. These soils are prone to squeezing and can experience substantial settlement post-excavation, necessitating careful monitoring of pore water pressures and potential soil consolidation around tunnelling structures [196, 198]. The literature emphasizes the identification of effective tunnelling strategies through the use of slurry-tunnelling methods to manage ground deformations that often occur in soft clays [198,

199]. Understanding the interaction dynamics of soft clays is essential for predicting and mitigating the risks associated with tunnel stability.

5. Silts

Silts exhibit relatively low cohesion and are often considered tricky to work with in tunnelling scenarios due to their susceptibility to excessive deformation [200]. The poor drainage capacity of silt can result in significant pore pressure build-up, complicating excavation processes and increasing the risk of instability [136]. Mitigation strategies, including careful selection of tunnelling methods and continuous monitoring of ground conditions, are necessary to manage the challenges posed by silt-laden ground [141, 201].

The distinct types of soil behaviour observed during tunnelling necessitate specific approaches tailored to each soil type's characteristics. Understanding these behaviours, particularly in terms of stability, consolidation, and interaction with tunnel structures, is pivotal for the efficient and safe execution of tunnelling projects. Key considerations include the application of appropriate engineering practices to manage groundwater influences, excavation stability, and long-term settlement risks.

Engineering Implications

Soil classification plays a pivotal role in various engineering applications, particularly in the realm of geotechnical and tunnel engineering. Understanding soil properties directly informs critical decisions regarding excavation methods, support systems, groundwater control, settlement risk management, and ground improvement techniques.

The classification of soil is fundamental in determining the appropriate excavation method, which can include open-face excavation, Earth Pressure Balance (EPB) Tunnel Boring Machines (TBMs), slurry TBMs, or Sequential Excavation Method (SEM) [202, 203]. For instance, EPB TBMs are particularly effective in cohesive soils, as they can maintain stability using the earth's natural pressure, while slurry TBMs are designed for softer, more unstable ground conditions where the risk of collapse is higher [202, 203]. Zhang et al. [203] suggested the necessity of tailoring excavation strategies to the soil type to mitigate construction challenges, emphasizing the interaction between ground properties and excavation methodologies.

Soil characteristics influence not just the type of excavation method but also dictate the timing and design of support systems. Mair [204] outlined the significance of assessing ground movements and tunnel lining loads, which are influenced by soil behaviour during excavation. The real-time assessment of soil properties can allow for timely modifications to the support system, such as adjusting the pressure of the support in response to changing soil conditions [204, 205].

The management of groundwater is critical in excavation projects, particularly in areas with high water tables or permeable soils. Soil classification informs the necessity and design of dewatering systems and hydraulic barriers. Rattia et al. [40] indicated the importance of estimating ground settlements induced by TBM operation, which includes accounting for groundwater influences. Poor groundwater management can lead to excess pore pressures, subsequently increasing the risk of settlement and affecting the stability of excavation sites [40].

Settlement risk is directly correlated with soil type, and managing this risk involves understanding the specific properties of the soils involved. Bhuiyan et al. [206] provided a detailed review of ground grouting techniques as a means to mitigate settlement risks, particularly in loose or unstable soils. Grouting can bolster the soil's load-bearing capacity, thus reducing potential settlement [206]. Advanced modelling approaches, such as those discussed by Hartwig and Moreira [207], enable engineers to predict and manage ground deformation effectively.

When excavation reveals challenging soil conditions, ground improvement techniques, such as grouting, freezing, or soil mixing. may be required. Zhao et al. [202] emphasized the role of ground improvement in constraining ground deformation during tunnelling, particularly in mixed-face and challenging terrain types. Grouting, specifically, has been shown to enhance the overall stability of foundations and mitigate the risks associated with tunnelling activities in sensitive environments, demonstrating its relevance as a sustainable and economically viable solution for soil improvement [206]. Furthermore, Liang highlighted the benefits of integrating advanced predictive models for the optimization of grouting practices, ensuring better performance during excavation [208].

Overall, the classification of soil is integral to the foundation of successful engineering outcomes in excavation projects. From determining excavation methods and timing of support systems to managing groundwater and settlement risks, the implications are far-reaching. Effective ground improvement techniques further enhance project feasibility and safety, underscoring the importance of accurate soil classification.

Faults, Joints, Fractures, and Bedding Planes

Faults, joints, fractures, and bedding planes are collectively referred to as geological discontinuities, or natural planes of weakness within soil and rock masses that strongly control tunnel behaviour. While intact rock strength is important, these discontinuities usually govern how the ground deforms, fails, and transmits water when a tunnel is excavated.

Faults

Faults are fractures in the Earth's crust along which there has been measurable movement due to tectonic forces. In tunnelling, faults are among the most significant geological hazards because they often contain crushed, sheared, or altered material with much lower strength than the surrounding rock. Fault zones may include gouge (clay-like material), breccia (broken rock fragments), and highly fractured rock, all of which can behave more like soil than rock.

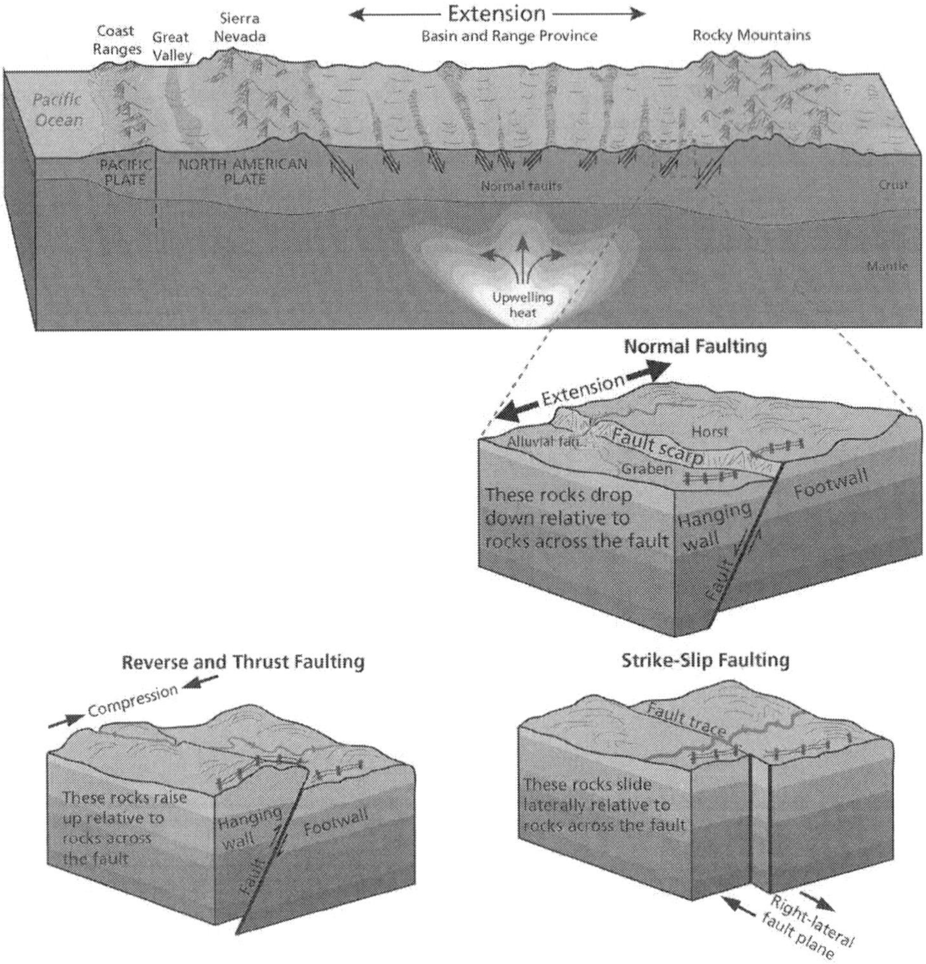

Figure 29: Graphic of Basin and Range extension, normal faults, and other fault types. National Parks Gallery, Public Domain, via Picryl.

Figure 29 explains how different tectonic stress regimes create different types of faults and how these faults deform the Earth's crust—an understanding that is critical for tunnelling, slope stability, and groundwater prediction.

At the top of the image, a large regional cross-section shows the Basin and Range Province of western North America. Here, the Earth's crust is being pulled apart by extensional tectonic forces. As the crust stretches, it thins and breaks along normal faults, producing alternating uplifted blocks (horsts) and down-dropped basins (grabens). Heat rising from the mantle contributes to weakening the crust, promoting faulting. This tectonic setting creates highly fractured ground with frequent faults—often challenging conditions for tunnelling due to variable strength and groundwater pathways.

The normal faulting diagram (upper right inset) illustrates what happens under extension. The block of rock above the fault plane, called the hanging wall, moves downward relative to the block below the fault plane, known as the footwall. This downward movement creates fault scarps at the surface and can form basins that fill with sediments such as alluvial fans. In tunnelling, normal faults often introduce crushed rock, reduced strength, and high permeability zones.

The reverse and thrust faulting diagram (lower left) shows deformation under compressional stress, where the crust is being pushed together. In this case, the hanging wall moves upward relative to the footwall. Thrust faults are low-angle reverse faults and are common in mountain belts. These faults can place older rock over younger rock and generate high in-situ stresses. For tunnels, reverse and thrust faults may cause squeezing ground, high stress concentrations, and significant deformation.

The strike-slip faulting diagram (lower right) represents faulting under shear stress, where crustal blocks slide horizontally past each other with little vertical movement. The image shows right-lateral motion, meaning the opposite side of the fault moves to the right when viewed from one side. Strike-slip faults, such as major plate boundary faults, can create linear valleys, offset rivers, and zones of intense fracturing. In tunnelling, they often pose seismic risk and unpredictable ground conditions.

Across all diagrams, the concepts of hanging wall, footwall, fault plane, and fault trace are illustrated, showing how movement direction defines fault type. From an engineering perspective, the image highlights why faults are critical geological features: they disrupt rock continuity, reduce strength, concentrate groundwater flow, and introduce abrupt changes in ground behaviour.

Faults commonly act as pathways for groundwater, leading to sudden inflows, high pore pressures, or loss of face stability. When a tunnel intersects a fault, load paths are disrupted, deformation increases, and support requirements rise sharply. Engineering responses often include reduced advance lengths, staged excavation, additional reinforcement, drainage, or ground improvement.

Joints

Joints are natural cracks or separations in rock where there has been little or no movement parallel to the fracture plane. They are ubiquitous in rock masses and typically occur in sets with consistent orientations. Joints control block size, shape, and stability, making them a primary factor in rock mass behaviour.

In tunnelling, unfavourably oriented joints can daylight into the excavation, creating unstable wedges in the crown or sidewalls. Joint properties such as spacing, persistence, roughness, infilling, and water pressure determine whether blocks remain interlocked or slide. Well-interlocked, rough joints can provide good stability, while smooth or clay-filled joints significantly reduce shear strength.

Figure 30: Joints in granite (Cadillac Mountain Granite, Silurian-Devonian boundary times, ~419 Ma; Thunder Hole, Mt. Desert Island, Maine, USA. James St. John, CC BY 2.0, via Wikimedia Commons.

Fractures

The term fractures is a general descriptor that includes joints, cracks, and small breaks in rock that may not form well-defined joint sets. Fractures increase rock mass permeability and deformability and reduce overall strength by breaking the rock into smaller blocks.

Highly fractured rock masses often behave poorly in tunnelling, even if the intact rock is strong. Excavation in such ground can lead to ravelling, overbreak, and face instability, requiring immediate and systematic support such as shotcrete, rock bolts, or steel ribs.

Figure 31: Fractured Metamorphosed Sedimentary Layers. Wing-Chi Poon, CC BY-SA 2.5, via Wikimedia Commons.

Bedding Planes

Bedding planes are planar features formed during the deposition of sedimentary rocks. They represent boundaries between layers of differing material properties and often act as planes of weakness. Bedding planes introduce anisotropy, meaning the rock mass behaves differently depending on the direction of loading or excavation.

Figure 32: Bedding and cleavage. Rudi Winter, CC BY-SA 2.0, via Wikimedia Commons.

In tunnelling, bedding planes that dip toward the tunnel opening can create planar sliding or wedge failures, especially when combined with joints or faults. Bedding planes may also control groundwater flow, allowing water to travel laterally along layers and enter the tunnel at specific horizons.

Engineering Significance in Tunnelling

In tunnelling, geological discontinuities such as faults, joints, fractures, and bedding planes are significant determinants of ground behaviour and stability. These features control how ground fails around an excavation site, influencing the occurrence of instability in the form of wedge failures, planar sliding, or rotational collapse. The interaction between these discontinuities and the rock mass is vital since tunnel stability is often more influenced by their geometry and condition than by the strength of the intact rock mass itself. For instance, research shows that weak geological structures such as foliation in schistose rock can lower stiffness in directions normal to the discontinuities and promote instability in tunnels, particularly in anisotropic conditions [168, 209]. The interplay of discontinuities also directly affects tunnel deformation; as certain joint planes become preferential pathways for failure, this leads to significant stability concerns during excavation [210].

Discontinuities significantly reduce the effective shear strength of the rock mass when compared to intact material. Factors such as weak infill, smoothness of surfaces, weathering, or groundwater presence at discontinuity planes can enable blocks of rock to shift or separate under relatively low stress. This phenomenon necessitates that engineers account for the decreased strength when selecting excavation methods and designing both primary and permanent support systems. Documented instances highlight that rock mass behaviour during tunnelling can exhibit strong dependency on the mechanical properties of these discontinuities, particularly under varying stress conditions such as those induced by excavation [211, 212]. As such, employing analytical models to estimate potential deformation and instability during tunnelling is essential [213].

Furthermore, geological discontinuities play a significant role in groundwater behaviour, often acting as preferential flow paths. The presence of faults and joint networks can channel substantial amounts of water into an excavation, consequently increasing pore pressures and reducing effective stress, which complicates construction and poses risks to stability. Sudden inflows from fractured zones can lead to severe safety issues if not anticipated adequately [214]. In complex geological environments, understanding groundwater dynamics is critical in managing the associated risks during excavation projects [215].

Additionally, the presence of discontinuities introduces considerable variability in ground conditions over short distances. A tunnel may unexpectedly transition from competent rock to highly fractured or faulted material, necessitating rapid alterations in excavation methods and support strategies. This inherent variability underscores the importance of observational methods and flexible design approaches in tunnelling operations [216]. It is vital that engineering geology plays a central role in identifying the characteristics of these discontinuities—such as their orientation, spacing, persistence, and condition—prior to excavation. Such information is crucial for adapting excavation sequences, selecting suitable support types, and effectively managing instability and groundwater risks, thus ensuring safety and efficiency in tunnel construction [217, 218].

Case Study Scenario: Influence of Geological Discontinuities on Tunnel Stability

Project context

A 6.5 km long twin-tube rail tunnel is constructed through a mountainous region to provide a new intercity rail link. The alignment passes predominantly through metamorphic rock, mainly schist and gneiss, at depths ranging from 150 to 450 m below ground surface. The tunnel is excavated using the Sequential Excavation Method (SEM/NATM) due to anticipated geological variability.

Geological setting

Pre-construction investigations identify a strongly foliated schist with multiple joint sets and occasional fault zones intersecting the tunnel alignment at oblique angles. The foliation dips steeply toward the tunnel axis in several sections, while faults are infilled with clay-rich gouge and exhibit high groundwater permeability. Although laboratory tests indicate high intact rock strength, the rock mass classification (RMR and Q) varies widely due to discontinuity spacing, orientation, and condition.

Observed behaviour during excavation

During excavation of the top heading, significant instability develops in one section where foliation planes daylight into the tunnel crown and sidewalls. Wedge failures occur at the face and shoulders, despite the intact rock strength being high. The failures are traced to smooth, mica-rich foliation surfaces with low shear resistance, confirming that discontinuity geometry—not intact strength—controls stability. Tunnel convergence measurements show asymmetric deformation, with greater displacement occurring in directions normal to the foliation.

In a nearby faulted zone, excavation intersects a fractured rock mass with clay-filled joints. Upon breakthrough, a sudden inflow of groundwater occurs along the fault plane, increasing pore pressures and triggering local face instability and slaking of weakened rock. Temporary closure of the heading is required while drainage holes and additional support are installed.

Engineering response and adaptive design

In response, the excavation sequence is modified. Advance lengths are reduced, and systematic face dowels are installed to reinforce the face and control wedge formation. Shotcrete thickness is increased locally, lattice girders are added, and rock bolts are reoriented to intersect unfavourably oriented discontinuities. Drainage measures are enhanced ahead of the face to reduce water pressure in faulted zones. Continuous monitoring of deformation and groundwater inflow informs further adjustments to support design in accordance with the observational method.

Groundwater Conditions and Hydrostatic Pressure

Groundwater conditions constitute a vital aspect of tunnelling, affecting the stability and safety of construction processes. The presence of water in soil, rock pores, and fractures significantly influences ground strength, stability, and overall construction safety. Groundwater behaviours are contingent upon characteristics such as permeability, connectivity, and the hydrogeological conditions of the region [219, 220]. As tunnelling activities commence, they disrupt the natural stress equilibrium within the ground, forcing groundwater to respond to these alterations, often becoming a dominant factor influencing subsurface behaviour [221].

Hydrostatic pressure, the pressure exerted by stationary groundwater, escalates with depth due to the weight of the water column above. This pressure acts uniformly in all directions and contributes to pore water pressure in undisturbed ground. The increase in hydrostatic pressure diminishes effective stress in the surrounding soil and rock, rendering them less capable of resisting shear forces during tunnelling operations [222, 223]. Effective stress is foundational in understanding ground behaviour, particularly in water-saturated conditions, as it relates directly to tunnel stability [224].

When excavation occurs, the surrounding material experiences a decrease in total stress around the tunnel opening, while groundwater pressure may initially remain stable. This discrepancy can lead to various forms of instability, such as face instability, base heave, or inward ground movement. In soils and weaker rock types, high hydrostatic pressures can cause phenomena such as running ground and sudden flows, while in fractured rock, rapid water flow through joints can create localized zones of weakness [138, 225].

The variability of groundwater conditions is inherently linked to the geological context. For instance, in coarse-grained soils like sands and gravels, water movement is less impeded, allowing pore pressures to quickly dissipate. Conversely, fine-grained soils, such as clay, exhibit low permeability, leading to slower pore pressure dissipation and time-dependent behaviour, particularly post-excavation [226, 227]. In consolidated rock masses, groundwater dynamics are predominantly dictated by the fracture network, with intact rock showing minimal influence [228].

Controlling hydrostatic pressures during tunnelling is imperative. Engineers employ strategies such as drainage techniques, including the installation of probe holes and drainage pipes, as well as depressurization protocols and grouting methods to mitigate water ingress [229]. Earth Pressure Balance (EPB) and slurry Tunnel Boring Machines (TBMs) are also utilized to maintain pressures at the tunnel face, thus safeguarding effective stress levels within the surrounding materials [230].

Figure 33: Concrete slab constructed to seal off the groundwater from penetrating the Track A approach structure. MTA Capital Construction Mega Projects, CC BY 2.0, via Wikimedia Commons.

A comprehensive understanding of groundwater behaviour is crucial for anticipating inflow rates and assessing stability risks. By employing sound hydrogeological assessments, engineers can design adequate support systems, thereby protecting adjacent infrastructure [231]. Failure to effectively manage groundwater can precipitate excessive deformation, flooding, ground loss, and enduring durability challenges. Therefore, groundwater conditions and their management are integral elements of successful tunnel construction and design [232, 233].

Geological Risk Identification and Interpretation

Geological risk identification and interpretation are essential components in the safe and economical design and construction of tunnels. The inherent uncertainties associated with subsurface conditions necessitate a systematic approach to identify, interpret, and manage these risks effectively.

Geological risk refers to the uncertainties and potential adverse outcomes resulting from ground conditions that deviate from those assumed during the design phase. These uncertainties arise because direct observation of subsurface conditions is not possible, and variability is often spatially dependent [234, 235]. Hence, tunnelling projects must focus on comprehensive geological assessments rather than attempting to eliminate geological risks entirely.

The identification of geological risks starts during the planning and alignment selection stages, where regional geology, tectonic settings, and historical data are reviewed to pinpoint significant hazards like fault zones and weak rock units [235, 236]. Such desk studies contribute to the initial geological model and clarify uncertainties needing further exploration [234]. For example, the use of digital tools in developing geological models has proven beneficial, as evidenced by projects like Crossrail 2 in London, which effectively utilized geographic information system (GIS) technology to identify and assess geological hazards [236].

Figure 34: View towards the west of the Crossrail construction site with the Connaught Tunnel entrance, as seen from the pedestrian bridge between Connaught Road and Factory Road in Silvertown, London Docklands. Kleon3, CC BY-SA 4.0, via Wikimedia Commons.

Following the initial assessment, site investigations are conducted to refine the geological model. These investigations typically include boreholes, core logging, and geophysical surveys, with the goal of mitigating uncertainty surrounding subsurface conditions [237]. Engineering geologists analyse data from these investigations to identify risks such as highly fractured zones and unfavourable discontinuities, emphasizing the importance of assessing both average and adverse scenarios that may affect construction behaviour [238].

Interpreting geological risks involves translating geological data into expected engineering behaviours—essentially assessing how geological features can impact stability, deformation, groundwater inflow, and overall constructability [239]. For instance, faults are interpreted as risks due to their potential for reduced shear strength and increased permeability, while rock fabric may indicate directional instability [240].

This process requires an understanding of how different construction methods interact with geological conditions. The risks associated with a given ground condition can vary significantly depending on whether excavation is executed via drill-and-blast methods or using a TBM [241].

A crucial outcome of geological risk interpretation is the formulation of a Geotechnical Baseline Report (GBR), which formalizes the expected ground conditions [41]. This report serves as a contractual and technical reference during construction, delineating anticipated ground conditions from unforeseen circumstances, thus aiding in dispute resolution and risk allocation [242].

Geological risk management is not a static process but evolves throughout the construction phase. The observational method is vital, utilizing techniques such as continuous face mapping and groundwater monitoring to identify risks as they arise [234]. This ongoing interpretation facilitates timely adjustments to excavation and support strategies, supporting efficient and safe construction practices [243, 244].

Effective geological risk identification and interpretation do not erase uncertainties; rather, they enable their transformation into manageable risks. By understanding and anticipating geological behaviours, tunnel construction projects can enhance safety, control costs, and cultivate predictable performance outcomes despite inherent subsurface variability.

Example Case Study: Geological Risk Identification and Interpretation in an Urban Rail Tunnel

Project overview

A 9 km underground metro extension is proposed beneath a dense urban area, passing below historic buildings, utilities, and a river crossing. The tunnel is planned at depths of 20–40 m and will be constructed using a combination of Earth Pressure Balance (EPB) TBMs and short sections of mined tunnels near stations. From the outset, geological risk identification and interpretation are recognised as critical to project success due to the sensitivity of surface structures and the variable subsurface conditions.

Early risk identification (planning and alignment stage)

During the planning phase, desk studies review regional geology, historical borehole records, old mining maps, and previous infrastructure projects. These studies identify several potential geological hazards, including interbedded sands and clays, buried river channels, possible faulted zones, and high groundwater levels near the river crossing. GIS-based geological modelling is used to integrate historical data with modern mapping, highlighting zones of increased uncertainty along the alignment. These early findings inform alignment optimisation, avoiding known high-risk zones where possible and flagging others for targeted investigation.

Site investigation and refinement of the geological model

A staged ground investigation program is carried out, including deep boreholes, continuous core logging, in-situ testing, piezometer installation, and geophysical surveys. Engineering geologists interpret the results and identify key risks, such as locally thick layers of loose sand beneath the water table, lenses of soft clay with high compressibility, and a previously unmapped fault zone filled with silty material. Although average ground conditions appear manageable, adverse scenarios—such as sudden transitions from stiff clay to loose sand—are identified as potential triggers for face instability and excessive settlement.

Geological risk interpretation into engineering behaviour

The identified geological features are translated into expected engineering responses. Loose, water-bearing sands are interpreted as posing a high risk of face instability and groundwater inflow, requiring pressurised excavation. Soft clays are interpreted as a settlement risk due to consolidation and long-term deformation. The fault zone is interpreted as a combined mechanical and hydrogeological risk, with reduced shear strength and increased permeability likely to cause localised deformation and inflow. These interpretations directly influence the choice of EPB TBMs, face pressure ranges, advance rates, and lining design.

Integration with construction methods and GBR development

The interpreted risks are formalised in a Geotechnical Baseline Report (GBR), which defines the expected ground conditions, groundwater pressures, and allowable ranges of variability. The GBR establishes baseline assumptions for TBM performance, support requirements, and settlement control measures. Conditions outside these baselines are clearly defined as unforeseen, providing clarity for contractual risk allocation and reducing the likelihood of disputes during construction.

Construction-phase risk management (observational method)

As tunnelling progresses, continuous face mapping, TBM operational data, settlement monitoring, and groundwater measurements are used to compare actual conditions with the geological baseline. In one section, higher-than-expected sand content and water inflow are encountered near the river crossing. Because this risk had been identified and interpreted during design, predefined mitigation measures—such as increased face pressure, reduced advance length, and supplementary grouting—are implemented quickly, preventing loss of ground and limiting surface settlement.

Ground Investigation Methods (Boreholes, Geophysics, Sampling)

Ground investigation methods are used to characterise subsurface conditions before and during tunnelling so engineers can predict ground behaviour, identify hazards, and design appropriate excavation and support systems. Because the ground cannot be observed directly, investigations combine direct methods (boreholes and sampling) with indirect methods (geophysics) to reduce uncertainty and manage geological risk.

Boreholes

Boreholes are the primary method of direct subsurface investigation. They provide point-specific information on soil and rock types, stratigraphy, groundwater conditions, and material properties along the tunnel alignment.

In soil, boreholes allow in-situ testing such as Standard Penetration Tests (SPT), Cone Penetration Tests (CPT), vane shear tests, and permeability tests. These tests help estimate strength, density, stiffness, and compressibility. In rock, boreholes provide core samples, enabling assessment of rock type, weathering, fracture frequency, Rock Quality Designation (RQD), and groundwater inflow zones.

Standard Penetration Test (SPT)

The Standard Penetration Test (SPT) is a critical in-situ testing method widely used in geotechnical engineering. Originally developed for soil investigations, this test is particularly valuable for assessing the properties of soils needed for tunnelling and foundation designs. The SPT involves driving a split-spoon sampler into the ground using a drop hammer and recording the number of blows required to achieve a specified penetration depth, providing a quantitative value known as the SPT N-value.

The SPT is performed within a borehole, where a split-spoon sampler is driven into the soil. A standard hammer, dropped from a fixed height, delivers blows to drive the sampler through the last 300 mm of the borehole. The number of blows recorded is the SPT N-value, which is indicative of soil density for granular soils and consistency for cohesive soils [245, 246]. This correlation is vital, as the N-value is often employed to estimate important geotechnical properties such as shear strength and bearing capacity [247].

The SPT N-value is a fundamental parameter that can indicate the density of granular soils and the consistency of fine-grained soils. There are established correlations between the N-value and various soil properties, including shear strength and compressibility, making it an essential tool for designing foundations and assessing excavation conditions in tunnel constructions [246]. For instance, it aids in evaluating the stability of tunnel faces and predicting potential settlements during and after excavation processes [248].

Despite its widespread usage, the SPT has limitations that industry practitioners must consider. Factors such as the efficiency of hammer energy, borehole disturbance, and the technique employed by the operator can significantly impact the results obtained from the test [247]. As a result, correction factors are often necessary to ensure an accurate representation of sub-surface conditions. It is also essential to note that the reliability of SPT results tends to be higher in sands and gravels, while correlations can be less reliable when evaluating very soft clays or coarse soils, where the underlying assumptions may not hold true [249].

The Standard Penetration Test remains a cornerstone of geotechnical investigation due to its relative simplicity, cost-effectiveness, and the critical data it provides regarding soil properties necessary for foundation and tunnelling design. However, engineers must apply caution and consider the limitations of the SPT when interpreting results to ensure that the resulting data accurately informs project specifications and safety evaluations.

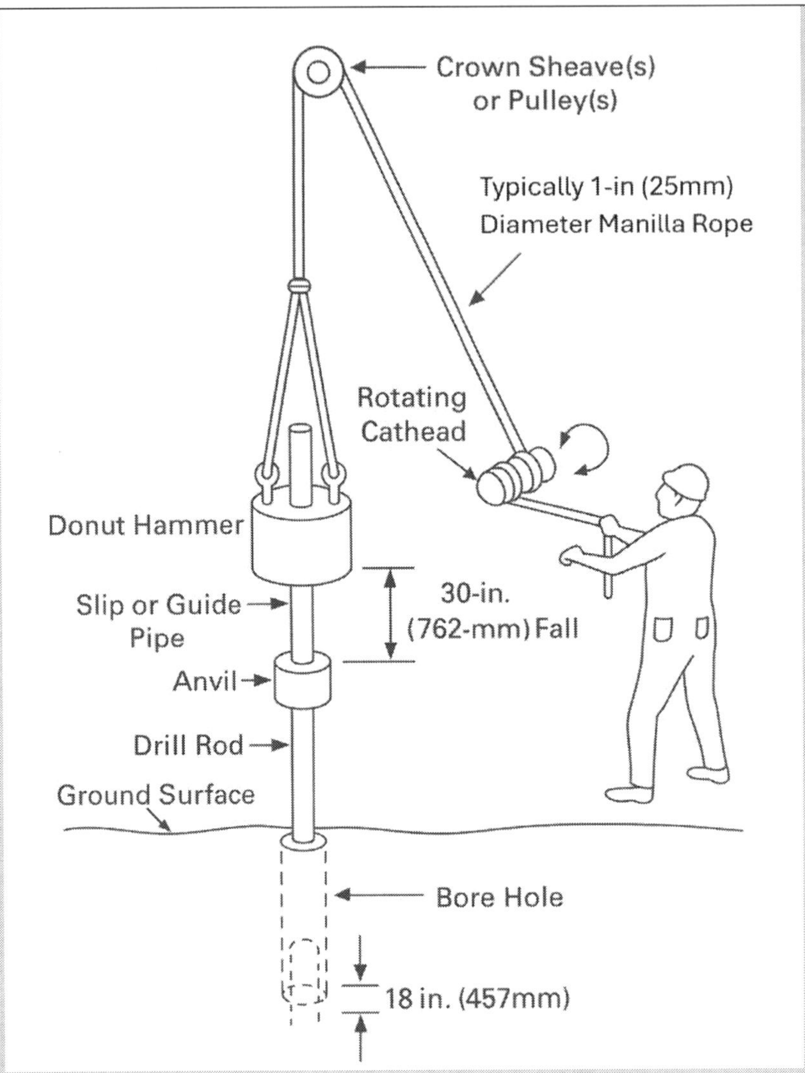

Figure 35: Standard Penetration Test Arrangements. Adapted from Sarker & Abedin (2015) [250].

Figure 35 illustrates the Standard Penetration Test (SPT) arrangement, a widely used in-situ test for assessing the strength and density of soils during borehole investigations.

At the ground surface, a borehole has been drilled to the required test depth. Inside the borehole sits the drill rod, which transmits impact energy downward to the sampler. At the lower end of the rod is the SPT sampler (anvil and split-spoon sampler), positioned at the test depth where soil resistance is to be measured.

The driving energy for the test is provided by a donut hammer, a standardized weight that is repeatedly raised and dropped. The hammer falls freely through a fixed height of 30 inches (762 mm), guided by a slip or guide pipe to ensure consistent alignment. The hammer impacts the anvil, which transfers the force directly to the sampler and into the surrounding soil.

A manila rope—typically 1 inch (25 mm) in diameter—is looped over crown sheaves or pulleys at the top of the rig. An operator rotates the cathead, winding the rope to lift the hammer and then allowing it to fall under gravity. This simple mechanical system ensures repeatable hammer drops, although variations in technique can influence energy efficiency.

During the test, the sampler is driven into the soil in increments. The total penetration measured is 18 inches (457 mm). The number of hammer blows required to drive the sampler through the final 12 inches (300 mm) of penetration is recorded as the SPT N-value. The first 6 inches are considered a seating drive and are not included in the N-value.

The recorded N-value provides an indication of soil density or consistency. Low blow counts suggest loose sands or soft clays, while high blow counts indicate dense sands or stiff to hard clays. These results are commonly correlated with parameters such as shear strength, bearing capacity, and settlement behaviour, making the SPT a foundational tool in geotechnical and tunnelling investigations.

Cone Penetration Test (CPT)

The Cone Penetration Test (CPT), originally developed in the 1950s in the Netherlands, is an essential in-situ geotechnical investigation method for determining soil properties and defining stratigraphy with a high level of precision. Its evolution from the Dutch cone test has led to it becoming one of the most widely accepted methods of soil investigation globally due to its efficiency, repeatability, and ability to generate continuous subsurface profiles, allowing engineers to make informed decisions based on detailed geotechnical data [251, 252].

The Cone Penetration Test (CPT) involves pushing a steel cone with a known geometry into the ground at a constant rate, usually without drilling. Sensors in the cone measure cone tip resistance, sleeve friction, and often pore water pressure (CPTu).

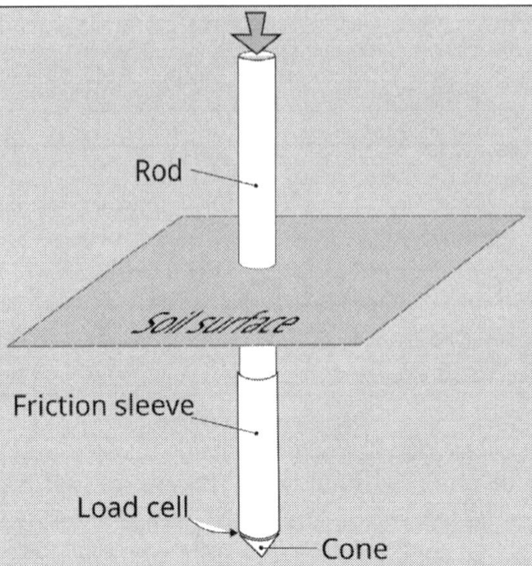

Figure 36: Cone penetrometer. Lusilier, CC BY-SA 3.0, via Wikimedia Commons.

CPT provides continuous, high-resolution profiles of soil behaviour with depth, making it extremely valuable for identifying thin layers, soft zones, and stratigraphic changes. In tunnelling, CPT data are used to estimate undrained shear strength in clays, relative density in sands, soil classification, and settlement potential. Because CPT does not require boreholes, it is faster and often more repeatable than SPT, though it is limited in very coarse gravelly soils or hard ground where penetration is difficult.

CPT involves pushing an instrumented steel cone into the ground at a uniform rate, typically between 1.5 and 2.5 cm/s. The cone, affixed to rods, is advanced using hydraulic rams supported by ballasted trucks or ground anchors. Unlike other impact-based tests, CPT is static, providing smoother and more uniform measurements of soil resistance. The cone's design, with standard cross-sectional areas of 10 cm^2 or 15 cm^2, influences the test's resolution—smaller cones yield higher stratigraphic resolution, essential for identifying thin soil layers [251, 253]. Technological advancements have led to the creation of ultra-miniature cones that cater to specialized applications, demonstrating CPT's adaptability [254].

The introduction of friction sleeves and the capacity to separate resistance components has substantially enhanced the test's usefulness. Initially, equipment measured only total penetration resistance; however, advancements allowed for separate measurements of cone tip and sleeve friction, providing more critical insights into soil type and strength, particularly in cohesive soils [252, 254]. Furthermore, the introduction of the piezocone test (CPTU), which integrates pore water pressure measurements, has refined CPT's applicability, improving soil classification and offering insights into drainage conditions, which is particularly valuable in geotechnical projects [255-257].

One of the primary advantages of CPT over traditional testing methods, such as the Standard Penetration Test (SPT), is its ability to generate continuous soil profiles at tight intervals, enhancing the identification of soil layer boundaries and the detection of weak zones—vital in complex environments or urban tunnelling [258]. The CPT process yields near-continuous measurements that not only enhance stratigraphic precision but also help evaluate geotechnical properties like hydraulic conductivity and shear strength through empirical correlations [259-261].

Modern CPT systems can incorporate various sensors to measure additional soil characteristics, such as seismic velocities, resistivity, or chemical properties. For instance, seismic CPT (SCPT) can measure shear and compression wave velocities to facilitate dynamic analyses, crucial for assessing liquefaction potential in seismic zones [262, 263]. Furthermore, the integration of magnetometers in CPT probes is noteworthy, particularly when operating in areas previously affected by conflict, as they can detect buried ferrous objects [256, 262].

CPT protocols are governed by international standards, including guidelines from ASTM and the International Society for Soil Mechanics and Geotechnical Engineering (ISSMGE), which ensure consistency and reliability across different geotechnical investigations [264]. Thanks to its accuracy, speed, low costs, and capability to incorporate diverse testing equipment, CPT has become a cornerstone in modern geotechnical investigations—particularly beneficial for tunnelling, foundation design, and risk assessment in varying soil environments [251, 265].

Vane Shear Test

The vane shear test is specifically designed to measure the undrained shear strength of soft to very soft cohesive soils, such as clays. A four-bladed vane is inserted into the soil and rotated at a controlled rate until the soil fails. The torque required to cause failure is used to calculate shear strength.

This test is particularly valuable in tunnelling through soft clays, where accurate assessment of undrained strength is critical for evaluating face stability, base heave risk, and short-term deformation. Vane shear tests can also be used to measure sensitivity, which indicates how much strength is lost when clay is remoulded. The test is not suitable for sands or stiff, fissured clays.

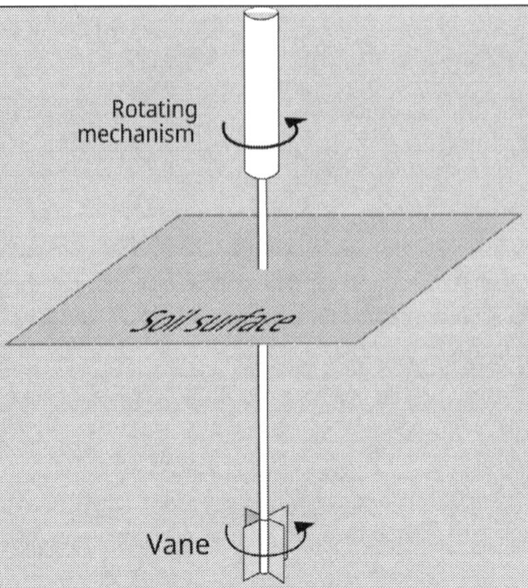

Figure 37: Schematic diagram of a shear vane. Lusilier, CC BY-SA 3.0, via Wikimedia Commons.

The shear vane test is an essential in-situ geotechnical test primarily used for quantifying the undrained shear strength of cohesive soils, notably soft to firm clays. Introduced in the late 1940s, the method remains a vital tool in conditions where traditional sampling methods can be challenging due to soil disturbance or where a clear assessment of short-term soil strength is needed under undrained conditions. The method serves as a direct measure of soil properties, thereby providing critical data that influences construction and engineering practices [266, 267].

The test involves using a shear vane, which comprises a straight rod with four thin rectangular vanes arranged cruciformly at one end. This device is inserted into the soil to a typical depth of around 500 mm without rotation. The vane is then rotated slowly, generally between 6 to 12 degrees per minute. As the vane rotates, it mobilizes shear resistance along a cylindrical surface in the surrounding soil, and the torque required for this rotation is measured using a torque gauge attached to the rod. This maximum torque value is converted into undrained shear strength, expressed in kilonewtons per square meter (kN/m^2), according to pre-established calibration factors [267, 268].

Different dimensions of vanes can be employed based on soil strength, with larger vanes used for softer clays and smaller ones for stronger clays. This variance allows for a broad applicability of the vane test across different types of cohesive soils [269, 270]. The British Standards (BS 1377) outline the procedures for conducting the shear vane test, ensuring that tests remain reliable and consistent across civil engineering applications in the UK [266].

One of the prominent advantages of the shear vane test is its ability to evaluate soil sensitivity. After conducting an initial test, the soil can be remoulded by rotating the vane before retesting, a process that reveals information about the soil's resilience to disturbance, which is crucial during excavation or tunnelling projects [266, 267]. This sensitivity assessment helps in identifying clays that exhibit significant strength reduction upon disturbance, assisting engineers in formulating strategies for effective excavation [270].

The test's portability and rapid execution make it exceptionally advantageous for field investigations in soft cohesive soils, especially when conventional boring or advanced in-situ testing methods are impractical [266, 267, 270]. While the shear vane test is predominantly associated with clay soils, it has been successfully adapted to assess various materials, including some construction materials and geotechnical applications, enhancing risk assessments in specific contexts [268, 271].

However, the shear vane test is primarily limited to measuring undrained shear strength, which necessitates its use alongside other investigative techniques to create a comprehensive profile of soil behaviour and stability [266, 272]. The correlation between in-situ values obtained from the shear vane test and laboratory tests highlights the necessity for careful interpretation in complex geological settings where soil behaviour may vary based on factors such as consolidation history and anisotropic conditions [273, 274].

Permeability Tests

Permeability tests measure how easily water flows through soil or rock, which is essential for assessing groundwater inflow and pore pressure behaviour in tunnelling.

In soils, common in-situ permeability tests include falling-head and constant-head tests performed in boreholes, as well as pumping tests for larger-scale assessment. In rock, permeability is often measured using packer tests, where sections of a borehole are sealed and pressurised to measure water loss.

Permeability data are used to predict groundwater inflow rates, assess the effectiveness of drainage or grouting, and determine whether pressurised excavation methods are required. Highly permeable ground may require face pressure control or ground improvement, while low-permeability soils may exhibit slow pore pressure dissipation and time-dependent behaviour.

Boreholes are also used to install instrumentation such as piezometers and extensometers to monitor groundwater levels and deformation over time. Their main limitation is that they provide information only at discrete locations, meaning important features such as faults or cavities may be missed if borehole spacing is too wide.

Geophysical Methods

Geophysical investigations are indirect methods that infer subsurface conditions by measuring physical properties such as seismic velocity, electrical resistivity, or electromagnetic response. These methods provide continuous or areal coverage, helping to identify variability between boreholes.

Common geophysical techniques in tunnelling include seismic refraction and reflection, which are used to estimate rock depth, rippability, and fracture zones; electrical resistivity tomography (ERT), which is effective for identifying groundwater, weak zones, and clay-rich materials; and ground-penetrating radar (GPR), which can detect shallow voids, utilities, or near-surface features.

Seismic refraction is based on measuring the travel time of seismic waves as they pass through the ground. A controlled energy source, such as a hammer blow or small explosive, generates seismic waves that travel through soil and rock layers. As these waves encounter materials with different stiffness and density, their velocities change. When a wave reaches a faster, denser layer—typically bedrock—it refracts along that interface and returns energy back to surface sensors (geophones). By analysing the arrival times of these refracted waves, engineers can estimate the depth to bedrock, layer thicknesses, and seismic velocities. These velocities are strongly correlated with material stiffness and strength, making refraction particularly useful for assessing rock depth, rippability, and the degree of weathering. Zones of fractured or weak rock typically exhibit lower seismic velocities, allowing them to be identified as potential instability or excavation-difficulty zones.

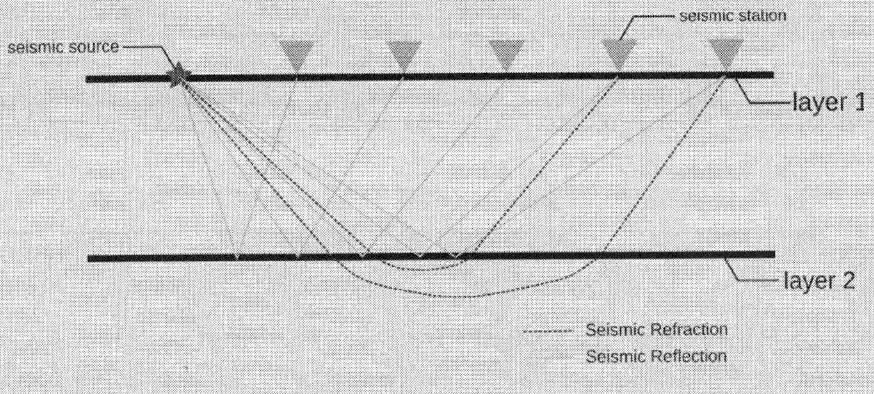

Figure 38: Seismic Refraction and Reflection. Louiskmn, CC BY-SA 4.0, via Wikimedia Commons.

Seismic reflection uses the same basic wave-generation principle but relies on waves reflecting back from subsurface interfaces rather than refracting along them. Reflected waves are recorded by arrays of geophones and processed to create detailed images of subsurface layering. Seismic reflection is capable of much higher resolution than refraction, particularly at greater depths, and is well suited to identifying complex geological structures such as faults, folded strata, and deep fracture zones. In tunnelling projects, reflection surveys are often used to map major structural features ahead of excavation and to identify variations in rock quality that could affect alignment, excavation method, or support design. However, reflection surveys are more data-intensive and require more sophisticated processing and interpretation than refraction surveys.

Electrical resistivity tomography (ERT) is a geophysical method that measures the electrical resistance of subsurface materials. Electrodes are placed in the ground along a survey line, and electrical current is injected while resulting voltage differences are measured. Because different materials conduct electricity differently, resistivity values can be used to infer subsurface conditions. Saturated soils and fractured rock filled with groundwater generally have low resistivity, while dry, intact rock exhibits high resistivity. Clay-rich materials also show low resistivity due to their high ion content and surface conductivity. By processing the data, ERT produces two- or three-dimensional images of resistivity variations, making it particularly effective for identifying groundwater zones, weak or weathered rock, faults, and clay-rich layers. In tunnelling, ERT is often used to assess groundwater inflow risks and to locate weak zones that may require pre-support or ground improvement.

Figure 39: Deployment of a permanent Electrical resistivity tomography profile on a longitudinal section of an active landslide. Ekrem Canli, CC BY-SA 3.0, via Wikimedia Commons.

Ground-penetrating radar (GPR) uses high-frequency electromagnetic waves transmitted into the ground from a surface antenna. When these waves encounter changes in material properties—such as voids, buried objects, or changes in moisture content—they are reflected back to the receiver. GPR is best suited to shallow investigations, typically within a few metres of the surface, depending on soil type and moisture conditions. It is highly effective for detecting voids, buried utilities, near-surface fractures, and construction features such as old foundations or cavities. GPR performs best in dry, resistive materials such as sand, gravel, and rock, but its effectiveness is reduced in clay-rich or highly conductive soils, where signal attenuation is high.

Figure 40: Using ground-penetrating radar equipment. The Official CTBTO Photostream, CC BY 2.0, via Wikimedia Commons.

Geophysics is particularly valuable for locating fault zones, karst features, buried channels, and transitions between soil and rock. However, geophysical results are interpretative and must be calibrated against borehole and sampling data to avoid misleading conclusions.

Sampling

Sampling provides material for laboratory testing and visual inspection, allowing engineers to quantify mechanical and hydraulic properties.

In soils, disturbed samples are used for classification tests such as grain size distribution and Atterberg limits, while undisturbed samples (e.g. Shelby tubes) are required for strength, compressibility, and permeability testing. In rock, core samples are logged for lithology, fracture characteristics, and weathering, and tested for properties such as uniaxial compressive strength, tensile strength, elastic modulus, and durability.

Sampling also helps identify problematic materials such as swelling clays, collapsible soils, organic deposits, or abrasive rock, which can significantly affect excavation methods and equipment selection.

Integration and Engineering Use

No single investigation method can fully characterise subsurface conditions on its own. Effective ground investigation depends on integrating information from boreholes, geophysical surveys, and laboratory testing into a single, coherent geological and geotechnical model. By combining direct subsurface observations with indirect measurements and material property data, engineers develop a more reliable understanding of ground behaviour than would be possible using any individual technique in isolation.

This integrated ground model underpins critical engineering decisions throughout the life of a tunnelling project. It is used to select appropriate excavation methods, such as the type of tunnel boring machine, the application of the Sequential Excavation Method, or the use of drill-and-blast techniques, based on anticipated ground conditions. The model also informs the design of both temporary and permanent support systems by identifying expected load paths, deformation mechanisms, and potential instability zones.

In addition, the integrated investigation framework allows engineers to assess groundwater inflow potential and hydrostatic pressure risks, which are central to tunnel stability and construction safety. Predictions of ground deformation and surface settlement are derived from this model, supporting the protection of nearby structures and infrastructure. The investigation outcomes further contribute to the establishment of geotechnical baselines for construction contracts, enabling clearer risk allocation and more effective management of uncertainty.

Ground investigation is rarely a one-off activity completed during design. Instead, it is typically progressive and continues throughout construction. Additional probing, face mapping, instrumentation, and targeted geophysical surveys are often undertaken as excavation advances, allowing the geological and geotechnical model to be refined as new information becomes available and ensuring that construction methods and support systems remain aligned with actual ground conditions.

Boreholes provide direct, high-quality point data; geophysics reveals spatial variability between investigation points; and sampling enables detailed laboratory testing of material properties. Together, these ground investigation methods form the foundation of safe, economical, and predictable tunnelling by transforming subsurface uncertainty into manageable engineering knowledge.

Chapter 4

Tunnel Boring Machine (TBM) Types and Selection

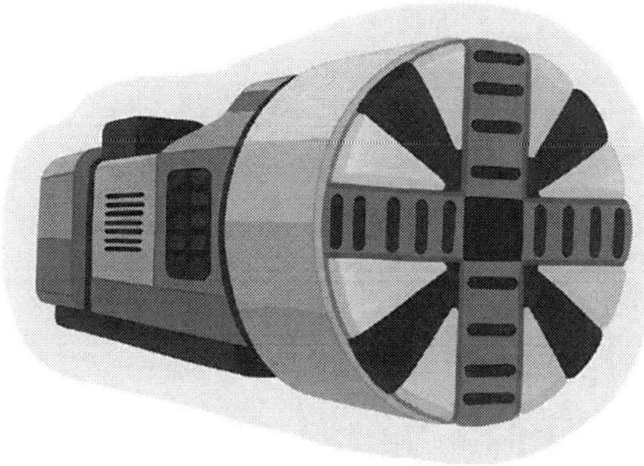

This chapter builds on the foundational overview of mechanised tunnelling by examining the principal types of Tunnel Boring Machines (TBMs) and the engineering logic that governs their selection. As modern tunnelling projects increasingly traverse complex geological, hydrogeological, and urban environments, the choice of TBM has become a decisive factor influencing safety, construction risk, cost, programme certainty, and environmental performance. There is no universally optimal machine; instead, TBM selection represents a careful balance between ground behaviour, tunnel geometry, groundwater conditions, alignment constraints, and external environmental and regulatory pressures.

The chapter first introduces the major TBM categories—open (hard-rock) TBMs, shielded TBMs, Earth Pressure Balance (EPB) machines, slurry and Mixshield TBMs, dual-mode and variable-density systems, and microtunnelling and pipe-jacking machines—explaining their operating principles and the conditions for which each is best suited. Real-world case studies are used throughout to illustrate how specific geological and project constraints directly influence machine choice in practice.

The latter sections shift from machine description to systems-level decision-making, presenting TBM selection as an integrated engineering process rather than a purely mechanical choice. By linking ground conditions, tunnel diameter, length and alignment, groundwater pressure, and environmental constraints, the chapter provides a structured framework for understanding why particular TBM types are selected on major infrastructure projects. This approach reflects contemporary tunnelling practice, where successful outcomes depend on aligning machine capability with the full spectrum of geotechnical, operational, and environmental risks encountered along the tunnel alignment.

Tunnel Boring Machines – Overview

A Tunnel Boring Machine (TBM) represents a cutting-edge approach to tunnel excavation, merging efficiency and precision in construction that is vital for modern infrastructure projects. TBMs are mechanized systems employed for excavating tunnels through varied geological conditions such as rock, soil, or mixed grounds, functioning continuously and in a controlled manner [275, 276].

Mechanism and Functionality of TBMs

Central to the operation of a TBM is a rotating cutterhead designed to break ground using either disc cutters for rock or advanced cutting tools for soil. This integrated process encompasses excavation, ground support, and spoil removal, minimizing surface disturbance and ground vibration. By transporting the excavated material via conveyors or slurry systems and utilizing thrust mechanisms to propel forward, TBMs effectively adapt to different ground conditions, operating as either open machines in solid rock scenarios or shielded, pressurized units in softer, water-saturated grounds [19, 277]. The latter significantly mitigates displacement around the excavation site, resulting in more stable structures and smoother tunnel walls, which in turn reduces the requirement for heavy lining [65, 278].

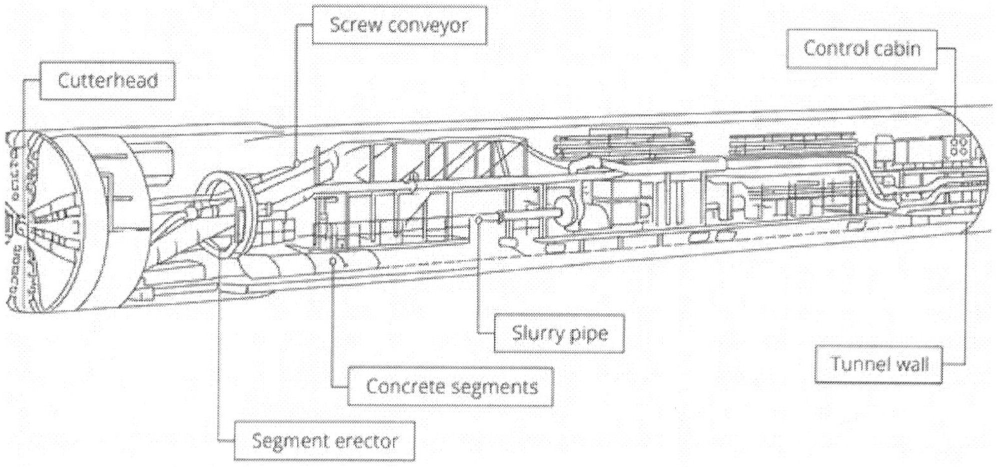

Figure 41: Tunnel Boring Machine Diagram.

Engineering Advantages and Challenges

The advantages offered by TBM technology are multifaceted. Continuous excavation allows for enhanced precision in tunnel geometry, leading to improved structural efficiency and less material overbreak compared to conventional drill-and-blast methods [279, 280]. Moreover, the operational consistency of TBMs translates into higher productivity rates, particularly noted during large projects like the Gotthard Base Tunnel, where exceptional performance and cut efficiency have been recorded [19, 276]. Nevertheless, the deployment of TBMs is accompanied by significant initial investment costs related to design, manufacturing, and assembly, making them economically viable primarily for long tunnel projects where costs can be amortized over greater distances [281].

Historical Development of TBMs

The evolution of TBMs traces back to the development of tunnelling shields in the 19th century, an innovation by Marc Isambard Brunel for the Thames Tunnel [65]. Over the years, numerous pioneers such as Henri Maus and Thomas English contributed to the mechanization of tunnelling, albeit with limitations in power and reliability [282]. By the late 20th century, advancements enabled TBMs to incorporate hydraulics, electronics, and automated systems, vastly amplifying their capabilities and applications in extensive infrastructure projects [283, 284].

Figure 42: The giant cutter head of tunnel boring machine (TBM) 'Kathleen' is lowered into the shaft at Barangaroo, ready to be assembled ahead of the historic start of the first rail crossing under Sydney Harbour. John Cowper from Sydney, Australia, CC BY 2.0, via Wikimedia Commons.

Performance and Adaptability of Modern TBMs

Contemporary TBMs are designed to operate efficiently in challenging conditions, achieving impressive advance rates of several hundred meters per week under favourable conditions [279]. This adaptability is critical, as geological variability can significantly affect tunnelling performance. Issues such as equipment failure or adverse ground conditions may lead to decreased efficiency, necessitating robust operational strategies and continuous monitoring to maintain productivity [280, 285, 286].

TBMs signify a monumental leap in tunnelling technology, providing unparalleled efficiencies and capabilities in modern engineering, particularly for large, urban infrastructure projects. Their ability to conduct precise excavation under complex conditions makes them a vital tool in contemporary civil engineering.

Open TBMs (Hard Rock)

Open Tunnel Boring Machines (often called open TBMs or hard rock TBMs) are mechanised excavation systems designed specifically for tunnelling through competent to moderately fractured rock where the tunnel can remain stable without continuous full-face support. Unlike shielded TBMs used in soft or water-bearing ground, open TBMs operate without a closed shield and rely on the inherent strength of the surrounding rock mass for stability during excavation.

At the front of an open TBM is a large rotating cutterhead fitted with disc cutters. These cutters apply concentrated normal forces to the rock surface, inducing tensile cracking and progressive chipping as the cutterhead rotates. The broken rock (muck) falls through openings in the cutterhead and is collected by buckets or scrapers, then transferred via conveyor belts or muck cars to the rear of the machine. Because excavation is continuous, open TBMs can achieve very high advance rates in favourable ground conditions.

A defining feature of open TBMs is their gripper system. Hydraulic grippers press laterally against the tunnel sidewalls, providing a reaction force that allows the cutterhead thrust cylinders to push forward into the rock. This system eliminates the need to react against a lining or shield and makes open TBMs highly efficient in strong rock. However, it also means that sufficient rock competence is required to withstand gripper pressures without excessive damage or instability.

S-210 was one of four large Herrenknecht hard-rock gripper tunnel boring machines used in the construction of the Gotthard Base Tunnel. Launched from the south portal at Bodio, the machine advanced northward in the eastern tube and was known to the tunnelling crews by the nickname "Sissi." Its role formed part of the southern excavation strategy for what would become the world's longest railway tunnel.

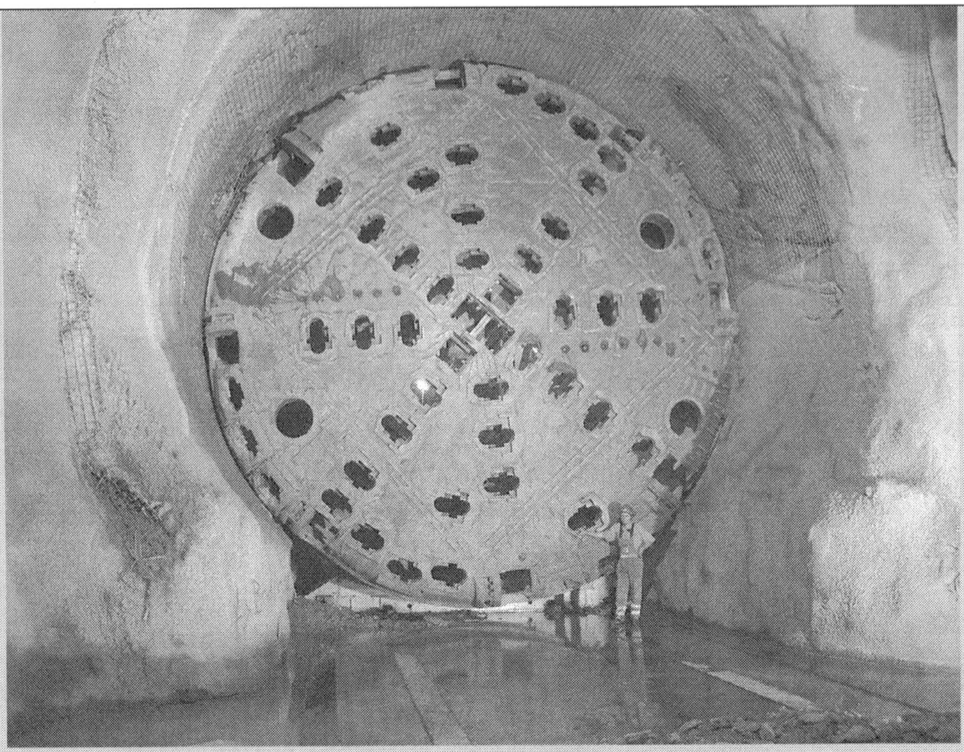

Figure 43: S-210 Tunnel boring machine used to excavate the eastern tube of the Gotthard Base tunnel between Bodio and Faido, Canton Ticino, Switzerland. Cooper.ch 16:58, 19 September 2006 (UTC), CC BY-SA 2.5, via Wikimedia Commons.

The machine's principal assignment was the Bodio–Faido section in the Canton of Ticino. This southern lot was predominantly excavated using a gripper TBM where tunnel geometry and ground conditions permitted. In parallel, other underground structures along the same alignment—such as multifunction stations, cross passages, and complex cavern geometries—were constructed using conventional drill-and-blast methods, reflecting the need to adapt excavation techniques to varying spatial and geological constraints.

As an open hard-rock gripper TBM, S-210 reacted its thrust forces through hydraulic side grippers that pressed directly against the tunnel walls, rather than against a segmental lining or closed shield. This configuration made the machine highly efficient in competent rock capable of sustaining high gripper pressures. It also allowed higher advance rates and simpler machine architecture compared to pressurised TBMs, provided that ground conditions remained stable and water inflow was manageable.

In practical operation, excavation was carried out using a rotating cutterhead fitted with disc cutters, which fractured the rock through compressive and tensile stress concentration. Forward motion occurred through a cyclic process: thrust and rotation produced penetration into the rock, followed by re-gripping and another thrust stroke. Although the thrust cycles were discrete, the excavation process itself was effectively continuous. Excavated material was transported away from the face via conveyor systems to the rear logistics trains and then out of the tunnel. Because the machine did not provide continuous shield support, primary ground support—including rock bolts, mesh, shotcrete, and occasionally steel ribs—was installed close behind the cutterhead as required by ground conditions.

Publicly available project data describe S-210 as a very large and powerful machine by hard-rock standards. It was manufactured by Herrenknecht, had an installed cutterhead power reported at approximately 3,500 kW, and bored tunnels with a diameter generally reported in the range of 8.8 to 9.4 metres, depending on the specific drive and reporting source. These specifications placed it among the most capable hard-rock TBMs operating at the time of construction.

Geologically, the Bodio–Faido drives passed through a range of Alpine rock types, including gneiss, granite, and slate. While these lithologies are generally well suited to gripper TBM excavation, the extreme overburden of the Gotthard massif introduced high in-situ stresses. In zones of weaker or foliated rock, these stresses led to squeezing behaviour and localized instability, posing significant challenges for machine operation and support design.

One notable early challenge occurred shortly after startup, when S-210 encountered squeezing ground conditions. These conditions imposed high loads on the machine structure, reduced clearance between the TBM and surrounding rock, and necessitated changes to excavation strategy. Adjustments included shorter advance lengths, increased support intensity, modified operating parameters, and, where necessary, ground treatment measures such as grouting. These responses highlighted the sensitivity of open TBMs to stress-driven deformation in deep tunnels.

In tunnelling terms, S-210 is significant because it exemplifies both the strengths and limitations of large open hard-rock gripper TBMs. The machine achieved high production rates in competent ground while also confronting the realities of deep Alpine tunnelling, including squeezing and spalling. Its performance underscores the importance of close integration between geological prediction and face mapping, careful control of TBM operating parameters, and the rapid installation of appropriate primary support to manage ground response effectively.

Ground support in open TBM tunnels is installed behind the cutterhead, typically within the backup system. Support measures may include rock bolts, mesh, steel ribs, or spot shotcrete, depending on the degree of jointing and stress conditions. In good-quality rock, support requirements may be minimal, while in more fractured or stressed rock, systematic reinforcement is often required. Because the excavation face is fully exposed, support installation relies on careful geological mapping and rapid response to changing ground conditions.

Open TBMs are particularly well suited to long tunnels in hard rock, such as water conveyance tunnels, hydropower headrace tunnels, mining access tunnels, and base tunnels through mountainous terrain. They offer excellent alignment control, smooth tunnel profiles, and low overbreak compared to drill-and-blast methods. The circular excavation produced by the TBM also results in efficient load transfer and favourable long-term stability.

Despite their advantages, open TBMs have limitations. They are poorly suited to weak, squeezing, or highly water-bearing ground, where face instability or excessive deformation may occur. Sudden transitions from competent rock to faulted or crushed zones can significantly disrupt progress and require temporary measures such as forepoling, spiling, or short shield extensions. As a result, thorough geological investigation and realistic assessment of ground variability are critical when selecting an open TBM.

Major Examples of Open Tunnel Boring Machines

Gotthard Base Tunnel TBMs (Switzerland)

The Gotthard Base Tunnel was excavated using multiple large Herrenknecht gripper TBMs, including machines such as S-210 (“Sissi”), S-211, S-229, and S-230. These machines bored through deep Alpine gneiss and granite under extreme overburden, demonstrating both the productivity and geological sensitivity of open TBMs.

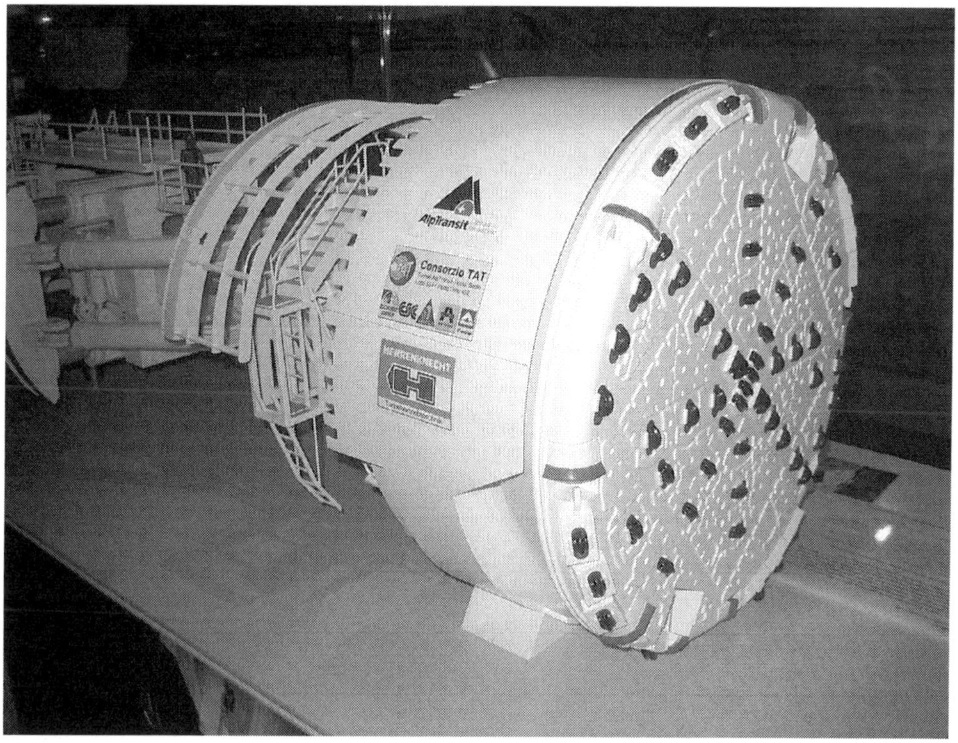

Figure 44: Model of the S-210 Tunnel boring machine used to excavate the southern part of the Gotthard Base Tunnel, Canton Ticino, Switzerland. Cooper.ch 16:58, 19 September 2006 (UTC), CC BY-SA 2.5, via Wikimedia Commons.

Key features:

- Diameter: ~8.8–9.4 m
- Rock types: gneiss, granite, schist
- Challenges: squeezing ground, spalling, high stress

<u>Lötschberg Base Tunnel TBMs (Switzerland)</u>

The Lötschberg Base Tunnel employed open hard-rock TBMs for long sections of its twin-tube alignment. The project highlighted the suitability of gripper TBMs for long rail tunnels in competent Alpine geology.

Key features:

- Long continuous drives
- High advance rates in massive rock
- Combined TBM and drill-and-blast excavation

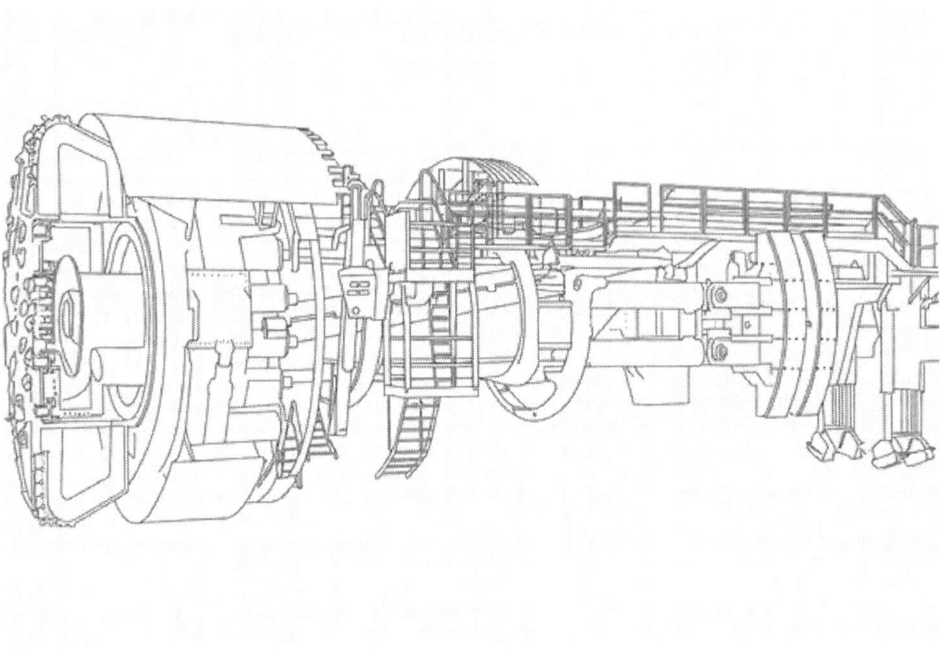

Figure 45: Diagram of Lötschberg Base Tunnel TBM.

Niagara Tunnel Project – "Big Becky" (Canada)

The Niagara Tunnel was excavated using "Big Becky," one of the largest hard-rock TBMs ever built at the time.

Key features:

- Manufacturer: The Robbins Company
- Diameter: 14.4 m
- Application: Hydropower water conveyance
- Rock type: Dolostone and shale

This project demonstrated the scalability of open TBMs for large-diameter water tunnels.

Queens Water Tunnel No. 3 – Stage 1 (USA)

Stage 1 of Queens Water Tunnel No. 3 used open TBMs to excavate deep tunnels beneath New York City through competent metamorphic rock.

Key features:

- Deep alignment (>150 m below surface)
- Competent schist and gneiss
- Minimal surface impact despite urban setting

Hallandsås Tunnel TBMs (Sweden)

The Hallandsås Tunnel initially used open TBMs, though the project became notable for groundwater inflow issues, illustrating the limitations of open TBMs in fractured, water-bearing rock.

Lessons learned:

- Open TBMs are highly sensitive to groundwater
- Geological prediction is critical
- Ground improvement may be required

Typical Applications

The operational effectiveness of open or hard-rock tunnel boring machines (TBMs) is best understood through their application in real-world scenarios where geological conditions and project requirements are well-aligned, as exemplified above. Several notable examples exist where such machines have proven their worth, significantly influenced by factors such as tunnel purpose, length, alignment, and surrounding ground conditions.

Open TBMs are frequently employed in deep Alpine and mountain base tunnels under conditions of competent rock mass and considerable overburden. Projects like the Gotthard Base Tunnel and the Lötschberg Base Tunnel in Switzerland exemplify this use. These tunnels traverse robust rock types such as granite and gneiss, which effectively support the TBM's operations under high cover. The rock masses in these areas can withstand the lateral forces exerted by the TBM grippers, while manageable groundwater inflows contribute to favourable excavation conditions. Extended, relatively straight tunnel alignments enhance continuous excavation capabilities, allowing open TBMs to achieve high advance rates over distances exceeding tens of kilometres [287].

For instance, the S-210 TBM, known as "Sissi," at the Gotthard Base Tunnel, exemplifies the effectiveness of open TBMs in such challenging environments. These machines not only produce efficient circular tunnel profiles that ensure long-term stability but also minimize the need for segmental linings where rock quality is optimal [287]. Liu et al. emphasize that the correct design and selection of TBMs lead to substantial improvements in excavation rates, particularly in robust geological setups [287].

Open TBMs are equally significant in the construction of hydropower headrace and water conveyance tunnels, where long, smooth excavations are imperative for hydraulic efficiency. The Niagara Tunnel Project and various Alpine hydropower schemes in regions like Norway and Switzerland provide clear cases where open TBMs excel. These systems are often located in competent rock, which allows for high-quality circular excavations that minimize hydraulic losses and improve flow efficiency [288, 289].

Projects of this nature often feature extensive tunnel lengths, where reduced overbreak translates into economic benefits through the lower demand for excessive lining. The "Big Becky" TBM, utilized in the Niagara Tunnel Project, is a prominent example of how open TBMs can be successfully scaled for major water conveyance tasks, showcasing their ability to reduce construction timelines through continuous excavation [288].

In urban settings, open TBMs are applicable for excavating deep rock tunnels when operating entirely within competent bedrock. An illustrative example is the Queens Water Tunnel No. 3 in New York City, where the machines manage to minimize surface settlement risks and control groundwater inflows effectively [289]. The geological stability provided by uniform rock conditions means that open TBMs can operate with minimized vibrations and disturbances compared to conventional drill-and-blast methods. This capability affords excellent alignment control beneath sensitive urban infrastructure, which is critical in densely populated areas [290].

Moreover, in hard-rock mining regions, open TBMs are extensively used for infrastructure tunnels, including conveyor and haulage paths. Their capacity to create smooth tunnel profiles is particularly beneficial for optimizing ventilation and material transport while ensuring higher productivity across long distances [291]. The TBMs used in these contexts demonstrate a clear advantage over conventional excavation methods, reflecting the synergy between machine capabilities and geological constraints [291].

Despite their advantages, open TBMs are not universally applicable. Their deployment is typically avoided in soft ground conditions, heavily faulted rock, or scenarios with high, uncontrollable groundwater pressures [292]. Transitioning between varying soil and rock types or conditions requiring confined face stability also restricts the effectiveness of open TBMs. In such situations, alternative solutions like shielded TBMs or standard excavation methods are preferred to ensure safe and controlled tunnelling processes [292].

Table 2: Summary Table — Where and When Open TBMs Are Used.

Project Type	Typical Location	Geological Conditions	Why Open TBM
Alpine base tunnels	Switzerland, Austria	Gneiss, granite, high stress	Long drives, high productivity

Project Type	Typical Location	Geological Conditions	Why Open TBM
Hydropower tunnels	Alps, Scandinavia	Massive crystalline rock	Smooth hydraulic profile
Deep urban tunnels	New York, Boston	Competent bedrock	Low disturbance
Mining infrastructure	Canada, Australia	Hard, stable rock	Permanent access tunnels

The deployment of open TBMs is significantly dictated by the interplay between geological conditions and project specifications. Their successful application in deep tunnels, hydropower projects, urban environments, and mining contexts demonstrates their versatility and efficiency. However, it is crucial to assess the suitability of these machines against prevailing ground conditions to optimize performance and minimize risk during excavation.

Shielded TBMs

Shielded tunnel boring machines (TBMs) are mechanized excavation systems crucial for constructing tunnels in unstable ground conditions. Their design incorporates a rigid steel cylindrical shield that provides continuous support to the surrounding ground while excavation occurs, making them indispensable in challenging geological scenarios such as soft soils, mixed ground, water-bearing conditions, and urban environments [57, 293, 294]. Unlike open, gripper-type TBMs, shielded machines operate within this protective envelope, ensuring that ground movements and settlements are meticulously controlled, a necessity in densely populated areas [15].

Figure 46: WIRTH 529 Telescopic Shielded TBM. Tambo at Hungarian Wikipedia., CC BY-SA 3.0, via Wikimedia Commons.

The core operational principle of a shielded TBM revolves around maintaining stability at the face of the excavation. The cutterhead located at the front excavates the ground, while the shield prevents any collapse and limits surrounding ground deformations [295]. The forward thrust required for the machine's movement is generated by hydraulic thrust cylinders that push against the tunnel lining, which is typically made of precast concrete segments. This arrangement not only supports the tunnel immediately after excavation but also forms a permanent structural lining [296].

The methodology for ground support at the excavation face varies significantly depending on the nature of the soil. In competent ground conditions, the TBM may operate with a partially open face; however, in less stable environments, like soft or saturated soils, a fully supported face utilizing controlled pressure is essential [65]. The pressure counters the forces exerted by the earth and groundwater, thus preventing inflow and maintaining stability [297]. The removal of excavated material employs different systems tailored to ground conditions, including belt conveyors for rocky substrates and slurry circuits for softer soils [298].

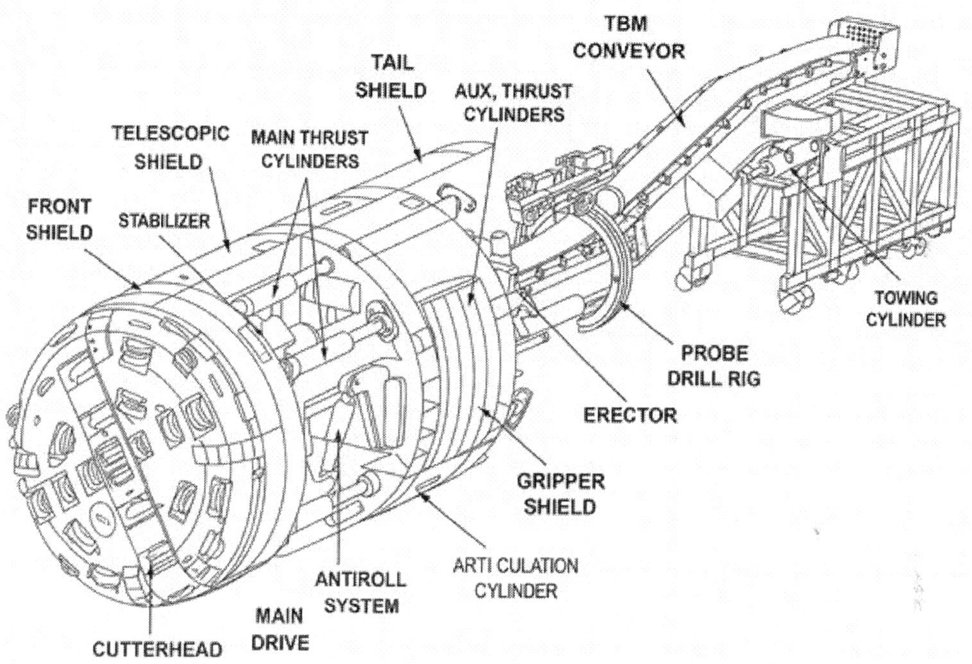

Figure 47: Double Shield (DS) TBM Diagram.

Among the shielded TBMs, single-shield machines are frequently employed in ground conditions that are fractured or squeezing, where traditional open gripper TBMs cannot provide the required support. These machines push directly against the installed lining rings, which might result in slower operational speeds but significantly improves stability under difficult conditions [296].

Conversely, double-shield TBMs carry the versatility of both open and shielded operations, enabling them to efficiently navigate competent rock while simultaneously installing linings. This dual capability significantly enhances productivity in long tunnels with variable geological characteristics. Nonetheless, their complexity and increased cost necessitate thorough planning and justification in project execution [282].

The interaction of shielded TBMs with varying geological formations presents numerous challenges. For instance, TBMs can face difficulties when operating in squeezing ground conditions where excessive convergence could result in machine entrapment. This situation often arises during protracted stoppages that allow ground displacement to impact the shield negatively [295, 299]. Advanced predictive analyses and simulations, such as those detailed in recent studies, play a critical role in understanding these dynamics and improving TBM design and operation [10, 15].

Implementing effective soil conditioning techniques has shown promise in enhancing the operational capabilities of EPB TBMs, especially in less stable environments such as weathered granite soils [300]. These techniques improve extraction efficiency and face stability by modifying the behaviour of the soil during the excavation process [301].

Shielded TBMs are particularly advantageous in urban metro and rail tunnels due to their ability to operate beneath existing buildings and utilities with minimal surface disruption. Beghoul and Demagh [57] emphasize that these machines effectively control ground movements, which is essential in shallow tunnels where settlement could impact above-ground structures. Additionally, the use of Earth Pressure Balance (EPB) TBMs, as described by Tyrer et al. [302], supports safe excavation in heterogeneous environments, reinforcing the need for precise geological understanding for effective pressure management.

The technology is crucial for constructing road tunnels beneath densely populated urban areas or rivers, allowing for stable excavation in water-rich conditions. Yubero et al. [62] provide an analysis of EPB tunnelling through soft, deltaic deposits, demonstrating the effectiveness of shielded TBMs in minimizing local settlements during operations. Such applications necessitate the timely installation of segmental linings, ensuring durability and structural integrity, which remains a significant concern during design and execution phases.

Shielded TBMs excel in mixed-face conditions—areas where soil and rock intermingle—due to their flexibility in managing varying material properties without compromising performance. The study by Lu [303] highlights the capabilities of these machines in transitioning between varying geological features, ensuring effective tunnelling even in challenging conditions.

A key advantage of shielded TBMs is their steel shield, which provides continuous ground support, significantly reducing the risk of tunnel collapse. This feature enables precise control over face stability, addressing high groundwater pressures effectively [304]. The machinery can maintain controlled face pressures, preventing excessive ground deformation. This stability is crucial in urban environments where unexpected geological changes can occur [295].

The installation of segmental linings during excavation enhances structural continuity and ensures a predictably shaped tunnel geometry. The study by Mirković [63] highlights the importance of maintaining optimal face pressures to manage ground settlements effectively, contributing to long-term performance and safety. The machines are designed to facilitate immediate lining installation, securing the tunnel shortly after passage, which minimizes risks associated with construction schedules.

Despite their numerous advantages, shielded TBMs entail high capital and operational costs due to their complex mechanical systems and extensive support requirements. The logistical demands of segment supply and spoil removal can complicate construction efforts [305]. This aspect often makes the deployment of TBMs less appealing in scenarios where traditional boring methods might suffice.

Furthermore, shielded TBMs can be sensitive to abrupt geological shifts, such as sudden transitions to hard rock or encountering unforeseen obstructions, which can result in operational inefficiencies. Zhou and Yang [306] note that in uniformly strong rocks, shielded TBMs may be less efficient compared to their open gripper counterparts. This emphasizes the necessity for thorough pre-construction geological investigations to ensure proper machine selection and operational parameters.

Examples of Shield TBMs

Single-Shield TBMs

(Fractured or squeezing rock where grippers cannot be used reliably)

Brenner Base Tunnel – Austria / Italy

Single-shield TBMs are used in sections of the Brenner Base Tunnel where rock quality is poor or highly fractured.

Figure 48: TBM excavation (advance) in the Ahrental–Pfons exploration tunnel. BBT SE, CC BY-SA 4.0, via Wikimedia Commons.

Why single-shield TBMs were used:

- Faulted and squeezing rock
- Inability of rock mass to sustain gripper forces
- Need for immediate segmental lining

Double-Shield TBMs

(Long tunnels with variable rock quality)

<u>Gotthard Base Tunnel – Switzerland</u>

In addition to open gripper TBMs, double-shield TBMs were used in sections with variable rock quality to maintain productivity while retaining the ability to push off the lining when conditions deteriorated.

Why double-shield TBMs were used:

- Alternating competent and fractured rock
- Long drives requiring high productivity

- Flexibility to switch between gripper and shield mode

Road and Urban Tunnels (Shielded TBMs)

WestConnex – Australia

Shielded TBMs were used to excavate major motorway tunnels beneath Sydney, controlling settlement and groundwater inflow in urban ground.

Earth Pressure Balance (EPB) TBMs

Earth Pressure Balance (EPB) tunnel boring machines (TBMs) represent a significant technological advancement in underground construction, specifically designed for tunnelling in soft ground and mixed-face conditions. These machines operate effectively where the geological formations are unstable and where groundwater pressure must be meticulously controlled. Their design is particularly advantageous in urban environments, where minimizing ground disturbance is crucial for the safety of existing infrastructure, such as buildings and transportation systems.

The primary function of an EPB TBM is to maintain equilibrium between earth pressures at the tunnel face by utilizing the excavated material itself. As the cutterhead works through the soil, it breaks up material which is retained in a pressurized working chamber located directly behind the cutterhead. This innovative method allows operators to manage the volume and consistency of the excavated material to stabilize the pressure within the chamber equal to that of the surrounding ground and groundwater [293, 302]. Effective pressure balance is essential for preventing face collapse, negotiating adverse ground movements, and managing groundwater inflow.

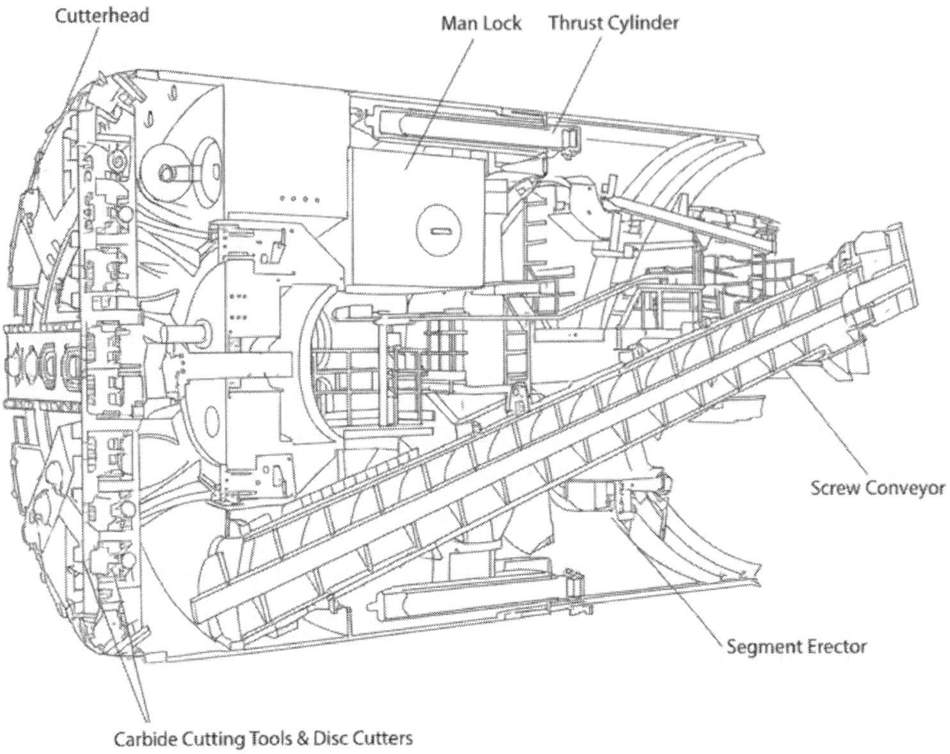

Figure 49: Diagram of an Earth Pressure Balance (EPB) tunnel boring machine.

To enhance tunnelling performance, the excavated soil is conditioned with additives such as polymers or foam. This conditioning transforms the soil into a manageable, paste-like consistency, allowing it to support the tunnel face while obviating uncontrolled inflow of water [307]. The optimized soil conditions promote a stable working environment, contributing significantly to the safety and efficiency of EPB operations.

Spoil removal is conducted through a screw conveyor, a critical component that maintains pressure control. By adjusting operational parameters such as extraction rate and torque, operators can finely tune the pressure within the working chamber [308]. Forward motion of the TBM is achieved through thrust cylinders that engage with previously installed segmental linings, providing immediate structural support and ensuring continuous excavation [308, 309].

EPB TBMs excel in environments characterized by clays, silts, sandy clays, and mixed-face conditions where cohesive soil properties can be effectively managed. Their design enables them to operate efficiently in areas with low to moderate permeability and manageable groundwater pressures [310, 311]. However, they exhibit reduced performance in highly permeable soils, such as clean sands and gravels, and are ill-suited for hard, abrasive rock

[312]. As a result, careful site assessments and soil characterizations are imperative for successful tunnelling.

From an engineering standpoint, EPB TBMs provide numerous benefits, including high face stability, effective control of surface settlement, and the capability for continuous excavation combined with immediate installation of permanent linings. These attributes make them the preferred choice for constructing urban metro systems and other critical infrastructure projects [313, 314].

Despite their versatility, EPB TBMs are also accompanied by challenges. Their performance can deteriorate significantly in high-permeability soils, and they require continuous monitoring and skilled operation to adjust to abrupt changes in geological conditions. Additionally, the operational complexity and associated costs are generally higher than alternative tunnelling methods, such as open TBMs, particularly in challenging geological formations [294, 307].

Earth Pressure Balance TBMs are sophisticated, shielded machines designed for maintaining stability during tunnel excavation in sensitive geological conditions. Their ability to utilize conditioned excavated soil to balance earth and water pressures underscores their essential role in urban tunnelling projects where minimizing ground disturbance is critical.

Figure 50: The TBM-EPB S610 "Adéla" tunnelling shield, together with its twin TBM-EPB S609 "Tonda," is one of the two shield machines that excavated the metro route from Vypich

toward Ruzyně (from Dejvická to Vypich). Jiří Janíček, CC BY-SA 3.0, via Wikimedia Commons.

Major Urban Metro Systems (Soft Ground & Mixed Face)

London Underground – United Kingdom: Earth Pressure Balance (EPB) TBMs have been extensively used on London Underground projects from the 1990s to the present, including the Jubilee Line Extension and the Crossrail/Elizabeth line stations and running tunnels. These projects were constructed predominantly through London Clay and mixed alluvial deposits, with a high sensitivity to groundwater conditions. In addition, strict settlement limits were imposed due to the presence of historic buildings and dense urban development above the tunnels. EPB TBMs were selected because they allowed precise control of face pressure, enabling excavation to proceed with minimal ground movement and very low surface settlement.

Sydney Metro – Australia: The Sydney Metro programme, delivered between 2017 and 2023 through the Metro Northwest and City & Southwest projects, made extensive use of EPB TBMs. The tunnels passed through highly variable mixed-face conditions, including Hawkesbury sandstone, shale, and alluvial deposits, as well as beneath a dense urban environment and harbour crossings. EPB TBMs were chosen for their ability to manage changing ground conditions while maintaining stable face pressures. As a result, safe excavation was achieved beneath buildings, roads, utilities, and sensitive transport corridors with effective settlement control.

Prague Metro – Czech Republic: During the extension of Prague Metro Line A in the 2010s, EPB TBMs were employed to manage variable soils and weathered rock under a high groundwater table. The project utilised two EPB machines, TBM-EPB S610 "Adéla" and S609 "Tonda." These machines were well suited to the complex geological conditions encountered, enabling controlled settlement and a stable advance rate despite frequent transitions between soil and weaker rock.

<u>Subaqueous and River Crossings</u>

Channel Tunnel Rail Link – United Kingdom: Between 1998 and 2007, EPB TBMs were used on the Channel Tunnel Rail Link in soft chalk marl formations containing groundwater. The long tunnel drives required consistent and reliable face pressure control to prevent instability and water inflow. EPB machines provided the necessary pressurisation at the face, ensuring safe and predictable excavation under these conditions.

Highway 99 Tunnel – United States: The Highway 99 Tunnel in Seattle, constructed between 2013 and 2017, employed the EPB TBM "Bertha," with a diameter of 17.45 metres. The tunnel passed through saturated glacial soils beneath downtown Seattle, presenting significant geotechnical and urban constraints. Although the project faced notable operational challenges, it demonstrated the capability of EPB technology at very large diameters and confirmed its suitability for soft, water-bearing ground beneath dense cities.

Road and Rail Tunnels in Soft Ground

Rome Metro Line C – Italy: Rome Metro Line C, under construction since 2007, has relied heavily on EPB TBMs due to the city's archaeologically sensitive environment and the presence of soft volcanic soils with groundwater. EPB machines allowed tunnelling to proceed with minimal disturbance to historic structures and subsurface remains, while maintaining strict control over ground movement.

Delhi Metro – India: Since the early 2000s, EPB TBMs have become the standard tunnelling method for the Delhi Metro. The network is constructed predominantly through alluvial soils and silty sands with a high groundwater table. EPB machines were selected for their effectiveness in controlling face stability and groundwater pressures, enabling rapid expansion of the metro system beneath a densely populated city.

Manufacturer-Specific EPB TBM Examples

Herrenknecht EPB TBMs: Herrenknecht has supplied numerous EPB TBMs worldwide, including S-series machines used on the Prague Metro and large-diameter EPB TBMs deployed on projects such as Crossrail and Sydney Metro. These machines are recognised for their advanced pressure control systems, ability to operate at large diameters, and adaptability to mixed-ground conditions.

Hitachi Zosen EPB TBMs: Hitachi Zosen is best known in the EPB context for delivering the "Bertha" TBM used on the Seattle Highway 99 Tunnel. The machine exemplifies the company's capability in manufacturing very large EPB TBMs for complex urban megaprojects, particularly where soft ground and groundwater pressures dominate design considerations.

Slurry / Mixshield TBMs

Slurry and Mixshield TBMs are specifically designed for high-risk environments characterized by unstable, water-bearing, and coarse-grained soils. These machines utilize a dual method of managing excavation by creating a pressurized face that helps counterbalance earth and groundwater pressures, making them ideal for subaqueous crossings and deep urban tunnelling projects [315-317].

The operational foundation of slurry TBMs is based on a pressurized working chamber filled with a bentonite-based slurry. This slurry serves multiple functions: it stabilizes the tunnel face by transmitting pressure, facilitates the transportation of excavated material, and maintains effective groundwater control [318-320]. As the cutterhead excavates material, the slurry prevents face collapse and minimizes ground deformation, ensuring a safe and efficient tunnelling process [318, 321].

Moreover, Mixshield TBMs have evolved from conventional slurry machines to include greater operational flexibility. They are capable of switching between different modes of operation depending on the geological conditions. In high-permeability soils, they often operate in full slurry mode, while in mixed or transitional conditions, they can adapt to density-controlled or semi-EPB modes [322, 323]. This adaptability enhances their effectiveness when tunnelling through varying soil types without compromising stability or performance [320, 324].

The components of slurry and Mixshield TBMs are engineered to ensure effective excavation and support. Key components, such as the cutterhead and excavation chamber, are hermetically sealed to maintain a constant pressure regime [319, 320]. The slurry circuit includes pumps that transport the excavated slurry to the surface for separation, allowing the reconditioned slurry to be cycled back to the cutterhead. This closed-loop system is crucial for maintaining pressure balance during operations, thus preventing groundwater inflow and ensuring the safety of the excavation [325, 326].

The surface separation plant typically includes cyclones, screens, and centrifuges to manage excavated materials, allowing for rapid adjustment of slurry density and viscosity, which enhances the TBM's operational performance in complex ground conditions [323, 327]. Additionally, the steel shield surrounding the TBM provides vital structural support, while thrust cylinders facilitate machine advancement by pushing against a ring of precast concrete lining segments [316, 326].

Slurry and Mixshield TBMs excel in environments with high groundwater pressure and unstable soils, such as alluvial deposits, river sediments, and mixed face conditions involving soil and rock interfaces [315, 317]. Their ability to maintain stable face pressure in permeable soils differentiates them, particularly in contexts where Earth Pressure Balance (EPB) TBMs may encounter rapid drainage issues [317, 324].

From an engineering perspective, these machines provide numerous advantages:

- **Enhanced Face Stability**: They offer exceptional support in soft-ground conditions and ensure effective groundwater control, which is vital for urban projects where ground settlement must be minimized [321, 328].

- **Flexibility to Geological Changes**: The ability to fine-tune operational parameters according to changing geological conditions helps maintain a stable excavation environment [318, 326].
- **Predictable Tunnel Geometry**: Continuous, circular excavation results in predictable tunnels, thereby reducing project risks [316, 323].

Despite their many benefits, slurry and Mixshield TBMs also face significant challenges, including high upfront costs and operational complexities related to slurry handling and management [317, 328]. Their application is most justified in projects where groundwater management and face stability are paramount concerns.

Below are representative real-world examples of slurry and Mixshield tunnel boring machines that illustrate where, when, and why this technology is selected. Together, they demonstrate how slurry-based TBMs are chosen in response to challenging ground conditions, high groundwater pressures, and stringent settlement control requirements.

One of the most prominent examples is the Chek Lap Kok–Tuen Mun Road Tunnel in Hong Kong, constructed between 2016 and 2019. This project used a Herrenknecht Mixshield TBM known as S-880 "Qin Liangyu," with an excavation diameter of approximately 17.6 metres, making it one of the largest TBMs ever built. The tunnel passes beneath the seabed through highly permeable marine sands, gravels, and reclaimed land under significant hydrostatic pressure. These conditions require precise and reliable face-pressure control that EPB machines cannot consistently achieve. The Mixshield system enabled stable excavation beneath the sea, effectively controlled groundwater inflow, and limited surface settlement, demonstrating the capability of slurry technology at extreme scale.

Figure 51: Tuen Mun–Chek Lap Kok Link site view (July, 2019). Wpcpey, CC BY 4.0, via Wikimedia Commons.

Another landmark application is the Øresund Fixed Link between Denmark and Sweden, constructed from 1995 to 2000. Slurry TBMs were used to excavate subaqueous tunnels beneath the Øresund Strait, where ground conditions included chalk, limestone, sands, and high groundwater pressures. Continuous face support and effective control of water inflow were essential in this permeable marine environment. Slurry TBMs provided the required stability, allowing safe tunnelling beneath the seabed as part of a combined bridge–tunnel transport connection.

In China, slurry and Mixshield TBMs have been widely deployed on the Shanghai Metro and associated urban infrastructure projects from the early 2000s to the present. Shanghai is built on very soft, water-saturated alluvial soils with high groundwater pressures and extremely low tolerance for settlement due to dense urban development. Under these conditions, slurry-based machines became the preferred solution for deep metro tunnels and river crossings, delivering excellent settlement control and predictable performance beneath sensitive surface structures.

The Thames Tideway Tunnel in London, constructed between 2016 and 2022, provides another example of slurry TBM application. Slurry machines were selected to excavate beneath the River Thames and its floodplain through water-bearing sands, gravels, and chalk. The high permeability and groundwater pressures along the alignment exceed the reliable operating range of EPB machines. Slurry TBMs maintained face stability and controlled groundwater inflow, enabling safe construction beneath one of the world's most constrained and sensitive urban river corridors.

Figure 52: A picture of the back of the Thames Tideway Tunnel Boring Machine Rachel named after Rachel Parsons, first President and founder of the Women's Engineering Society. The machine is 90m long. The image shows the shaft down to the Thames Tideway Tunnel, the tunnel opening, a smaller tunnel connecting to another site above and a crane moving material down to the machine. MyrtleGal, CC BY-SA 4.0, via Wikimedia Commons.

In Singapore, Mixshield TBMs have been used extensively on major metro projects such as the Downtown Line and the Thomson–East Coast Line during the 2010s and 2020s. Singapore's geology is highly variable, comprising marine clays, sands, weathered rock, and high groundwater pressures beneath urban and coastal areas. Mixshield machines were chosen for their ability to switch between slurry-dominated and density-controlled operating modes, allowing stable excavation through rapidly changing ground conditions while maintaining tight settlement control.

A further example is the Elbe River crossings in Hamburg, Germany, developed during the late twentieth century. Slurry shield TBMs were used to tunnel beneath the river through clean sands and gravels under high groundwater pressure. In this environment, slurry support was essential to maintain face stability and prevent excessive settlement. The slurry system enabled controlled excavation beneath critical port facilities and river infrastructure.

Across these projects, certain manufacturers have become closely associated with slurry technology. Herrenknecht Mixshield TBMs have been used on major projects such as the Chek Lap Kok–Tuen Mun Tunnel, Singapore metro lines, and numerous European river crossings. Slurry TBMs produced by Hitachi Zosen and Kawasaki have also been widely applied on Japanese and Asian subaqueous tunnels and large-diameter urban infrastructure projects.

Taken together, these examples show that slurry and Mixshield TBMs are typically selected when tunnelling involves clean sands and gravels with high permeability, high groundwater pressures, river, harbour, or seabed crossings, dense urban environments with extremely low settlement tolerance, or mixed ground conditions where EPB pressure control is unreliable.

Slurry and Mixshield TBMs represent advanced technologies explicitly designed for tunnelling in challenging geological scenarios. Their unique operational characteristics, such as pressurized face support and adaptability to different excavation conditions, make them the preferred choice for a wide range of projects, from urban tunnelling to subaqueous crossings. While they require meticulous management and incur higher costs than some alternatives, their advantages often outweigh these challenges in critical tunnelling scenarios.

Dual-Mode and Variable-Density Machines

Dual-mode and variable-density TBMs have emerged as sophisticated solutions designed to address the complexities of mixed ground conditions encountered in urban and infrastructure projects. Dual-mode TBMs are engineered to seamlessly switch between Earth Pressure Balance (EPB) mode and slurry mode, depending on ground conditions. This capability allows for an adaptive tunnelling strategy that enables continuous operation under varying geological scenarios without halting excavation [329, 330]. EPB TBMs excel in cohesive or fine-grained soils, creating a stable pressure plug at the tunnel face, while slurry TBMs are better suited for highly permeable conditions where hydraulic pressure stability is crucial [316]. The development of dual-mode machines reflects a direct response to the limitations of traditional machines, which often operate effectively within narrower ranges of geological conditions [329].

> The development of multi-mode soft-ground tunnel boring machines represents a response to one of the most persistent challenges in mechanised tunnelling: highly variable ground conditions combined with groundwater pressure. Early TBMs were optimised for narrow ranges of geology—either cohesive soils (EPB) or permeable sands and gravels (slurry). However, many modern infrastructure projects encounter frequent transitions between soils, rock, mixed faces, and water-bearing formations. This drove the first attempts, beginning in the 1980s, to create machines capable of operating in more than one support mode. Over time, these ideas evolved into today's mature Variable Density TBM technology [331].
>
> The first true multi-mode TBM was developed in the 1980s for a project with alternating stable rock, mixed faces, and water-bearing soils. While technically ambitious, this early machine demonstrated that combining EPB and slurry principles in one system could be feasible. Subsequent refinements over several decades, particularly in face pressure control, slurry management, and machine automation, led to the Variable Density TBM

concept, which allows smooth, controlled transitions between modes without mechanical intervention [331].

A major milestone in multi-mode TBM development was the Socatop (A86 West Tunnel) project in Paris, completed around 2000. This 10 km road tunnel formed the final link of the A86 ring road and passed through highly variable geology. Approximately 60% of the alignment was well suited to EPB operation, while the remaining 40% required slurry support due to permeable, water-bearing soils. To address this, an 11.56 m diameter TBM was designed that could operate in both EPB and slurry modes. This was the first project to demonstrate that a single TBM could successfully switch between the two face support concepts in long, associated geological sections [331].

From an operational perspective, both EPB and slurry shields rely on a filled excavation chamber and controlled face pressure, but they differ fundamentally in how that pressure is generated and regulated. In slurry mode, pressure is controlled hydraulically—often using an air bubble separated from the excavation chamber by a submerged wall—while in EPB mode, pressure is controlled mechanically through the balance between advance rate and screw conveyor extraction. The Socatop TBM accommodated both systems by installing parallel muck transport arrangements within the large-diameter machine. This included both a slurry circuit and a screw conveyor system, with additional equipment such as a movable jaw crusher to handle coarse material during slurry operation. Although mechanically complex, the project proved that such hybridisation could be justified when geological variability was sufficiently extreme [331].

For many years, Socatop remained a unique solution. However, increasing urbanisation, environmental sensitivity, and geological uncertainty renewed interest in adaptable TBM systems. A significant step forward occurred with the Port of Miami Tunnel project, completed in 2013. Here, an EPB TBM was adapted to operate in a Water-Controlled Process (WCP) mode when excavating beneath Biscayne Bay. The ground comprised porous sands and limestone with high permeability, but environmental constraints discouraged the use of a full slurry system due to the risk of slurry loss into the marine environment. The solution was a hybrid system in which excavated material could be hydraulically transported via a slurry pipeline after passing through a crusher, while still retaining the core EPB configuration. This WCP mode represented an intermediate step toward full Variable Density TBM technology [331].

The Variable Density TBM concept emerged from these experiences as a fully integrated solution for mixed and uncertain ground. Unlike earlier multi-mode machines, Variable Density TBMs are designed to transition continuously between EPB-like and slurry-like behaviour without mechanical modification of the excavation chamber or backup systems. They achieve this by controlling the density, viscosity, and composition of the chamber filling material. The machine can operate with classic slurry support, classic EPB support, or intermediate high-density suspensions that are too thick for conventional slurry operation but too fluid for traditional EPB. Crucially, face pressure remains fully controlled throughout these transitions, eliminating the need for chamber interventions [331].

This technology was first deployed at scale on the Klang Valley MRT Line 1 in Kuala Lumpur. The tunnels passed through Kenny Hill formation and Kuala Lumpur Limestone, which is notorious for highly erratic karst features, cavities, and sudden voids beneath soft soils. Six Variable Density TBMs with diameters of 6.62 m were used over approximately 8.6 km of tunnel. In zones where standard slurry would rapidly drain into karst cavities, the machines operated with high-density slurry prepared at dedicated surface mixing plants. This allowed face pressure to be maintained even in highly permeable or voided ground, significantly reducing the risk of collapse or surface settlement [331].

The success in Kuala Lumpur led to further applications of Variable Density TBMs worldwide. On Lima Metro Line 2, a 10.21 m diameter Variable Density TBM was selected to deal with coarse gravels, sands, silts, and clays under groundwater. The machine incorporated a dual screw conveyor system and a slurryfier box with an integrated crusher, allowing seamless switching between EPB and slurry logistics. Similar technology was applied in Hong Kong's Shatin to Central Link, where extremely heterogeneous ground—including decomposed granite, boulders, marine deposits, and fill—required continuous adaptability, especially under shallow cover [331].

Variable Density TBMs were also used on the Forrestfield Airport Link in Perth, Australia, where tunnels passed beneath airport runways in variable sands and cemented layers under groundwater. Maintaining face stability and controlling settlement were critical, and the machines were operated continuously in closed, pressurised modes with the ability to adjust between EPB and slurry behaviour as conditions changed [331].

When operating in EPB mode, these machines utilize conditioned spoil to manage face pressure through the regulation of screw conveyor discharge rates. In contrast, during slurry operation, the chamber is filled with bentonite slurry, allowing for hydraulic control of face pressure via a circulation system [329]. This operational flexibility is especially beneficial in areas where tunnelling transitions occur, such as from cohesive clays to mixed soil–rock faces [5].

Variable-density TBMs, often associated with Mixshield technology, differ fundamentally from dual-mode TBMs in that they maintain a constant suspension density within the excavation chamber, adjusting parameters such as viscosity in response to geological changes. This capacity allows for a dynamic response to ground conditions, mimicking the behaviour of both EPB and slurry systems depending on immediate requirements [316, 332]. The adaptability inherent in variable-density TBMs significantly enhances operational efficiency in mixed or heterogeneous soils, which are prevalent in urban areas and complex geological scenarios [316].

Additionally, these machines utilize sophisticated monitoring systems to oversee and adjust excavation parameters in real-time, mitigating risks associated with the unpredictability of ground behaviour [333, 334]. Their ability to fine-tune operational parameters instantaneously allows for improved face stability and settlement control, essential for preserving the integrity of surrounding infrastructure [300].

Both dual-mode and variable-density TBMs are designed with several critical features that support their advanced operational capabilities. These include a sealed and pressurized excavation chamber, adaptable cutterheads for various soil types, slurry circulation systems, and integrated monitoring and control systems that facilitate precise management of excavation pressures [5, 329]. Such features are vital for ensuring stability in sensitive environments, particularly where groundwater and soil interaction play crucial roles [335].

Despite their advantages, both dual-mode and variable-density TBMs come with challenges, including higher capital costs and complex operational logistics, particularly relating to slurry management and treatment [336]. Their deployment necessitates thorough pre-excavation geological investigation and skilled operation to navigate the intricacies of construction in unpredictable ground conditions. Real-time monitoring of excavation parameters is critical, as it heavily influences the success of tunnelling operations [5].

The applicability of dual-mode and variable-density TBMs is prominent in urban metro systems, subaqueous tunnels, and projects under various geological uncertainties [336, 337]. Their principal advantages include enhanced adaptability to changing conditions, improved stability during excavation, reduced risk of operational failures, and more favourable settlement control outcomes. These features collectively contribute to greater reliability in complex tunnelling projects, minimizing delays and cost overruns associated with conventional methods [316, 329].

Dual-mode and variable-density TBMs represent significant progress in the field of mechanized tunnelling. Their design reflects an advanced understanding of the need for flexibility in varying ground conditions, enabling successful navigation through complex geological environments. As urbanization continues to increase and infrastructure demands grow, these hybrid machines are likely to play a crucial role in efficient and effective tunnelling projects.

Microtunnelling and Pipe-Jacking Machines

Microtunnelling and pipe-jacking are integral trenchless construction methods designed for the installation of small- to medium-diameter tunnels with enhanced precision and minimal surface disruption. These technologies are particularly essential in urban environments or ecologically sensitive areas where conventional excavation poses risks, including safety hazards and significant social or environmental impacts.

Microtunnelling is a remote-controlled tunnelling process that employs a microtunnel boring machine (MTBM) to excavate, remove spoil, and construct tunnels without human presence within the tunnel itself. This method is initiated from a launching shaft and proceeds toward a reception shaft while being conducted under strict surface control. The remote operation is particularly advantageous in hazardous ground conditions or beneath existing infrastructure, significantly reducing exposure risk to labourers [338, 339].

Figure 53: Slurry Pipe Jacking Micro Tunnelling Shaft. Sabino Freitas Correa, CC BY-SA 4.0, via Wikimedia Commons.

The primary feature distinguishing microtunnelling from traditional methods is its implementation of full-face pressure control mechanisms. This is achieved using a slurry mixture that stabilizes both earth and groundwater pressures, thus maintaining equilibrium during excavation in challenging geotechnical environments [340]. Precision guidance systems, typically utilizing lasers, allow for millimetre-level accuracy in both the alignment and gradient of the excavated path. Such meticulous control is pivotal for ensuring optimal hydraulic performance in applications like gravity-fed pipelines, where even minute misalignments can lead to significant operational inefficiencies [338, 340].

Pipe-jacking serves as a complementary technique whereby the tunnel lining is installed simultaneously with the excavation process. As the MTBM progresses, pre-fabricated pipes (composed of materials such as reinforced concrete or steel) are jacked forward utilizing hydraulic jacks from the launch shaft, thereby forming the tunnel's permanent lining [338,

341]. This simultaneous installation eliminates the need for a separate lining phase, streamlining the construction process.

The jacking forces exerted on the pipe string must be adequately managed, often necessitating the design of systems that can endure high axial loads. For lengthy jacking operations, the placement of intermediate jacking stations may be crucial to minimize friction and manage peak load situations effectively. Additionally, to enhance performance, lubricants like bentonite slurry are often injected into the spaces surrounding the pipes, which further reduces resistance against surrounding soil [341, 342].

Microtunnelling technologies generally adopt two primary methods for managing face support and spoil removal: slurry microtunnelling and earth pressure or mechanical systems. Slurry systems dominate globally due to their efficacy in soft or water-saturated soils, whereas mechanical systems are typically confined to scenarios with simpler geotechnical demands [343]. The choice of cutting tools on an MTBM varies per site-specific conditions and includes diverse implements like scrapers or disc cutters, which are essential for sustaining excavation rates and minimizing wear in abrasive environments [343].

Figure 54: Breakthrough of the TBM cutterhead at the end of a drive in soft rock, inside the reception pit/trench. UgoLazzarini, CC BY-SA 4.0, via Wikimedia Commons.

Microtunnelling and pipe-jacking find extensive applications in constructing sewer systems, stormwater drainage, utility conduits, and various urban infrastructure projects. Their capacity to operate in compact urban environments and through difficult ground conditions positions them as superior alternatives to conventional excavation methods [338, 341]. The method's advantages include negligible surface disruption, high alignment accuracy, enhanced groundwater management, and reduced labour risks; thereby minimizing potential environmental and social repercussions associated with surface excavation [340, 343].

Despite its advantages, microtunnelling is accompanied by several limitations, such as considerable upfront investments in specialized machinery and shaft construction. Furthermore, once tunnelling commences, flexibility in altering alignments is severely restricted, and significant challenges arise when dealing with exceptionally hard or abrasive geological materials [341]. Additionally, a skilled workforce is essential for managing the sophisticated equipment and ensuring the continuous monitoring required to mitigate project risks effectively [338, 342].

TBM Selection Criteria

Selecting the appropriate Tunnel Boring Machine (TBM) is a critical engineering decision that directly affects safety, cost, programme, and environmental impact. TBM selection is never based on a single factor; instead, it results from balancing several interrelated criteria.

Ground Conditions

Ground conditions are critical determinants in the selection of Tunnel Boring Machines (TBMs) due to their significant influence on excavation stability and efficiency. The characteristics of the ground—including its strength, structure, permeability, and overall variability—are paramount in ensuring that the tunnel face and excavation perimeter maintain stability without requiring additional support [344]. As different geological formations present unique challenges, proper TBM selection is crucial to mitigate risks associated with ground behaviour during tunnelling [345].

For tunnels situated in competent hard rock formations, such as granite or gneiss, open (gripper) TBMs are often favoured. The inherent stability of these rock types allows them to support the tunnel structure, thus enabling the TBM's grippers to efficiently push against the rock, facilitating high advance rates [346, 347]. Gripper TBMs effectively use the rock's strong natural support, which significantly enhances productivity in these stable environments [348].

In contrast, environments characterized by fractured rock or squeezing ground require more specialized TBM types, such as single-shield or double-shield TBMs. These machines are designed to provide continuous support while allowing for mechanized excavation, thus addressing the instability caused by ground fractures or excessive squeezing. The use of these TBMs helps to avert risks such as tunnel collapse, which is often exacerbated in zones where ground conditions are unpredictable [344, 349]. The dual functionality of shielding and excavation provided by these TBMs makes them suitable for such complex environments [350].

When tunnelling through soft ground conditions—such as clays, silts, and sands—shielded TBMs are essential. The design of these machines includes mechanisms to minimize the risk of ground loss and collapse, which is critical when dealing with softer, less stable geological formations [350]. Specifically, Earth Pressure Balance (EPB) TBMs are adept at managing excavated soil, conditioning it to create a pressure plug, which is vital in maintaining stability during the excavation process [351].

On the other hand, in cases involving clean sands, gravels, or highly permeable soils, slurry or Mixshield TBMs are preferred. These machines utilize a slurry to counteract high groundwater pressures while excavating, which is crucial for maintaining operational efficiency in such shifting conditions [352, 353].

Highly variable or mixed-face ground conditions demand flexible TBM designs, such as dual-mode or variable-density TBMs. These machines are adept at managing rapid transitions between soil and rock, alongside fluctuating groundwater regimes. The ability to switch modes in response to changing ground conditions allows for greater operational adaptability and reduction of downtime due to unexpected geological changes [344, 354]. As such, thorough ground investigation is necessary before operation, as erroneous assumptions regarding ground behaviour can lead to significant underperformance of the chosen TBM [355, 356].

The selection of a TBM is fundamentally influenced by the ground conditions encountered during a tunnelling project. Competent hard rock supports open TBMs, while fractured and squeezing ground conditions necessitate the use of shielded machines such as single or double-shield TBMs. Soft grounds require shielded TBMs, particularly EPB TBMs for effective management of excavated soil, whereas slurry TBMs address the challenges posed by permeable soils. Finally, in highly variable conditions, dual-mode TBMs offer the necessary flexibility for successful tunnelling. Comprehensive ground investigations are fundamental to guide these selections and ensure project success.

Tunnel Diameter

The diameter of a tunnel significantly dictates the choice of tunnelling methods and machine types utilized in construction. This relationship is observed through varying classifications of diameters, each associated with specific construction challenges and machine adaptations. The typology of tunnelling, from microtunnelling to large-diameter tunnelling, further illustrates how mechanical and engineering considerations evolve with an increase in tunnel diameter.

Small Diameter Tunnels (≈0.3–3 m): Microtunnelling and pipe-jacking machines are predominantly used for small-diameter tunnels, typically ranging from 0.3 to 3 meters. These machines are well-suited for the installation of utilities and drainage systems under roads or buildings due to their effectiveness in constrained environments and minimal surface disruption. The adoption of these technologies facilitates precision and control, which is crucial for maintaining safety in urban settings [339]. Pipe-jacking has been widely recognized as a trenchless method of choice for these applications, especially under varying ground conditions that would otherwise complicate open-cut trenching.

Medium Diameter Tunnels (≈3–6 m): For medium diameters, approximately between 3 to 6 meters, the tunnelling landscape transitions towards methods appropriate for metro, rail, and road infrastructures. In this category, Earth Pressure Balance (EPB), slurry, and shielded Tunnel Boring Machines (TBMs) are often the preferred choices, contingent on ground conditions. The operational flexibility of EPB and slurry TBMs permits their use under challenging geological configurations where stability must be closely monitored [357, 358]. Research indicates that advancements in manufacturing and operational criteria have expanded the applicability of these machines in urban tunnelling projects, allowing for deeper excavations and complex routing [322].

Large Diameter Tunnels (≈6–18+ m): As the diameter escalates, particularly for large tunnels ranging from 6 to over 18 meters, the engineering complexity increases dramatically. This category often accommodates multi-lane road tunnels, service tunnels, or large rail tunnels, with shielded TBMs dominating applications in soft ground scenarios. The logistical demands, including thrust forces and face pressure management, intensify considerably with growing diameter sizes [359, 360].

In soft ground, shielded TBMs are advantageous, offering a means to control face pressure effectively while minimizing risks associated with tunnelling operations [361]. In contrast, large open TBMs are primarily deployed in competent rock, where the stability of the surrounding geology can be assured [319]. Comprehensive studies illustrate that as tunnel diameters increase, the need for enhanced segmental lining designs and thrust mechanisms becomes vital due to the pressing challenges of managing excavation forces and ground interactions [362, 363].

Furthermore, the influence of buoyancy and uplift increases with diameter—crucial factors that dictate the design and operational strategies of large-diameter tunnels [364]. Safety protocols and structural integrity thus become paramount, necessitating advanced engineering practices and material innovations tailored for large-scale projects.

The diameter of tunnels undeniably shapes both the types of machinery employed and the feasibility of construction projects. From microtunnelling methods for small diameters to sophisticated shielded TBMs for large tunnels, each category possesses distinct engineering requirements and operational complexities. Understanding these dynamics is crucial for effective project planning and execution in diverse geological environments.

Length and Alignment

The decision to employ mechanised tunnelling, specifically utilizing Tunnel Boring Machines (TBMs), is significantly influenced by the length and geometry of the tunnel. This analysis will explore how these factors dictate the feasibility of mechanised approaches and the optimal type of TBM to be used.

Mechanised tunnelling via TBMs is particularly advantageous in long, continuous drives that extend several kilometres. These machines enhance productivity, improve consistency, and reduce overbreak compared to traditional drill-and-blast methods. For instance, it has been documented that efficient operational parameters of TBMs lead to a more predictable excavation process, minimizing disruptions in urban settings where stability and adherence to schedules are critical [365, 366]. In urban areas, the small disturbance caused by TBM operations aligns with modern construction demands, making them a preferred option [365].

The accuracy of TBMs in handling large volumes of earth and their ability to maintain a controlled tunnelling environment supports confidence in their use for lengthy projects. This aligns with findings that assert TBMs have become the preferred choice for extensive tunnelling projects because they balance feasibility and cost-effectiveness, especially in complex urban contexts [316, 367].

Conversely, in shorter tunnels or where the alignment varies significantly, the utility of TBMs may be less economically justified. Mobilisation costs associated with TBMs are substantial, as these machines must be transported and set up at the site, which can be inefficient for shorter projects. Research indicates that for tunnels with limited lengths or where flexibility is required, conventional excavation methods may provide a better financial return [368]. Notably, studies have highlighted that specialised mobile machinery becomes less beneficial in contexts requiring quick adjustments and rapid deployment, underscoring the economic trade-off of using TBMs in variable geometries [368, 369].

When considering the geometry of the tunnel design, straight or gently curved alignments present an ideal scenario for TBMs, particularly when utilizing open and double-shield machines. Such configurations allow these machines to operate at optimal performance, facilitating efficient material excavation and ground support management [294, 370]. The operational principles of TBMs are well-suited to these alignments, maximizing productivity and reducing the risk of geological disturbances.

However, tight curves, frequent cross-passages, and complex geometries impose limitations on TBM usage. Such scenarios might necessitate the use of smaller TBMs, segmented drives, or even conventional excavation methods. For example, complex urban environments often demand flexible tunnelling solutions that larger TBMs cannot provide due to their size and operational constraints [371].

To effectively navigate these challenges, employing a smaller TBM or adjusting to traditional methods can mitigate the risks associated with complicated tunnelling environments. This adaptability reflects a growing recognition of the need for tailored approaches in mechanised tunnelling projects, heavily influenced by site conditions and operational requirements [368, 372].

The justification for mechanised tunnelling and the selection of the appropriate TBM type are intricately linked to the length and geometry of the tunnel. Long, straight drives favour the use of TBMs, enhancing both productivity and consistency. In contrast, shorter tunnels with complex alignments often require alternative approaches due to the high costs and operational limitations associated with TBMs. This variability necessitates a nuanced understanding of each project's requirements to optimize tunnelling strategies effectively.

Water Pressure

The selection of Tunnel Boring Machines (TBMs) is a crucial phase in the planning of tunnelling projects, as groundwater conditions significantly influence their performance and suitability. Understanding these groundwater conditions helps determine the most appropriate TBM type to mitigate risks associated with tunnelling.

In areas where low groundwater pressure exists, Open TBMs or Earth Pressure Balance (EPB) machines can be viable solutions if the inflows are manageable. EPB TBMs have a pressure balance mechanism that can efficiently handle minor water inflows by maintaining stable face pressure, thereby minimizing the risk of face instability [307]. EPB machines are particularly effective in cohesive soils and can be adapted to control water inflow through careful soil conditioning and face pressure management [307].

As groundwater pressure becomes moderate, the choice of TBM must ensure that soil conditioning techniques effectively control the excavation conditions. Enhanced soil conditioning can reduce the permeability and adjust the pressure transmitted at the face, thus ensuring stability. Studies indicate that appropriate use of conditioning agents can significantly improve the effectiveness of EPB TBMs by managing the behaviour of the soil during excavation [300, 373]. Cheng et al. emphasize the importance of this hydromechanical control, which allows maintaining the stability of tunnel faces in moderate groundwater situations [374].

In situations where groundwater pressure is high, including subaqueous tunnelling, Slurry or Mixshield TBMs are preferred. These machines utilize a hydraulic pressure transmission system that provides reliable support at the face, particularly in permeable soils [319]. The dynamic application of slurry pressure helps balance the geological forces acting on the tunnel face, making slurry TBMs essential in high-pressure conditions that could otherwise lead to face collapse and significant operational challenges [306]. The importance of proper assessment of groundwater conditions to prevent issues like face collapse and flooding cannot be overstated, as inadequate control can lead to disastrous outcomes, including blowouts and surface settlements [36, 375].

Failure to adequately assess and control groundwater pressures can lead to severe consequences during TBM operations. Such oversight is a major predictor of failure, which can include flooding and substantial ground subsidence [375]. On the other hand, proper pre-excavation grouting and continuous monitoring systems are critical in managing groundwater during TBM excavation, as these prevent disruptions by ensuring that expected groundwater conditions align closely with real-time operational demands [376]. Thus, thorough groundwater assessment is a non-negotiable aspect of selecting the appropriate TBM.

Groundwater conditions play a decisive role in the selection of TBMs. Different groundwater pressures dictate the suitability of specific types of TBMs, from Open TBMs and EPB machines to Slurry and Mixshield TBMs, depending on the operational environment. The efficient management of groundwater, through techniques such as soil conditioning and pre-excavation grouting, is imperative for the successful completion of tunnelling projects without facing operational risks such as face collapse, blowouts, and flooding.

Environmental Constraints

The selection of Tunnel Boring Machines (TBMs) in construction projects is increasingly influenced by environmental and social factors, particularly in urban environments and sensitive settings. This influence extends beyond mere engineering considerations, encapsulating a spectrum of regulatory and ecological constraints that shape the optimal choice of TBMs.

In urban environments, TBMs, particularly shielded types such as Earth Pressure Balance (EPB) and Slurry TBMs, are preferred due to their minimized impact on surrounding communities. These machines exhibit superior settlement control, which is crucial to avoiding damage to adjacent structures and infrastructure [365, 377]. The operational mechanisms of shielded TBMs allow for reduced vibration and noise, key factors in densely populated areas where disturbances are heavily scrutinized by local communities and regulatory bodies [78, 365]. This is critical as the risk of damage to existing buildings from construction-induced settlements is a significant concern in urban settings, thus making the choice of TBMs as much a social consideration as an engineering one [78, 377].

Furthermore, the regulatory landscape often necessitates the adoption of technologies that align with public health and safety requirements. The environmental benefits of employing TBMs that ensure minimal disruption are compelling in contexts where public approval is essential for project advancement [378, 379]. As urbanization increases, stakeholders are more sensitive to the need for preserving community quality of life, thus reinforcing the preference for less invasive tunnelling technologies.

When tunnelling beneath rivers, harbors, wetlands, or contaminated sites, the choice of TBMs is dictated not merely by technical specifications but also by stringent environmental regulations aimed at protecting groundwater and surrounding ecosystems. Slurry TBMs, for example, are often mandated in these environments due to their ability to prevent groundwater loss and mitigate environmental contamination risks [319, 380]. These machines operate under conditions where maintaining hydrostatic pressure is critical; thus, their design inherently incorporates mechanisms to reduce ecological footprints [374, 380].

In sensitive ecosystems, closed-face TBMs with controlled spoil handling systems are favoured for their capacity to minimize environmental risks and handle potentially hazardous materials safely [319, 380]. The methodical control over spoil and the maintenance of pressure within the excavation site are paramount to prevent any adverse effects on surface and subsurface environments [319, 380]. This emphasizes the dual focus on engineering efficacy and environmental stewardship in TBM selection processes.

The operational principle of TBMs also allows for significant limitations on surface disruption—a key advantage in urban and environmental contexts. By reducing the need for extensive open excavation, TBMs preserve roadways, buildings, and public services and minimize the ecological footprint of construction projects [381, 382]. This aspect aligns with contemporary urban planning goals and environmental sustainability efforts, making TBM selection not just a matter of technical feasibility but also a regulatory mandate influenced by community needs and ecological considerations [380, 383].

As such, the inclusion of environmental constraints often eliminates certain technically viable options from consideration, establishing a framework where regulatory compliance becomes as crucial as engineering performance metrics [382, 383]. This shifting paradigm underscores a transformative approach within the tunnelling industry, where decisions are increasingly made at the intersection of engineering, environmental science, and social responsibility.

The decision-making process regarding Tunnel Boring Machine selection is becoming appreciably complex, as urban and sensitive settings demand a broader view that encompasses environmental and social factors. The adoption of shielded TBMs and slurry systems in these contexts reflects the industry's evolving understanding of operational impacts on both surrounding infrastructures and local ecosystems. Thus, TBM choice is increasingly characterized by a regulatory framework that seeks to harmonize engineering excellence with community and environmental considerations.

Systems-Level Decision

The selection of a Tunnel Boring Machine (TBM) is a complex, systems-level decision that integrates various factors critical to successful tunnelling projects. This decision-making process encompasses geology, hydraulics, tunnel geometry, construction logistics, and environmental considerations. Each of these factors plays a pivotal role in ensuring safety, performance predictability, and environmental sustainability during tunnel excavation.

Geological conditions are foundational to TBM selection, determining how the face must be supported during excavation. Research indicates that adverse geological conditions can necessitate specific support mechanisms to mitigate risks such as ground settlement or sudden water inflows [30, 384]. For instance, the pressure of underground water significantly influences excavation methods, necessitating pressurized excavation techniques to manage high water inflow rates effectively [385]. Proper geological assessments allow for the identification of risk levels associated with groundwater, directly impacting TBM choice.

The geometry of the tunnel, including its diameter, significantly affects the feasibility and type of TBM employed. A larger diameter may necessitate a distinct TBM design due to mechanical limitations and construction logistics [386]. Additionally, tunnel length and alignment are critical determinants of economic viability; longer and more complex alignments increase costs and require more sophisticated TBMs to ensure efficient operation and safety [387]. For instance, closed-type shield TBMs are noted for their efficacy in managing ground settlement in urban areas, showcasing their advantage over traditional excavation methods in maintaining site integrity [30].

Construction logistics also play a vital role in TBM selection, influencing both the method employed and the machine's operational capabilities. The scheduling of TBM operation must consider factors such as surrounding infrastructure, project timelines, and resource availability [387]. Moreover, environmental constraints significantly shape acceptable construction methods. The integration of sustainability principles into TBM selection ensures compliance with regulatory frameworks and minimizes adverse environmental impacts [388]. For example, studies on tunnelling practices highlight the necessity of addressing environmental risks during the selection process to ensure the protection of surrounding ecosystems [389].

A successful TBM selection must balance geological factors, hydraulic pressures, tunnel geometry, construction logistics, and environmental constraints. Studies underscore that the interplay of these variables determines the effectiveness and safety of the tunnelling operation. A well-chosen TBM can facilitate safe excavation, predictable performance, and lower environmental impact [388]. Proper evaluation of these elements, as evidenced in various case studies, leads to better management practices and sustainable outcomes in tunnelling projects [384, 387].

Table 3: Tunnel Boring Machine (TBM) Selection Criteria Matrix.

Selection Criterion	Key Factors Considered	Typical Ground / Project Conditions	Preferred TBM Types	Engineering Rationale / Implications
Ground Conditions	Strength, structure, permeability, variability, squeezing behaviour	Competent hard rock (granite, gneiss)	Open (Gripper) TBMs	Rock mass provides natural support and gripper reaction, enabling high advance rates and efficient excavation
		Fractured rock, fault zones, squeezing ground	Single-shield or Double-shield TBMs	Shield provides continuous support where rock cannot reliably sustain gripper forces
		Soft ground (clays, silts, sands)	Shielded TBMs (EPB)	Excavated soil can be conditioned to form a pressure

Selection Criterion	Key Factors Considered	Typical Ground / Project Conditions	Preferred TBM Types	Engineering Rationale / Implications
				plug, maintaining face stability
		Clean sands, gravels, highly permeable soils	Slurry / Mixshield TBMs	Hydraulic pressure transmission controls face stability and groundwater inflow
		Mixed-face or highly variable ground	Dual-mode / Variable-density TBMs	Ability to adapt excavation mode to rapidly changing geology and groundwater regimes
Tunnel Diameter	Structural demand, thrust force, lining design, buoyancy	Small (≈0.3–3 m)	Microtunnelling / Pipe-jacking	High positional accuracy and minimal surface disruption for utilities and drainage
		Medium (≈3–6 m)	EPB, Slurry, Shielded TBMs	Common for metro, rail, and road tunnels in urban environments
		Large (≈6–18+ m)	Shielded TBMs (soft ground) / Open TBMs (hard rock)	Increased thrust, face pressure, and lining demands favour shielded systems in soft ground
Length and Alignment	Drive length, curvature, cross-passages, complexity	Long, continuous drives (km-scale)	TBMs (all types depending on ground)	High productivity, consistent excavation, reduced overbreak
		Short tunnels or highly variable alignment	Conventional excavation or small TBMs	TBM mobilisation costs may outweigh benefits
		Straight or gently curved alignments	Open and Double-shield TBMs	Optimal performance and efficiency

Selection Criterion	Key Factors Considered	Typical Ground / Project Conditions	Preferred TBM Types	Engineering Rationale / Implications
		Tight curves, complex geometry	Small TBMs or drill-and-blast	Large TBMs lack flexibility in constrained geometry
Water Pressure	Groundwater level, permeability, inflow risk	Low groundwater pressure	Open TBMs or EPB TBMs	Manageable inflows without full hydraulic face support
		Moderate groundwater pressure	EPB TBMs	Soil conditioning and pressure control stabilise face
		High groundwater pressure / subaqueous tunnels	Slurry / Mixshield TBMs	Hydraulic pressure transmission prevents face collapse and flooding
Environmental Constraints	Settlement limits, vibration, noise, contamination, regulation	Dense urban areas	Shielded TBMs (EPB, Slurry)	Excellent settlement control, low vibration and noise
		Rivers, harbours, wetlands	Slurry / Mixshield TBMs	Prevent groundwater loss and environmental contamination
		Contaminated ground	Closed-face TBMs with controlled spoil handling	Minimise exposure and environmental risk
		Strict surface disruption limits	All closed-face TBMs	Reduce need for open excavation and protect infrastructure
Systems-Level Integration	Geology, hydraulics, geometry, logistics, environment	Complex, high-risk projects	TBM selected via integrated assessment	TBM choice balances safety, predictability, cost, and environmental impact

Chapter 5

TBM Components and Systems

Tunnel Boring Machines (TBMs) represent some of the most complex and highly integrated construction systems used in modern civil engineering. Far from being a single excavation device, a TBM operates as a coordinated assemblage of mechanical, hydraulic, electrical, and digital subsystems that must function in precise synchronisation to safely and efficiently construct tunnels through a wide range of ground conditions. The performance of a TBM is therefore not determined by any single component, but by the way its systems interact with each other and with the surrounding ground.

This chapter provides an overview of the major components and systems that make up a TBM, explaining both their individual functions and their collective role within the excavation process. It examines how cutting tools, thrust systems, shields, spoil-handling mechanisms, power and control systems, and support infrastructure are integrated into a continuous tunnelling operation. Particular emphasis is placed on the relationship between machine design and ground behaviour, highlighting how TBM systems are selected, configured, and controlled to manage face stability, ground deformation, wear, and operational risk.

Understanding TBM components at a systems level is essential for engineers, operators, and project managers involved in tunnel design and delivery. Decisions made during TBM selection, design, and operation—such as cutterhead configuration, thrust control philosophy, spoil-removal method, or automation strategy—have direct implications for productivity, safety, settlement control, and overall project outcomes. This chapter establishes the technical foundation needed to appreciate those decisions and to interpret TBM performance in real-world tunnelling conditions.

By presenting the TBM as an integrated underground production system rather than a collection of isolated parts, this chapter sets the framework for deeper discussion of excavation mechanics, ground–machine interaction, and operational control in subsequent sections.

TBM Components and System Overview

A Tunnel Boring Machine (TBM) is a highly integrated system composed of mechanical, hydraulic, electrical, and digital control components that operate together to excavate ground, stabilise the tunnel, remove spoil, and install permanent lining in a continuous and controlled process. Rather than acting as a single piece of equipment, a TBM functions as a mobile underground factory in which each subsystem performs a specialised role. The effectiveness of TBM tunnelling depends on the precise coordination of these subsystems, as the performance of any one component directly affects overall safety, productivity, and ground response.

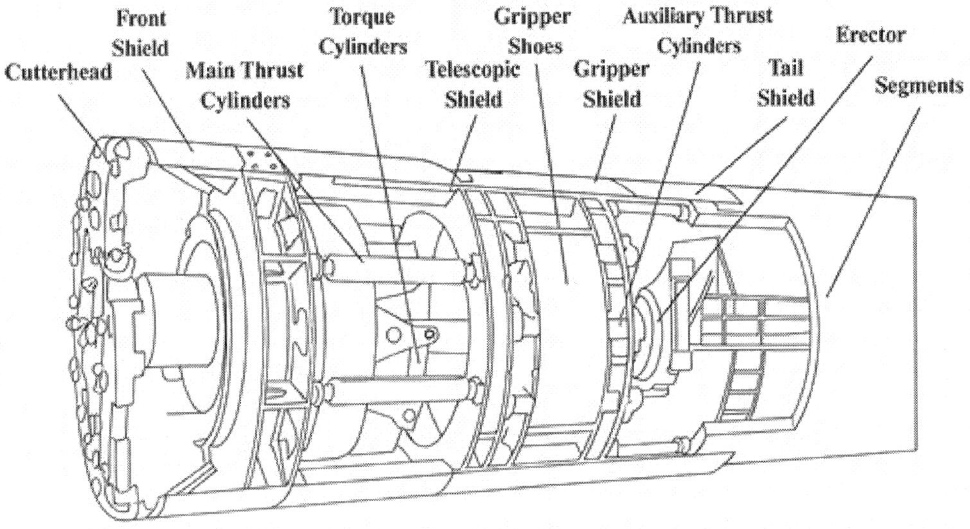

Figure 55: Key Components of a TBM.

At the front of the machine is the cutterhead, which performs the primary excavation function. The cutterhead is a large rotating steel structure fitted with cutting tools selected to suit the anticipated ground conditions. Disc cutters are used in hard rock to induce compressive fracturing, while scrapers, picks, or teeth are used in soft soils and mixed ground. In variable geology, combinations of cutting tools may be installed. As the cutterhead rotates, it breaks the ground into fragments small enough to be handled by the machine, with openings in the cutterhead allowing the excavated material to pass into the excavation chamber or collection system immediately behind it.

Directly behind the cutterhead lies the excavation chamber, which plays a critical role in maintaining face stability. The function of this chamber varies depending on the TBM type. In open (gripper) TBMs, the face remains open and stability is provided by competent rock. In Earth Pressure Balance (EPB) TBMs, the chamber is filled with conditioned excavated soil that balances earth and groundwater pressures. In slurry or Mixshield TBMs, the chamber is filled with pressurised bentonite slurry that hydraulically supports the tunnel face. Regardless of configuration, the excavation chamber is fundamental to preventing face collapse, limiting ground movement, and controlling water inflow, particularly in soft or water-bearing ground.

Shielded TBMs incorporate a steel cylindrical shield that encloses the excavation chamber and working area. This shield provides temporary ground support during excavation, protects personnel and equipment, and limits deformation of the surrounding ground until the permanent lining is installed. In contrast, open TBMs do not include a full-length shield, although they may incorporate partial shields or protective canopies to provide limited protection in less stable sections.

Forward movement of the TBM is achieved through the thrust system. Hydraulic thrust cylinders generate the force required to push the cutterhead into the ground. In open hard-rock TBMs, thrust is reacted through hydraulic grippers that press against the tunnel walls. In shielded TBMs, thrust is reacted against previously installed segmental lining rings. Thrust force, cutterhead torque, and penetration rate are continuously monitored and adjusted to suit ground behaviour, ensuring efficient excavation while avoiding excessive loads on the machine or the surrounding ground.

Efficient removal of excavated material is essential for continuous TBM operation. The muck removal system varies according to TBM type and ground conditions. Open TBMs and some EPB machines typically use belt conveyors or muck cars to transport spoil away from the face. EPB and variable-density machines commonly use screw conveyors to control material extraction and maintain face pressure. Slurry and Mixshield TBMs transport spoil hydraulically via slurry pipelines to a surface separation plant, where excavated solids are removed and the slurry is conditioned and recycled back to the face.

In shielded TBMs, permanent tunnel support is installed using a segment erector located immediately behind the shield. The segment erector places precast concrete lining segments to form a complete ring that provides long-term structural support, seals against groundwater, and serves as the reaction surface for the thrust cylinders. The lining installation cycle is closely synchronised with excavation to ensure continuous progress while maintaining ground stability.

Open hard-rock TBMs rely on a gripper system to provide thrust reaction. Hydraulic side grippers press laterally against the tunnel walls, allowing the machine to push forward without reacting against a lining. This arrangement enables very high advance rates in competent rock but depends entirely on the strength and integrity of the surrounding rock mass. Where the rock cannot reliably sustain gripper forces, shielded machines are required.

Behind the main TBM body is a series of backup gantries that support ongoing operations. These gantries house electrical transformers, power distribution equipment, hydraulic pumps, cooling systems, ventilation and dust suppression equipment, control rooms, and storage for lining segments and consumables. Together, the backup systems ensure continuous operation and provide a controlled, safe working environment underground.

Modern TBMs are equipped with advanced guidance, control, and monitoring systems that allow precise alignment control and real-time assessment of machine and ground performance. Laser guidance and gyroscopic navigation systems maintain tunnel geometry, while sensors continuously measure thrust, torque, vibration, pressure, and machine position. These data feed into automated control systems that allow operators to adjust operating parameters rapidly in response to changing ground conditions.

All TBM components operate as a tightly integrated system. The cutterhead excavates the ground while the excavation chamber stabilises the face. Thrust cylinders advance the machine, spoil is removed through dedicated transport systems, the shield provides temporary support, and lining segments are installed immediately behind the excavation zone. Backup systems supply power, ventilation, and operational control. This continuous, coordinated process enables safe, efficient, and predictable tunnel construction with minimal surface impact and controlled ground behaviour.

Cutterhead Design and Tooling

The cutterhead is the primary excavation component of a Tunnel Boring Machine (TBM) and plays a decisive role in excavation performance, ground stability, tool wear, and overall project risk. Its design and tooling must be carefully matched to the expected geology, abrasivity, strength, structure, and variability of the ground. A well-designed cutterhead not only breaks the ground efficiently but also controls muck flow, supports face stability, and protects the machine from excessive loads and damage.

Figure 56: TBM Cutterhead. MTA Construction & Development Mega Projects, CC BY 2.0, via Flickr.

At its core, the cutterhead must apply sufficient energy to fracture or cut the ground while allowing excavated material to pass through into the excavation chamber or muck collection system. To achieve this, the cutterhead design balances several competing requirements: structural strength, open area for material flow, tool accessibility for maintenance, compatibility with face-support systems, and resistance to wear and impact. Poor cutterhead design can lead to low penetration rates, clogging, excessive vibration, tool damage, or face instability.

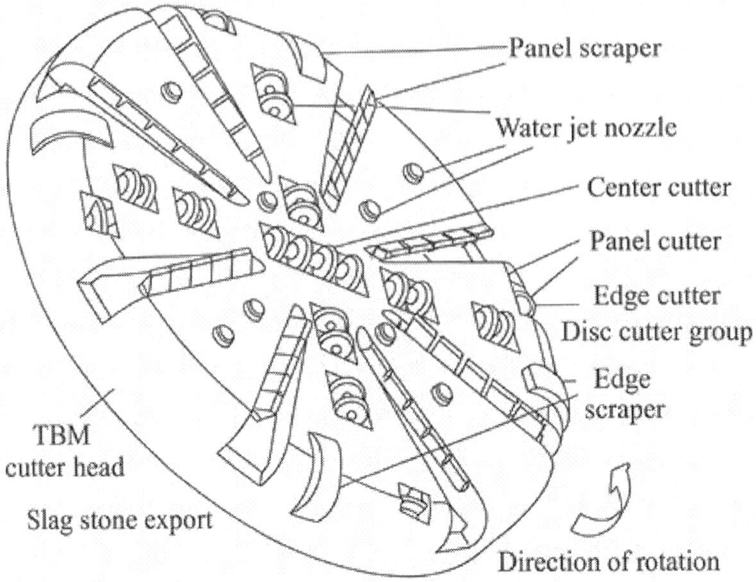

Figure 57: Example Cutterhead design (with a high-pressure water jet nozzle). Adapted from Wang et al. (2020) [390].

In hard rock tunnelling, cutterheads are typically robust, stiff structures equipped with disc cutters. These cutters apply high normal forces to the rock surface, inducing compressive stress and tensile cracking that causes the rock to spall and chip. Hard-rock cutterheads generally have a lower open area to maintain stiffness and transmit thrust efficiently, as the rock fragments are relatively coarse and free-flowing.

In soft ground and mixed ground conditions, such as those encountered by EPB and slurry TBMs, cutterheads are designed with a much higher open area. This allows soil, sand, or slurry-mixed spoil to flow easily into the excavation chamber, supporting pressure control at the face. These cutterheads are typically fitted with scrapers, cutting knives, picks, or teeth rather than disc cutters, although mixed-ground designs may incorporate both tool types.

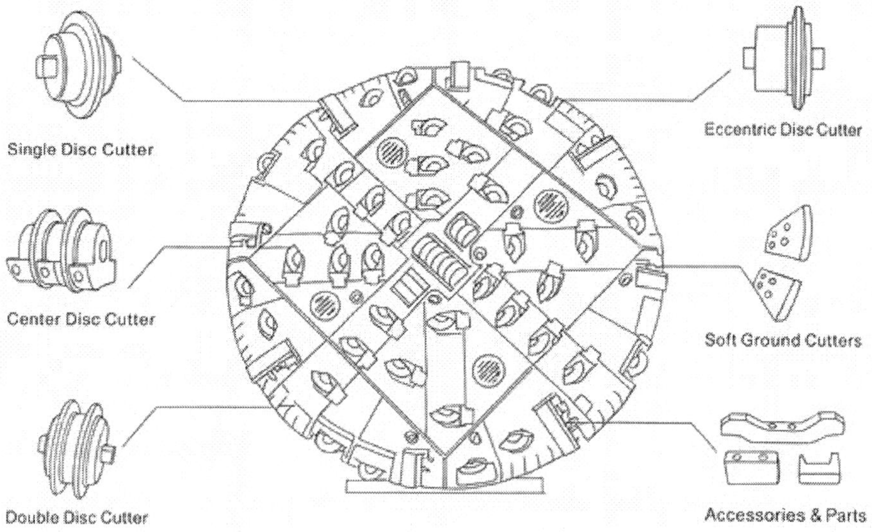

Figure 58: Disc Cutter example. Adapted from Lithian [391].

Disc cutters are the dominant cutting tools for hard rock TBMs. They consist of hardened steel rings mounted on bearings, allowing them to roll as they press into the rock. Cutter diameters commonly range from about 13 to 19 inches, with larger cutters used in very hard or abrasive rock. The spacing, layout, and load distribution of disc cutters are critical: too wide a spacing reduces efficiency, while too close a spacing increases wear and energy consumption.

Disc cutter performance depends on rock strength, abrasivity, jointing, and in-situ stress. High stresses can improve cutting efficiency through stress-assisted spalling, while highly abrasive rocks accelerate cutter wear. Cutterhead designers carefully model cutter forces and wear rates to ensure acceptable maintenance intervals and predictable performance.

In soft soils such as clays, silts, and sands, excavation relies on cutting, shearing, and scraping rather than fracturing. Cutterheads are fitted with scrapers, knives, chisels, and drag picks that shave material from the face and promote controlled flow into the chamber. These tools are often replaceable and designed to work in conjunction with soil conditioning agents such as foam, polymers, or bentonite.

In cohesive soils, tool geometry is selected to minimise clogging and adhesion, while in sandy soils the emphasis is on durability and resistance to abrasion. The cutterhead layout must also support uniform pressure distribution across the face to avoid localised collapse or excessive ground movement.

Many real-world tunnels encounter mixed-face conditions, where soil and rock appear simultaneously at the tunnel face. In these cases, cutterheads are equipped with a combination of disc cutters and soft-ground tools. The arrangement is carefully planned so that hard inclusions such as boulders or rock ribs can be broken effectively, while softer materials continue to flow smoothly.

Mixed-ground cutterheads often incorporate reinforced structures, replaceable wear plates, and adaptable tool zones to cope with abrupt geological transitions. These designs are particularly common in urban tunnelling, river crossings, and karstic environments.

The proportion of open area in the cutterhead is a critical design parameter. A higher open area improves spoil flow and pressure control in EPB and slurry machines but reduces structural stiffness. A lower open area increases stiffness and cutting efficiency in rock but limits material flow. Designers must therefore tailor the cutterhead geometry to both the excavation method and the face-support mechanism.

In pressurised TBMs, the cutterhead also acts as part of the face-support system. Its geometry influences how pressure is transmitted to the ground, how material is mixed in the chamber, and how stable the face remains during excavation and stoppages.

Cutterheads are exposed to extreme mechanical wear and impact. To manage this, they are fitted with wear plates, hardfacing, and sacrificial components that protect the main structure. Tool change strategy is a key operational consideration: in open TBMs, cutters can often be changed at the face, while in pressurised machines, tool changes may require compressed-air interventions or hyperbaric work, significantly increasing cost and risk.

Wear Plates

Wear plates are replaceable steel plates mounted on areas of the TBM that are exposed to high abrasion, erosion, or impact. They are commonly installed on cutterhead surfaces, cutterhead spokes, the excavation chamber, screw conveyor inlets, slurry inlets, and transition zones where muck changes direction or velocity.

These plates are manufactured from abrasion-resistant steels (such as hardened alloy steels) and are designed to be thicker and more wear-resistant than the underlying structural steel. Their primary function is to shield the load-bearing structure of the TBM from direct wear. When wear plates reach the end of their service life, they are removed and replaced, restoring protection without requiring repair of the main structure.

From an engineering perspective, wear plates provide a controlled and predictable wear mechanism. Rather than allowing uncontrolled thinning of structural components—which could lead to deformation or failure—wear is concentrated in components specifically designed to be consumed and replaced.

Hardfacing

Hardfacing is a surface treatment process in which a layer of very hard, wear-resistant material is welded or deposited onto exposed steel surfaces. Unlike wear plates, hardfacing does not add a separate component; instead, it enhances the surface properties of an existing part.

In TBMs, hardfacing is commonly applied to cutterhead surfaces, buckets, scrapers, crusher housings, slurry pipes, and screw conveyor flights. The deposited materials often contain carbides (such as chromium or tungsten carbides) that provide exceptional resistance to abrasion, erosion, and gouging.

Hardfacing is particularly valuable in areas where installing removable wear plates is impractical due to geometry, space constraints, or load transfer requirements. It extends component life and reduces the frequency of repairs, but once worn away it must be reapplied—often requiring welding work in confined or pressurised environments.

Engineering trade-offs are important here: while hardfacing significantly improves wear resistance, it can also increase brittleness, making components more susceptible to cracking under impact if not correctly specified.

Sacrificial Components

Sacrificial components are parts intentionally designed to wear out or fail before more critical or expensive components are damaged. In TBMs, these include items such as cutter rings, scraper tips, picks, teeth, wear bars, and crusher liners.

These components are relatively easy to replace and are expected to be consumed as part of normal operation. Their design philosophy is similar to an electrical fuse: they protect the system by failing first. For example, disc cutter rings are sacrificial elements that wear against the rock while protecting the cutter hub and bearing assembly. Likewise, sacrificial wear bars on cutterhead openings prevent erosion of the main cutterhead body.

Sacrificial components allow maintenance to be planned, modular, and cost-effective, rather than reactive and structural. Their wear rates are often predicted during design based on rock abrasivity, expected penetration rates, and cutter loads.

Effective wear protection is not just a durability issue; it directly affects safety, programme, and cost. Excessive or unexpected wear can force unplanned stoppages, hyperbaric interventions, or even early machine withdrawal. Conversely, well-designed wear systems enable predictable maintenance intervals, reduce downtime, and support consistent advance rates.

In high-abrasivity ground, wear protection strategy can be a deciding factor in TBM selection and cutterhead design. It must be based on realistic ground investigation data, including abrasivity indices, grain size, mineralogy, and groundwater conditions. As a result, cutterhead and tooling design strongly influence not only excavation performance but also safety, maintenance planning, and project programme.

From an engineering perspective, cutterhead design and tooling form the interface between the machine and the ground. They translate thrust and torque into controlled excavation while directly influencing face stability, advance rate, energy consumption, and wear. Successful TBM projects depend heavily on matching cutterhead design to geological conditions, supported by thorough ground investigation, realistic wear modelling, and contingency planning for adverse ground.

Disc Cutters, Drag Bits, and Scrapers

The cutterhead of a Tunnel Boring Machine (TBM) is equipped with cutting tools specifically selected to match the mechanical behaviour of the ground. Among the most common tools are disc cutters, drag bits, and scrapers. Each works on a different physical principle and is suited to particular ground conditions. Correct selection and arrangement of these tools is critical to excavation efficiency, wear control, and face stability.

Disc Cutters

Disc cutters are the primary cutting tools used for hard rock tunnelling. They consist of hardened steel rings mounted on bearings, allowing them to rotate freely as they roll across the rock surface.

Figure 59: Disc cutter from the tunnel boring machine (TBM) "Åsa" used to dig and line the Hallandsås tunnel. Erik Lundin, CC BY-SA 3.0, via Wikimedia Commons.

Rather than cutting or grinding the rock, disc cutters work by inducing compressive stress. As the cutterhead rotates and thrust is applied, each disc presses into the rock, creating a crushed zone beneath the disc. When adjacent cutter tracks overlap, tensile cracking develops between them, causing chips or slabs of rock to break away. This mechanism is known as compressive–tensile fracturing and is highly efficient in strong, brittle rock such as granite, gneiss, and basalt.

Disc cutters are extremely robust and can withstand very high loads, but they are also heavy and expensive. They are considered sacrificial components: the cutter ring gradually wears and must be replaced once it reaches a wear limit. Cutter size (typically 13–20 inches), spacing, and layout are carefully designed to suit rock strength, abrasivity, and expected penetration rates.

Typical use:

- Open (gripper) TBMs in hard rock
- Single- and double-shield TBMs in competent or fractured rock

Drag Bits

Drag bits are fixed cutting tools that shear or plough through the ground rather than crushing it. They usually consist of steel bodies with hardened edges or carbide inserts and do not rotate independently like disc cutters.

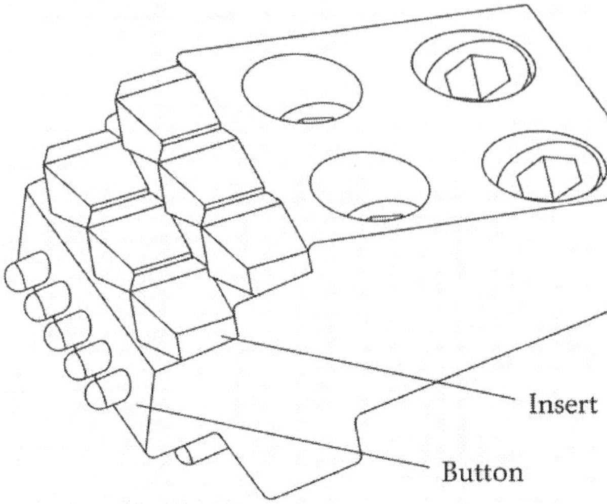

Figure 60: Diagram of a Drag Bit.

As the cutterhead turns, drag bits scrape through the soil, breaking it by shear failure. This mechanism is effective in soft, cohesive soils such as clays and silts, where compressive fracturing is unnecessary and inefficient. Drag bits require much lower thrust and torque than disc cutters, making them ideal for soft-ground machines.

However, drag bits wear rapidly in abrasive or coarse materials and are unsuitable for hard rock. They are often combined with other tool types in mixed-ground cutterheads.

Typical use:

- EPB TBMs in clay and silt
- Soft-ground and mixed-face conditions
- Low-strength soils with limited coarse particles

Scrapers

Scrapers are also fixed tools, but their primary role is not just cutting—it is material collection and control. They scrape loosened soil or rock fragments from the tunnel face and guide them through cutterhead openings into the excavation chamber.

In soft ground, scrapers assist drag bits by gathering excavated soil and maintaining a uniform flow of material into the chamber. In EPB machines, this is particularly important because controlled material flow is essential for maintaining face pressure. In mixed or transitional ground, scrapers help manage partially broken material and prevent blockages at the cutterhead.

Scrapers are generally simpler and lighter than drag bits and experience significant wear, especially in abrasive soils. As a result, they are often treated as sacrificial wear components.

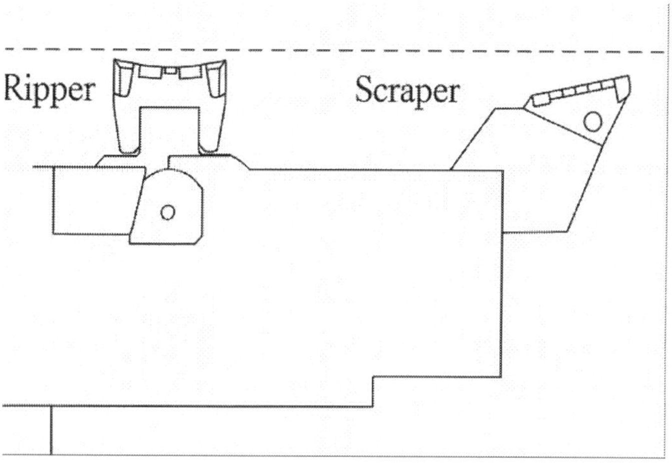

Figure 61: Diagram of a ripper and scraper.

Typical use:

- EPB TBMs
- Slurry and Mixshield TBMs
- Soft ground and mixed-face conditions

> Scrapers and rippers are both fixed, non-rotating cutting tools commonly fitted to soft-ground and mixed-ground TBM cutterheads, but they are designed to perform distinctly different mechanical functions. While both contribute to excavation, they act on the ground in different ways and are selected to address different material behaviours.
>
> Scrapers are primarily material-handling tools rather than aggressive cutting tools. They are relatively broad, shovel- or blade-shaped elements mounted on the cutterhead to collect loosened soil or fragmented ground and guide it through the cutterhead openings into the excavation chamber. Any cutting action they provide is secondary to their role in controlling and directing muck flow.
>
> In operation, scrapers penetrate only minimally into the ground and work mainly on material that has already been loosened by other tools. Their design promotes smooth and uniform spoil movement, which is especially important in pressure-controlled

systems. By helping regulate the rate and consistency of material entering the chamber, scrapers play a key role in maintaining stable face pressure in EPB TBMs and in reducing the risk of clogging or blockages at the cutterhead.

Because of this function, scrapers are most commonly used in EPB and slurry or Mixshield TBMs operating in soft soils, silts, clays, and mixed ground. They are particularly important in environments where predictable spoil flow and pressure balance are more critical to face stability than aggressive cutting performance.

Rippers, by contrast, are designed for active and aggressive ground break-up. They are narrow, pointed, tooth-like tools that concentrate cutting force over a small area, allowing them to penetrate deeply into firm or stiff materials. Their primary role is to tear, fracture, and disaggregate ground that is too strong or bonded for scrapers or drag bits to handle efficiently.

Rippers achieve high penetration depths and are effective in breaking stiff clays, cemented soils, weathered rock, and weak rock bands. They are often fitted with carbide tips to improve wear resistance and durability. The fragments and chunks created by rippers are then collected and transported by scrapers or other material-handling tools on the cutterhead.

These tools are typically used on mixed-ground TBMs and EPB machines encountering transitional conditions, such as interfaces between soil and rock, cemented sands, or weak mudstone and shale. Rippers bridge the gap between soft-ground tools and hard-rock disc cutters, providing an effective solution where the ground is too stiff for purely soft-ground tooling but not hard enough to justify full hard-rock cutting systems.

In practical TBM operation, scrapers and rippers function as complementary tools rather than alternatives. Rippers act first at the tunnel face, penetrating and tearing the ground to break it into fragments where the material is stiff, bonded, or partially rock-like. Scrapers then follow by collecting these loosened fragments and guiding them through the cutterhead openings into the excavation chamber, ensuring controlled and continuous spoil flow.

In Earth Pressure Balance (EPB) machines, this combined action is particularly important because stable operation depends on maintaining a consistent pressure plug within the excavation chamber. Rippers ensure that the ground can be effectively disaggregated, while scrapers regulate how that material enters the chamber, helping to stabilise face pressure and prevent sudden pressure losses or surges.

In mixed-ground conditions, the use of both tools allows the cutterhead to respond to abrupt changes in material stiffness without the need to change machine type or significantly alter operating parameters. Cutterheads intended for variable ground conditions are therefore typically designed with alternating arrangements of rippers and scrapers across the face, balancing cutting efficiency, wear distribution, torque demand, and controlled muck flow.

From an engineering perspective, the correct balance between scrapers and rippers is critical. Using scrapers where rippers are required can result in poor penetration, elevated torque, and an increased risk of clogging at the cutterhead. Conversely, overusing rippers in ground where scraping and material control should dominate can lead to over-excavation, unstable face conditions, and unnecessarily high wear rates and energy consumption.

For these reasons, the ratio and spatial arrangement of scrapers to rippers are carefully engineered for each project. Design decisions are informed by detailed assessments of soil and rock strength, cohesion, cementation, abrasivity, and groundwater conditions, ensuring that the cutterhead delivers efficient excavation while maintaining face stability and predictable machine performance.

Combined Use in Practice

Modern Tunnel Boring Machine (TBM) cutterheads are designed with sophisticated configurations that combine various tool types to effectively manage complex ground conditions encountered during excavation. This paper explores the specific tool arrangements utilized across different geological scenarios, including hard rock, soft ground, and mixed-ground environments, detailing how these configurations contribute to continuous excavation and operational efficiency.

Hard rock TBM cutterheads predominantly utilize disc cutters, recognized for their effectiveness in breaking through hard rock formations. This tool type is carefully engineered to handle substantial forces during operation. The design minimizes the use of auxiliary tools, such as scrapers, which primarily aid in muck flow control. The arrangement of disc cutters on the cutterhead is crucial; it is often optimized for superior rock penetration and wear resistance [392-394]. Studies have shown that the thrust and torque requirements for hard rock tunnelling can be predicted based on the cutterhead's design and material properties [394, 395]. The focus on disc cutters allows for enhanced productivity and reliability in hard rock environments [394].

Conversely, soft-ground TBM cutterheads predominantly rely on drag bits and scrapers. Drag bits are particularly suited for softer materials and aid in the effective excavation of less cohesive ground conditions, while scrapers help manage the movement of muck and ensure efficient removal from the excavation site [313, 396]. The effectiveness of soft-ground cutterheads hinges on the optimal arrangement of these tools, which must be designed to minimize soil disturbance and maintain stability during excavation [397]. The dynamic response of the cutterhead is heavily influenced by the operational conditions, making it vital to understand the geological context when selecting tool configurations [390].

Mixed-ground scenarios present unique challenges that necessitate a hybrid approach in cutterhead design, often integrating combinations of disc cutters, drag bits, and scrapers. These mixed-ground cutterheads must be capable of transitioning between soil types without compromising excavation efficiency [395]. The strategic arrangement and spacing of these tools are engineered to maintain continuous operation even in the event of individual tool wear or failure. This redundancy in tool design allows TBMs to adapt swiftly to changing geological conditions, thereby minimizing downtime and enhancing overall project viability [383].

The engineering design of TBM cutterheads not only involves selecting appropriate tools but also encompasses their arrangement, spacing, and the incorporation of redundancy. Achieving the ideal layout is essential for maximizing effective cutting forces and minimizing complications such as excessive wear or failure during the excavation process [398, 399]. Various studies underline the significance of computational models in predicting wear patterns and optimizing tool placement on cutterheads under differing geological conditions [393, 394].

In addition, the coupling of innovations such as high-pressure water jets with traditional cutting methods has been found to further enhance performance in challenging extraction environments [390]. As TBM technology advances, the continuous refinement of cutterhead designs will provide new techniques for tackling increasingly complex ground conditions encountered in modern tunnelling projects.

Engineering Significance

The choice between disc cutters, drag bits, and scrapers is a critical factor influencing various operational parameters in construction and tunnelling activities, particularly in the context of tunnel boring machines (TBMs). This selection not only affects performance metrics like penetration rate and advance speed but also plays a significant role in energy consumption, torque demand, maintenance frequency, face stability, spoil flow, and overall project costs.

1. Penetration Rate and Advance Speed

The penetration rate (ROP) and advance speed of a TBM are significantly influenced by the type of cutting tool employed. Disc cutters have been identified as effective for hard rock conditions, providing better control over torque and facilitating higher penetration rates due to their design and operational dynamics [400]. Research indicates that the weight on bit (WOB) and cutter specifications directly correlate with performance metrics, revealing that optimizing these parameters can lead to substantial increases in ROP [401]. For instance, a comparative study by Chakeri et al. [402] demonstrates that different cutter designs exhibit varying efficiencies under specific geological conditions, affecting both penetration and advance speed.

2. Energy Consumption and Torque Demand

The energy consumption and torque demand are intrinsic to the operational efficiency of TBMs using different cutting tools. The performance parameters such as torque and energy consumption relate directly to the design and arrangement of the cutting tools on the cutterhead [403]. When examining various cutterhead configurations, it becomes evident that inappropriate selection can lead to excessive torque and energy inefficiencies, which ultimately increases operational costs and project delays [401, 404]. Effective cutterhead design can mitigate these issues by ensuring a balanced load distribution across the cutting tools, significantly influencing energy use [402].

3. Tool Wear and Maintenance Frequency

The wear of cutting tools is a pivotal consideration in tunnelling operations, as it dictates the maintenance frequency and operational costs. Disc cutters, for instance, exhibit specific wear patterns that can be quantified and modelled to predict their lifespan under different operational conditions [402]. Alwaar et al. [405] discussed how vibration-assisted rotary drilling (pVARD) can extend the life of disc cutters by reducing wear rates, thereby minimizing maintenance interventions. Research underlines that frequent maintenance due to tool wear can lead to increased downtime and costs, emphasizing the need for careful selection based on expected geological conditions [402].

4. Face Stability and Spoil Flow

The stability of the face during tunnelling operations is crucial, as it affects spoil flow and overall safety. Different cutting tools have varying implications on face stability; for example, drag bits may not provide adequate support in unstable formations, whereas properly configured disc cutters can enhance stability by extracting material in a controlled manner [402]. Studies indicate that the type of cutter influences not only the immediate performance but also the long-term geological interactions at the tunnel face, impacting spoil management and flow characteristics [404].

5. Overall Project Cost and Downtime

Ultimately, the choice of cutting tools carries significant weight in determining overall project costs and downtime. Poor tool selection can lead to unplanned maintenance and operational inefficiencies, driving up total expenses beyond initial estimates [406]. The alignment of cutterhead design with geological parameters is essential to minimize costs and improve efficiency, highlighting the synergy between geological investigation data and cutter design [404]. The integration of these considerations into the planning process is fundamental for successful project execution.

Main Bearing and Seals

The main bearing in Tunnel Boring Machines (TBMs) is a critical component designed to withstand extreme mechanical stresses while facilitating the operation of the cutterhead. It not only supports the cutterhead but also plays a significant role in the transmission of thrust and torque, essential for a TBM's effective function in tunnel construction. Given the hostile environments in which TBMs operate—characterized by exposure to water, soil, and abrasive materials—the seal system that protects this bearing is equally important for ensuring operational reliability and longevity.

The main bearing serves as the interface between the rotating cutterhead and the stationary shield, imposed with a complex mix of loading conditions. TBMs must manage significant axial loads from thrust cylinders that propel the machine forward, as well as radial loads due to uneven cutting forces encountered in heterogeneous ground materials [407]. Recent studies emphasize the necessity for TBM bearings to endure bending moments caused by eccentric loading and tool wear, particularly since large TBMs can carry loads exceeding tens of thousands of tonnes [407, 408].

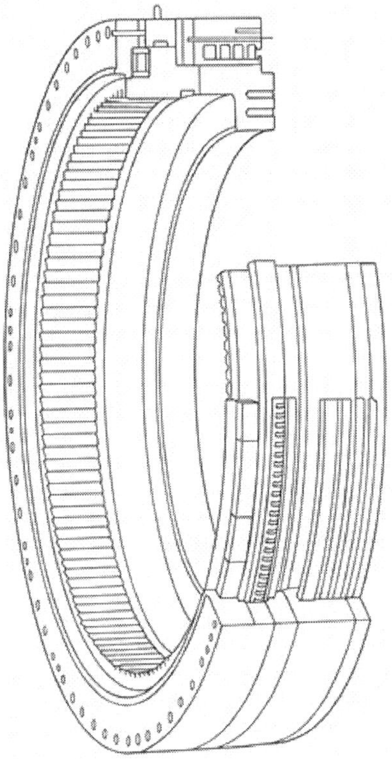

Figure 62: The slewing bearing is the link between the pinion drives and the cutter head.

Modern TBMs typically employ large-diameter multi-row roller bearings tailored to these demanding conditions. For example, triple-row roller bearings are prevalent in earth pressure balance (EPB) and slurry TBMs, while double-row designs are more common in smaller or harder rock applications [316]. These bearings are custom-engineered for optimal low-speed, high-load performance rather than the high-speed operations characteristic of most industrial bearings [408].

The functionality of the bearing heavily relies on an effective seal system to protect against contamination. Seals are engineered to resist high external water pressures—which can reach several bars in deep underground scenarios—while also providing barriers against abrasive materials and chemicals from conditioning agents like bentonite slurry [316]. Failure of the seal system poses one of the most significant risks in TBM operations, as contamination can lead to bearing failure, which is costly and time-consuming to rectify [407, 409].

Typically, multi-stage seal systems are implemented to provide redundancies. These may include multiple radial lip seals, mechanical face seals for high-pressure conditions, and pressurized grease zones to maintain a protective barrier against external contaminants [316]. This robustness is critical for ensuring operational integrity, particularly in challenging tunnelling environments, such as those encountered with slurry TBMs, where robust sealing measures are paramount [316, 407].

Continuous and effective lubrication is vital for both the main bearing's performance and the seals' integrity. Centralized lubrication systems deliver controlled grease applications to roller elements while simultaneously supporting seal functions [407, 408]. Modern TBMs incorporate sophisticated condition-monitoring technologies, including embedded temperature sensors and vibration analytics, to provide real-time diagnostics regarding bearing health [407, 408].

Monitoring systems are critical for early detection of potential issues, allowing for timely interventions before failures arise. They track parameters such as temperature variations, torque transmission efficiency, and seal pressure status [407, 409]. Such predictive maintenance strategies enhance overall system reliability and extend operational lifespan, reinforcing the vital importance of the main bearing and seals in the overall design and engineering of TBMs.

The requirements for bearing and seal systems vary significantly across different types of TBMs. For instance, open (gripper) TBMs generally face lower groundwater pressures; nonetheless, they encounter high mechanical loads [316]. Conversely, EPB TBMs must contend with soil conditions that demand bearings and seals robust enough to handle associated clogging and pressure fluctuations [410]. Among the most challenging conditions are those experienced by slurry TBMs, which necessitate sophisticated seals capable of functioning under elevated pressures and abrasive conditions, making seal performance a critical risk factor in these operations [316, 409].

The main bearing of a TBM acts as a critical nexus for facilitating cutterhead rotation while supporting heavy loads from thrust, torque, and bending forces. The accompanying seal system is essential for safeguarding the bearing from the harsh tunnelling environment, underscoring the interdependence of both systems in maintaining TBM reliability, safety, and efficiency during complex underground projects. As technology advances, further innovation in bearing design, seal robustness, and condition monitoring will play an indispensable role in enhancing TBM performance, especially in challenging tunnelling scenarios.

Thrust Cylinders

TBM thrust cylinders are the hydraulic actuators that provide the forward force required to push a Tunnel Boring Machine (TBM) into the ground. They translate hydraulic pressure into controlled axial thrust, allowing the cutterhead to penetrate the ground while the machine advances in a stable, predictable manner. Along with the cutterhead drive and main bearing, the thrust system is a core load-bearing element that directly governs penetration rate, advance control, and overall excavation safety.

Each thrust cylinder extends to press against a reaction surface, generating axial force that advances the TBM. The load path differs by machine type. In shielded TBMs (EPB, slurry, Mixshield), thrust is reacted through the previously installed segmental lining rings. In open (gripper) hard-rock TBMs, thrust is reacted through side grippers bearing against the tunnel walls. The combined thrust from all cylinders must overcome cutting resistance at the face, friction along the shield, and any additional resistance from ground deformation or steering corrections.

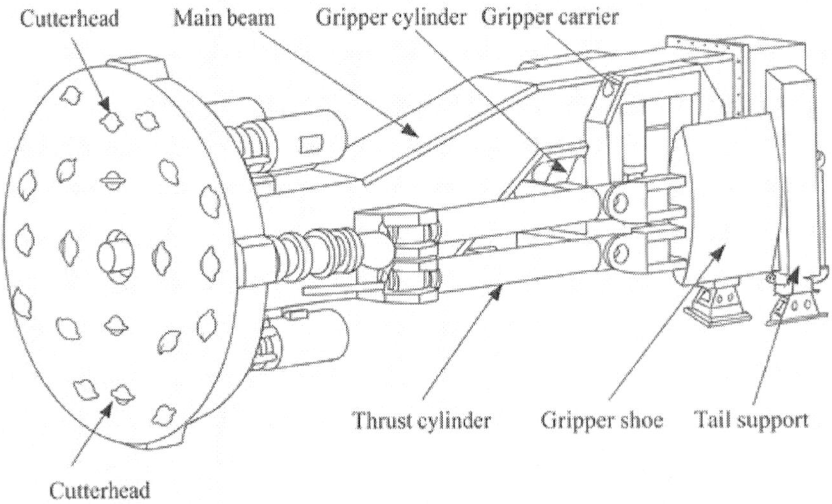

Figure 63: TBM example Thrust Cylinder arrangement. Adapted from Chen et al. (2022) [411].

Thrust systems typically consist of multiple large-diameter hydraulic cylinders arranged symmetrically around the TBM circumference. Large urban TBMs may have 20–30 or more cylinders, each capable of delivering hundreds to thousands of tonnes of force. Cylinders are grouped into control zones so operators can vary thrust distribution locally, an essential capability for steering, correcting alignment, and managing asymmetric ground resistance.

Hydraulic power units supply pressurised fluid to the cylinders, with proportional valves enabling fine control of extension, retraction, and force. Operators adjust total thrust and individual cylinder groups to match ground behaviour and cutterhead performance. Thrust, penetration rate, cutterhead torque, and face pressure are monitored continuously; these parameters are tuned together to avoid overloading tools, inducing excessive settlement, or stalling the machine.

In shielded machines, thrust cylinders push directly on the most recently erected lining ring. This dual role advances the TBM while simultaneously holding the lining in compression against the surrounding ground until annular grouting is completed. Because thrust loads are transmitted through the lining, segment design, joint configuration, and allowable jacking forces are critical. Over-thrusting can crack segments or damage gaskets; under-thrusting can reduce advance efficiency or compromise face stability.

Differential thrust is a primary steering mechanism. By increasing force on one side of the machine and reducing it on the opposite side, operators induce controlled yaw or pitch corrections. This is coordinated with articulation joints (where fitted) and guidance systems (laser/gyro) to maintain tight tolerances on line and grade, especially important in urban and utility tunnels.

Thrust cylinders operate in harsh environments with high loads, repetitive cycles, and potential contamination. Robust cylinder barrels, hardened rods, high-performance seals, and protective bellows are used to resist abrasion, corrosion, and slurry or soil ingress. Redundancy is built in: if one cylinder is isolated for maintenance, others can often maintain reduced but continuous operation.

Modern TBMs instrument thrust cylinders with pressure transducers and stroke sensors to provide real-time force and position data. Abnormal trends, such as rising thrust at constant penetration, can indicate tool wear, face instability, increased friction, or ground changes. Safety systems limit maximum thrust to protect the machine, lining, and ground.

Thrust cylinders are not merely "pushers"; they are precision actuators central to excavation control. Their sizing, arrangement, and control philosophy must be matched to ground conditions, TBM type, tunnel diameter, and lining design. Well-designed thrust systems enable efficient advance, accurate steering, and safe interaction with the ground and lining, while poor thrust management can lead to segment damage, excessive settlement, or stalled drives.

The following are examples that show how thrust cylinders act as precision control systems rather than simple "pushers," and what happens when they are well designed—or poorly managed—in real tunnelling contexts.

Example 1: Soft ground urban tunnelling — segment protection and settlement control

In EPB metro tunnelling through soft clay, thrust cylinders must be carefully sized and controlled to match both the soil strength and the segmental lining capacity. For example, in a large urban EPB TBM, total thrust is distributed across multiple cylinder groups so that no single segment or joint is overloaded during advance. Operators deliberately limit peak jacking forces while maintaining steady penetration by coordinating thrust with cutterhead torque and chamber pressure.

When this balance is achieved, the TBM advances smoothly with minimal surface settlement and without cracking segments. If thrust is applied too aggressively or unevenly, the lining can suffer radial cracking, joint opening, or gasket damage, which compromises watertightness and may increase settlement behind the shield.

Example 2: Mixed ground conditions — steering through asymmetric resistance

In mixed-face ground, where one side of the cutterhead is cutting soil and the other is cutting rock, resistance across the face is uneven. Thrust cylinders are used selectively to compensate for this imbalance. Operators increase thrust on the softer side and reduce it on the harder side to prevent the TBM from drifting off line.

Here, thrust cylinders function as steering actuators, working in combination with articulation joints and guidance systems. If thrust distribution is poorly managed, the machine can yaw or pitch uncontrollably, leading to over-steering corrections, higher shield friction, and increased wear on cutters and seals.

Example 3: High groundwater pressure — maintaining face stability

In slurry or Mixshield TBMs operating under high water pressure, thrust must be matched precisely to slurry face pressure. If thrust advances too quickly without sufficient pressure control, the face can unload, risking instability or slurry loss. Conversely, insufficient thrust can cause the cutterhead to stall while slurry pressure builds excessively.

A well-designed thrust system allows fine incremental advance, maintaining equilibrium between hydraulic face support and mechanical penetration. Poor thrust control in these conditions has led on some projects to face blowouts, slurry escape, or sudden settlement, even when the TBM itself was mechanically capable.

Example 4: Long drives — managing friction and preventing stalls

In long shield drives, friction between the shield and ground increases progressively as the tunnel length grows. Thrust cylinders must therefore be sized with sufficient reserve capacity, but also controlled to avoid sudden force spikes. Operators may gradually increase thrust over distance while maintaining constant penetration rates.

Where thrust capacity is underestimated or poorly managed, TBMs can experience progressive loss of advance, culminating in a stalled machine that cannot overcome accumulated friction. Recovery often requires lubrication improvements, thrust redistribution, or costly stoppages—demonstrating that thrust design is a long-term systems decision, not just a peak-force calculation.

Example 5: Hard rock tunnelling — interaction with grippers and rock mass

In open hard-rock TBMs, thrust cylinders work in conjunction with side grippers that react against the tunnel walls. Thrust must be coordinated with gripper force so that the rock mass is not overstressed or crushed. In competent rock, this allows very high advance rates with excellent efficiency.

In fractured or blocky rock, however, excessive thrust can cause local rock failure at gripper pads, leading to loss of reaction force and reduced advance. This is one of the reasons projects in weaker rock often switch from open gripper machines to single- or double-shield TBMs, where thrust is reacted through the lining instead.

Example 6: Poor thrust management — cumulative project impacts

Projects that suffer from poor thrust control philosophy often experience a cascade of issues rather than a single failure. Excessive thrust can damage segments, which then increases water ingress; increased water inflow raises friction; higher friction requires more thrust; and eventually the TBM stalls or settlement exceeds allowable limits.

These cases illustrate that thrust cylinders are part of a feedback-controlled system involving ground response, face support, lining performance, and machine behaviour. Treating thrust as a simple pushing force rather than a precision-controlled parameter is a common root cause of TBM underperformance.

Screw Conveyor and Slurry Circuits

Screw conveyors and slurry circuits are the two principal systems used to remove excavated material (muck) from a Tunnel Boring Machine while maintaining stable pressure at the tunnel face. Although both systems serve the same essential function—controlled spoil removal and pressure regulation—they operate on fundamentally different mechanical and hydraulic principles and are closely linked to TBM type and ground conditions. Screw conveyors are most commonly associated with Earth Pressure Balance (EPB) and variable-density machines, whereas slurry circuits are integral to slurry and Mixshield TBMs. In modern hybrid machines, both systems may be installed and selectively used as ground behaviour changes along the alignment.

A screw conveyor consists of a rotating helical auger enclosed within a steel tube that extracts excavated material from the pressurised excavation chamber and discharges it into the tunnel behind the TBM. In EPB machines, the screw conveyor performs a critical dual role: it not only transports muck but also acts as a key pressure-control device. By carefully regulating the screw's rotational speed and the opening of the discharge gate, operators control the volume of material leaving the chamber, thereby balancing earth and groundwater pressures at the tunnel face. In this system, the excavated soil itself functions as the pressure-transmitting medium, forming a stable plug within the screw conveyor.

Screw conveyor systems rely on conditioned spoil with appropriate plasticity and cohesion. Conditioning agents such as foam, polymers, and water are added to modify the soil so it behaves as a semi-plastic mass capable of transmitting pressure without flowing uncontrollably. Engineering design focuses on precise discharge control, pressure-resistant housings near the chamber, wear-protected flights and liners, and close integration with face pressure monitoring systems. When properly matched to the ground, screw conveyors provide effective and controllable excavation performance.

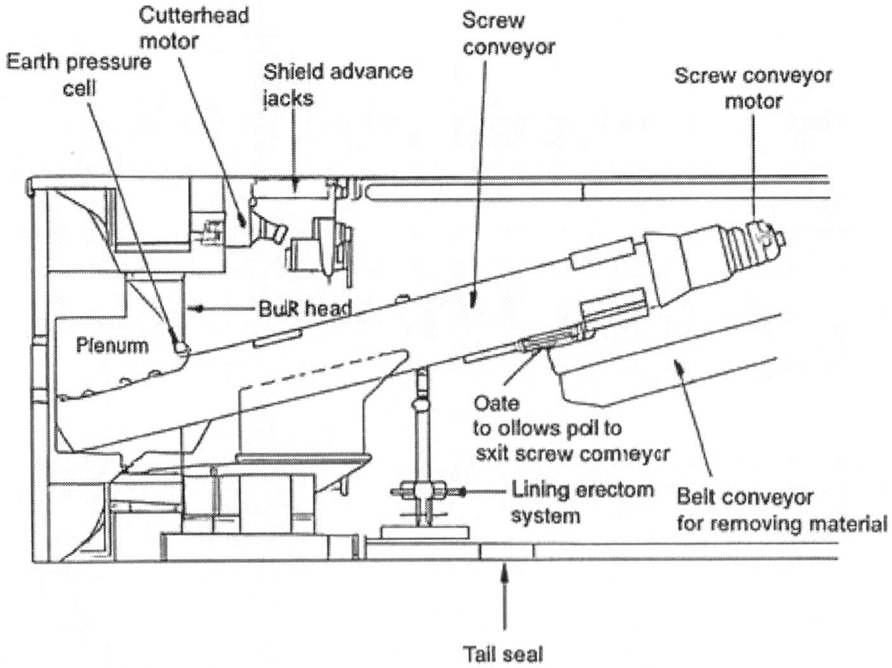

Figure 64: Earth Pressure Balance (EPB) Tunnel Boring Machine showing Screw Conveyor.

Figure 64 illustrates the screw conveyor system within an Earth Pressure Balance (EPB)–type Tunnel Boring Machine, showing how the screw conveyor links the pressurised excavation chamber at the tunnel face to the downstream spoil-handling system. The screw conveyor is not simply a transport device; it is a primary pressure-control mechanism that governs face stability.

In the diagram, the screw conveyor runs diagonally from the excavation chamber (plenum) behind the cutterhead toward the rear of the shield. Its inlet is directly connected to the pressurised chamber, while its outlet discharges material onto a belt conveyor for removal from the tunnel. This positioning is critical because it places the screw at the boundary between pressurised excavation and atmospheric tunnel conditions.

At the front of the machine, the cutterhead excavates the ground and deposits the material into the plenum (excavation chamber). In EPB operation, this chamber is filled with conditioned spoil that transmits earth and groundwater pressure to the tunnel face. The screw conveyor extracts material from this chamber in a controlled manner.

The diagram labels an earth pressure cell, which monitors the pressure inside the chamber. This pressure is directly influenced by the rate at which the screw conveyor removes material. If too much spoil is extracted, pressure drops and the face can become unstable; if too little is removed, pressure rises and can cause ground heave or excessive thrust loads.

The screw conveyor motor, shown at the rear end of the screw, drives the helical flights inside the conveyor casing. By adjusting the rotation speed of the screw and the opening of the discharge gate, operators precisely regulate:

- The volume of material leaving the chamber
- The pressure retained within the plenum
- The balance between excavation rate and shield advance

In effect, the excavated soil inside the screw forms a pressure plug, allowing pressure to be transmitted along the screw while still enabling controlled discharge. This is the defining principle of EPB tunnelling.

The diagram shows a gate at the outlet of the screw conveyor, just before material transfers onto the belt conveyor. This gate provides an additional level of control, allowing fine adjustment of discharge resistance. Together with screw speed, it ensures that pressure decays gradually rather than suddenly, preventing blow-outs or uncontrolled inflow.

Once material passes the gate, it is no longer pressurised and can be safely transported by the belt conveyor through the backup system to the surface or muck handling area.

The screw conveyor works in coordination with the shield advance jacks and the lining erector system, both shown in the diagram. As the shield advances forward, thrust cylinders push against the previously installed lining while the screw removes spoil at a rate that maintains face pressure equilibrium. This synchronisation ensures:

- Stable face conditions
- Controlled ground deformation
- Safe installation of lining segments

Figure 64 highlights why screw conveyors are central to EPB TBM performance. They simultaneously:

- Control face pressure
- Regulate spoil flow
- Isolate pressurised ground from the open tunnel
- Enable continuous excavation

Poor screw conveyor control can lead to pressure loss, surface settlement, clogging, or excessive torque demand. For this reason, screw design (diameter, pitch, wear protection) and operational control are treated as critical engineering parameters, not secondary components.

The primary advantages of screw conveyors lie in their relative mechanical simplicity and reduced surface infrastructure requirements compared with slurry systems. They are particularly effective in clays, silts, and mixed soils where conditioned spoil can reliably form a pressure plug. However, their effectiveness diminishes in clean sands, gravels, or highly permeable soils, where material drains too freely to sustain face pressure. In such conditions, pressure loss, clogging, or uncontrolled groundwater inflow may occur unless hybrid solutions or slurry-based methods are employed.

A slurry circuit, by contrast, uses a pressurised bentonite slurry to support the tunnel face and hydraulically transport excavated material to the surface. The excavation chamber is completely filled with slurry, which transmits pressure uniformly to the face regardless of spoil characteristics. As excavation progresses, spoil mixes with the slurry and is pumped through pipelines to a surface separation plant, where solids are removed using cyclones, screens, and centrifuges. The cleaned slurry is then returned to the TBM in a closed-loop system. In Mixshield TBMs, face pressure is often controlled using a compressed air cushion separated from the slurry by a submerged wall, enabling very precise pressure regulation under high groundwater loads.

Slurry systems rely on hydraulic pressure rather than soil behaviour, making face support largely independent of spoil consistency. This characteristic makes slurry circuits particularly reliable in permeable, water-bearing ground. Their core components include a pressurised excavation chamber, slurry feed and return pipelines, high-capacity pumps, a surface slurry treatment plant, and continuous monitoring of slurry density, viscosity, and flow conditions.

The major advantage of slurry circuits is their ability to maintain face stability and groundwater control in clean sands, gravels, marine deposits, and subaqueous tunnelling conditions, where EPB systems often struggle. They typically achieve very low settlement and high levels of safety in challenging hydrogeological environments. However, these benefits come at the cost of increased complexity, higher capital and operating expenses, and significant logistical demands. Slurry plants require substantial surface space, power, water management, and skilled operation, and they are generally less efficient in dry, cohesive clays where hydraulic transport offers limited advantage.

Modern TBM designs increasingly integrate both screw conveyor and slurry handling capabilities. Variable-density and dual-mode TBMs commonly extract material through a screw conveyor while allowing downstream handling to switch between dry conveyor transport and hydraulic slurry transport. In these machines, the screw conveyor continues to control chamber pressure, while material can either be discharged onto a belt conveyor in EPB mode or diverted into a slurryfier or crusher box for hydraulic transport. This arrangement allows gradual transitions between EPB-like and slurry-like behaviour without stopping the drive or entering the excavation chamber.

From an engineering perspective, both screw conveyors and slurry circuits are fundamental to face stability, settlement control, and advance efficiency. The choice between them is driven primarily by ground permeability, groundwater pressure, and spoil behaviour rather than by machine preference alone. A poor match between spoil-handling system and ground conditions can lead to pressure loss, excessive settlement, clogging, reduced advance rates, and increased wear and downtime. Consequently, spoil removal systems are treated as a core design decision in TBM selection, rather than as secondary or interchangeable components.

Shield and Articulation Systems

Central to the functionality of TBMs are the TBM shield and articulation systems. These components provide essential structural and operational capabilities that enable a TBM to navigate unstable ground conditions while minimizing ground deformation and maintaining alignment necessary for successful tunnel construction. The following discussion synthesizes relevant literature to comprehensively detail the roles of these systems in TBM operation.

The TBM shield is a robust steel cylindrical structure designed to encase the cutterhead and excavation chamber. Its primary purpose is to offer temporary support to surrounding ground during the excavation phase, particularly in soft or fractured scenarios where soil cannot remain stable without confinement. The shield's design prevents ground collapse, limits deformation and surface settlement, and facilitates installation of the tunnel lining [302, 335]. This functionality is emphasized particularly in Earth Pressure Balance machines, which utilize the shield to create a pressurized excavation chamber against varying earth and hydrostatic pressures [294, 302].

In soft-ground conditions, the TBM shield's structural integrity becomes even more critical. The shield's form allows it to cope with significant axial and lateral loads while resisting external pressures [295, 412]. The requirement for robust design elements in these applications is exemplified by findings from studies on tunnelling in Istanbul, where the performance of EPB TBMs effectively managed ground settlements [413, 414].

Single-shield TBMs incorporate a continuous shell and advance by thrusting against pre-installed segmental linings, providing stability in mixed or fractured ground conditions [415]. This method, while slower than methods using open TBMs, significantly enhances ground stability [335]. Conversely, double-shield TBMs, which have two shield sections, adapt their thrust reaction modes depending on ground conditions, thereby increasing operational flexibility and advance rates in favourable geology while still accommodating challenging conditions [416, 417].

The articulation system is pivotal for enabling TBMs to navigate curved pathways, critical for urban tunnelling projects where geographical and infrastructural constraints impose such design needs. By connecting the shield's sections via articulation joints, the TBM can adjust its angle of operation slightly, facilitating precision steering without the need for modifications to the machine itself [302, 418]. This engineering finesse minimizes the chances of exceeding alignment tolerances or causing undue surface deformation, both of which can have significant ramifications in densely populated areas [413, 414].

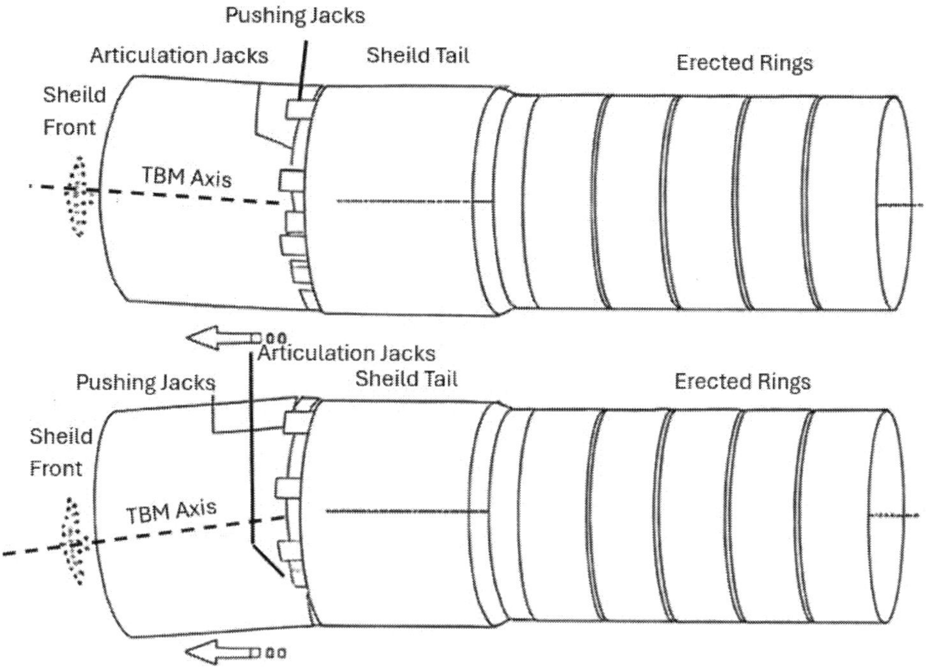

Figure 65: TBM Articulation Jacks.

In the context of Earth Pressure Balance and slurry TBMs, articulation becomes even more critical due to inherent pressures and the frequent curvatures encountered [286, 417]. Articulation coordination with thrust systems ensures that ground pressures are balanced effectively while avoiding excessive wear or damage to both the lining and the TBM itself [10, 417].

Recent advancements in TBM articulation control have involved the use of hydraulic actuators to enable precise movement between machine sections [418, 419]. These innovations enhance the capability of articulation systems, allowing for real-time adjustments based on geological feedback, thus improving operational efficiency while ensuring safety and stability in complex tunnelling environments.

The combination of sophisticated TBM shield and articulation systems is essential in contemporary tunnelling practices, particularly in urban and geologically challenging environments. These systems not only offer temporary ground support and control but also facilitate alignment and precision steering, enabling TBMs to operate effectively under adverse conditions. Their integrated performance significantly impacts the success of tunnelling operations while minimizing potential risks associated with ground instability and deformation.

Backup Gantries

Backup gantries are the trailing structural platforms that follow immediately behind the main body of a Tunnel Boring Machine (TBM). Together, they form the machine's mobile support system—often described as an "underground factory"—housing the equipment, services, and logistics required to keep excavation, lining installation, and spoil removal operating continuously. While the cutterhead and shield perform the excavation itself, it is the backup gantries that enable sustained, controlled, and efficient tunnelling over long distances.

The primary role of the backup gantries is to support uninterrupted TBM operation by supplying power, hydraulics, ventilation, materials, and control functions to the cutterhead and shield. They also provide protected working areas for personnel and form the functional interface between the TBM and the completed tunnel behind it. In long drives, the reliability and capacity of the backup system are just as critical to overall performance as the cutting capability of the TBM itself.

Structurally, backup gantries are arranged as a series of steel-framed platforms coupled together behind the TBM. Depending on tunnel diameter and machine complexity, a TBM may have only a few gantries or more than a dozen. Each gantry has a defined role and is designed to fit within the tunnel envelope while allowing safe access for inspection, maintenance, and material handling. The gantries are typically supported on wheels or skids that run on the tunnel invert or rails and are connected to the TBM by articulated couplings to accommodate curves and alignment changes.

One of the most important functions of the backup gantries is housing the electrical power systems. These include transformers, switchgear, variable-frequency drives, and power distribution boards. High-voltage electricity is supplied from the surface and stepped down on the gantries to power cutterhead motors, thrust cylinders, screw conveyors, slurry pumps, and auxiliary equipment. Because any power interruption immediately halts excavation, electrical reliability and redundancy are essential design considerations.

Hydraulic systems are another core component mounted on the backup gantries. Hydraulic power units supply pressurised oil to thrust cylinders, articulation systems, grippers (where fitted), segment erectors, and other actuators. These systems include pumps, reservoirs, filters, coolers, and monitoring instrumentation, and they must operate continuously under high loads and elevated temperatures typical of underground tunnelling environments.

Ventilation and environmental control systems are also routed along the backup gantries. Ventilation ducts, fans, and air-handling equipment supply fresh air to the TBM face and working areas while removing dust, heat, and exhaust gases. In pressurised TBMs, the backup system interfaces with compressed-air installations used for face access or pressure management, ensuring safe working conditions throughout the tunnel.

Spoil transport is another critical function supported by the backup gantries. In EPB and hard-rock TBMs, this typically involves belt conveyor systems that receive material from the screw conveyor or cutterhead and transport it rearward along the gantries. In slurry and Mixshield TBMs, the backup carries slurry pipelines, pumps, valves, and instrumentation that convey excavated material hydraulically to the surface separation plant. Any disruption to these systems directly limits advance rates.

For shielded TBMs, backup gantries also handle lining segments and consumables. Precast concrete segments, gaskets, bolts, and other materials are stored and transported along the gantries using rail-mounted cars, multi-service vehicles, or overhead cranes. These systems deliver segments efficiently and safely to the segment erector, ensuring that lining installation remains synchronised with excavation.

Modern TBMs include operator cabins and control rooms mounted on the backup gantries. From these locations, operators monitor and control cutterhead rotation, thrust, face pressure, spoil removal, guidance, and safety systems. Sensor data from across the TBM and backup gantries is collected and analysed in real time, allowing rapid responses to changing ground conditions, equipment performance, or emerging risks.

Backup gantries are tightly integrated with the TBM excavation cycle. As the cutterhead advances, thrust and face pressure are coordinated with spoil removal systems on the gantries, while segment supply, erection, and grouting are synchronised with shield advance. Any bottleneck within the backup system, such as conveyor downtime, slurry pump failure, or segment delivery delays, immediately constrains progress, regardless of how effectively the cutterhead is cutting.

The configuration of backup gantries varies significantly with TBM type. Open (gripper) TBMs generally have simpler backups focused on power supply, ventilation, and conveyor systems. EPB TBMs add screw conveyor discharge handling, soil conditioning systems, and segment logistics. Slurry and Mixshield TBMs require the most complex backup arrangements, incorporating high-capacity slurry pumps, extensive pipework, separation interfaces, and sophisticated monitoring and control equipment.

From an engineering and project-delivery perspective, backup gantries are central to productivity, safety, and reliability. Well-designed backup systems support high utilisation rates, efficient logistics, and rapid response to maintenance or ground changes. Conversely, poorly designed or under-capacity backups can severely restrict advance rates, increase downtime, and elevate safety risks, even when the cutterhead and shield are well matched to the ground.

Power, Hydraulic, and Electrical Systems

Tunnel Boring Machines (TBMs) represent a sophisticated and integral technology in modern tunnelling practices, with their power, hydraulic, and electrical systems forming the primary operational infrastructure enabling a range of essential functions such as excavation, steering, face control, spoil handling, and lining installation. The performance and reliability of these systems are critical, particularly when operating under extreme loads and harsh conditions underground. Together, they convert energy from the surface into highly controlled mechanical actions at the cutterhead and throughout the machine.

The power supply and distribution system of a TBM is pivotal for driving its various functions. Typically, electrical power is delivered from the surface through high-voltage cables, which range from 6.6 kV to 33 kV depending on project specifics such as tunnel length and machine size. This high-voltage transmission is essential for minimizing electrical losses over long distances [420]. Step-down transformers are employed on backup gantries to reduce voltage to acceptable levels for motors and ancillary equipment, enabling a wide range of machinery, including cutterhead drive motors, hydraulic power units, screw conveyors, ventilation fans, and segment erectors, to function effectively [421].

The power demand on TBMs can be both substantial and highly variable. For instance, large-diameter machines might require several megawatts just to drive the cutterhead motor, with demand fluctuating based on ground hardness, penetration rate, and cutter wear. To manage these varying demands, TBM power systems are designed with features such as peak load accommodation, redundancy, and robust protection mechanisms. This design philosophy is crucial in preventing unplanned shutdowns that could delay excavation. The integration of hydraulic systems is particularly noteworthy, as they deliver the significant forces necessary for functions like thrust and steering [422].

In TBMs, the electrical system can be compared to a nervous system responsible for distributing power while also facilitating monitoring and control functions. This system includes various components like switchboards, motor control centres, variable-frequency drives (VFDs), and control cabinets. VFDs are particularly important as they provide precision in motor speed and torque control, adapting operations like cutterhead rotation to changing geotechnical conditions in real-time. Adopting controlled acceleration and deceleration reduces mechanical stress, thereby enhancing the longevity of the TBM components.

Furthermore, TBMs are equipped with extensive sensor networks that collect real-time data relating to thrust force, torque, temperature, pressure, and machine alignment, which is displayed in operator cabins and control rooms. This data-driven approach allows for proactive adjustments to be made by operators, thereby mitigating potential issues during tunnelling operations [421].

Hydraulic systems in TBMs provide high-force actuation that is indispensable for various tasks, including thrust, alignment, and gripper operations on hard-rock TBMs. A hydraulic power unit on the backup gantries generates pressurized oil through electrically driven pumps, distributing it through hoses and pipework to different actuators [422]. For instance, thrust cylinders are powered hydraulically to push the TBM forward, while articulation cylinders are used for steering.

Moreover, hydraulic efficiency is maintained through comprehensive design principles that emphasize filtration and cooling systems to minimize contamination and manage heat generated from high-pressure operations. Clean oil is vital as it prevents damage to crucial components. The integration of hydraulic functions also ensures redundancy and load management, with hydraulic systems often interlinked with electrical control systems for coordinated operation [422].

The integration of power, electrical, and hydraulic systems forms a single, coordinated network essential for TBM operation. Electrical systems drive hydraulic pumps and control motors based on sensor feedback, creating a feedback loop that ensures stable excavation in ever-changing underground environments. The complexity of the TBM systems grows with the machine's specifications; for example, while open-gripper TBMs are relatively simple, Earth Pressure Balance (EPB) TBMs incorporate additional controls for soil conditioning and face pressure, thereby increasing operational intricacies [421].

TBM power, hydraulic, and electrical systems are inherently interdependent, with their effectiveness directly affecting the capability, reliability, and safety of operations. Properly designed systems are essential for ensuring rapid advancements in tunnelling while minimizing mechanical wear and promoting safe operational conditions. As tunnelling projects become increasingly complex and demanding, understanding these systems will be integral to successful engineering and project delivery strategies.

Control and Automation Systems

TBM control and automation systems are the intelligence layer of a Tunnel Boring Machine, integrating sensors, software, and actuators to manage excavation, steering, face support, spoil removal, and lining installation in a coordinated and safe manner. These systems transform a TBM from a collection of powerful mechanical components into a responsive, adaptive tunnelling platform capable of operating efficiently in highly variable ground and groundwater conditions.

At the core of TBM control is a distributed control architecture, typically based on programmable logic controllers (PLCs) networked across the cutterhead, shield, and backup gantries. Each subsystem—cutterhead drives, thrust cylinders, screw conveyors or slurry pumps, articulation, segment erectors, and auxiliary services—has local control modules that communicate with central PLCs. This architecture provides robustness and redundancy, ensuring that critical functions remain operational even if part of the system is disrupted.

> Programmable Logic Controllers (PLCs) are the core real-time control devices used to operate and coordinate Tunnel Boring Machines (TBMs). They function as the industrial "brains" of the machine, continuously reading sensor inputs, executing control logic, and commanding mechanical, hydraulic, and electrical actuators. In TBM operations, PLCs

are fundamental to achieving safety, precision, automation, and operational reliability in demanding underground environments.

A PLC is a ruggedised industrial computer designed specifically for deterministic, real-time control. Unlike general-purpose computers, PLCs are built to operate continuously around the clock, tolerate vibration, dust, moisture, and electrical noise, and respond predictably within milliseconds. They are also designed to fail safely under fault conditions. TBMs rely on PLCs because tunnelling requires precise, repeatable, and fail-safe control in conditions where human reaction time alone is insufficient to manage rapidly changing loads, pressures, and ground behaviour.

PLCs operate through a continuous loop known as the scan cycle, which runs many times per second. During the input scan, the PLC reads signals from a wide range of sensors, including thrust force transducers, cutterhead torque and speed encoders, hydraulic pressure sensors, face pressure instrumentation for EPB or slurry systems, slurry density and flow meters, and temperature, vibration, and alignment sensors. These inputs provide a real-time picture of both machine performance and ground response.

Once inputs are read, the PLC executes the programmed control logic. This logic determines how fast motors should run, how much hydraulic pressure should be applied, whether operating limits are being approached or exceeded, and whether alarms or shutdowns are required. Following this, the PLC updates its outputs, sending commands to variable-frequency drives, hydraulic valves, thrust cylinders, screw conveyors or slurry pumps, segment erectors, and auxiliary systems. Each scan cycle concludes with housekeeping and diagnostic tasks, including communications management, fault checking, and redundancy verification, before the loop repeats. This rapid, continuous cycle allows the TBM to react predictably to changing operating and ground conditions.

In TBM operations, PLCs control virtually every critical function. For excavation, they regulate cutterhead rotation speed, motor torque limits, penetration rate, and anti-stall or overload protection, automatically adjusting machine behaviour based on ground resistance and cutter performance. For thrust and advance control, PLCs manage thrust cylinders by balancing total thrust against ground resistance, applying differential thrust for steering, protecting lining segments from overload, and coordinating thrust with cutterhead advance. In EPB and slurry machines, thrust control is closely integrated with face-pressure control.

Face pressure management is one of the most safety-critical PLC functions. In EPB TBMs, PLCs coordinate thrust rate, cutterhead advance, screw conveyor speed, and soil conditioning injection to maintain a stable pressure plug in the excavation chamber. In slurry and Mixshield TBMs, PLCs regulate slurry pressure, control pumps and valves, manage compressed-air cushions where used, and prevent pressure drops or blowouts. Accurate, continuous pressure control is essential to avoid face instability, excessive settlement, or groundwater inflow.

PLCs also play a central role in steering and guidance. They process data from laser guidance systems, gyroscopes, and total stations, and use this information to control articulation cylinders, apply differential thrust, and support ring taper selection in coordination with operators. This enables millimetre-level control of tunnel line and grade, even over long drives and curved alignments.

For lining installation, PLCs automate or semi-automate segment erector movements, monitor loads during segment placement, verify ring completion, and enforce interlocks that prevent unsafe operations. This automation improves erection speed, repeatability, and safety. Similarly, PLCs coordinate spoil removal systems by regulating screw conveyor speed and torque, conveyor belts, slurry pumps and valves, and interfaces with surface separation plants, ensuring that spoil removal remains synchronised with excavation to avoid pressure loss or overfilling.

Safety, interlocks, and emergency functions are deeply embedded in PLC logic. PLCs enforce operating envelopes and hard limits, manage emergency stop systems, trigger responses to fire detection, initiate overpressure or overload shutdowns, and control access and maintenance interlocks. Many TBM PLC systems are designed to fail safe, meaning that loss of signal, power, or communication defaults the machine to a safe condition.

PLCs are programmed using standardised industrial languages defined in IEC 61131-3. These typically include Ladder Logic, which resembles traditional relay circuits and remains widely used; Function Block Diagrams, which are common for process control such as pressure, flow, and speed regulation; Structured Text, a high-level, code-like language used for complex logic; and Sequential Function Charts, which are well suited to step-by-step processes such as startup sequences or segment erection cycles. TBM control systems usually combine several of these languages within a single control architecture.

The control philosophy of TBM PLC systems is modular and structured. Programs are divided into subsystems such as thrust, cutterhead, EPB pressure control, slurry circuits, and segment erection. State machines define operating modes such as startup, boring, ring build, shutdown, and emergency states, while closed-loop controllers—often using PID algorithms—regulate pressure, speed, and flow. This modular structure supports safe operation, easier troubleshooting, and future upgrades.

Before tunnelling begins, PLC logic is thoroughly tested through simulations and factory acceptance tests. On site, commissioning activities fine-tune parameters to match actual ground conditions, and control limits and alarm thresholds are progressively adjusted as experience is gained during the drive. This staged approach reduces risk and improves performance as tunnelling progresses.

TBMs rarely rely on a single PLC. Instead, they use a distributed control system architecture, with multiple PLCs located on the cutterhead, shield, and backup gantries. These are connected by high-speed industrial communication networks such as

Profibus, Profinet, or Ethernet/IP, and critical systems are often provided with redundant CPUs and power supplies. Integration with SCADA systems and human–machine interfaces allows operators to visualise data, trends, and alarms in real time.

PLCs do not replace operators; rather, they support them. Operators set target parameters such as advance rate, face pressure, and torque limits, monitor trends and alarms, override automation when necessary, and apply engineering judgment during unusual ground conditions. PLCs handle the fast, repetitive, and safety-critical actions that humans cannot reliably perform in real time.

From an engineering and risk perspective, PLCs are critical to TBM projects because they enable stable excavation in variable ground, reduce human error, protect expensive mechanical components, improve advance rates and consistency, and generate high-quality operational data for optimisation and forensic analysis. Many major TBM incidents are ultimately linked not to mechanical failure, but to control logic, parameter selection, or sensor interpretation.

In summary, Programmable Logic Controllers form the central control and automation backbone of modern Tunnel Boring Machines. By continuously reading sensor data, executing real-time control logic, and commanding mechanical and hydraulic systems, PLCs maintain stable excavation, precise alignment, and safe operation. Programmed using industrial control languages and integrated into distributed architectures, they enable TBMs to function as intelligent, adaptive underground production systems capable of operating safely and efficiently in some of the most challenging engineering environments in the world.

A dense network of sensors feeds real-time data into the control system. These sensors measure thrust force, cutterhead torque and speed, penetration rate, vibration, temperature, hydraulic pressures, face pressure, slurry density and flow (where applicable), electrical loads, and machine position and orientation. Guidance systems using lasers, gyroscopes, or total stations continuously track line and grade. Together, these inputs create a comprehensive picture of both machine behaviour and ground response at the tunnel face.

Control algorithms use this data to regulate TBM operation through closed-loop control. For example, cutterhead speed and torque are adjusted automatically to maintain a target penetration rate without overloading cutters or drives. Thrust force is modulated to match ground resistance while protecting the lining or grippers. In EPB and slurry TBMs, face pressure is continuously balanced by coordinating thrust, cutterhead advance, and spoil extraction via the screw conveyor or slurry circuit. This continuous feedback and adjustment are essential for maintaining face stability and controlling settlement.

Automation plays a major role in managing repetitive and safety-critical tasks. Segment erection, for instance, is often semi-automated, with control systems guiding the erector to the correct position, verifying segment orientation, and monitoring loads during placement. Grouting systems behind the lining can be automated to deliver the correct volume and pressure based on advance rate and ground conditions. Conveyor systems, slurry pumps, and conditioning agent injection are likewise coordinated automatically with excavation progress.

Modern TBMs increasingly incorporate advanced automation and decision-support tools. Trend analysis and performance dashboards allow operators and engineers to identify early signs of cutter wear, clogging, abnormal vibration, or seal degradation. Alarm systems prioritise warnings based on risk, drawing attention to conditions that could threaten safety or machine integrity. Some systems include adaptive control logic that modifies operating parameters automatically when ground conditions change, reducing reliance on manual intervention.

Human–machine interfaces (HMIs) are a critical element of TBM control systems. Operator cabins on the backup gantries provide real-time visualisation of machine status, ground response, and alignment, typically through multiple screens displaying key parameters, trends, and alarms. Although automation handles much of the routine control, skilled operators remain essential for interpreting data, making strategic decisions, and responding to unusual or unexpected conditions.

The level of automation varies with TBM type and project risk. Open (gripper) TBMs focus automation on thrust control, cutterhead performance, and guidance, while EPB machines add sophisticated face-pressure and spoil-flow control. Slurry and Mixshield TBMs require the most advanced automation, as they must manage hydraulic face support, slurry density and flow, separation-plant interaction, and often compressed-air interfaces under high groundwater pressure. In these machines, automation is central to safety and feasibility.

From an engineering and project-delivery perspective, TBM control and automation systems are critical to productivity, safety, and predictability. Well-designed systems reduce human error, enable rapid response to changing conditions, and allow consistent operation over long drives. Poorly integrated or poorly understood control systems, by contrast, can lead to instability, excessive wear, or major incidents despite sound mechanical design.

Chapter 6

Tunnel Boring Operations

Tunnel boring operations represent the point at which planning, design, and construction preparation converge into active underground excavation. Once a Tunnel Boring Machine (TBM) is launched, its performance is governed not only by mechanical capability, but by the integration of systems, procedures, and operational discipline that control how the machine interacts with the ground. This chapter examines the full operational lifecycle of a TBM drive, from site establishment and launch preparation through to steady-state boring, steering, and advance control.

The chapter begins by addressing site setup and launch shaft construction, recognising these works as the physical and operational foundation of mechanised tunnelling. Decisions made at this stage—relating to shaft geometry, ground support, logistics, and utilities—directly influence assembly efficiency, launch accuracy, and long-term tunnel alignment. The discussion highlights the launch shaft as a critical interface between surface construction and underground excavation, where errors are difficult to correct once boring begins.

Building on this foundation, the chapter explores TBM assembly, commissioning, and start-up procedures. These transitional phases are treated as engineered processes rather than routine construction tasks, with emphasis placed on system integration, verification under load, and controlled interaction with ground and groundwater conditions. Particular attention is given to start-up as a high-risk phase, where disciplined procedures are essential to establish stable face support, protect critical components, and set reliable operational baselines.

The chapter then examines the normal boring cycle and its key elements—thrust application, cutterhead rotation, spoil excavation, and muck removal—showing how these functions must operate as an integrated system. Concepts of advance rate, penetration control, and operational efficiency are introduced to illustrate how productivity is balanced against ground stability, machine wear, and risk management. Steering and guidance systems are discussed in detail, highlighting their role in maintaining alignment, protecting the lining, and controlling ground response over long drives.

Finally, the chapter contrasts continuous and cyclic modes of operation, demonstrating how different tunnelling methods influence machine design, ground behaviour, and project risk profiles. Through this systems-level perspective, tunnel boring is presented not as a single mechanical action, but as a coordinated sequence of engineering decisions and operational controls that collectively determine the safety, accuracy, and success of mechanised tunnelling projects.

Site Setup and Launch Shaft Construction

Site setup and launch shaft construction are the foundational phases of any Tunnel Boring Machine (TBM) project. Before excavation can begin, the site must be prepared to safely receive, assemble, and operate the TBM, while the launch shaft (and associated launch chamber) provides the structural and geometric starting point for the tunnel drive. Errors or shortcuts at this stage can cascade into alignment problems, delays, safety risks, or even machine damage once tunnelling begins.

Site setup establishes the physical, logistical, and environmental conditions required for continuous TBM operation. This phase transforms a surface site into a controlled interface between surface infrastructure and underground excavation.

Figure 66: On May 29, 2012, the tunnel boring machine shown here completed mining the third East Side Access tunnel to be built in Queens. This photo shows the machine being reassembled as crews prepared it for the tunnel mining operation. Metropolitan Transportation Authority, CC BY 2.0, via Flickr.

Key elements of site setup include establishing safe access and exclusion zones, preparing working platforms, installing utilities, and creating space for TBM assembly and material handling. Power supply is a major early activity, with high-voltage electrical connections installed to support TBM drives, pumps, ventilation, and site services. Water supply and drainage systems are also required for cooling, dust suppression, slurry preparation (where applicable), and general construction needs.

Equally important is logistics planning. Areas must be allocated for segment storage, muck handling, slurry treatment plants (for slurry or Mixshield TBMs), workshops, control rooms, and offices. Traffic management plans are implemented to manage deliveries and spoil removal without disrupting surrounding communities. In urban environments, noise, vibration, dust, and working-hour restrictions are addressed early to ensure regulatory compliance and community acceptance.

The launch shaft is the vertical or inclined excavation from which the TBM is assembled and launched. Structurally, it must withstand earth and groundwater pressures while providing a stable, accurately aligned base for the TBM. Functionally, it serves as the transition point between surface construction and underground mechanised tunnelling.

Figure 67: TBM Launch Cradle at Tunnel B/C. Metropolitan Transportation Authority, CC BY 2.0, via Flickr.

Launch shafts vary widely in size and depth depending on tunnel diameter, TBM type, and local ground conditions. They must be large enough to accommodate the TBM shield, backup gantries, cranes or lifting systems, and working space for assembly and commissioning. Tolerances are tight: the shaft base and launch frame must be constructed to millimetre-level accuracy to ensure correct line and grade from the very first ring.

Shaft Construction Methods

The method used to construct a launch shaft depends primarily on ground conditions, groundwater level, shaft depth, and surrounding constraints.

In soft ground or water-bearing soils, shafts are often constructed using diaphragm walls (slurry walls) or secant pile walls, which provide continuous structural support and groundwater cut-off. Once the perimeter walls are complete, the shaft is excavated internally, typically with temporary strutting or ground anchors installed progressively as depth increases.

In more competent ground or rock, shafts may be excavated using conventional drill-and-blast or mechanical excavation, with rock bolts, shotcrete, or concrete lining providing support. In some cases, caisson or sunk-shaft methods are used, particularly where space is limited.

Figure 68: TE11 Stevens MRT construction. Nethaniel, CC BY-SA 4.0, via Wikimedia Commons.

Figure 68 depicts the Tunnel Boring Machine (TBM) launch shaft and associated site setup works for the TE11 Stevens MRT Station on Singapore's Thomson–East Coast Line during construction. It captures a critical transition point in the project, where surface civil works give way to mechanised underground tunnelling.

At the centre-right of the image is a deep TBM launch shaft. This large excavation is the point from which the TBM is assembled and launched into the ground. The shaft is supported by diaphragm (slurry) walls with extensive internal steel strutting, a construction approach well suited to Singapore's soft, water-bearing soils. Blue steel walers and struts brace the walls to resist earth and groundwater pressures, while the depth of the excavation indicates that the shaft serves deep running tunnels rather than only a shallow station box.

To the left of the shaft is the TBM assembly area. The cylindrical components and the prominent ring are major TBM elements, such as shield sections, cutterhead assemblies, or main bearing components. These parts are transported to site in sections, assembled at surface level, and later lowered into the shaft for final underground assembly. Temporary roofing covers parts of this area, protecting sensitive mechanical and electrical equipment from weather exposure during assembly.

The image also shows extensive heavy lifting operations. Several large crawler cranes are positioned around the site, handling TBM components, segment moulds, and heavy temporary works. This concentration of cranage reflects the TBM assembly and launch preparation phase, which is among the most logistically complex and risk-intensive stages of a tunnelling project due to the size, weight, and precision requirements of the components involved.

Segment handling and broader site logistics are clearly visible across the site. Precast concrete tunnel lining segments, or segment cages and moulds, are stacked in designated storage areas. These segments will later be transported underground and erected to form the permanent tunnel lining as the TBM advances. The presence of internal haul roads, laydown areas, and controlled access zones demonstrates that this is a fully developed TBM construction site, rather than a simple station excavation.

The surrounding hoarding and noise barriers highlight the environmental and urban constraints governing the project. Stevens is located in a dense, high-value residential area, where strict limits apply to settlement, vibration, and noise. These constraints strongly influence construction methods and help explain the use of shielded TBMs—such as Earth Pressure Balance or slurry machines—designed to minimise ground movement and environmental impact.

Overall, Figure 68 represents the TBM launch preparation stage of construction. This includes completion of the launch shaft, delivery and surface assembly of TBM components, installation of temporary works and lifting systems, and readiness for TBM lowering, commissioning, and launch. From an engineering perspective, the image illustrates the critical interface between surface construction and underground mechanised tunnelling, the scale of infrastructure required before any tunnelling can begin, and why launch shafts are major technical and risk focal points. It also underscores that urban MRT tunnelling is as much about logistics, systems integration, and risk management as it is about excavation itself.

Groundwater control is a critical consideration throughout shaft construction. Dewatering systems, cut-off walls, or base plugs may be required to prevent inflow and uplift at the shaft base, especially for deep shafts below the water table.

Base Slab and Launch Frame Construction

At the bottom of the shaft, a heavily reinforced concrete base slab is constructed. This slab must carry the full weight of the TBM during assembly and resist thrust forces during initial excavation. Embedded rails, anchor points, or a launch cradle are cast into the slab to position the TBM accurately.

Figure 69: Installing rebar for the walls and base slab for the final tunnel approach structure. MTA Capital Construction Mega Projects, CC BY 2.0, via Wikimedia Commons.

The launch frame or cradle supports the TBM shield during startup, guiding the cutterhead into the ground and reacting early thrust forces before the first lining rings are installed. Its geometry is critical: any misalignment at launch can persist throughout the drive and require continuous correction, increasing wear and risk.

TBM Assembly and Commissioning in the Shaft

Once the shaft and base slab are complete, the TBM is delivered in sections and assembled within the launch shaft. Assembly typically proceeds from the cutterhead and shield backward through the backup gantries. This phase requires careful sequencing, heavy lifting operations, and strict safety controls.

After mechanical assembly, electrical, hydraulic, and control systems are connected and tested. Commissioning includes dry testing of drives and actuators, calibration of guidance systems, and verification of safety interlocks. For EPB or slurry TBMs, the excavation chamber, screw conveyor or slurry circuits, and pressure-control systems are tested before the face is opened.

Launch Operations and Initial Excavation

Launching the TBM is a controlled and high-risk operation. The cutterhead is slowly advanced into the ground while face support systems are activated to maintain stability. Initial thrust forces are carefully managed, as the TBM has not yet formed a complete reaction system against installed lining rings.

During this phase, ground behaviour, machine response, and alignment are closely monitored. Adjustments to thrust, cutterhead speed, face pressure, and conditioning are common as the TBM transitions from static assembly to steady-state tunnelling.

TBM Assembly and Commissioning

TBM assembly and commissioning are critical transitional phases that connect design and manufacture with live tunnelling operations. During these stages, a TBM is transformed from a series of large mechanical subassemblies into a fully integrated underground production system. The quality and discipline applied during assembly and commissioning have a direct influence on safety, reliability, and performance during the early stages of the tunnel drive.

TBMs are too large and complex to be transported as a single unit, so they are delivered to site as major subassemblies. Assembly typically takes place either at the surface adjacent to the launch shaft or partially within the shaft itself, depending on space constraints, shaft depth, and lifting capacity. The objective of the assembly phase is to reconstruct the TBM precisely to its design geometry, align all mechanical interfaces, and integrate power, hydraulic, and control systems so the machine is structurally and functionally complete before launch.

Figure 70: Borehead – TBM Ulrikke, assembly in Arna in September 2015. Svein Harkestad, CC BY-SA 4.0, via Wikimedia Commons.

Before assembly begins, the launch shaft must be fully prepared. This includes completion of permanent and temporary structural works, watertight base slabs or thrust blocks capable of resisting launch forces, and installation of working platforms, access systems, and utilities. Heavy lifting equipment such as crawler cranes, gantries, or shaft cranes must be in place, along with power, ventilation, drainage, and safe access arrangements. At this stage, the shaft effectively becomes the TBM's first operational station and must accommodate high loads, complex logistics, and elevated construction risk.

Assembly normally starts with the forward components of the TBM. The cutterhead, main bearing, and forward shield sections are installed and aligned first. Cutting tools and wear protection are fitted, and tight tolerances are verified to ensure concentricity and balance. Any misalignment introduced at this stage can lead to vibration, uneven cutter wear, or accelerated bearing damage once excavation begins, making precision essential.

Following this, the main drive system and thrust arrangement are installed. Drive motors, gearboxes, and thrust cylinders are mounted and mechanically connected to the cutterhead and shield. This stage ensures correct torque transmission, accurate alignment of thrust forces along the machine axis, and full integration with the hydraulic and electrical systems that control advance and steering.

Figure 71: Welding In Tail Seal Brushes On The Yard Lead TBM. Metropolitan Transportation Authority, CC BY 2.0, via Flickr.

Once the main body of the TBM is assembled, the backup gantries are installed sequentially behind the shield. These gantries house the electrical distribution systems, transformers, hydraulic power units, control rooms, ventilation equipment, spoil handling systems, and segment storage and handling equipment. Each gantry is mechanically, electrically, and hydraulically connected, forming a continuous mobile production system that supports excavation and lining installation throughout the drive.

The final stage of assembly is full system integration. Mechanical, electrical, hydraulic, and control subsystems are interconnected and checked for compatibility. This includes cable pulling and termination, hydraulic hose installation and pressure testing, and integration of PLCs, sensors, communication networks, and monitoring systems. Only once these systems are fully connected can commissioning begin.

Major Assembly Stages

1. Launch Shaft Readiness

Before TBM assembly begins, the launch shaft must be structurally complete, watertight, and equipped with:

- Base slabs or thrust blocks to react launch forces
- Temporary works and working platforms

- Lifting systems (crawler cranes, gantries, or shaft cranes)
- Utilities such as power, ventilation, drainage, and access

The shaft effectively becomes the TBM's first "station" and must tolerate high loads and operational risk.

2. Shield and Cutterhead Assembly
The cutterhead, main bearing, and forward shield sections are usually assembled first. This includes:

- Mounting the cutterhead onto the main bearing
- Installing cutting tools and wear protection
- Aligning the cutterhead and shield to tight tolerances

Any misalignment at this stage can cause vibration, uneven wear, or bearing damage once excavation begins.

3. Main Drive and Thrust System Installation
Drive motors, gearboxes, and thrust cylinders are installed and mechanically connected. This stage ensures:

- Correct torque transmission to the cutterhead
- Accurate thrust alignment with the machine axis
- Full integration with hydraulic and electrical systems

4. Backup Gantry Assembly
The backup gantries are assembled sequentially behind the TBM shield. These gantries house:

- Power distribution and transformers
- Hydraulic power units
- Control rooms and operator cabins
- Ventilation, slurry or conveyor systems
- Segment handling and storage equipment

Each gantry is mechanically and electrically connected, forming a continuous mobile production system.

5. System Integration
All subsystems—mechanical, electrical, hydraulic, and control—are connected and checked for compatibility. This stage includes:

- Cable pulling and termination
- Hydraulic hose installation and pressure testing

- Integration of PLCs, sensors, and communication networks

Commissioning is the structured process that verifies the assembled TBM performs as intended before excavation starts. Its purpose is to test, calibrate, and validate all systems under controlled conditions, reducing the risk of early failures, delays, or unsafe operation once the machine begins boring.

Commissioning typically begins with staged energisation of the electrical and control systems. Power distribution and protection systems are tested, PLC logic and safety interlocks are verified, emergency stop functions are checked, and sensors and instrumentation are calibrated. Dry testing, carried out without excavation, confirms that motors, drives, and control responses behave as designed.

Hydraulic system commissioning follows, with circuits flushed, filled, and pressurised. Thrust cylinders, articulation systems, grippers (where fitted), segment erectors, and auxiliary actuators are tested under load. Pressure, flow, temperature, and oil cleanliness are closely monitored to prevent early damage to sensitive components.

Mechanical functional testing then confirms the behaviour of rotating and moving parts. The cutterhead is rotated at low speed to assess vibration, alignment, torque transmission, and bearing lubrication and seal performance. Spoil handling systems such as screw conveyors, slurry pumps, conveyors, and segment handling equipment are tested under controlled conditions to confirm reliable operation.

For EPB and slurry TBMs, commissioning includes face pressure and spoil handling trials. These tests validate pressure control logic, screw conveyor or slurry circuit response, and the interaction between advance rate and muck removal. This phase is particularly critical in settlement-sensitive or high groundwater environments, where pressure instability can have immediate surface consequences.

Guidance and steering systems are then calibrated. Laser guidance, gyroscopes, and total station systems are aligned to the tunnel reference geometry, and articulation and differential thrust responses are verified. This ensures the TBM can maintain accurate line and grade control from the start of the drive.

In many projects, commissioning concludes with a short trial bore or soft-ground launch. This allows verification of excavation performance under real conditions, fine-tuning of control parameters, and operator familiarisation with machine behaviour. Trial boring often reveals minor adjustments that can be addressed before full production begins.

TBM assembly and commissioning are collaborative processes involving TBM manufacturer specialists, contractor engineers and operators, designers and geotechnical experts, and client or independent certifiers. Formal checklists, hold points, inspections, and sign-offs are used to confirm readiness at each stage and to ensure accountability.

Key Commissioning Activities

1. Electrical and Control System Testing
Commissioning begins with energising the TBM in stages:

- Power distribution and protection systems are tested
- PLC logic, interlocks, and emergency stop systems are verified
- Sensors and instrumentation are calibrated

Dry testing (without excavation) confirms that motors, drives, and control logic respond correctly.

2. Hydraulic System Commissioning
Hydraulic circuits are flushed, filled, and pressurised. Testing includes:

- Thrust cylinders and articulation systems
- Grippers (for open or double-shield TBMs)
- Segment erectors and auxiliary actuators

Pressure, flow, temperature, and cleanliness are monitored closely to prevent early component damage.

3. Mechanical Functional Testing
The cutterhead is rotated at low speed to:

- Check vibration and alignment
- Confirm torque transmission
- Verify bearing lubrication and seal integrity

Screw conveyors, slurry pumps, conveyors, and segment handling systems are tested under controlled conditions.

4. Face Pressure and Spoil Handling Trials
For EPB and slurry TBMs, commissioning includes simulated or initial low-risk operation to validate:

- Face pressure control logic
- Screw conveyor or slurry circuit response
- Interaction between advance rate and spoil removal

These tests are crucial for settlement-sensitive or high groundwater projects.

5. Guidance and Steering Calibration
Laser guidance, gyroscopes, and total station systems are aligned to the tunnel reference geometry. Articulation and thrust steering responses are verified to ensure:

- Accurate line and grade control
- Predictable steering behaviour

6. Trial Boring / Soft Ground Launch
In many projects, commissioning culminates in a short trial bore. This allows:

- Verification of excavation performance
- Fine-tuning of control parameters
- Operator familiarisation with machine behaviour

This stage often reveals minor adjustments needed before full production boring begins.

From an engineering and risk perspective, these phases are among the most critical in any TBM project. Errors are easiest and least costly to correct before excavation begins, while early failures can cause months of delay and substantial financial impact. Experience shows that many serious TBM incidents stem not from fundamental design flaws, but from inadequate assembly checks, rushed commissioning, or poorly tuned control systems. Properly executed assembly and commissioning therefore provide the foundation for safe, reliable, and efficient tunnelling.

Start-up Procedures

TBM start-up procedures are the structured sequence of checks, tests, and controlled actions carried out when a Tunnel Boring Machine transitions from a stationary, commissioned state into active excavation. Start-up is one of the highest-risk operational phases of a TBM drive because the machine begins interacting with real ground, groundwater, and in-situ stresses for the first time. For this reason, start-up is deliberately slow, methodical, and heavily monitored.

The start-up procedures for Tunnel Boring Machines (TBMs) are crucial in ensuring that these machines, along with their supporting systems, can engage safely and predictably with the ground. The primary aim of these procedures is to confirm the operational readiness of mechanical, electrical, hydraulic, and control systems under load conditions, rather than relying solely on no-load commissioning tests. It is essential to verify that all systems function correctly under operational stresses to uncover any potential issues that could impact performance or safety during excavation [35, 40, 423].

A critical objective of the TBM start-up is to establish stable face support conditions prior to full excavation advance. This involves maintaining appropriate pressures, whether earth pressure, slurry pressure, or natural rock stability. Proper face support is vital to avoiding collapses, excessive ground movements, or uncontrolled water inflow, which could destabilize the tunnel during operations [40, 424, 425]. Research emphasizes that maintaining an optimum range of face support pressure can significantly mitigate ground deformations and enhance pile responses during tunnelling activities [35, 424].

Furthermore, operational alignment, thrust reaction, and steering behaviour are essential factors that need verification during start-up procedures. By systematically calibrating these parameters, operators ensure that the TBM accurately responds to control inputs and follows the intended design line [335]. The alignment checks also play a crucial role in preventing deviations that could compromise the tunnel's structural integrity.

Start-up procedures are not only about operational readiness but also about protecting critical TBM components such as the cutterhead, main bearing, seals, and tunnel lining from early damage. By introducing thrust, rotation, and excavation forces in a gradual manner, mechanical stresses are kept within safe thresholds, allowing for careful monitoring of the machine's operational behaviour [423, 425]. The complexity of the TBM mechanics necessitates that operators use precise control mechanisms and dynamically adjust parameters based on real-time observations, ensuring the longevity and functional reliability of the machine [425, 426].

The start-up phase establishes an operational baseline for the entire tunnelling process. Any errors or hurried decisions can lead to severe consequences such as face instability, excessive settlement, and premature failure of key components [335, 423]. A thorough understanding and execution of start-up procedures is treated as a critical engineering process rather than merely an operational step. Errors at this stage can lead to significant repercussions not only for the TBM but also for the safety of adjacent structures and personnel [40, 423].

Pre-Start Readiness Checks

Before any excavation begins, the TBM must be declared "ready to bore." This involves coordinated sign-off across several systems.

Mechanical Readiness

- Cutterhead tools installed and inspected
- Main bearing lubrication active and pressures verified
- Thrust cylinders aligned and functional
- Segment erector, backup gantries, and conveyors operational

Electrical and Control Readiness

- Power supply stable and redundant feeds available

- PLCs, safety interlocks, and emergency stop systems tested
- Sensors (pressure, torque, thrust, temperature, alignment) calibrated

Hydraulic Readiness

- Hydraulic circuits flushed and pressurised
- Oil cleanliness verified
- Cooling systems operational

Ground and Shaft Readiness

- Launch shaft sealed and capable of resisting thrust and pressure
- Ground treatment (if required) completed
- Initial lining rings or thrust reaction blocks in place

No start-up proceeds without formal approval that all systems are operational and safe.

Establishing Initial Face Support

The establishment of initial face support is a crucial step in the operation of Tunnel Boring Machines (TBMs), significantly influencing the safety and efficiency of tunnelling operations. This paper aims to outline the methodologies employed in establishing and maintaining face support across different TBM types, emphasizing the importance of stabilization before any forward movement occurs.

Face support is critical for preventing tunnel face collapses and managing excessive ground movements. Prior to the active excavation by a TBM, it is essential to stabilize the tunnel face to mitigate risks associated with geological unpredictability, as emphasized by Chung et al. [5], who highlight that unforeseen ground conditions can lead to significant construction delays and increased costs. Establishing stable conditions during the initial phase allows for safe operations, confirming the adequacy of support pressures prior to advancement.

Methods of Face Support

Earth Pressure Balance (EPB) Machines: In the case of Earth Pressure Balance TBMs, practitioners fill the excavation chamber with conditioned soil and adjunct materials such as foam or polymers to form a pressure plug that supports the tunnel face. The gradual increase of chamber pressure until it matches the calculated earth and groundwater pressures is essential in stabilizing the face. This technique has been detailed in studies addressing the mechanics of bentonite slurry in front of TBMs, indicating that proper pressure management is vital for preventing ground movement and ensuring structural integrity [318, 323, 427]. The interaction between the sludge and surrounding earth conditions must be continuously monitored to affirm stability [323].

Additionally, soil conditioning techniques significantly enhance the functionality of EPB machines by promoting the reliability of face support through rheological adjustments of the slurry. Research elucidates that various conditioning agents, particularly those tailored for specific soil types, can optimize tunnelling performance by minimizing clogging and enhancing excavated material transport [301, 428].

Slurry and Mixshield TBMs: In contrast to EPBs, Slurry and Mixshield TBMs utilize a bentonite slurry, which acts as hydraulic face support. The system's design requires careful pressurization to balance the given earth loads effectively. According to Zhou et al. [306], the management of slurry pressure is essential, with precise adjustments necessary to maintain equilibrium and prevent issues like water spewing, which can jeopardize the excavation process. Studies underscore that the engineered slurry mix must be adaptable to specific geological conditions to ensure successful tunnelling operations while maintaining face stability [429].

Open TBMs: Open TBMs that operate in competent rock rely primarily on the natural stability of the rock mass itself. Prior to the activation of the cutterhead, proper inspections of the tunnel face are conducted to confirm stability and drainage conditions. This verification is paramount, as any unanticipated groundwater inflows must be handled within manageable limits to avoid destabilization [302]. Monitoring of the tunnel face during static conditions ensures a safe transition into active excavation practices.

Continuous monitoring of face pressure is essential, which can be achieved through specific instrumentation and routine visual assessments. This comprehensive oversight ensures no advancement of the TBM occurs until all parameters indicate stability [430, 431]. As emphasized by Huang et al. [432], understanding the dynamic pressures at play, particularly in relation to geomechanical factors, is vital for maintaining operational safety throughout the excavation phase.

The process of establishing initial face support is paramount in TBM operations, with methodologies varying significantly across different machine types. The emphasis on soil conditioning, precise pressure management, and continuous monitoring collectively contribute to effective risk mitigation strategies. Such practices not only enhance the safety and efficiency of tunnelling projects but also ensure that operations adhere to the highest engineering standards.

Initial Cutterhead Rotation (No Advance)

Before commencing excavation, the operational protocols for Tunnel Boring Machines (TBMs) dictate that the cutterhead is initially rotated without applying thrust. This controlled rotation is crucial as it enables thorough testing of the mechanical systems under motion while the TBM remains stationary, significantly reducing exposure to active excavation forces [433, 434]. By facilitating a non-invasive assessment of the machine's functioning, this protocol enables early identification of potential issues, enhancing the overall safety and maintenance of the TBM during subsequent excavation phases.

During the initial stages of rotation, it is essential that the cutterhead operates at a very low speed. This precautionary step minimizes loading conditions on the machine and permits engineers and operators to closely monitor machine behaviour without the additional stresses introduced by thrust application [434, 435]. Critical parameters such as cutterhead torque, vibration levels, main bearing temperature, and seal performance are carefully monitored through the TBM's integrated control systems and instrumentation [436, 437].

This phase of operation is critical because any anomalous indications, such as unusual noise, excessive vibration, unexpected temperature rise, or pressure fluctuations, trigger immediate safety protocols, including shutting down the system for further investigation [438]. This reflects best practices in engineering, where proactive measures are taken to mitigate risks before moving into full excavation mode.

The continuous monitoring of key operational parameters during this low-speed rotation phase is integral to ensuring the cutterhead, main bearings, drive systems, and seals are performing correctly. Detecting issues early is crucial for both efficiency and safety in tunnelling operations [387]. Should any abnormalities be detected, they can be addressed prior to the TBM engaging in ground cutting, thus significantly reducing the likelihood of incurring damage or operational failure when the machine begins its excavation activities [439].

The ability to correlate operational disturbances with mechanical performance under variably compressive stresses during initial rotation further reinforces the critical nature of this phase. Studies have shown that issues such as severe vibrations or thermal anomalies can lead to long-term inefficiencies or catastrophic failures if not addressed promptly [383, 434].

Controlled Initial Advance

When operating a Tunnel Boring Machine (TBM), particularly during the initial advancement phase following stable cutterhead rotation, the transition from stationary to active excavation must be handled with precision. This phase involves carefully controlled increments of advancement to mitigate risks to both the TBM systems and the surrounding ground.

Once the stability of the cutterhead rotation is confirmed, the TBM begins to advance cautiously. This deliberate incremental approach is essential to ensure a smooth transition while minimizing risks associated with machine operation and ground instability. Operators manage critical parameters such as thrust force and cutterhead rotation speed, maintaining total thrust below design limits to prevent excessive stress on the TBM structure, thereby ensuring that operational responses remain predictable and manageable [387, 434].

The thrust application is methodically introduced via thrust cylinders, which are pressurized gradually to avoid unintended steering responses. This controlled thrust prevents significant differential thrust, thus averting alignment deviations and potential risks during the excavation process [440]. Real-time assessment of operational actions and mechanical responses during this phase is vital for confirming predictable behaviour before increasing production rates [441].

During the initial advancement phase, it is crucial to monitor the excavation rate closely. The cutterhead speed, torque, and thrust are incrementally adjusted to ensure that operators can assess and address the cutting behaviour adequately. Synchronization between spoil removal systems and the advance rate is another critical component to maintaining stable face conditions, thereby preventing potential disruptions in the excavation process [387, 434]. For Earth Pressure Balance (EPB) TBMs, the screw conveyor speed is meticulously matched to the advance rate to avoid overfilling the excavation chamber or compromising face pressure. Similarly, for slurry TBMs, consistent slurry pump flow rates and density are fundamental to maintaining hydraulic support at the tunnel face [434, 442].

The use of real-time data and monitoring systems enables operators to gauge how the ground interacts with the TBM during cutting and maintain a cautious yet effective excavation strategy [440, 443].

Monitoring the reaction of the ground to both cutting and mechanical responses is a defining aspect of this phase. The stability of the ground and the TBM must be evaluated continuously to avert complications, which could arise from geological variability or unexpected interactions between the TBM and the surrounding soil [444, 445]. Operators utilize various data-driven methodologies to analyse operational data and enhance TBM efficiency and safety during this decisive phase of tunnelling [379].

The cautious approach during the initial advancement phase of TBM operation, characterized by gradual thrust application, controlled excavation parameters, and continuous ground monitoring, is pivotal in ensuring operational safety and efficiency. This methodology aids in preventing alignment issues and maintaining stable excavation conditions, critical for the successful and effective operation of tunnel boring machines in complex geological environments.

Steering and Alignment Verification

During the early stages of advance, the TBM's steering response is carefully tested to confirm that the machine can accurately follow the designed tunnel alignment. Guidance systems, including laser guidance, gyroscopes, and total station measurements, are used to verify that the TBM is maintaining the correct line and grade as it begins to move forward.

Small, controlled adjustments are then made using the articulation cylinders and differential thrust to assess the steering sensitivity and response of the machine. These corrections are deliberately minor, allowing operators to understand how the TBM reacts to steering inputs without introducing unnecessary stress into the system.

At the same time, the lining installation process is checked by verifying ring geometry and taper selection. This ensures that segment orientation and ring build support the intended alignment and do not introduce cumulative deviation. Any misalignment detected at this stage is corrected immediately, when thrust forces are still low and adjustments can be made easily and safely, preventing larger corrections and increased risk later in the drive.

Progressive Ramp-Up to Normal Boring

The process of gradually increasing operational parameters during the ramp-up phase of Tunnel Boring Machines (TBMs) is crucial for achieving stability and efficiency in tunnelling operations. This phase involves careful monitoring and adjustment of cutterhead speed, thrust force, penetration rate, and other parameters to optimize machine performance while ensuring minimal ground disturbance.

Once stable excavation and machine behaviour have been confirmed, the ramp-up process begins gradually. The cutterhead rotation speed is incrementally increased, which aligns with the machine's cutting performance capabilities and the geological response from the ground. The study by Samadi et al. emphasizes the assessment of geotechnical parameters such as cutterhead torque and thrust force, which directly influence the operational stability of Earth Pressure Balance Tunnel Boring Machines (EPB-TBM) in different soil types [440].

Simultaneously, thrust force and penetration rates must also be incremented step by step, allowing for a gradual adaptation to the ground's behaviour [311, 440]. Monitoring of pressure stability, spoil flow, and surface settlement is essential during this phase, as evidenced by research indicating the significant impact of operational parameters on ground disturbance [446].

Fine-tuning face pressure based on ongoing observations of ground behaviour is a critical component of the ramp-up phase. The adjustments made alleviate risks such as excessive settlement or instability. For EPB machines, maintaining an effective pressure plug is achieved through the careful adjustment of soil conditioning additives, which is crucial for controlling face stability [447, 448]. In slurry TBMs, it is necessary to modify the slurry's density, viscosity, and flow rates, as these factors play a pivotal role in maintaining consistent face support throughout excavation [447].

Research by Yang and Yin elucidates the complexities involved in the coupled hydro-mechanical interactions during tunnelling, reinforcing the importance of optimal slurry properties in ensuring operational efficiency and stability [319]. The recommendation for maintaining an effective pressure environment is reiterated by findings on the need for adjustments based on geological conditions [448].

The duration of the ramp-up phase can vary significantly, influenced by factors such as ground variability, groundwater conditions, and the overall risk tolerance of the project. Several studies suggest that this cautious, controlled phase may extend over multiple lining rings or days depending on the specific conditions encountered [311, 449]. The importance of a gradual approach is underlined, as successful transitions to full production boring are contingent on establishing a stable and repeatable operating environment [311, 440, 446].

Research indicates that in urban tunnelling projects, ramp-up durations can vary substantially, with the most effective strategies tailored to the specific context of the project at hand [446, 450]. For instance, findings show that urban projects often require particularly cautious ramp-up strategies to mitigate risks associated with surface settlement [311].

The ramp-up process for TBMs is integral to ensuring effective and stable tunnelling operations. The gradual increase in operational parameters—spurred by real-time monitoring and strategic adjustments—yields a balance between machine performance and geological challenges. The documented phases emphasize the need for a nuanced understanding of equipment behaviour, geological response, and operational metrics to navigate the complexities of underground excavation effectively.

Monitoring and Decision-Making During Start-up

Throughout the start-up phase, real-time operational data is closely scrutinised to confirm that the TBM, ground, and support systems are interacting as expected. Key parameters monitored include the relationship between thrust and penetration rate, fluctuations in cutterhead torque, and the stability of face pressure within the excavation chamber. These indicators provide early insight into ground behaviour, cutting efficiency, and the effectiveness of face support.

In parallel, external monitoring systems track ground movement and surface settlement, while water inflow rates are observed to identify any unexpected changes in groundwater conditions. Together, these data streams allow engineers to assess risk in real time and verify that excavation remains within safe limits. Start-up is frequently paused at predefined checkpoints to allow detailed engineering review and decision-making before advancing to full production, ensuring that any issues are addressed while corrective actions are still straightforward and low-risk.

Differences by TBM Type

Open (Gripper) TBMs: For open or gripper TBMs, the start-up procedures are primarily focused on confirming the stability of the surrounding rock at the tunnel face, ensuring the reliability of the gripper mechanisms, and managing groundwater inflows. This type of TBM relies on the strength of the surrounding rock to provide the necessary thrust and stability; therefore, verifying the rock's ability to withstand applied loads is crucial. Inadequate assessment can lead to excessive deformation or uncontrolled water ingress, which poses significant operational risks [14, 451].

Research indicates that the gripping-thrusting mechanism in open TBMs plays a critical role during the start-up process. Effective performance of this mechanism is necessary for maintaining continuous and efficient tunnelling, which highlights the need for thorough preliminary checks to ascertain rock stability prior to initiating excavation [347, 452]. Engineering studies emphasize that close observation and adjustments are vital during these initial interactions with the ground to mitigate potential failure modes [453, 454].

Earth Pressure Balance (EPB) TBMs: In contrast, EPB TBMs require a different approach during start-up, centring on the formation and maintenance of a stable pressure plug within the excavation chamber. This pressure plug is crucial for preventing ground settlement and face instability, especially in soft ground conditions. Start-up procedures for EPB TBMs closely monitor several factors, including the effectiveness of soil conditioning, chamber pressure response, and the behaviour of the screw conveyor. An improper formation of the pressure plug can lead to severe operational issues such as face collapse and ground movements [7, 322].

The dynamic response of the EPB system to pressure fluctuations directly impacts tunnel stability and excavation efficiency, making it imperative to establish precise control mechanisms from the outset [319]. As identified in research, understanding the complex interactions between the TBM and the surrounding soil is essential for effective operation, signalling a critical focus on maintaining optimal conditions during early operations [7, 435].

Slurry and Mixshield TBMs: For slurry and Mixshield TBMs, the start-up process emphasizes achieving and maintaining hydraulic pressure balance at the tunnel face. This involves meticulous control of parameters such as slurry density and pump flow rates. In the case of Mixshield machines, air-cushion pressure also plays a vital role in stabilizing the face. Insufficient management of hydraulic balance can quickly destabilize the face pressure, leading to operational hazards [319].

Research highlights that maintaining an adequate separation and circulation of slurry through surface separation plants is equally critical. If this process fails, it could rapidly lead to operational instability, necessitating careful monitoring throughout the start-up activities [455, 456]. The controllable variables during the start-up phase are pivotal, and adherence to specified operational parameters can significantly reduce the risk of encountering face instability, which might jeopardize the tunnelling project [319, 453].

Application Example

The following is a summary of the Start-up Procedures for the Akkerman 720 Series II Tunnel Boring Machine, based on the Operation – TBM Start-Up Procedure and Launching the TBM sections of the manufacturer's operator manual [457].

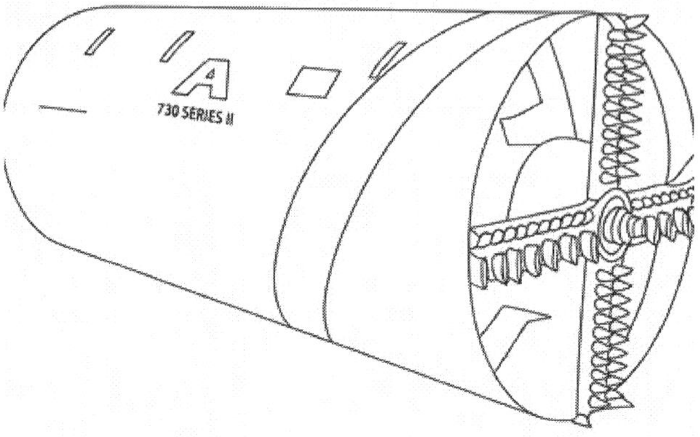

Figure 72: Diagram of the Akkerman 720 Series II Tunnel Boring Machine.

The Akkerman 720 Series II start-up procedure reflects a conservative, safety-focused philosophy typical of pipe-jacking TBMs. Rather than rapid engagement, the process emphasises controlled rotation, verified spoil handling, confirmed lubrication, and precise steering alignment. These steps reduce the risk of early mechanical damage, conveyor incidents, misalignment, or loss of control during the initial interaction between the TBM and the ground, setting a stable baseline for the subsequent jacking and excavation process.

The Akkerman 720 Series II Tunnel Boring Machine (TBM) is a purpose-built microtunnelling or pipe-jacking machine designed for installing small to medium diameter pipelines and conduits with high precision and minimal surface disruption. It's part of Akkerman's Series II family of earth pressure microtunnelling equipment, commonly used for utility installations such as sewer, water, drainage, stormwater, and other underground infrastructure beneath roads, railways, rivers, and urban areas where open-cut excavation is not feasible.

Key characteristics and engineering aspects of the Akkerman 720 Series II:

The 720 Series II TBM is a guided, steerable, remote-controlled machine that excavates and installs pipe simultaneously, typically in a trenchless microtunnelling or pipe-jacking application. It combines excavation, spoil removal, steering, and guidance into a single integrated system, allowing the operator to perform controlled tunnelling from a launch shaft to a reception shaft.

Unlike full-size mechanized TBMs used for metro or highway tunnels, this type of machine is tailored for more confined, shallow installations and usually operates with a hydraulic jacking frame that pushes the TBM and the pipe string forward in a coordinated advance. The shield at the front protects the tunnel face and working area, while guidance systems maintain alignment and grade with high accuracy.

The 720 Series II includes:

- **Cutting head and excavation chamber:** Fitted with tooling suited for soft soils, clays, silts, sands, or mixed ground typically encountered in utility corridors.
- **Steering and guidance system:** High-precision laser or gyro guidance allows the operator to maintain strict compliance with line and grade tolerances, often measured in millimetres over long drives.
- **Hydraulic power and jacking systems:** Robust hydraulic units provide the thrust needed to advance the machine and pipe string while transmitting reaction forces safely back to the launch shaft.
- **Spoil removal system:** Typically involves screw conveyors or slurry circuits (depending on configuration) to remove excavated material while maintaining face stability and pressure balance.
- **Control and monitoring:** Integrated electrical and hydraulic controls allow the operator to monitor torque, thrust, advance rate, alignment, and other key parameters in real time.

The Akkerman 720 Series II is especially well-suited to:

- **Urban utility installations:** Installing stormwater, sewer, potable water, or telecom conduits beneath roads, pavements, rail lines, or buildings with minimal surface disruption.
- **River or creek crossings:** Where surface excavation would disturb sensitive environments or watercourses.
- **Confined or restricted launch sites:** The machine's size allows assembly and operation from relatively small launch shafts.
- **Projects requiring high accuracy:** Situations that demand precise line and grade control, such as tie-ins to existing pipelines or gravity drainage systems.

Start-Up and Operational Behaviour

Start-up of the 720 Series II is methodical and deliberate, as outlined in the operation manual [457]. Before forward advance, the machine's cutterhead, conveyor, steering, guidance, hydraulic, and safety systems must be tested under controlled conditions. Operators gradually introduce rotation and jacking thrust to ensure that all components perform under load before actual ground cutting begins. Early cuts are made at low penetration rates, with continuous monitoring of torque, vibration, alignment, and spoil handling to validate system response.

The machine's design supports incremental advancement with careful coordination of thrust, cutter rotation, steering corrections, and spoil removal to reduce the risk of face instability, settlement, or mechanical overload—especially important in mixed soils, when near existing infrastructure, or where groundwater pressures are variable.

Start-up Purpose and Context: The Akkerman 720 Series II start-up procedure establishes safe, controlled operation of the boring head, conveyor, steering, and support systems before the machine is launched into the ground. The procedure is designed to verify mechanical integrity, hydraulic function, lubrication, spoil handling, and operator safety under live operating conditions, while minimising risk during the critical transition from stationary to advancing operation.

Pre-Start Safety and System Checks: Before any rotation or conveying begins, the operator must confirm that all safety systems are operational. This includes verifying that emergency stop circuits function correctly, the operator can evacuate quickly in an emergency, and all personnel are clear of moving components. The manual emphasises strict adherence to lockout/tagout principles and awareness of pinch points, rotating equipment, and suspended loads.

Lubrication systems are checked at this stage. The bearing oil lubrication pump, seal grease pump, and scavenging pump must be operating correctly, as these systems protect the cutterhead bearing and seals from early damage during rotation and loading.

Initial Boring Head and Conveyor Operation: The start-up sequence begins with controlled activation of the boring head. The operator engages the boring head control and gradually adjusts rotation speed as required. Cutterhead rotation is deliberately kept conservative during start-up to observe torque behaviour, vibration, and general mechanical response.

The conveyor system—belt or screw, depending on configuration—is then operated using the dedicated conveyor control. Conveyor speed is adjusted incrementally, with the manual stressing that the operator must remain seated in the normal operating position at all times while the cutterhead is rotating. Safety chains must be secured, and the conveyor safety valve (cutterhead drive dump valve) must be tethered and tested to prevent uncontrolled movement or injury.

Conveyor Positioning and Safety Controls: The conveyor is moved into the operating position so that excavated material will fall correctly into the dirt bucket. The manual warns explicitly that improper conveyor positioning or unsecured safety chains can result in severe injury or fatality. Continuous awareness of conveyor movement is mandatory during this phase.

After verification, the conveyor may be returned to the maintenance position prior to launch, depending on the stage of preparation and coordination with the pipe jacking system.

Steering System Preparation and Laser Guidance Setup: Before launch, the steering cylinders are fully extended and then retracted by approximately 64 mm (2.5 inches). This establishes correct steering geometry and ensures forward pull is maintained on the conveyor lifting cables and safety chains throughout the advance stroke. Failure to maintain this pull can cause the conveyor to disengage from the cutter bar centre pin, creating a serious safety hazard.

The laser guidance system is then set to the required line and grade. The laser beam must have a clear path to the cutter bar target bolt, and operators are warned never to stare into the laser beam due to the risk of severe eye injury. The manual notes that some operators deliberately steer slightly upward (typically 13–19 mm) before launch to compensate for the tendency of the TBM to dip as it enters the ground, with the exact adjustment dependent on ground conditions.

Transition to Launching: Only after the full TBM start-up procedure has been completed is the machine cleared for launching into the ground. The manual explicitly states that the start-up procedure must be performed before launch and that launching operations are coordinated with the 5000 Series II pipe jacking system and associated manuals.

Normal Boring Cycle

The normal boring cycle of a Tunnel Boring Machine (TBM) reflects a systematic and interdependent series of operations that are essential for the effective excavation of tunnel sections. It encompasses thrust application, cutterhead rotation, spoil excavation, and muck removal. Although specific operations may vary according to the type of TBM—be it open, Earth Pressure Balance (EPB), slurry, or microtunnelling—the fundamental cycle remains consistent. Integrating these components is vital for maintaining stability, maximizing productivity, and ensuring safety during tunnelling operations.

Thrust Application

Thrust application serves as the primary force propelling the TBM forward, executed through hydraulic thrust cylinders. These cylinders extend to create force against the tunnel lining in shielded TBMs, grippers against rock in open TBMs, or jacking frames in pipe-jacking systems. It is critical that thrust is applied steadily during boring as abrupt increments can lead to mechanical stress or instability. The thrust must be calibrated carefully concerning ground resistance, ensuring efficient penetration while preserving the integrity of the cutterhead and tunnel linings [335, 458].

Differential thrust application becomes particularly crucial in curved tunnelling alignments, where varying thrust across the cutterhead assists in maintaining the desired trajectory while preserving machine stability. This continuous adjustment of thrust, informed by real-time data on penetration depth and cutterhead torque, forms a feedback loop that regulates overall machine performance [335, 459].

Cutterhead Rotation

In concert with thrust application, the rotation of the cutterhead plays a pivotal role in the excavation process. The speed at which the cutterhead rotates depends on the nature of the material being excavated; harder materials necessitate reduced rotation speeds and increased thrust, while softer materials allow for higher speeds. An efficient balance between thrust and rotation rates is essential; excessive thrust without adequate rotation risks damaging the cutting tools, and high rotation with insufficient thrust can reduce the efficiency of the cutting operation [281, 313].

Modern TBMs often employ variable-frequency drives, enabling dynamic adjustments in cutterhead speed and torque, thereby enhancing responsiveness to shifting ground conditions. Closed-loop control systems further optimize cutting performance by ensuring consistent operational parameters [453].

Spoil Excavation

Once the cutterhead has engaged and sheared the ground, spoil excavation occurs at the tunnel face. The effectiveness of this process is dictated by the type and design of the cutting tools. For example, disc cutters primarily apply compressive forces, while rippers or picks are more effective for disaggregating mixed or fractured materials. In pressurized TBMs such as EPB and slurry types, maintaining face pressure during excavation is crucial, requiring careful correlation between spoil removal and excavation rates to counteract external loads from the surrounding earth and groundwater [285, 427]. Regular assessment of the spoil's characteristics, such as grain size and moisture content, is essential for adjusting properties and conditioning agents used during tunnelling [300, 333].

Muck Removal

Muck removal is the concluding aspect of the boring cycle, involving the transportation of excavated spoil away from the working face. Each TBM type employs different muck removal systems: screw conveyors in EPB and variable-density TBMs; hydraulic pumping in slurry TBMs; and mechanical conveyors in open TBMs. The synchronization of muck removal with thrust and cutterhead operations is fundamental; rapid removal may result in loss of face pressure, while slow removal can lead to overfilling and increased operational demands [460].

Integrated Cycle Control

For optimal TBM operation, thrust application, cutterhead rotation, spoil excavation, and muck removal must function cohesively. Advanced control systems integrate these elements, utilizing real-time sensor data to dynamically balance operations [281, 461, 462]. The TBM advances incrementally, executing the normal boring cycle ring by ring, ensuring that ground stability is maintained while minimizing surface disruption and maximizing excavation efficiency.

The execution of this cycle is not only a technical endeavour but also a critical factor in ensuring safety and efficiency in tunnelling operations. The cumulative effects of precise management of these processes underpin the operational success of tunnel boring machines and contribute to the overall reliability of underground construction projects.

Advance Rates and Penetration Control

Advance rates and penetration control describe how fast a Tunnel Boring Machine (TBM) progresses and how efficiently it converts thrust and cutterhead rotation into excavation. Together, they are fundamental performance indicators in mechanised tunnelling and play a central role in balancing productivity with ground stability, machine wear, and operational safety.

The advance rate refers to the forward progress of the TBM over time and is typically expressed in metres per hour or metres per day, often averaged over a shift or full working day. Advance rate is not governed by a single parameter; rather, it reflects the combined effect of ground conditions, TBM type and size, cutterhead performance, spoil removal efficiency, lining installation time, and the project's operational strategy and risk constraints.

In practice, a distinction is made between instantaneous advance rate and net or average advance rate. The instantaneous rate represents the speed achieved during active cutting, while the net rate includes all stoppages for ring building, maintenance, inspections, and delays. As a result, a TBM may achieve high cutting speeds during boring but still record modest daily progress if downtime is significant.

Penetration is the core cutting efficiency metric and describes how much the TBM advances for each revolution of the cutterhead, commonly measured in millimetres per revolution. It provides a direct indication of how effectively thrust and torque are being translated into ground breakage. High penetration indicates that cutters are engaging the ground efficiently and that thrust and rotation are well matched to ground strength. Conversely, low penetration suggests issues such as insufficient thrust, inappropriate cutterhead speed, worn or unsuitable tools, or unfavourable ground behaviour such as clogging, squeezing, or hard inclusions.

Penetration is primarily controlled by the interaction between thrust force, cutterhead rotation speed, and the strength and abrasivity of the ground. Increasing thrust, within safe limits, generally increases penetration, while increasing rotation speed without a corresponding increase in thrust often reduces it. Excessive thrust without adequate rotation can lead to high wear, vibration, and damage to cutters, bearings, or the machine structure. Each ground type therefore has an optimal operating window in which penetration is maximised without overstressing either the TBM or the surrounding ground.

The way penetration is managed varies by TBM type. In hard-rock TBMs, penetration is governed by disc cutter load and spacing, with operators targeting a penetration per revolution suited to the rock strength. Too little penetration leads to cutter polishing and poor efficiency, while excessive penetration increases cutter wear, vibration, and bearing loads. In EPB TBMs, penetration is closely linked to face pressure and soil conditioning; poorly conditioned soil can significantly reduce penetration even if thrust is increased, and screw conveyor performance and chamber pressure strongly influence achievable cutting efficiency.

For slurry and Mixshield TBMs, penetration must be carefully balanced against slurry pressure and flow capacity. Excessive penetration can destabilise the tunnel face if the slurry support system cannot respond quickly enough, so penetration is often deliberately conservative in highly permeable or subaqueous ground. In microtunnelling and pipe-jacking TBMs, penetration is constrained by jacking force capacity and pipe strength, making precise control especially critical due to tight tolerances and high axial loads in the pipe string.

Modern TBMs manage advance rate and penetration through closed-loop control systems that rely on real-time feedback from thrust cylinders, cutterhead torque and speed sensors, penetration measurements, face pressure instrumentation, and spoil flow monitoring. Operators typically set target penetration values or thrust limits and allow the control system to adjust rotation speed and advance rate automatically, intervening when data trends indicate instability, inefficiency, or emerging risk.

Advance rates are often deliberately limited by non-cutting factors. In urban environments, settlement risk may dictate conservative penetration and advance rates. Groundwater pressure control, lining installation cycle times, monitoring requirements, and regulatory constraints can also cap achievable progress, regardless of the machine's cutting capability.

Maximising advance rate does not necessarily maximise overall project performance. Overly aggressive penetration can lead to face instability, excessive settlement, segment damage, premature cutter and bearing wear, and unplanned shutdowns. In many urban or high-risk projects, the optimal approach is a stable, repeatable, moderate advance rate that minimises downtime and risk, resulting in more reliable overall progress.

Steering and Guidance

TBM steering and guidance are the systems and operational processes that keep a Tunnel Boring Machine accurately following the designed tunnel alignment while maintaining ground stability and protecting the lining. Together, these systems allow a TBM—often hundreds of metres long and weighing thousands of tonnes—to be controlled with millimetre-level precision as it advances through complex and variable underground conditions.

The primary purpose of steering and guidance is to ensure that the TBM progresses along the correct line, level, and orientation defined by the tunnel design. Accurate steering is essential to meeting strict geometric tolerances for rail, road, and service tunnels, avoiding clashes with existing infrastructure, and preventing small deviations from accumulating into major alignment errors over long drives. It also plays a key role in protecting the tunnel lining from uneven loads and in controlling ground deformation and settlement. Even minor misalignments, if not corrected early, can compound over distance and result in costly rectification or long-term operational constraints.

Guidance systems are responsible for determining the TBM's actual position and orientation relative to the design alignment. Modern TBMs rely on a combination of technologies to achieve high accuracy and reliability. Laser guidance systems are the most common, using a fixed laser set up in the tunnel behind the TBM to project a beam along the designed tunnel axis. Targets mounted inside the machine detect the laser position, allowing real-time calculation of horizontal and vertical deviations. Gyroscopic navigation systems are often used alongside lasers or as a backup, particularly in long tunnels, curved alignments, or locations where line-of-sight is interrupted. These systems measure angular changes independently of external references. Periodic total station surveys are used to verify alignment and recalibrate the guidance system, ensuring long-term accuracy and confidence in the data. Together, these tools continuously answer the fundamental question of where the TBM is relative to where it should be.

While guidance systems measure position, steering systems physically change the TBM's direction. Steering methods vary depending on the TBM type, but all rely on controlled asymmetry—deliberately creating small differences in force or geometry across the machine. In shielded TBMs such as EPB, slurry, and Mixshield machines, steering is achieved through a combination of articulation systems and differential thrust. Articulation systems divide the shield into front and rear sections connected by hydraulic cylinders, allowing the cutterhead to be angled slightly relative to the installed lining. Differential thrust is applied by increasing thrust force on one side of the machine relative to the other, gently pushing the TBM in the desired direction.

In open or gripper TBMs, steering is achieved mainly by varying gripper forces and thrust distribution against the rock, sometimes assisted by limited articulation or cutterhead steering features. This approach depends heavily on rock competence and predictable ground behaviour, as the surrounding rock provides both reaction forces and stability.

Steering is closely linked to lining installation and geometry. In segmentally lined tunnels, alignment control is supported through careful ring taper selection, where slightly tapered rings are installed to introduce controlled curvature, and through segment orientation, which influences how the lining responds to thrust and steering forces. Poor coordination between steering inputs and lining geometry can lead to segment cracking, gasket damage, or gradual alignment drift that becomes difficult to correct later in the drive.

During excavation, steering adjustments are made continuously and incrementally rather than as abrupt corrections. Operators and control systems apply very small articulation angles, often only fractions of a degree, and use minimal differential thrust to avoid overcorrection. The TBM's response is monitored over several rings before further adjustments are made. This gradual approach maintains stable excavation conditions, minimises ground disturbance, and protects both the machine and the lining from excessive stress.

Modern TBM steering is heavily supported by automation and control systems. PLC-based systems process guidance data in real time, display deviation trends and steering recommendations to operators, and enforce limits on articulation angles and differential thrust to keep operations within safe envelopes. These systems can prevent steering actions that might overstress the lining or shield. Despite this automation, steering remains a collaborative process, with experienced operators interpreting data and making engineering judgments, particularly in variable or challenging ground conditions.

Ground conditions and pressure regimes strongly influence steering behaviour. In soft or mixed ground, steering must be carefully coordinated with face pressure control to avoid inducing settlement or instability. In squeezing ground or highly variable geology, steering responses may be delayed or asymmetric, requiring cautious interpretation of guidance data. High groundwater pressure, changing face resistance, or cutter wear can also reduce steering efficiency, making continuous monitoring essential.

From an engineering perspective, steering and guidance are not merely geometric controls but critical risk management tools. Effective steering maintains tunnel geometry within tight tolerances, minimises ground movement and settlement, reduces wear on cutters, shields, and seals, and protects the lining from uneven loading. Many tunnelling problems—such as segment damage, excessive thrust demand, unexpected settlement, or loss of alignment—can often be traced back to poor steering control or delayed corrective action.

As a practical example, the following provides a system-level explanation of how the Akkerman 720 Series II pipe-jacking TBM steering and guidance systems work, based directly on the Operator's Manual and standard Akkerman pipe-jacking practice [457].

Overview: Steering and Guidance Philosophy

The Akkerman 720 Series II is an articulated, hydraulically steered pipe-jacking TBM designed for high-accuracy microtunnelling. Its steering and guidance system combines:

- Laser-based alignment guidance
- Articulated steering joint with hydraulic cylinders
- Operator-controlled hydraulic valves and pressure monitoring

- Roll control using torque/dirt wings

Unlike large shield TBMs, the Akkerman 720 relies on continuous fine steering adjustments made at the machine head while thrust is applied from the jacking frame behind the pipe string. Accuracy is achieved through small, frequent corrections, not large directional changes.

Guidance System: Knowing Where the TBM Is

Laser Guidance System

The primary guidance system is a laser sight and target arrangement, consisting of:

- A laser mounted in the launch shaft, aligned precisely to the design line and grade
- A laser target inside the TBM, visible to the operator from the control seat

The laser beam represents the theoretical tunnel axis. As the TBM advances, the operator continuously observes the position of the beam relative to the target to determine:

- Horizontal deviation (left/right)
- Vertical deviation (up/down)

The manual stresses that frequent checking of the laser is critical, as even small angular errors can result in large positional offsets over distance.

"A good survey is critical for proper line and grade." (Operating Guidelines)

Temperature control inside the pipe is also important, as thermal gradients can cause laser refraction and apparent misalignment.

Steering System: How Direction Is Changed

Articulated Steering Joint

The 720 Series II uses a steering joint between the front and rear sections of the TBM, controlled by three sets of hydraulic steering cylinders:

- Left steering cylinders
- Right steering cylinders
- Top steering cylinders

These cylinders allow the front section (cutterhead) to be angled slightly relative to the pipe string, creating a controlled bend that redirects the TBM.

At start-up, all steering cylinders are set to approximately 50% stroke (about 2.5 in / 63.5 mm) to allow correction in all directions.

Steering Controls and Operator Inputs

Manual Hydraulic Control

Steering is performed using dedicated steering control levers on the TBM's hydraulic control valve bank. The operator applies corrections by extending or retracting specific cylinders:

- **Steer Up**
  Extend left and right cylinders equally, or retract top cylinders
- **Steer Down**
  Extend top cylinders, or retract left and right cylinders
- **Steer Left**
  Extend right cylinder and retract left cylinder
  (optionally extend top cylinder at half the correction)
- **Steer Right**
  Extend left cylinder and retract right cylinder
  (optionally extend top cylinder at half the correction)

These combinations allow smooth three-dimensional steering without twisting or overstressing the machine.

Pressure Control and Structural Protection

Steering Pressure Limits

Steering force is carefully limited to protect both the TBM and the pipe string:

- Normal steering corrections: ≤ 3,000 psi
- Absolute maximum during advancement: 4,800 psi

Exceeding these limits risks:

- Hydraulic component failure
- Structural damage to the TBM
- Excessive pipe-to-ground friction

Steering cylinder pressures are displayed on dedicated gauges (left, top, right), allowing the operator to see real-time load response.

The manual explicitly warns that large or rapid steering corrections increase jacking forces, because the front and rear sections are no longer parallel.

Steering Strategy: Small Corrections Over Distance

Akkerman explicitly instructs operators to:

- Make only minor steering adjustments
- Apply corrections gradually over several feet of advance
- Avoid abrupt directional changes

This approach:

- Reduces jacking force buildup
- Prevents pipe damage
- Minimises ground disturbance
- Maintains predictable machine behaviour

The system is designed for continuous correction, not episodic realignment.

Roll Control: Keeping the TBM Level

Torque Wings and Dirt Wings

In addition to directional steering, the 720 Series II includes torque wings or dirt wings to control TBM roll:

- Torque wings: non-directional fins that resist rolling
- Dirt wings: directional fins that counteract rotation caused by cutterhead torque

If roll exceeds approximately ¼–½ inch (6–13 mm), the operator extends the wings hydraulically until the machine returns to level.

This is critical because roll errors can:

- Distort steering response
- Cause uneven cutter wear
- Affect laser alignment accuracy

Interaction with the Pipe Jacking System

Unlike self-propelled TBMs, the Akkerman 720:

- Does not generate its own thrust

- Is pushed forward by external jacking cylinders acting on the pipe string

Steering therefore occurs while the pipe is being pushed, meaning:

- Steering resistance directly affects jacking force
- Excessive steering increases friction along the pipe
- Poor steering discipline can stall the drive

This is why the manual emphasises coordination between the TBM operator and pump unit operator during steering corrections.

TBM steering and guidance combine precise measurement with controlled mechanical response. Guidance systems continuously determine the TBM's position relative to the design alignment, while steering systems use articulation, differential thrust, and lining geometry to gently and predictably correct its path. Together, they enable massive underground machines to follow complex alignments with remarkable accuracy, making safe, efficient, and reliable mechanised tunnelling possible even in the most demanding ground conditions.

Continuous vs Cyclic Operation

Tunnelling technologies encompass various methods that impact ground interaction, machine design, and operational dynamics. Among these, continuous and cyclic operation stand out due to their distinct approaches to excavation, spoil management, and ground support installation, influencing aspects such as productivity, operational risk, and machine suitability for different environments.

Continuous Operation in Tunnelling

Continuous operation refers to the method where Tunnel Boring Machines (TBMs) excavate, remove spoil, and install ground support in an overlapping manner, allowing for steady, uninterrupted advancement. Machines like Earth Pressure Balance (EPB) TBMs, slurry TBMs, and some double-shield hard-rock TBMs utilize this approach effectively. The process assures controlled pressure management and steady face advance, which significantly minimizes ground settlement and enhances stability during tunnelling operations, particularly in urban areas where ground control is critical [35, 62].

The continuous method presents several advantages, including high productivity and reduced cyclic stress on the surrounding geology, which lowers the likelihood of detrimental ground movements. The machinery can operate at a higher average daily advance rate, making it particularly advantageous for long tunnels where interruptions for inspections or maintenance are minimized [177, 311]. Continuous systems' reliance on closed-loop control systems further stabilizes excavation conditions by continuously monitoring factors like thrust, torque, and face pressure [62]. However, the setup requires substantial capital investment and comprehensive support systems, which could limit flexibility in short or interrupted tunnelling projects [57, 463].

Cyclic Operation in Tunnelling

Cyclic operation contrasts with continuous methods by executing excavation in discrete cycles. Machines in this category, including microtunneling TBMs and drill-and-blast systems, undergo repeated stopping and starting: they excavate over a short distance, halt for spoil removal and lining installations, and then resume. This approach provides flexibility and positional accuracy ideal for small-diameter tunnels or projects with specific geometric constraints [464, 465].

The cyclic method is characterized by its ability to manage operational timelines and costs profitably. Unlike continuous methods, cyclic systems can adapt to various environmental conditions and smaller projects where extensive tunnelling equipment is impractical. Yet, this operational framework comes with limitations; average advance rates tend to be lower due to operational stoppages, and there is a heightened sensitivity to soft ground conditions due to the periodic stress relief at the face [177, 466]. Thus, effective management of operator timing and adherence to procedural discipline becomes critical to mitigate the effects of thrust fluctuations and ground deformation [293].

Key Engineering Differences

Significant engineering distinctions between the two operations include their impacts on ground response and overall tunnelling dynamics. Continuous operation maintains relatively constant stress conditions at the tunnel face, while cyclic operation introduces intermittent loading and potential stress relaxation, particularly detrimental in soft ground scenarios. The choice between continuous and cyclic methodologies extends beyond mere equipment selection; it encompasses considerations related to settlement risk, advance predictability, and thrust demands [335, 461, 467].

The decision to employ continuous or cyclic operation should be made on a systems-level basis, as both approaches cater to different tunnelling needs influenced by environmental conditions, project scope, and financial considerations. Continuous operation is optimal for longer, settlement-sensitive tunnels that require stable conditions, while cyclic methods provide necessary flexibility for shorter or small projects. Careful evaluation of these operational modes enables better management of risks associated with tunnelling and enhances project outcomes.

Chapter 7

Excavation Mechanics and Cutting Theory

Mechanised tunnelling relies fundamentally on the controlled interaction between cutting tools and ground materials. For Tunnel Boring Machines (TBMs), excavation performance, tool wear, energy efficiency, and face stability are all governed by the mechanics through which forces are transferred from the machine into rock or soil and converted into material failure and removal. A rigorous understanding of excavation mechanics and cutting theory is therefore essential for interpreting TBM behaviour, predicting performance, and making informed design and operational decisions.

This chapter develops the mechanical foundations that underpin TBM excavation, focusing on how thrust, torque, cutter geometry, and cutterhead motion interact with the physical properties of rock and soil. Unlike conventional excavation, TBM cutting is not a single mechanism but a combination of fracture-dominated processes in hard rock and shear-dominated, plastic flow processes in soft ground. These mechanisms determine whether excavation proceeds efficiently through tensile cracking and chipping, or inefficiently through crushing, grinding, and excessive deformation.

The chapter begins with rock cutting mechanics, establishing the fracture-controlled nature of hard-rock excavation and the critical role of disc cutter indentation, stress concentration, and crack propagation. It then examines the working principles of TBM disc cutters, including cutter arrangement, spacing, and load distribution, highlighting how these factors influence chip formation, specific energy, and cutter wear. Dynamic effects such as vibration, torque fluctuation, and load redistribution are also addressed, as they play a decisive role in operational stability and component life.

Subsequent sections extend the discussion to soft-ground excavation, where soil cutting and plastic flow dominate. Here, excavation efficiency depends less on fracture initiation and more on mobilising controlled shear failure and maintaining favourable rheological conditions for face support and spoil transport. The chapter links cutting mechanics directly to cutterhead torque–thrust relationships, wear mechanisms, penetration rate theory, and the influence of rock strength and abrasivity.

By integrating fundamental mechanics with practical TBM behaviour, this chapter provides the theoretical framework needed to understand why TBMs perform as they do in different ground conditions. It establishes the mechanical basis for later chapters on penetration rate prediction, cutter consumption, machine selection, and performance optimisation, ensuring that subsequent empirical and analytical models are grounded in first-principles understanding of excavation physics.

Rock Cutting Mechanics

The process of rock cutting is fundamental to various excavation techniques, particularly in the operation of Tunnel Boring Machines (TBMs), roadheaders, and drilling equipment. Understanding the mechanics of rock cutting involves examining the physical processes through which a cutting tool applies force to the rock, induces fractures, and subsequently removes material in the form of chips or fragments. This understanding is crucial for enhancing the efficiency, performance, and safety of excavation projects.

Mechanics of Rock Cutting

Fracture-Dominated Process

Rock cutting fundamentally differs from metal machining in that it is a fracture-dominated process. Typically, intact rock exhibits brittleness, heterogeneity, and anisotropy, which leads to failure mechanisms based on cracking rather than yielding or plastic deformation [68]. Effective rock cutting relies predominantly on the controlled initiation and propagation of fractures, rather than on continuous crushing, thereby emphasizing the need for precise control over the energy imparted to the material [468].

Cutting Actions

The process of rock cutting involves three closely coupled actions:

1. **Force Application**: The cutter exerts force on the rock's surface, initiating stress concentration beneath the cutter tool [469].
2. **Stress Development**: This applied force results in highly concentrated stresses in the rock mass, leading to crack initiation along pre-existing weak planes [470].

3. **Fracture Propagation**: These stresses facilitate the propagation of cracks, ultimately forming detachable chips from the rock surface [468].

The efficiency of rock cutting can thus be assessed by how effectively this mechanical energy is utilized to promote tensile cracking instead of being lost to crushing, vibration, or heat [471].

Material Properties and Cutting Efficiency

Key material properties significantly influence the rock's response to cutting. For instance, uniaxial compressive strength determines the force needed to initiate crushing, while tensile strength governs the initiation of cracks and chip formation. Notably, tensile strength is usually much lower than compressive strength [472]. Additionally, brittleness is a critical factor, as brittle rocks tend to break more effectively during cutting, whereas ductile rocks absorb energy without achieving clean fractures [473].

Natural discontinuities such as joints and fractures also play an essential role in chip size and cutting efficiency [474]. Consequently, while intact rock strength is a defining factor for cuttability, the orientation and spacing of fractures can be even more consequential.

TBM Cutting Dynamics

In hard-rock TBMs, the mechanics of cutting are primarily performed by disc cutters that engage the rock surface in a rolling motion. The dynamics of this interaction include the establishment of a crushed zone directly beneath the cutter tip, which is crucial for chip formation when tensile cracks propagate from this zone [287]. Parameters such as cutter load, penetration per revolution, and cutter spacing directly influence the cutting process's effectiveness [475].

Failure Modes

Two predominant rock failure modes are identified in the context of cutting:

- **Chipping**: This is the desired mode characterized by tensile fracture, yielding larger chips and lower specific energy requirements.
- **Crushing**: An inefficient mode where compressive failure leads to excessive fines and increased cutter wear, requiring more energy [476].

Successful TBM operations strive to maintain cutting within the chipping regime to optimize performance and reduce wear.

Specific Energy and Operational Considerations

Specific energy, defined as the energy required to excavate a unit volume of rock, serves as an essential indicator of cutting efficiency [477]. Efficient performance correlates with low specific energy values, indicating effective cutting techniques and equipment conditions.

Conversely, increases in specific energy can signal issues such as worn cutters or suboptimal operational parameters [475].

The rock's response to cutting also varies, influenced by cutter impacts with uneven surfaces and resulting vibrations [478]. Excessive vibrations not only accelerate wear of cutter bearings but can also hinder overall excavation efficiency, necessitating adjustments to operational parameters to mitigate these effects [479].

Working Principle of TBM Disc Cutters in Tunnel Engineering

Tunnel Boring Machines excavate rock by converting thrust and rotational energy into controlled rock fracture at the tunnel face. The cutterhead, fitted with multiple disc cutters, is the component that directly interacts with the rock. Disc cutters do not grind the rock away; instead, they induce cracking and chipping through concentrated contact forces [480]. Understanding how these cutters work is fundamental to understanding TBM performance, wear behaviour, and advance rates.

Force and Rock-Breaking Mechanism of TBM Disc Cutters

Each disc cutter breaks rock through a combination of forces generated by the TBM's thrust system and cutterhead rotation [480].

The normal force is the dominant force in rock breakage. It is applied perpendicular to the tunnel face as the TBM thrust cylinders push the cutterhead forward. This force presses the cutter ring into the rock surface, creating a crushed zone directly beneath the cutter. The magnitude of this force largely determines penetration depth and the ability to initiate fractures, especially in strong, massive rock.

The tangential force is generated by cutterhead torque as the cutter rolls across the face. This force acts parallel to the cutting direction and enables the cutter to roll rather than slide, which significantly reduces energy loss and wear. The rolling motion allows tensile cracks to propagate efficiently between adjacent cutter tracks, which is essential for effective chipping.

A smaller lateral force develops due to cutter geometry and rotational effects. This force acts toward the centre of the cutterhead but is generally minor compared to normal and tangential forces. While usually negligible in cutting efficiency, lateral forces influence bearing loads and cutter stability.

As the cutterhead rotates, each disc cutter simultaneously rolls on its own axis and revolves around the cutterhead axis. This motion creates concentric grooves in the rock. Tensile cracks form between these grooves, and when cracks intersect, rock chips detach. The resulting muck consists of chips, fragments, and fines, with the proportions depending on rock strength, brittleness, cutter spacing, penetration, and thrust [480].

Strategic Arrangement of TBM Cutters on the Cutterhead

The performance of disc cutters depends not only on individual cutter design but also on how cutters are arranged across the cutterhead. Cutters are positioned to ensure balanced force distribution, which minimises uneven loads on the cutterhead structure and main bearing. Spiral or quasi-spiral layouts are commonly used to distribute cutting forces evenly and avoid torque imbalance or vibration [480].

Rock breaking is designed to occur sequentially. Cutters closer to the centre initiate fractures, while adjacent and trailing cutters exploit those fractures to complete chip formation. Slight phase offsets between cutter paths improve fracture interaction and reduce the amount of energy spent on pure crushing [480].

Different cutter types are used at different positions. Single disc cutters are common across most of the face for consistent wear. Double or triple cutters may be used near the centre where space is limited and cutter speed is lower. Gauge cutters at the periphery shape the tunnel diameter and experience the highest wear due to greater rolling speed and contact with intact rock, requiring reinforced designs.

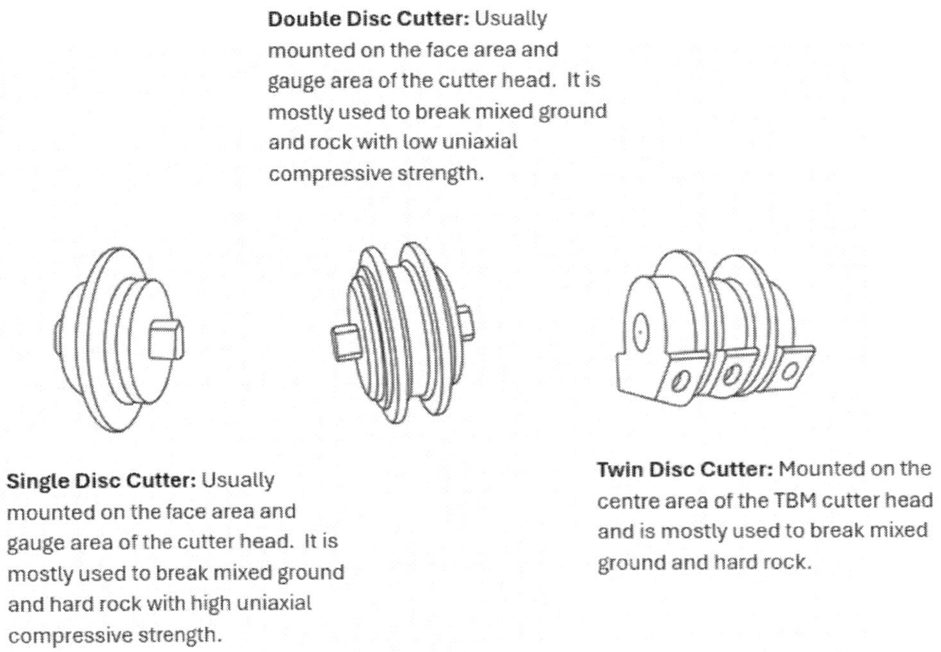

Figure 73: Disc Cutter types.

Cutter spacing is a critical design parameter. Typical spacing ranges from about 50 to 120 mm, with closer spacing used in strong, massive rock and wider spacing in weaker or fractured rock [480]. Spacing can be fixed, with thrust adjusted to suit ground conditions, or varied through cutterhead design changes for mixed geology. The goal is always to ensure crack interaction between adjacent cutters.

Managing Wear and Tear in TBM Disc Cutters

Disc cutters operate under extreme mechanical and abrasive conditions, making wear management a central aspect of TBM operations.

Wear monitoring is essential, especially for gauge cutters that experience high contact stress and speed. Regular inspections identify ring wear, bearing degradation, cracking, or deformation before failure occurs.

Debris and muck management plays an indirect but important role. Efficient removal of excavated material reduces re-crushing of fragments under the cutterhead, which accelerates wear and increases energy consumption.

Planned cutter replacement is critical. Worn cutters lead to uneven load distribution, increased vibration, higher torque demand, and accelerated wear of adjacent cutters and bearings. Modular cutter mounting systems allow cutters to be replaced during planned interventions.

Operational parameter control, adjusting thrust, torque, and penetration, can significantly extend cutter life. Excessive thrust in hard rock increases bearing stress, while insufficient penetration causes inefficient grinding and rapid wear [480].

Effective wear management can substantially reduce downtime and operating costs while maintaining stable advance rates.

Future Trends in TBM Cutter Technology

Disc cutter technology continues to evolve in response to demands for higher productivity, reduced downtime, and improved sustainability.

One major trend is sensor integration, with cutters increasingly equipped to monitor temperature, vibration, and load. This enables predictive maintenance and early detection of bearing or ring failure.

Advanced materials and coatings are being developed to improve wear resistance while reducing friction and energy consumption. These include improved alloy steels, surface treatments, and composite materials.

Automation and intelligent control are also expanding. Cutter data is increasingly integrated into TBM control systems, allowing thrust, rotation, and penetration to be adjusted automatically in response to cutter performance and ground conditions.

These developments aim to improve reliability, reduce manual interventions, and optimise cutting efficiency in increasingly complex geological environments [480].

Key Aspects of TBM Disc Cutters and Their Impact

TBM disc cutters function by converting thrust and rotation into controlled tensile rock fracture rather than grinding. Their effectiveness depends on cutter load, spacing, arrangement, and interaction with rock properties. Proper cutterhead design and operational control maximise chipping efficiency, minimise wear, and stabilise ground response. Advances in materials, sensing, and automation are further enhancing cutter performance and reliability, reinforcing the central role of disc cutters in modern mechanised tunnelling [480].

The mechanics of rock cutting encompass the application of forces that generate stresses capable of initiating fractures and producing chips in brittle rock. The successful execution of this process is rooted in an intricate balance between cutter load, spacing, penetration, and the material properties of the rock. Understanding these mechanics is vital for optimizing TBM performance, reducing wear on cutting tools, controlling vibrations, and achieving cost-effective excavation in hard-rock tunnelling projects.

Disc Cutter Indentation and Fracture Propagation

The process of rock excavation using Tunnel Boring Machines (TBMs) involves complex mechanical interactions that are primarily dictated by the physical mechanisms employed by disc cutters. Unlike grinding or abrasive techniques, disc cutters focus on inducing concentrated stresses that lead to brittle fracture, facilitating the efficient breaking of hard rock. Understanding these mechanisms is essential for enhancing TBM performance, including optimizing cutter spacing, thrust, penetration, energy efficiency, and wear rates.

The Mechanism of Rock Failure under Disc Cutters

When a disc cutter is engaged against the tunnel face, a thrust system applies a normal load that penetrates the rock surface, creating an indentation. This indentation depth plays a crucial role in the efficiency of the cutting process. According to Lin et al. [481], experiments indicate that this indentation is influenced by cutter load, cutter diameter, edge geometry, and rock strength. The stress field generated beneath the cutter edge is characterized by high compressive stresses, often surpassing the rock's uniaxial compressive strength, leading to the formation of a crushed zone where grains are pulverized [482] .

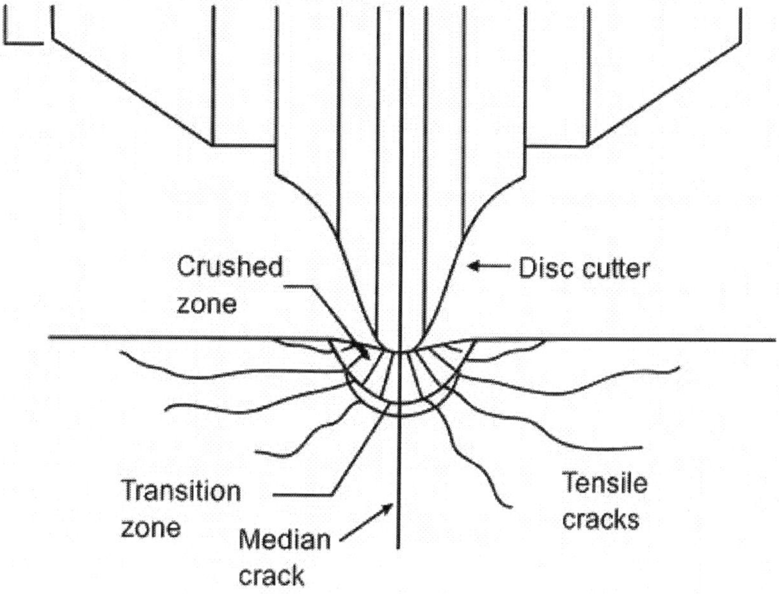

Figure 74: Development of a crushed zone and cracks under the disc cutter. Adapted from Cho et al. (2010) [483].

As the indentation progresses, radial and lateral tensile stresses develop around the perimeter of the crushed zone. It is these tensile stresses that facilitate crack initiation at the edges of the crushed zone and at natural weaknesses within the rock structure, such as microfractures and bedding planes [399]. This transition indicates the shift from a grinding mechanism dominated by crushing to an efficient fracture-driven excavation process.

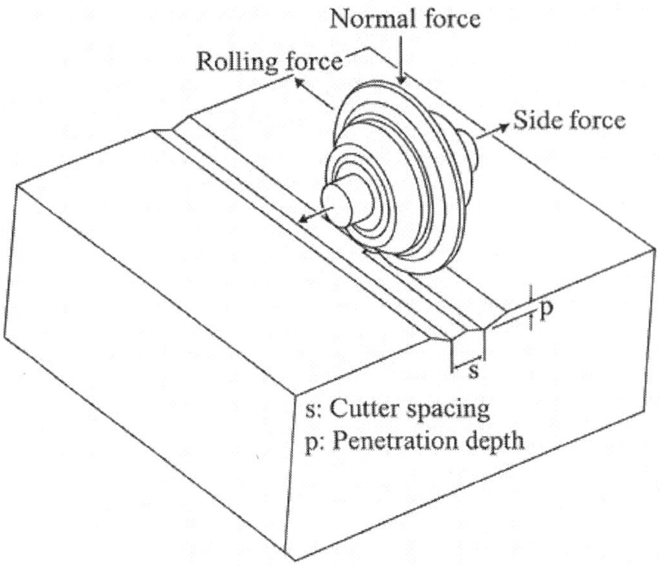

Figure 75: Idealized rock-cutting geometry and three-directional forces. Adapted from Cho et al. (2010) [483].

Indentation Depth and Its Effects on Cutting Efficiency

A critical factor in the optimization of cutter performance is the depth of indentation. Insufficient indentation depth results in stress levels too low to incite tensile cracking, leading to inefficient grinding behaviour rather than effective cutting [399]. Conversely, excessive indentation depth promotes crushing, thereby increasing energy consumption, vibrations, and cutter wear [484]. Optimizing the indentation thus requires finding a balance that minimizes crushing while promoting tensile fracture [484].

The crushed zone's dynamics, including its expansion and the development of tensile stresses, are essential considerations in maintaining efficient cutting operations [482]. Their research highlights how effective indentation can enhance chip formation while reducing energy requirements.

Dynamics of Fracture Propagation

Once cracks are initiated, they propagate outward due to the ongoing loading from the cutter and the release of elastic energy in the surrounding rock mass. The propagation path of these cracks is significantly influenced by the stress fields from neighbouring cutter indentations. If the spacing between cutters is too wide, cracks become isolated and each cutter must independently create a free surface, which leads to increased crushing and energy demands [469]. Optimal cutter spacing facilitates tensile cracks initiated by one cutter to be intercepted by those from adjacent cutters, enhancing chip detachment and formation [482, 485].

Rock properties also profoundly impact fracture propagation. For instance, in jointed or brittle rock, cracking is often more efficient compared to intact rock, where higher loads may be necessary to maintain crack growth [475]. The presence of natural fractures or discontinuities can guide the path of crack propagation, underscoring the significance of geotechnical assessments prior to TBM operations.

Dynamic Effects and Their Implications

The interaction between the disc cutter and the rock mass is inherently dynamic, leading to effects such as sudden load drops and stick-slip behaviour. These dynamics can result in vibrations that negatively impact cutting efficiency and the longevity of cutter components [472, 486]. Managing these variables requires careful adjustments to penetration depth and cutter rotation speed to avoid exceeding operational thresholds that compromise stability and predictability in TBM performance.

Disc cutter indentation and fracture propagation are integral to the efficient excavation of hard rock in TBM operations. The transition from crushing to tensile failure, influenced by variables such as indentation depth, spacing, and rock properties, is critical to maximizing the performance of TBMs. Effective management of these factors is essential to optimize cutting efficiency and to minimize energy consumption, cutter wear, and operational costs, thereby ensuring stable and predictable performance during tunnelling operations.

Example: Disc Cutter Indentation and Fracture Propagation in a Hard-Rock TBM Drive

Consider a hard-rock Tunnel Boring Machine excavating a 6.5 m diameter transport tunnel through moderately jointed granite with a uniaxial compressive strength (UCS) of approximately 180 MPa. The TBM is equipped with 19-inch disc cutters arranged in a spiral pattern on the cutterhead, with an average cutter spacing of 80 mm. During initial boring, the machine applies a thrust force that results in an average cutter load of 250 kN and a penetration rate of approximately 6 mm per revolution.

As each disc cutter is pressed into the tunnel face, the applied normal load creates a local indentation in the rock surface. Beneath the cutter edge, compressive stresses exceed the UCS of the granite, forming a crushed zone where mineral grains are compacted and microcracks develop. Surrounding this crushed zone, tensile stresses arise due to elastic rebound of the rock mass. These tensile stresses initiate subsurface cracks at the edges of the crushed zone, particularly along existing grain boundaries and microfractures within the granite.

As cutterhead rotation continues, adjacent disc cutters create overlapping stress fields. Tensile cracks generated beneath one cutter propagate laterally toward the indentation created by neighbouring cutters. Because the cutter spacing has been selected appropriately for the rock strength and brittleness, these cracks intersect, allowing a wedge-shaped rock chip to detach cleanly from the tunnel face. This chip formation indicates that the excavation process is operating in an efficient fracture-dominated regime rather than a crushing-dominated regime.

During this phase of the drive, monitoring data shows relatively low specific energy consumption and stable torque levels, confirming efficient cutting. Chip size observations at the conveyor reveal coarse, angular fragments rather than excessive fines, further indicating effective tensile fracture rather than grinding. Cutter wear rates remain within expected limits, and vibration levels are moderate and well controlled.

As the TBM encounters a short section of more massive, less fractured granite, penetration decreases despite maintaining the same cutter load. Operators respond by slightly increasing thrust to raise indentation depth while maintaining cutter spacing. This adjustment restores tensile crack initiation and propagation, re-establishing chip formation and preventing a shift into inefficient crushing. Conversely, when the TBM later enters a more heavily jointed zone, penetration increases naturally, allowing thrust to be reduced to limit vibration and unnecessary cutter wear.

This example demonstrates how disc cutter indentation initiates rock breakage through localized crushing, while fracture propagation—driven by tensile stresses and interaction between adjacent cutters—controls efficient chip formation. It highlights the critical importance of matching cutter load, spacing, and penetration to rock properties to maximise cutting efficiency, minimise energy consumption, and maintain stable TBM performance.

Soil Cutting and Plastic Flow

Tunnel Boring Machines (TBMs) are crucial in modern excavation technology, particularly for soft-ground tunnelling. This paper aims to synthesize the mechanisms of soil cutting and plastic flow involved during the excavation processes utilizing TBMs, particularly in the context of earth pressure balance (EPB) systems.

In the context of TBM excavation, soil cutting involves a different mechanism compared to hard rock excavation methods. While hard rock relies on brittle fracture, soft-ground TBMs predominantly employ shear failure and plastic deformation to remove material. This process is characterized by the ability of soils, particularly clays and silts, to undergo significant deformation without tensile failure [427, 467]. The dynamic between the soil mass and TBM cutting tools results in shear failure along well-defined slip planes, reflecting its cohesive or frictional properties, which subsequently informs the selection of cutting tools [431, 487].

The engagement of a TBM's cutterhead—through scrapers or picks—induces shear stresses. When the applied force surpasses the soil's shear strength, a shear plane manifests, resulting in material movement relative to the surrounding mass [373]. In cohesive soils like clays, this shear is more pronounced as materials tend to deform plastically as a unified mass, while in granular materials such as sands, shear occurs mainly through grain reorganization [467]. Consequently, managing the interface dynamics between the cutting tools and soil is critical to optimizing excavation performance and preventing issues like clogging.

> When a TBM cutterhead rotates, cutting tools such as scrapers or picks press into the soil and move relative to it. Rather than crushing the ground, these tools drag and push the soil sideways and forward. This motion generates shear stresses at the interface between the cutting tool and the soil mass. Shear stress is the force that attempts to make adjacent parts of the soil slide past one another, and it progressively builds up in front of and beneath the advancing tool as excavation continues.
>
> Every soil has a finite shear strength, governed by its cohesion, internal friction, and in-situ stress conditions. When the shear stress induced by the cutterhead exceeds this shear strength, the soil can no longer sustain its original structure. A shear plane then forms, marking a zone along which the soil yields and begins to move relative to the surrounding, less-disturbed ground. Once this shear plane develops, excavation proceeds through sliding and deformation rather than through fracture, which is the fundamental mechanism of soil cutting in TBM operations.
>
> In cohesive soils such as clays, soil particles are bonded together by electrochemical forces. When shear strength is exceeded, the material does not disintegrate into individual grains. Instead, it deforms plastically as a continuous mass. The clay behaves like a stiff paste, flowing, stretching, and reshaping while remaining coherent. This behaviour is advantageous for Earth Pressure Balance (EPB) TBMs because it allows the excavated soil to form a stable pressure plug. However, it also increases the risk of adhesion and clogging if the soil sticks to cutting tools or the walls of the excavation chamber.
>
> In granular soils such as sands, behaviour is markedly different. These materials have little to no cohesion, so shear failure occurs mainly through grain rearrangement and sliding rather than plastic flow. Individual particles move relative to one another instead of deforming as a single mass. As a result, sands tend to behave like a loose assembly of grains, making them prone to running ground, loss of face support, and uncontrolled inflow unless face pressure is carefully controlled. Plastic flow is limited, and the ability to transmit pressure is poor unless soil conditioning is applied.
>
> Because different soils fail and move in different ways, managing the interaction between cutting tools and the soil is a critical aspect of TBM operation. Poor control of this interface can lead to clogging—particularly in sticky clays—poor spoil flow into the chamber, instability of face pressure, increased torque and tool wear, and excessive ground movement or settlement at the surface. Effective management requires

appropriate tool selection, the use of soil conditioning agents such as foam, polymers, or water, and careful control of thrust, cutterhead rotation, and chamber pressure.

Overall, the passage highlights that soft-ground TBM excavation relies on inducing controlled shear failure rather than breaking material apart. Whether the soil responds through plastic deformation or grain rearrangement depends on whether it is cohesive or granular. Understanding and controlling this behaviour is essential for efficient, stable, and predictable tunnelling. In practical terms, successful soft-ground TBM performance depends far more on managing shear and plastic flow at the cutterhead–soil interface than on simply applying greater cutting force.

Upon the initiation of shear failure, soils transition into a plastic flow regime where they continue to deform without a notable increase in stress. The material flows around the cutting tools into the excavation chamber, a critical process for the successful operation of EPB TBMs where soil is utilized for face support [431]. Conditioning agents such as polymers or foams play a vital role in this transition, allowing granular soils to achieve sufficient plasticity to transmit pressure effectively while facilitating flow during excavation, thus avoiding excessive deformations [488].

Once the shear stress induced by the cutterhead exceeds the soil's shear strength, the soil reaches a yield condition. At this point, it does not suddenly collapse or fracture. Instead, it enters a plastic flow regime, meaning the soil can continue to deform and move while the applied stress remains relatively constant. In engineering terms, additional deformation occurs without a proportional increase in stress. This is a defining characteristic of plastic behaviour and distinguishes soil excavation from brittle rock cutting.

In this plastic flow state, the soil no longer resists movement as a rigid mass. Instead, it flows around the cutting tools, conforming to their shape and being guided through openings in the cutterhead into the excavation chamber. This controlled movement is not incidental—it is a required condition for EPB TBMs. In EPB machines, the excavated soil inside the chamber is deliberately used as the face-support medium, transmitting pressure back to the tunnel face to balance earth and groundwater pressures. For this to work, the soil must be plastic enough to deform and flow, yet cohesive enough to maintain pressure continuity.

Conditioning agents such as foams, polymers, or water are used to achieve and stabilise this plastic behaviour. Natural soils—especially granular soils like sands—do not naturally flow plastically. Instead, they tend to behave as loose particles with poor pressure transmission. Conditioning modifies the soil's rheological properties by reducing internal friction, increasing apparent cohesion, and binding particles together. This allows granular soils to behave more like a cohesive paste, capable of both flowing into the chamber and transmitting face pressure effectively.

At the same time, conditioning prevents excessive deformation and instability. Without sufficient plasticity, soil may flow uncontrollably, leading to loss of face pressure, running ground, or surface settlement. With excessive stiffness, soil may clog tools and the chamber. Proper conditioning therefore ensures a balanced plastic flow: the soil deforms and moves where required, but remains stable enough to support the face.

In essence, the passage explains that successful EPB TBM operation depends on deliberately driving the soil into a controlled plastic flow state after shear failure. This flow allows the soil to serve a dual role—both as excavated material and as a temporary structural support—while conditioning agents ensure that this behaviour is stable, predictable, and compatible with safe tunnelling.

Designing cutting tools for soft-ground TBMs necessitates an understanding of their interaction with the soil. Tools like scrapers are engineered to mobilize shear effectively while promoting smooth flows of material into the chamber [487]. Additionally, tools must not only penetrate but also initiate effective flow by managing the cohesive properties of the soil. Conditioning agents are essential to prevent the formation of lumps in cohesive materials, ensuring that the soil can flow smoothly into the operational areas of the TBM [301]. The interaction between tool geometry and soil rheology directly influences excavation efficiency and stability [427].

Natural soils do not consistently exhibit optimal flow conditions for TBM operations. As such, soil conditioning is employed to improve flow properties. This can include the injection of various conditioning agents to lower internal friction, augment cohesion, reduce adherence on cutting tools, and promote homogeneous flow [489]. Proper conditioning enhances the mechanical properties of soil, mitigating risks associated with water spewing during excavation operations [306]. In achieving controlled plastic behaviour of the excavated material, successful tunnelling is less contingent on brute cutting force but rather on the fluidity and stability of the material being excavated [301].

The excavation process notably impacts the surrounding geological structure, as stresses are redistributed around the tunnel face. When face pressure is inadequately balanced, it can lead to unwanted deformations such as settlement or collapse, highlighting the intricate relationship between excavation mechanics and ground response [490]. Unlike hard rock tunnelling, where localized failure occurs, the flow behaviour of soils can manifest as far-field deformations if not correctly managed, underscoring the importance of precision in pressure control during excavation [488].

All these factors establish that soil cutting and plastic flow mechanics are imperative for maintaining face stability and controlling settlement in TBM excavation operations. Poor management of these aspects can result in significant operational failures, often attributed not to insufficient excavation power but rather to ineffective control of plastic flow and accompanying conditions [491]. Therefore, understanding these mechanisms allows for enhanced operational safety and efficiency in urban underground construction.

Excavating with TBMs in soft ground relies fundamentally on the principles of shear failure and plastic flow rather than brittle fracturing. The implementation and optimization of cutting tools, soil conditioning, and careful management of excavation-induced pressures are key to successful tunnelling operations. Understanding these concepts equips engineers with the tools necessary for safe, efficient, and effective tunnelling in a diverse array of soil conditions.

Cutterhead Torque and Thrust Relationships

Tunnel Boring Machines (TBMs) utilize a combination of cutterhead torque and thrust forces as primary mechanical inputs to facilitate excavation and advance through various geological conditions. These two parameters are critical in determining penetration rates, cutting efficiency, and operational stability. A detailed understanding of their interaction is essential for effective TBM operation in both soft and hard ground conditions.

Thrust refers to the axial force exerted by the hydraulic cylinders of the TBM, which serves to push the cutterhead into the material being excavated. The relationship between thrust and penetration depth is critical; sufficient thrust is required to enable the cutting tools to engage effectively with the ground, initiating fracture in rock or shear failure in softer soils. Without adequate thrust, the ability of the cutters to penetrate is severely compromised, leading to inefficient excavation efforts [383, 492].

Research has shown that thrust not only dictates the level of force applied but also establishes the normal load on the cutting tools. As thrust increases, it directly correlates with penetration depth, but can also lead to higher cutter loads, necessitating adequate torque to maintain operational continuity [469, 492]. High thrust without corresponding torque may lead to stalling of the cutterhead under excessive resistance, resulting in operational inefficiencies [289, 469].

Torque is the rotational force that drives the cutterhead during operations, enabling it to rotate against the resisting forces encountered at the excavation front. It is essential for overcoming friction and cutting resistance. The torque demand can sharply increase based on the material properties and configuration of the cutter tooling. If the torque available is insufficient relative to the resistance met, it may lead to cutterhead rotation stalling or significant vibration, which can cause damage over time [287, 436, 493].

Studies indicate that the torque utilized in TBM operations is a reflection of the geological conditions encountered, higher resistance materials necessitate greater torque for effective cutting. Thus, monitoring torque can serve as a diagnostic tool for operational assessment; for example, a sudden increase in torque may indicate the encounter of harder rock or worn cutting tools [289, 493]. Moreover, the effective functioning of the TBM relies on maintaining a balance between thrust and torque to achieve optimal penetration without leading to excessive wear or operational hazards [383, 492].

The interplay between thrust and torque in TBM operations can be conceptualized as a system where thrust creates penetration and torque sustains cutting under that load. Efficient operation occurs when these parameters are actively balanced: increased thrust should ideally correspond with an adjustable torque output to mitigate resistance. The failure to achieve this balance can result in mechanical failure or inefficiencies characterized by grinding rather than cutting during excavation [401, 494].

In soft ground conditions, thrust is crucial for mobilizing shear forces and inducing soil deformation, while torque helps manage the adherence of the soil to the cutting tools and the overall cutting process [495, 496]. Conversely, in hard rock situations, thrust affects crack initiation, while torque is an indicator of rock strength and cutter performance [497, 498]. This nuanced interaction necessitates that TBM operators continuously monitor both parameters to adjust operational strategies proactively and efficiently [492, 494].

Modern TBMs are often equipped with closed-loop control systems designed to maintain these torque and thrust parameters within optimal ranges. These systems adjust cutterhead speed and advance rates based on real-time data to prevent overloading and ensure efficient excavation [287, 494]. The continuous assessment of torque and thrust trends allows operators to detect issues, such as increased wear or anomalous geological conditions, enabling corrective actions before significant operational disruptions occur [289, 493].

The efficacy of a Tunnel Boring Machine is significantly governed by the interrelationship between cutterhead torque and thrust. Thrust governs the penetration of the cutting tools into the geological material, while torque enables effective cutting operations under load. A careful balance of these two mechanical forces is essential to prevent operational inefficiencies and mechanical failures. Understanding their dynamics is central to effective TBM design, operational protocols, and ultimately, the success of tunnelling projects across diverse geological environments.

Wear Mechanisms and Cutter Life

The degradation of cutting tools in Tunnel Boring Machines (TBMs) is a complex issue that significantly impacts the efficiency, safety, and cost-effectiveness of tunnelling projects. Cutter wear, which refers to the gradual loss of material from the cutting tools, is not merely a maintenance concern but a critical performance and risk driver that must be actively managed throughout the excavation process [499, 500]. Understanding the mechanisms of wear and lifecycle of cutters is essential for optimizing TBM operations.

Cutter Wear Mechanisms

Several primary wear mechanisms contribute to the degradation of TBM cutters:

1. **Abrasive Wear**: This is prevalent when cutters engage hard mineral grains such as quartz and feldspar, which scratch and grind the cutter surface [501, 502]. In hard rock environments, particularly those with a high Cerchar Abrasivity Index (CAI), abrasive wear is recognized as a life-limiting mechanism for disc cutters [501, 502].
2. **Fatigue Wear**: Repeated cyclic loading induces microcracking in cutter materials, which can lead to catastrophic failure [502, 503]. This phenomenon is especially noticeable in hard, massive rocks where cutter fatigue manifests from continuous impacts on uneven surfaces [500, 503].
3. **Impact and Chipping Damage**: This occurs when cutters strike hard inclusions or boulders, resulting in localized chipping and cracking of cutter rings [501, 504]. Such wear is typical in mixed ground conditions where geological variability is prevalent.
4. **Adhesive Wear**: In cohesive soils like clays, soil adhesion to cutter surfaces can increase torque, exacerbating fatigue and damage under high-pressure conditions [505]. Although the material loss may be minor, performance degradation can occur swiftly.
5. **Corrosive and Chemical Wear**: In geologically aggressive environments, such as acidic groundwater, chemical wear can occur, leading to corrosion of cutter materials, thus shortening their functional lifespan [68].

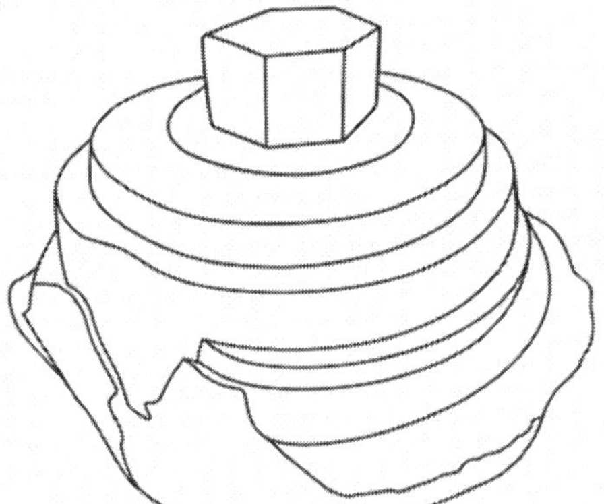

Figure 76: Worn TBM disc cutter.

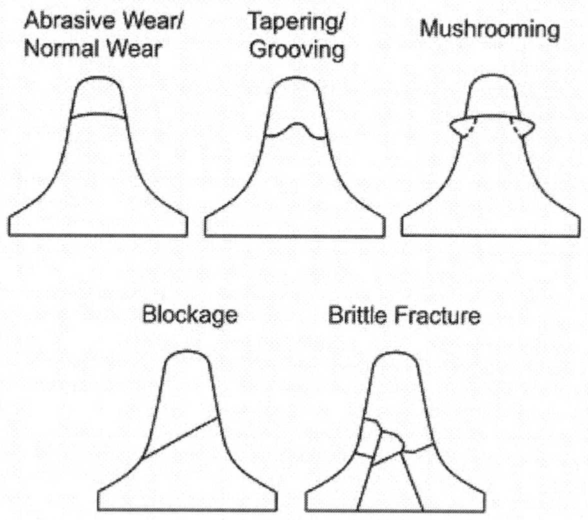

Figure 77: Basic visible wear shapes/types for TBM disk cutters. Adapted from Hamzaban et al. (2022) [506].

The Cerchar Abrasivity Index (CAI) is a standardised laboratory measure used to quantify how abrasive a rock is to steel tools. In tunnelling, mining, and drilling, it is one of the most important indices for predicting cutter wear, tool life, and maintenance demand on Tunnel Boring Machines (TBMs) and other excavation equipment. CAI plays a central role in assessing operational risk, estimating maintenance costs, and planning interventions during mechanised excavation.

CAI specifically measures the ability of a rock to wear steel through abrasion. It does not directly measure rock strength, hardness, or cuttability. Instead, it focuses on wear potential, which is critical for understanding how quickly cutting tools will degrade in service. This makes CAI particularly important for predicting disc cutter ring wear, drag tool and pick wear, drill bit consumption, maintenance frequency, downtime, and overall cost and schedule risk in tunnelling projects. Importantly, a rock can be relatively easy to excavate yet highly abrasive, or conversely very strong but only mildly abrasive, highlighting why CAI must be considered separately from strength-based indices.

The Cerchar Abrasivity Index is determined using a simple and repeatable scratch test. A hardened steel pin, typically with a 90-degree conical tip, is pressed onto a freshly broken rock surface under a constant normal force, usually 70 N. The pin is then drawn across the rock surface over a fixed distance, commonly 10 mm. After the test, the flat wear scar that forms on the tip of the steel pin is measured under a microscope. The diameter of this wear flat, expressed in millimetres, is reported as the CAI value.

In practical terms, the interpretation of the test is straightforward: greater wear on the steel pin corresponds to a higher CAI value and therefore a more abrasive rock. This direct relationship makes CAI a highly effective and widely used indicator for assessing tool wear risk and planning cutter management strategies in mechanised tunnelling.

Table 4: Typical CAI Classification.

CAI Range	Abrasivity Class	Practical Meaning
< 0.5	Very low	Minimal cutter wear
0.5 – 1.0	Low	Long cutter life
1.0 – 2.0	Medium	Moderate wear
2.0 – 3.0	High	Rapid cutter wear
> 3.0	Very high	Severe wear, high cost risk

Many hard-rock TBM projects become cutter-dominated once CAI exceeds about 2.0–2.5, regardless of rock strength.

The Cerchar Abrasivity Index (CAI) is critically important in TBM engineering because it directly influences cutter wear, machine design, operational decisions, and project risk. Unlike strength parameters, which govern whether rock can be cut, CAI governs how long cutting tools survive while doing so. For this reason, CAI is often a more decisive factor than uniaxial compressive strength when assessing the practicality and cost of mechanised tunnelling.

One of the most important roles of CAI is in cutter life prediction. CAI is among the strongest predictors of disc cutter ring consumption, gauge cutter replacement frequency, and cost per bored metre. In high-CAI ground, cutters may wear rapidly even when penetration rates are good and cutting appears efficient. This means that productivity measured purely by advance rate can be misleading, as frequent cutter changes can dominate downtime and cost.

CAI also has a major influence on TBM design and specification. Designers use CAI data to select appropriate cutter ring materials, including alloy composition and heat treatment, to resist abrasive wear. It affects cutter spacing, allowable cutter loads, bearing size, sealing requirements, and the need for additional wear protection on the cutterhead and face plates. In many cases, TBMs are sized and detailed for abrasivity rather than rock strength, because abrasive wear is often the life-limiting factor for cutters and bearings.

From an operational strategy perspective, high-CAI ground requires a more conservative approach. Penetration rates are often deliberately reduced, thrust limits are constrained, and cutter inspections are scheduled more frequently to prevent secondary damage. Advance rates are managed to protect tooling rather than maximised for short-term progress. Experience shows that aggressive cutting in abrasive ground usually increases overall cost by accelerating wear and downtime, rather than improving net productivity.

CAI is also central to cost and risk management. High abrasivity leads to increased spare cutter inventory requirements, longer and more frequent interventions, higher downtime risk, and greater schedule uncertainty. Many TBM cost overruns and programme delays can be traced back to underestimating CAI rather than misjudging rock strength. Accurate abrasivity assessment is therefore essential for realistic budgeting and risk allocation.

The magnitude of CAI values is controlled primarily by mineralogical and textural factors. Quartz content is the dominant influence, as quartz is highly abrasive to steel. Grain size also plays a significant role, with coarse-grained rocks often being more abrasive than fine-grained equivalents. Cementation type, degree of weathering, and contrasts in mineral hardness further affect abrasivity. As a result, rocks with similar strength can exhibit very different CAI values.

Typical high-CAI rocks include quartz-rich sandstones, quartzites, granites, gneisses, and some volcanic rocks. In contrast, low-CAI rocks generally include limestone, marble, chalk, and shale, unless the shale contains significant sandy or siliceous components. Understanding where a project's geology sits on this spectrum is essential for predicting cutter wear and managing TBM performance effectively.

Table 5: CAI vs Other Rock Indices.

Property	Measures	What it affects
UCS	Strength	Thrust and penetration
BTS	Tensile strength	Crack initiation
CAI	Abrasivity	Cutter wear
Brittleness	Fracture behaviour	Cutting efficiency

A low-UCS, high-CAI rock can be easy to excavate but extremely expensive to tunnel through.

Impact of Wear on Cutter Life

Cutter life, defined by the distance a cutter can excavate effectively before requiring replacement, is influenced by various factors:

- **Ground Conditions**: The abrasivity and composition of the rock or soil are critical determinants of cutter life. For instance, a weaker but highly abrasive rock may wear down cutters faster than a denser rock with lower abrasivity [393, 502].
- **Operational Parameters**: Critical operational factors include thrust per cutter, penetration per revolution, and overall cutterhead speed. Deviations from optimal settings can significantly accelerate cutter wear rates [499, 507].
- **Dynamic Effects**: Vibration levels induced during tunnelling can further impact cutter life by enhancing fatigue and increasing the likelihood of structural failures in cutterheads and bearings [499, 508].

Monitoring wear effectively through predictive analytics and data-driven approaches has gained traction, enabling operators to pre-emptively manage cutter life and performance [503, 509].

Management of Cutter Wear

Effective management strategies commence at the design phase, ensuring optimal cutter selection based on anticipated ground properties, coupled with a robust maintenance schedule [503, 510]. Real-time monitoring of wear against operational metrics, such as torque and distance bored, allows for timely interventions that minimize downtime and prevent catastrophic cutter failures [511, 512].

State-of-the-art methods, such as integrated sensor technologies for wear detection and predictive modelling, are emerging as leading methodologies for maintaining cutter efficiency [509, 511]. Additionally, managing cutter replacements proactively, especially for gauge cutters, has proven crucial since these components typically endure the highest wear rates [500, 512].

Cutter wear in TBMs is a complex phenomenon dictated by multiple mechanical and environmental factors. Its management is integral not only to ensuring effective tunnelling operations but also to sustaining project timelines and budgets. As such, ongoing research into wear mechanisms, alongside advancements in monitoring technologies, will continue to play a pivotal role in optimizing the performance of TBMs.

Penetration Rate Theory

Penetration rate theory plays a pivotal role in understanding how Tunnel Boring Machines (TBMs) advance through different substrates at varying speeds. This theory elucidates the interactions between machine inputs—thrust, torque, and cutterhead rotation—with excavation efficiency, cutter wear, vibration, and ground response. An advanced understanding of these interactions informs strategies to optimize both machine performance and excavation stability.

At its core, penetration rate is defined as the distance a TBM advances into the soil per unit of time, while penetration typically refers to the progress made per cutterhead revolution, usually quantified in millimetres per revolution. The theory emphasizes the significance of effective ground breakage as a function of the forces applied by the TBM. The effective application of these forces is crucial, as it influences the mode of failure in the ground, which ranges from tensile cracking in hard rock to shear failure in softer soils [513, 514].

The efficacy of penetration rate is fundamentally linked to the input parameters of the TBM. Increased thrust translates into deeper cuts, while the cutterhead's rotation speed dictates the frequency with which new material is engaged. The mechanical energy from the TBM must be optimally converted into ground breakage; otherwise, excess thrust can result in inefficient grinding, increased wear, and vibrations that jeopardize operational stability [433, 513]. Furthermore, ground properties such as strength, brittleness, and abrasivity significantly influence both the penetration efficiency and wear rates of the cutting tools [459, 515].

A critical parameter within penetration rate theory is Penetration per Revolution (PPR), which measures cutting efficiency. A higher PPR indicates a more effective engagement with the material, while a lower PPR suggests inefficient cutting, characterized by increased energy expenditure without corresponding advances [400, 513]. The theory posits that thrust serves as the primary driver of penetration, particularly in hard rock where it initiates crack propagation and chip formation [513, 516].

Rising thrust typically increases PPR to a point; however, this effectiveness plateaus, and excessive thrust can cause undesirable outcomes, such as increased wear rates and instability during operation [433, 469]. Torque, while traditionally viewed as a primary control variable, functions primarily as a response to the cutting resistance; rapid or erratic torque increases indicate deviations from optimal cutting conditions [517].

Specific energy is another essential component of penetration rate theory. It quantifies the energy required to excavate a unit volume of material, where efficient excavation corresponds to low specific energy [514]. The theory asserts that as penetration increases, specific energy decreases up to an optimal point. Beyond this point, issues such as crushing and excessive tool wear lead to a significant rise in energy costs [433, 516].

Ground properties greatly affect specific energy. For instance, rock strength, brittleness, and the degree of fracturing dictate the ease of penetration and thus energy consumption during excavation [459, 513]. The interplay between the dynamic responses of TBMs and the physical characteristics of the ground being excavated is vital in establishing optimal operating conditions [518].

The practical implementation of penetration rate theory in modern TBMs involves automated control systems that continuously adjust operational parameters based on real-time measurements of thrust and torque while monitoring the geological conditions encountered [399, 515]. Such systems aim to maintain stable efficiency during excavation, avoiding peaks in thrust or excessive rotation speeds that could lead to wear or operational failures [513, 519].

The following provides a worked, practical example that shows how penetration rate theory operates in a real TBM context.

Example: Applying Penetration Rate Theory in a Hard-Rock TBM Drive

Project Context

A hard-rock TBM is excavating a 7.5 m diameter water conveyance tunnel through moderately strong granite with the following characteristics:

- **Uniaxial Compressive Strength (UCS):** ~110 MPa
- **Rock mass condition:** Moderately fractured
- **Cerchar Abrasivity Index (CAI):** High (≈3.5)
- **TBM type:** Open hard-rock TBM with 17-inch disc cutters

The project team aims to optimise advance while controlling cutter wear and vibration.

Initial Operating Conditions

The TBM is initially operated conservatively:

- **Thrust:** 18 MN
- **Cutterhead speed:** 7.0 rpm
- **Measured penetration per revolution (PPR):** 2.0 mm/rev

Penetration rate (PR):

$$PR = PPR \times N = 2.0 \times 7.0 = 14.0 \text{ mm/min}$$

Operators observe:

- Low vibration
- Stable torque
- Low cutter wear
- But slower-than-expected advance

Applying Penetration Rate Theory

Based on penetration rate theory, engineers recognise that:

- Thrust is below the level required to consistently initiate tensile cracking.
- Cutting is partially dominated by grinding, not efficient chipping.
- Specific energy is higher than optimal.

Optimised Operating Adjustment

Thrust is increased incrementally while monitoring torque response:

- **New thrust:** 24 MN
- **Cutterhead speed:** Maintained at 7.0 rpm

Results after adjustment:

- **Penetration per revolution:** increases to 4.5 mm/rev
- **Torque:** increases smoothly but remains stable
- **Vibration:** remains within acceptable limits

New penetration rate:

$$PR = 4.5 \times 7.0 = 31.5 \text{ mm/min}$$

This nearly doubles the advance rate without introducing instability.

Why This Matches Penetration Rate Theory

This outcome illustrates several core principles from the theory:

1. **Thrust-driven penetration**
   Increased thrust pushed disc cutters deeper into the rock, enabling tensile crack initiation and chip formation.
2. **Torque as a response variable**
   Torque increased as expected but did not spike—indicating cutting remained in the efficient fracture regime.
3. **Improved specific energy**
   More rock volume was excavated per unit of energy, reducing grinding losses.
4. **Stable dynamic behaviour**
   Absence of torque spikes or vibration confirmed operation within the optimal penetration window.

What Happens If Thrust Is Increased Too Far

The team tests a higher thrust level:

- **Thrust:** 30 MN

Observed response:

- PPR increases marginally to 5.0 mm/rev
- Torque spikes intermittently
- Cutter temperature rises
- Vibration increases
- Cutter wear accelerates

Despite slightly higher penetration, specific energy increases, and cutter life drops sharply.

This confirms the penetration rate theory prediction:

Beyond the optimal point, additional thrust causes crushing, wear, and instability rather than useful penetration.

Operational Decision

Based on penetration rate theory, the TBM is operated at:

- **Thrust:** 23–25 MN
- **PPR:** ~4.3–4.6 mm/rev
- **Cutterhead speed:** 6.5–7.0 rpm

This balance delivers:

- High advance rate
- Controlled cutter wear
- Stable torque and vibration
- Predictable performance

Effective application of this theory ultimately influences the selection, design, and operational strategy for TBMs in various tunnelling projects, reinforcing the significance of balancing power input with actual performance metrics in the context of the geological environment [520]. Notably, failure to adequately adhere to the principles of penetration rate theory can lead to suboptimal performance characterized by low advance rates, high wear on tools, and increased operational risk [390, 496].

Penetration rate theory provides a comprehensive framework for understanding the complex interactions between TBM operational variables and geological factors. By emphasizing the importance of thrust, rotation speed, ground characteristics, and energy efficiency, the theory facilitates the optimization of tunnelling operations. Understanding its dynamics not only enhances the performance of TBMs but also ensures the stability and efficiency of excavation efforts.

Influence of Abrasivity and Strength

Tunnel Boring Machines (TBMs) operate in complex geological environments where the interaction between machine components and rock properties critically influences operational efficiency. Among the myriad of geological factors, abrasivity and strength stand out as fundamental properties that affect TBM performance in unique ways. Understanding how these properties influence excavation rates, cutter life, torque and thrust demands, and overall project risk is essential for effective tunnelling operations.

Rock Strength and Its Effects on TBM Performance

Rock strength, typically characterized by metrics such as uniaxial compressive strength (UCS), dictates the resistance encountered during the initial rock fragmentation phases when employing TBMs. It impacts how rock fails under applied loads, fundamentally influencing the force required for cutters to instigate fractures in the rock. Particularly, as rock strength increases, the force needed for the cutter to apply normal stress to induce a fracture also escalates, leading to significant challenges in TBM operations. High-strength formations often curtail penetration rates due to a predominance of fracture initiation failures over mere equipment performance [521, 522]. When rock strength is substantial, the needed thrust per cutter escalates, causing a rise in torque to maintain cutterhead rotation while cutting may shift towards less efficient crushing mechanisms if thrust exceeds optimal thresholds [499, 523, 524].

Conversely, in low-strength rock, easier penetration can be achieved, permitting higher advance rates and lower thrust demands. However, the benefits are contingent upon favourable abrasivity and rock mass structure conditions, as general mechanical characteristics of the surrounding rock also play a critical role in operational effectiveness [522, 525].

Uniaxial Compressive Strength (UCS) is a fundamental mechanical property used to describe the strength of intact rock. It represents the maximum compressive stress that a rock specimen can withstand before failure when loaded in a single, unconfined direction. In tunnelling and Tunnel Boring Machine (TBM) engineering, UCS is one of the most widely used parameters for assessing a rock's resistance to cutting, crushing, and penetration.

Physically, UCS reflects the inherent resistance of intact rock material to compressive loading. As a rock sample is compressed axially, existing microcracks within the material initially close, followed by the initiation and growth of new cracks. With increasing stress, these cracks propagate and coalesce into larger fractures, ultimately leading to failure. The stress level at which this failure occurs is recorded as the UCS. This value therefore captures the combined influence of mineral composition, grain bonding, cementation, and microstructural characteristics of the rock.

It is important to emphasise that UCS describes the strength of intact rock only, not the behaviour of the rock mass as a whole. In practice, a highly fractured or jointed rock mass may behave as relatively weak during excavation, even if laboratory tests show a high UCS for intact samples. This distinction is critical in tunnelling, where machines interact with the rock mass rather than isolated specimens.

UCS is determined through a standard laboratory compression test. A cylindrical rock core, typically prepared with a length-to-diameter ratio of about 2:1, is placed between two rigid platens in a compression testing machine. Axial load is applied at a controlled and constant rate until the specimen fails. The UCS is calculated by dividing the peak load at failure by the original cross-sectional area of the specimen, and the result is expressed in megapascals (MPa).

Based on UCS values, rocks are commonly classified into strength categories. Very weak rocks have UCS values below about 5 MPa, weak rocks range from roughly 5 to 25 MPa, medium-strength rocks from 25 to 50 MPa, strong rocks from 50 to 100 MPa, very strong rocks from 100 to 250 MPa, and extremely strong rocks exceed 250 MPa. Materials such as chalk and some shales typically fall into the lower strength categories, whereas granites, basalts, and quartzites commonly exhibit high UCS values.

In TBM excavation, UCS serves as a primary indicator of penetration resistance. As UCS increases, penetration per revolution generally decreases, higher thrust per cutter is required to initiate cracking, and torque demand rises to maintain cutterhead rotation. In hard, massive rock, UCS strongly influences whether disc cutters can generate efficient tensile cracking and chip formation, or whether cutting shifts toward inefficient crushing and grinding when thrust is excessive.

Despite its importance, UCS alone does not fully describe tunnelling conditions. TBMs interact with the rock mass, and factors such as joint spacing, joint orientation, weathering, and overall rock mass integrity can dominate excavation behaviour. For this reason, UCS is commonly used alongside parameters such as fracture spacing, integrity coefficients, or rock mass classification systems. A rock with a UCS of 120 MPa but closely spaced joints may, in practice, be easier to excavate than a massive, unjointed rock with a UCS of 80 MPa.

UCS also has several inherent limitations. It does not account for the presence of discontinuities, does not directly indicate abrasivity or cutter wear potential, does not capture time-dependent behaviours such as creep, and can be sensitive to sample quality and testing conditions. Consequently, UCS should be viewed as a baseline strength parameter rather than a complete predictor of TBM performance, and it must be interpreted in combination with other geological and operational factors to accurately assess tunnelling behaviour.

Abrasivity: Its Role in Tool Wear and Performance Degradation

In parallel with rock strength, abrasivity—defined as the capacity of rock to induce wear on cutting tools—significantly influences TBM effectiveness. Measured using the Cerchar Abrasivity Index (CAI), abrasivity does not correlate with how easily rock can be broken but rather defines how rapidly tool surfaces degrade during operation. High-abrasivity materials lead to drastically accelerated wear rates for cutter rings and associated components, often resulting in increased downtime due to the need for frequent replacements [507, 526]. Interestingly, abrasive rocks may initially allow satisfactory penetration rates; however, this performance can swiftly decline as cutter sharpness diminishes, necessitating increased energy inputs to maintain similar advance rates [527, 528].

The interplay between strength and abrasivity is encapsulated in the performance of a TBM, where one can encounter a rock type that is easy to break yet severely abrasive, promoting early operational efficiencies but ultimately leading to decreased long-term productivity due to rapid tool wear [68, 526].

Interdependence and Independent Assessment

It is crucial to underscore that while both properties influence TBM operations, they do so in fundamentally different ways and are not inherently correlated. Engineers must independently evaluate these characteristics to formulate accurate predictions regarding TBM performance [529, 530]. Instances of high-strength yet low-abrasivity rocks typically yield low cutter wear but are challenging to penetrate, necessitating a careful balance of thrust and operating conditions to optimize performance. Contrarily, low-strength, high-abrasivity rocks present a situation where operations may initially progress favourably but lead to substantial cutter consumption and associated costs over extended periods [393].

Chapter 8

Ground Support and Tunnel Lining Systems

Ground support and tunnel lining systems are fundamental to the safe, efficient, and durable construction of tunnels using Tunnel Boring Machines. As excavation progresses, the removal of in-situ ground stresses creates an immediate need for engineered systems that can stabilise the surrounding ground, control deformation, protect personnel and equipment, and ultimately ensure that the completed tunnel performs as intended over its full design life. In mechanised tunnelling, these requirements are addressed through a combination of temporary and permanent support systems that operate at different stages of construction and fulfil distinct but closely interrelated roles.

This chapter examines the principles, functions, and practical application of ground support and tunnel lining systems in TBM tunnelling. It begins by distinguishing between temporary support measures—installed to manage short-term instability and construction-stage risks—and permanent lining systems designed to carry long-term ground and groundwater loads while meeting structural, durability, and operational requirements. The chapter then explores the range of support systems commonly used in TBM projects, including rock bolts, shotcrete, steel ribs, face support measures, and segmental lining systems, highlighting how these elements interact as excavation advances.

Particular emphasis is placed on segmental tunnel linings, which have become the dominant permanent support system in modern shielded TBM tunnelling. The chapter describes their configuration, installation sequence, joint behaviour, sealing systems, backfilling and annular grouting processes, and the mechanisms by which loads are transferred from the ground to the lining over time. Through this integrated discussion, the chapter provides a framework for understanding how ground support and lining systems function not as isolated components, but as coordinated elements within a staged construction and load-transfer process that evolves from excavation through to long-term service

Temporary vs Permanent Support

In TBM tunnelling, ground support and tunnel lining systems are fundamental to maintaining excavation stability, controlling deformation, and ensuring the tunnel's long-term structural performance. These systems are generally classified as temporary support and permanent support, based on their purpose, design life, and role within the construction sequence. A clear understanding of the distinction between the two is essential for effective tunnel design, construction planning, and risk management.

The conceptual difference between temporary and permanent support lies primarily in purpose and time horizon. Temporary support is installed to stabilise the ground during excavation and construction, addressing the short-term instability risks created by the excavation process itself. Permanent support, by contrast, is designed to carry long-term loads and to meet structural, durability, and operational requirements throughout the tunnel's service life. Although temporary and permanent systems are often closely integrated in TBM projects, they are not interchangeable and are designed according to fundamentally different criteria.

Figure 78: Preparing Temporary Supports for Oil Static Lines Under LIRR Bridge at 43rd Street (12-21-2011). Metropolitan Transportation Authority, CC BY 2.0, via Flickr.

Temporary ground support is typically installed immediately behind the tunnel face or close to it, with the aim of controlling ground movement and maintaining safety until the permanent lining or long-term support becomes effective. Its primary functions are to prevent collapse or ravelling of the ground, control deformation and convergence, protect the TBM, crew, and equipment, and maintain face and crown stability as excavation progresses. These systems are usually designed for short-duration loading and often rely on the surrounding ground progressively stabilising as stresses redistribute.

Common forms of temporary support in TBM tunnelling include rock bolts or dowels, steel ribs or lattice girders, wire mesh, and shotcrete, either plain or fibre-reinforced. In poorer ground conditions, face bolts or spiles may be used, and in weak or squeezing ground, local steel sets may be required. In hard rock TBM tunnelling, temporary support is often installed after the TBM has passed, typically from the backup area using probe drilling and support installation. In soft ground or mixed ground conditions, temporary support may be required much closer to the face or integrated directly with the TBM shield. A key advantage of temporary support is its flexibility, allowing engineers to adapt the support system in response to changing geological conditions.

Figure 79: Drilling for Rock Dowel Installation. Metropolitan Transportation Authority, CC BY 2.0, via Flickr.

Rock Bolts or Dowels

Rock bolts and dowels stabilise the ground by reinforcing the rock mass rather than supporting it externally. They are installed into drilled holes and anchored mechanically, chemically (resin-grouted), or by friction. Once installed, they create a reinforced rock arch by tying together individual rock blocks and preventing them from moving independently. This increases the self-supporting capacity of the rock mass and limits dilation and loosening around the tunnel.

Bolts primarily work in tension and shear, transferring load from unstable outer zones into more competent rock deeper behind the excavation boundary.

Rock bolts are most effective in jointed but competent rock, where instability is caused by block movement rather than material weakness. In TBM tunnelling, they are commonly installed in hard rock sections behind the TBM, particularly where joint spacing, bedding planes, or faulting create potential for block fall-out. They are less effective in very weak or squeezing ground, where the rock mass cannot carry load even when reinforced.

Steel Ribs or Lattice Girders

Steel ribs and lattice girders provide external structural support to the tunnel perimeter. They act as load-bearing frames that resist ground pressure by forming a stiff ring. Loads from the surrounding ground are transferred onto the ribs and then redistributed along the tunnel axis and into adjacent support systems such as shotcrete.

Lattice girders are lighter and work compositely with shotcrete, while heavy steel ribs act as primary structural members.

These systems are used in weak, fractured, or squeezing ground where the rock mass cannot self-support even temporarily. In TBM tunnelling, steel ribs are often installed in short sections of poor ground, fault zones, or highly stressed areas. They are particularly valuable where deformation must be tightly controlled and where shotcrete or bolts alone are insufficient.

Wire Mesh

Wire mesh provides surface confinement rather than structural support. It prevents loose rock fragments or soil from falling out between bolts or behind shotcrete. The mesh distributes local loads and holds small blocks in place until other support systems mobilise.

Mesh does not carry significant ground pressure on its own but works in combination with bolts and shotcrete.

Wire mesh is used where ravelling or minor block fall-out is expected, such as in moderately fractured rock. In TBM tunnelling, it is commonly installed immediately after excavation, especially in crown and shoulder areas, to provide immediate protection for crews and equipment before shotcrete application.

Shotcrete (Plain or Fibre-Reinforced)

Shotcrete is sprayed concrete that forms a continuous lining on the tunnel surface. It works by providing early surface support, confinement, and load redistribution. Once hardened, shotcrete limits deformation, bonds loose material, and works compositely with bolts, mesh, or lattice girders.

Fibre-reinforced shotcrete improves toughness, crack control, and post-cracking strength, allowing it to absorb deformation without sudden failure.

Shotcrete is widely used in hard rock TBM tunnelling as the primary temporary support system. Plain shotcrete is suitable where deformations are small, while fibre-reinforced shotcrete is preferred in fractured, high-stress, or moderately squeezing ground. It is often applied shortly after the TBM passes, forming a stabilising shell that remains effective until permanent lining installation.

Face Bolts or Spiles (Poor Ground)

Face bolts and spiles are installed ahead of the tunnel face to reinforce the ground before excavation reaches it. They act as pre-support elements, improving face stability by increasing cohesion and stiffness in front of the TBM. This reduces the risk of face collapse, excessive deformation, or sudden inflow of material.

They work by forming a reinforced “umbrella” or dowelled zone that holds the ground together during excavation.

These systems are used in very poor, loose, or highly fractured ground, particularly near fault zones, shear zones, or mixed ground transitions. In TBM tunnelling, face bolts are critical where face stability cannot be guaranteed by the TBM alone, especially in hard-rock TBMs without full face confinement.

Local Steel Sets in Weak or Squeezing Ground

Local steel sets are heavy structural frames installed in sections where ground pressure is high and deformation is ongoing. Unlike bolts or shotcrete, steel sets provide passive resistance, carrying significant load as the ground converges. They are often combined with sliding joints or yielding elements to accommodate controlled deformation without collapse.

Steel sets are reserved for extreme conditions, such as squeezing ground, highly deformable rock, or zones of severe overstress. In TBM tunnelling, they are used sparingly and locally, often where other support measures have proven inadequate. Their installation is labour-intensive but essential for maintaining tunnel integrity in the worst ground conditions.

Permanent support, commonly referred to as the tunnel lining, is designed to provide long-term structural stability, durability, and functionality consistent with the tunnel's intended use. Its roles include carrying long-term ground and groundwater loads, providing a stable tunnel profile, ensuring watertightness where required, supporting operational systems such as track, roadway, or services, and meeting specified durability and design life requirements, often exceeding 100 years. Permanent support systems must satisfy structural, geotechnical, hydraulic, and durability criteria established during the design stage.

Typical permanent lining systems used in TBM tunnelling include precast concrete segmental linings, which are the most common, in-situ concrete linings used as secondary linings, composite systems combining segments with cast-in-place concrete, and steel liners for specialised applications. In shielded TBM tunnelling, segmental lining rings are installed directly behind the cutterhead, meaning that the permanent support is placed almost immediately after excavation, providing rapid long-term stability.

Figure 80: Precast concrete segment used in segmental tunnel linings. Maria Angela 93, CC BY-SA 4.0, via Wikimedia Commons.

Precast Concrete Segmental Linings (Most Common)

Precast concrete segmental linings consist of individual reinforced concrete segments that are assembled into circular rings inside the TBM shield immediately after excavation. The segments are manufactured off-site under controlled conditions and erected using a segment erector mounted within the TBM. Once a ring is completed, it forms a closed structural system capable of resisting ground pressure, groundwater loads, and TBM thrust forces.

Structurally, segmental linings work as a compression ring. Loads from the surrounding ground are transferred through the segments and joints, with forces distributed circumferentially and longitudinally along the tunnel. Gaskets installed between segments provide watertightness, while bolted or dowelled connections ensure ring integrity during erection and early loading. As grout is injected into the annulus between the lining and ground, the lining becomes fully supported and long-term load sharing is established.

Segmental linings are used in most shielded TBM projects, including EPB and slurry TBMs, and increasingly in hard rock TBM tunnels where immediate support is required. They are particularly well suited to long tunnels, high groundwater pressure, urban environments, and projects requiring strict control of deformation and settlement. Their predictability, speed of installation, and ability to act as both temporary and permanent support make them the dominant lining system in modern TBM tunnelling.

In-Situ Concrete Linings (Secondary Lining)

In-situ concrete linings are cast directly inside the excavated tunnel, typically after initial excavation and temporary support have stabilised the ground. In TBM tunnels, in-situ linings are most commonly used as secondary linings, placed after the tunnel has advanced and primary support—such as rock bolts, shotcrete, or segmental linings—has taken effect.

These linings work by forming a continuous, monolithic concrete shell that carries long-term loads and provides a smooth, durable tunnel profile. The lining is usually cast using formwork systems that advance along the tunnel, and reinforcement may be included depending on structural requirements. In many cases, the in-situ lining is designed to carry full long-term ground and water loads, while the primary support is partially or fully destressed over time.

In-situ concrete linings are commonly used in hard rock TBM tunnels where immediate permanent support is not required, allowing excavation to proceed with temporary support first. They are also used where a smooth internal profile is critical, such as in water conveyance tunnels, or where additional durability, fire resistance, or chemical resistance is required. In urban rail tunnels, in-situ linings are often used in combination with segmental linings to meet stringent operational or architectural requirements.

Composite Linings (Segmental Linings + Cast-in-Place Concrete)

Composite linings combine precast segmental linings with an additional cast-in-place concrete layer installed inside the segments. In this system, the segmental lining provides immediate structural support and watertightness, while the in-situ concrete layer enhances long-term structural capacity, durability, or functional performance.

Structurally, composite linings distribute loads between the segmental lining and the secondary concrete layer. The segmental lining typically carries construction-stage loads and initial ground pressure, while the cast-in-place lining may be designed to carry a portion or all of the long-term loads, depending on the design philosophy. The secondary lining can also protect the segments from aggressive environments and provide a smooth internal finish.

Composite linings are used where high durability, long design life, or enhanced fire and impact resistance is required. They are common in metro tunnels, road tunnels, and tunnels with strict serviceability criteria. This system is also adopted where segmental linings alone would be structurally sufficient but do not meet operational or maintenance requirements.

Steel Liners (Specialised Applications)

Steel liners consist of welded or bolted steel plates forming a continuous internal lining. Unlike concrete linings, steel liners are thin, flexible, and capable of withstanding high internal pressures. They rely on the surrounding ground or an external concrete shell for overall structural stability, while the steel itself primarily provides containment, impermeability, or resistance to internal pressure.

Steel liners work by acting as a pressure boundary rather than a load-bearing ground support system. In many cases, they are installed inside a concrete tunnel or embedded within a concrete lining, forming a composite system. Corrosion protection and long-term durability are critical design considerations.

Steel liners are used in specialised TBM applications such as pressure tunnels for hydropower, penstocks, oil and gas pipelines, or tunnels carrying aggressive fluids. They are also employed where absolute watertightness or resistance to high internal pressure is essential. Due to their cost and maintenance requirements, steel liners are rarely used for conventional transport tunnels.

In practice, temporary and permanent support systems often overlap in function, although their design intent remains distinct. For example, rock bolts may provide short-term stabilisation until a concrete lining is installed, shotcrete may control deformation while segmental linings carry long-term loads, and the TBM shield itself may provide temporary confinement before the installation of lining segments. In some hard rock TBM projects, the permanent lining may be installed significantly later, making temporary support critical for maintaining stability over extended construction periods. Conversely, in EPB or slurry TBM operations, the segmental lining frequently acts as the primary permanent support almost immediately, reducing the reliance on additional temporary systems.

The design philosophies for temporary and permanent support differ accordingly. Temporary support design is often observational and adaptable, based on short-term loading conditions and allowing for relatively higher deformation, with adjustments made during construction as conditions are observed. Permanent support design is deterministic and code-based, intended to resist full life-cycle loads while controlling deformation within strict limits and addressing durability and maintenance considerations. This distinction aligns closely with principles established in methods such as the New Austrian Tunnelling Method, although similar concepts apply equally in mechanised TBM tunnelling.

Incorrect separation or poor integration of temporary and permanent support can have serious consequences, including excessive deformation before lining installation, overstressing of segmental linings, increased surface settlement or damage, construction delays, and safety incidents. Successful TBM projects therefore clearly define what elements are carrying load at each stage, which systems are intended to provide long-term support, and how and when load transfer occurs between temporary and permanent systems.

Below are practical, project-style examples that illustrate how temporary and permanent support systems are applied in real TBM tunnelling scenarios.

Example 1: Hard Rock TBM Tunnel with Delayed Permanent Lining

(Hydropower headrace tunnel in jointed granite)

Temporary support used

- Rock bolts (3–5 m long, resin-grouted) installed from the TBM backup
- Wire mesh in the crown
- Fibre-reinforced shotcrete (50–100 mm thick)

Why temporary support was required

The intact granite had high UCS, but joint spacing varied significantly. Local block instability occurred immediately after excavation, particularly in the crown. The permanent lining was not installed until several hundred metres behind the TBM.

How it worked

Rock bolts tied unstable blocks into the surrounding rock mass, mesh prevented small block fallout, and shotcrete formed a stabilising shell. Together, these systems controlled deformation and ensured safety during the long construction period before lining installation.

Permanent support

- In-situ concrete lining cast later as a final structural lining

Key lesson

Temporary support carried *construction-phase loads for months*, while the permanent lining was designed only for *long-term service loads*.

Example 2: Urban EPB TBM Metro Tunnel

(Soft clay and silty clay under sensitive buildings)

Temporary support used

- TBM shield confinement
- Face pressure via conditioned soil
- Short face bolts locally in disturbed zones

Why temporary support was required

Soft cohesive clay exhibited plastic flow and risk of face instability. Ground movement had to be strictly controlled to prevent settlement beneath buildings.

How it worked

The TBM shield and pressurised excavation chamber provided continuous temporary support at the face. Where disturbed zones were encountered, face bolts were installed ahead of excavation to stiffen the ground.

Permanent support

- Precast concrete segmental lining installed immediately behind the cutterhead

Key lesson

In shielded TBM tunnelling, the TBM itself is a temporary support system, while the segmental lining becomes permanent support almost immediately.

Example 3: Mixed Ground TBM Tunnel through Fault Zone

(Sandstone transitioning into faulted shale)

Temporary support used

- Steel ribs with shotcrete infill
- Face spiles ahead of the TBM
- Heavy fibre-reinforced shotcrete

Why temporary support was required

The fault zone caused loss of confinement, ravelling, and uneven loading. Bolts alone were insufficient due to weak, crushed rock.

How it worked

Face spiles created a reinforced umbrella ahead of the TBM, reducing face instability. Steel ribs provided immediate structural resistance, while shotcrete locked the system together and controlled deformation.

Permanent support

- Cast-in-place concrete lining installed after clearing the fault zone

Key lesson

Temporary support systems can become structurally significant in extreme ground, even though they are not designed for long-term service.

Example 4: Squeezing Ground in Deep TBM Tunnel

(High-stress schist at depth)

Temporary support used

- Yielding steel sets with sliding joints
- Thick fibre-reinforced shotcrete
- Systematic rock bolting

Why temporary support was required

High in-situ stress caused time-dependent deformation (squeezing). Immediate confinement was needed, but the support also had to tolerate controlled deformation.

How it worked

Steel sets absorbed load as the ground converged, while yielding elements prevented sudden failure. Shotcrete and bolts worked together to control loosening.

Permanent support

- Reinforced concrete lining installed after ground deformation stabilised

Key lesson

Temporary support must sometimes be designed to deform, while permanent support is designed to resist deformation.

Example 5: Road Tunnel with Composite Lining System

(Moderate rock, high groundwater pressure)

Temporary support used

- Rock bolts and shotcrete
- Drainage measures

Permanent support

- Precast segmental lining
- Secondary cast-in-place concrete lining

Why both were needed

Temporary support controlled deformation immediately after excavation. The permanent lining system provided long-term load resistance and watertightness.

How load transfer occurred

As groundwater pressure increased, load gradually transferred from the temporary shotcrete shell to the permanent lining system.

Key lesson

Temporary and permanent supports may overlap in time, but their design intent remains distinct.

Example 6: What Happens When the Distinction Is Not Clear

(Common failure scenario)

Problem

Temporary shotcrete was assumed to be "good enough" and left in place too long without installing the designed permanent lining.

Outcome

- Excessive deformation
- Cracking of the shotcrete
- Overstressing of later-installed segmental lining
- Costly remedial works

Key lesson

Temporary support is not a substitute for permanent support, even if it appears stable initially.

These examples reinforce a fundamental principle of mechanised tunnelling: temporary support is primarily responsible for managing geotechnical and structural risk during excavation, while permanent support is designed to manage long-term risk over the full service life of the tunnel. Successful TBM projects therefore clearly define, at every stage of construction, which support system is carrying load, how loads are transferred as excavation progresses, and when responsibility transitions from temporary support measures to the permanent lining system.

Table 6: Summary Table of temporary and permanent support systems examples.

Scenario	Temporary Support Role	Permanent Support Role
Hard rock TBM	Stabilise jointed rock	Long-term structural lining
EPB metro tunnel	Face and ground confinement	Immediate segmental lining
Fault zone	Prevent collapse during advance	Restore long-term stability
Squeezing ground	Absorb deformation	Carry final loads
Composite lining	Control early deformation	Provide durability & watertightness

Segmental Lining Systems

Segmental lining systems are critically significant in mechanized tunnelling, particularly when utilizing tunnel boring machines (TBMs) such as Earth Pressure Balance (EPB) and slurry TBMs. These systems typically consist of precast concrete segments that are bolted and gasketed together, providing a robust structural solution for tunnel integrity and performance under various conditions.

The organization of segmental linings into rings is essential, as they are assembled immediately behind the TBM's advance. This immediacy is essential for maintaining stability and durability, which is a notable advantage over traditional lining systems. Once installed, these rings form a seamless structural lining, offering long-term stability and watertightness where necessary [531, 532]. Reinforced concrete has become standard for these segments due to its durability and minimal impact on structural integrity during segment installation [532].

Segmental linings are installed as part of a tightly coordinated, mechanised process that runs in parallel with TBM excavation. The installation sequence is designed to ensure that the ground is continuously supported, that the lining is correctly aligned and compressed, and that loads are safely transferred from the ground to the permanent structure. The process described below reflects standard practice for shielded TBMs such as Earth Pressure Balance (EPB) and slurry machines.

After the TBM completes excavation of one ring length—typically between about 1.2 and 2.0 metres—the machine stops advancing. At this point, the ground around the excavation is temporarily supported by the TBM shield, which prevents collapse and limits deformation while the lining is installed. The cutterhead remains at the face, maintaining face support pressure in soft ground or slurry-supported conditions.

Precast concrete segments are delivered from the backup system to the erector inside the shield. A hydraulic or electro-mechanical segment erector, usually a robotic arm with multiple degrees of freedom, picks up each segment individually. Segments are installed sequentially against the shield tail, following a predetermined order that ensures stability and correct geometry. The segments are carefully positioned to achieve the required alignment, joint contact, and gasket compression.

As each segment is placed, temporary bolts or dowels are installed to connect it to the previously installed segments. Circumferential bolts connect segments within the same ring, while longitudinal bolts connect the new ring to the ring installed previously. These connections provide immediate stability during erection, prevent segments from falling or rotating, and help control joint opening and ring geometry during the installation process.

The final piece installed in each ring is the key segment. This segment is typically wedge-shaped or slightly different in geometry so that, when inserted, it locks the ring into position and develops circumferential compression. Once the key segment is in place and bolted, the ring forms a complete structural unit capable of resisting ground pressure in combination with the surrounding ground and the TBM shield.

At the rear of the shield, tail seals—usually a system of steel brushes lubricated with grease—maintain a seal between the shield and the newly installed lining. These seals prevent soil, slurry, and grout from entering the shield and protect the erection area. Immediately after ring installation, grout is injected into the annular gap between the outside of the lining and the excavated ground. This annular grouting fills any voids created by over-excavation or shield taper, transfers load from the ground to the lining, and limits surface settlement.

Once grouting is complete and the ring has achieved initial stability, the TBM thrust cylinders push off the newly installed lining ring. This thrust allows the TBM to advance and excavate the next ring length. The process then repeats in a continuous cycle: excavate, support with the shield, erect segments, grout, and push forward.

Throughout installation, strict geometric and quality controls are applied. Ring alignment, segment fit, bolt torque, gasket compression, and grout volume and pressure are monitored to ensure structural performance and watertightness. Deviations can lead to uneven load distribution, increased leakage risk, or long-term durability issues, so precision is critical.

Structurally, segmental linings exhibit a combination of ring compression, bending resistance, and shear transfer across joints. This arrangement acts as a single unit capable of withstanding external ground loads and internal pressures, thus ensuring the tunnel's structural safety [533]. The capacity of these linings to distribute loads effectively through joint interactions and annular grouting further enhances their utility [534].

Ring Compression (Hoop Action)

Ring compression, commonly referred to as hoop action, is the primary load-carrying mechanism of a circular segmental tunnel lining. When ground pressure and groundwater pressure act radially and relatively uniformly around the excavation, compressive forces develop around the circumference of the lining. The tunnel ring therefore behaves in a manner analogous to a compression ring, transferring loads continuously around its perimeter.

Because concrete has high compressive strength, ring compression is an efficient and durable structural mechanism. The circular geometry of the lining promotes even redistribution of loads, which significantly limits the development of tensile stresses. Under ideal conditions—such as deep tunnels in uniform, competent ground—the lining response is dominated by compression, with bending effects remaining minimal.

The continuity of compressive forces around the ring depends on effective load transfer across segment joints. This transfer occurs through direct concrete-to-concrete contact, bolt pre-tension that clamps the segments together, and compression of elastomeric gaskets. Adequate gasket compression is therefore critical, not only to maintain watertightness but also to ensure uninterrupted force transmission around the circumference of the lining.

From an engineering perspective, ring compression minimises cracking risk, reduces reinforcement requirements, and underpins most design assumptions for deep bored tunnels. In practice, however, ground conditions are rarely perfectly uniform, and deviations from ideal loading activate additional structural mechanisms within the lining.

Bending Resistance (Flexural Behaviour)

Bending resistance becomes significant when loading around the tunnel perimeter is non-uniform. This may occur due to shallow cover with asymmetric overburden, variations in ground stiffness, construction-induced distortions, or localised thrust forces applied by TBM jacks. These influences cause the tunnel cross-section to deform from its ideal circular shape, a process known as ovalisation, which induces bending moments in the lining segments.

Structurally, each segment behaves as a reinforced concrete beam spanning between adjacent joints. Bending resistance is provided by the segment thickness, the longitudinal reinforcement, and the overall geometry of the segment, including any taper or variation in thickness. Segment joints act as semi-rigid connections, allowing limited rotation while still transmitting bending moments and maintaining overall ring integrity.

Bending behaviour has important design implications, as it governs reinforcement detailing, crack width control, and serviceability performance. Designers must explicitly account for bending effects in situations such as shallow urban tunnels, faulted ground, or mixed soil conditions, where non-uniform loading is expected to be significant.

Shear Transfer Across Joints

Shear forces develop in segmental linings as a result of differential ground movement, TBM thrust during ring erection, eccentric loading associated with bending, and longitudinal forces arising from alignment changes. If these forces are not adequately transferred across joints, relative displacement between segments may occur, leading to joint slip, step formation, or local cracking.

Shear resistance across segment joints is provided through a combination of mechanisms. Frictional resistance arises from normal compressive forces generated by ring compression and bolt pre-load. Mechanical interlock is achieved through features such as dowel bars, shear keys, or tongue-and-groove joint profiles, which physically limit relative movement between adjacent segments. Bolted connections contribute indirectly by maintaining joint closure and contact pressure, thereby enhancing both frictional and interlocking capacity.

Effective shear transfer is essential to maintain ring geometry, prevent relative displacement between segments, preserve waterproofing integrity, and support the long-term durability of the lining. Shear capacity is particularly critical during construction, when load conditions are often more uneven and less predictable than those experienced during long-term operation.

Integrated Structural Behaviour

In service, ring compression, bending resistance, and shear transfer act simultaneously rather than independently. Ring compression provides the primary resistance to ground and water pressures, bending resistance accommodates non-uniform loading and deformation, and shear transfer ensures compatibility and continuity across segment joints. A well-designed segmental lining balances these mechanisms through appropriate geometry, reinforcement detailing, joint configuration, and construction control, allowing the lining to function as a robust and redundant structural system under both construction-stage and long-term operational conditions.

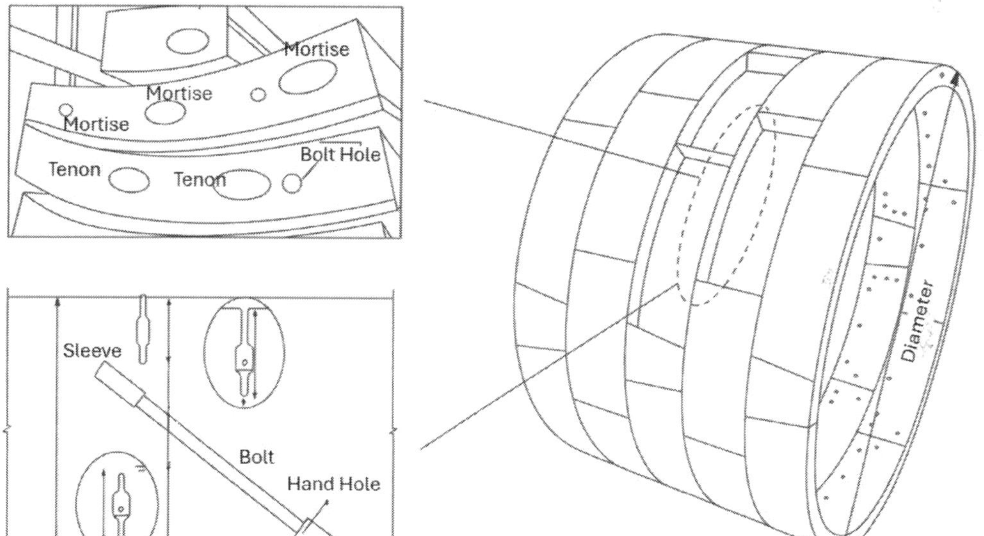

Figure 81: The segment structure and the detail of circumferential joint. Adapted from Gou et al. (2023) [535].

The installation process of segmental lining is mechanized and highly synchronized with the TBM operation. This allows for excavation and lining support to occur almost concurrently, significantly increasing the efficiency of the tunnelling process. According to studies, the TBM excavates a length equivalent to one ring, typically between 1.2 and 2.0 meters [533]. A segment erector is crucial in positioning the segments, securing them with circumferential and longitudinal bolts, and ultimately utilizing a key segment to complete the ring [536].

Precision in the assembly is vital; the segments must demonstrate tight dimensional tolerances to ensure correct fit and alignment. Additionally, the use of grout fills the annular space between the segments and the surrounding ground, facilitating load transfer and mitigating settlement risks [537].

From a design perspective, segmental linings must accommodate a variety of demands, including hydrostatic and ground-induced loads, operational traffic forces, and potential accidental impacts such as seismic activities [533]. The behaviour of joints in the segmental system is pivotal, as joint stiffness significantly influences the overall ring performance and divergence under load [538]. The design often employs analytical methods such as elastic ring theory and numerical modelling to predict performance and assess serviceability under varied conditions [539].

Hydrostatic pressures impose additional challenges, necessitating careful design for watertightness. The segmental linings must utilize low-permeability concrete combined with appropriately designed gaskets to counter potential leakage through the joints, which is often the primary point of failure [540].

Segmental lining systems offer numerous advantages including rapid construction, high-quality control from factory production, and exceptional watertightness. Their capacity to deliver immediate structural support is particularly beneficial in environments that require quick advancement and minimal disruption [533, 541]. However, challenges such as high upfront costs for manufacturing segments, the critical need for precise TBM tracking and erection accuracy, and sensitivity to construction tolerances are notable [534, 542].

Segmental lining systems are an essential component of modern TBM tunnelling, merging efficiency with advanced engineering practices to meet contemporary engineering needs. Their design and operational methodologies must be managed adeptly to ensure successful construction in a variety of environments.

Precast Concrete Segments

Precast concrete segments form the fundamental building blocks of segmental tunnel linings used in mechanised TBM tunnelling. They are factory-manufactured structural elements designed to be rapidly assembled behind the TBM, creating a continuous lining that is simultaneously load-bearing and watertight. The performance of these segments depends not only on concrete strength and reinforcement, but critically on geometric accuracy, tight construction tolerances, and reliable sealing systems. Unlike cast-in-place linings, precast segments must achieve structural continuity and long-term durability through precise fit and controlled interaction at their joints.

Each tunnel ring is assembled from several standard segments and one key segment that completes the ring. In plan, the segments are typically trapezoidal, allowing them to form a circular or near-circular geometry when assembled. The radial thickness of the segments is determined by structural capacity, durability requirements, and fire resistance considerations, while the axial width is governed by TBM stroke length, segment handling constraints, and erection efficiency. Depending on design requirements, segments may be uniform in thickness or locally thickened at bolt pockets, dowel locations, or load transfer zones. In tunnels with horizontal or vertical curvature, tapered rings are used, with subtle geometric adjustments that allow alignment changes without excessive joint opening or stress concentration.

Tight dimensional tolerances are essential to ensure that segments assemble correctly and that loads are distributed as intended. Control is required over segment length and width, radial thickness, circumferential angle, and the flatness and surface finish of joint faces. Even small deviations can accumulate around a ring, leading to misalignment, uneven gasket compression, or local stress concentrations. Poor tolerance control can also create erection difficulties, increase wear on the erector system, and raise the risk of segment damage during installation.

Accurate geometry and tolerance control have direct structural and construction implications. Uniform geometry ensures that ring compression is transferred consistently across joints, that bending moments are not unintentionally amplified by geometric eccentricities, and that shear forces are properly resisted through joint contact and mechanical interlock. From a construction perspective, tight tolerances reduce the need for corrective actions underground, improve erection speed and reliability, and minimise the risk of cracked segments or leakage paths developing during the service life of the tunnel.

Gaskets form the primary waterproofing system for segmental linings and are installed in grooves cast into the segment faces. They are designed to seal both circumferential joints between segments within a ring and longitudinal joints between adjacent rings. In addition to preventing water ingress, gaskets contribute to joint contact pressure and therefore play a secondary structural role in load transfer and joint stability.

The most commonly used gasket material is EPDM (ethylene propylene diene monomer) rubber, selected for its durability, elasticity, and resistance to chemical attack and ageing. Gasket systems may consist of a single continuous gasket around the segment perimeter, double-gasket arrangements for redundancy in high water pressure environments, or combined systems incorporating hydrophilic strips that swell when exposed to water. Hydrophilic elements are typically used as secondary seals, providing a degree of self-healing if minor leakage paths develop.

Effective sealing performance depends on achieving the correct level of gasket compression during ring assembly. This compression is controlled by the accuracy of segment geometry and joint flatness, the applied bolt tightening forces, and the precision of the erector during installation. Insufficient compression can result in leakage, while excessive compression may damage the gasket or increase joint stiffness beyond design assumptions. Gaskets are therefore designed to operate within a defined compression range that balances watertightness with long-term material performance.

Long-term durability is a critical requirement for gasket systems, as they must remain functional for the entire design life of the tunnel, often exceeding 100 years. This demands resistance to creep, stress relaxation, chemical attack from groundwater, and repeated load cycles caused by ground movement and operational effects. Proper storage, handling, and protection of gaskets during segment manufacture, transport, and erection are essential to avoid damage that could compromise sealing performance.

Overall, the effectiveness of precast concrete segments is governed by the interaction between geometry, tolerances, and sealing systems rather than any single feature in isolation. Precise geometry ensures correct load paths, tight tolerances provide uniform contact and controlled gasket compression, and well-designed gasket systems deliver durable watertightness. Together, these elements allow segmental tunnel linings to perform reliably under complex combinations of ground pressure, groundwater, construction loads, and long-term in-service demands.

The Role of EPDM Rubber in the TBM Process

EPDM (Ethylene Propylene Diene Monomer) rubber is a vital component in Tunnel Boring Machine tunnelling, particularly as a sealing material in segmental tunnel linings. This comprehensive analysis seeks to delineate the multifaceted roles EPDM rubber plays in the TBM process, highlighting its importance across various stages, from manufacturing to long-term performance in the field.

EPDM Gasket Preparation During Segment Manufacture: The incorporation of EPDM gaskets into precast concrete segments occurs during manufacturing. Each segment features a precisely formed gasket groove along the joint faces, which is essential for ensuring watertight performance. Upon demoulding and curing, gasket preparation involves either mechanical pressing or adhesive bonding of the EPDM gasket into the groove under controlled factory conditions. Such rigorous quality control is paramount; any defects, such as misalignment or cuts, jeopardize the gasket's capacity to continually resist groundwater ingress once installed in the tunnel [543, 544].

Handling and Protection During Transport and Storage: EPDM gaskets, once fitted, must be meticulously protected during transportation and storage. This protection is critical, as gaskets are susceptible to mechanical damage from stacking, impact, and abrasion. They should also be shielded from excessive heat and ultraviolet radiation, both of which can degrade rubber properties [545]. Utilizing protective covers and controlled stacking arrangements is a common practice to maintain the integrity of gaskets and ensure they arrive at the TBM site in optimal condition [543].

EPDM Gasket Behaviour During Ring Erection: During TBM operations, segments fitted with EPDM gaskets are installed immediately behind the cutterhead. The primary function of the gaskets comes into play as they make contact with adjacent segment gaskets. The controlled compression of the EPDM gasket as bolts are tightened plays a crucial role in ensuring a watertight seal, accommodating minor geometric tolerances [543, 544]. Properly executed compression facilitates the gasket's ability to fill surface irregularities, thereby creating a continuous sealing line resistant to both water pressure and gas ingress, reinforcing the construction's durability [543, 546].

Interaction with Bolting and Joint Geometry: The efficacy of EPDM gaskets is intricately linked to segment geometry, joint flatness, and the forces exerted by bolting. Bolts are essential as they maintain the clamping force, allowing for effective compression of the gasket [543]. Proper alignment and sufficiently flat joint faces help ensure that the compression along the gasket's length is uniform, thereby enhancing seal integrity. This interaction is crucial, especially in circumstances where the TBM thrust pushes the newly erected ring against previously installed segments [543].

Performance Under TBM-Induced Loads and Movements: During tunnelling operations, EPDM gaskets are subjected to various dynamic loading conditions. These include TBM thrust forces, ring build tolerances, and groundwater pressure acting shortly after installation. EPDM's elastic properties enable it to maintain contact pressure under these varying conditions, allowing the tunnel lining to accommodate construction-induced distortions while remaining watertight [543]. Such resilience under load is a compelling reason for its widespread use in TBM tunnelling applications [546].

Longitudinal Sealing and Ring-to-Ring Interaction: As the TBM progresses, the compression of each new ring against the previous one activates the EPDM gasket in the longitudinal joint, critical for resisting groundwater migration. The interaction between successive rings must ensure gasket compression is consistent, particularly along longitudinal joints where water ingress poses significant risks [543]. To bolster sealing efficacy, some designs include secondary hydrophilic strips that swell in contact with water, creating an added layer of protection against potential leaks [546].

Long-Term In-Service Performance: Post-construction, EPDM gaskets are subject to long-term stresses, including creep, stress relaxation, and chemical attack from groundwater, necessitating that they maintain their impermeability and elasticity throughout the tunnel's design life, often exceeding 100 years [543, 544]. The performance of these gaskets, therefore, heavily relies on their initial installation quality, underscoring the importance of meticulous handling and installation during the TBM process [546].

In the context of TBM tunnelling, EPDM rubber gaskets serve as critical functional components rather than mere waterproofing elements. Their performance hinges on numerous factors, including accurate manufacturing, careful handling, precise erection practices, and effective TBM operation. When executed correctly, EPDM gaskets provide the necessary watertight integrity and structural reliability required for the demanding conditions of mechanized tunnelling, ensuring longevity and durability from the initial excavation through to long-term service.

Segment Installation Sequence

The segment installation sequence is a critical aspect of Tunnel Boring Machine (TBM) operations, involving a meticulously coordinated process that ensures the structural integrity and stability of tunnel linings. Understanding this sequence provides insights into engineering methodologies and practices that optimize tunnelling performance.

Thrust and Face Stability

Once the TBM has completed a thrust stroke and excavation for a full ring, it halts its forward movement while maintaining ground support through applied face pressure, particularly in Earth Pressure Balance (EPB) or slurry TBM operations. The thrust cylinders must carefully redistribute thrust loads before any segment erection can begin, allowing selective retraction of the thrust cylinders without compromising the stability of the face, which is essential for preventing ground collapse during the installation phase [547, 548]. This redistribution creates the necessary space at the rear of the shield for segment installation, emphasizing the critical nature of thrust management within the tunnelling process [547].

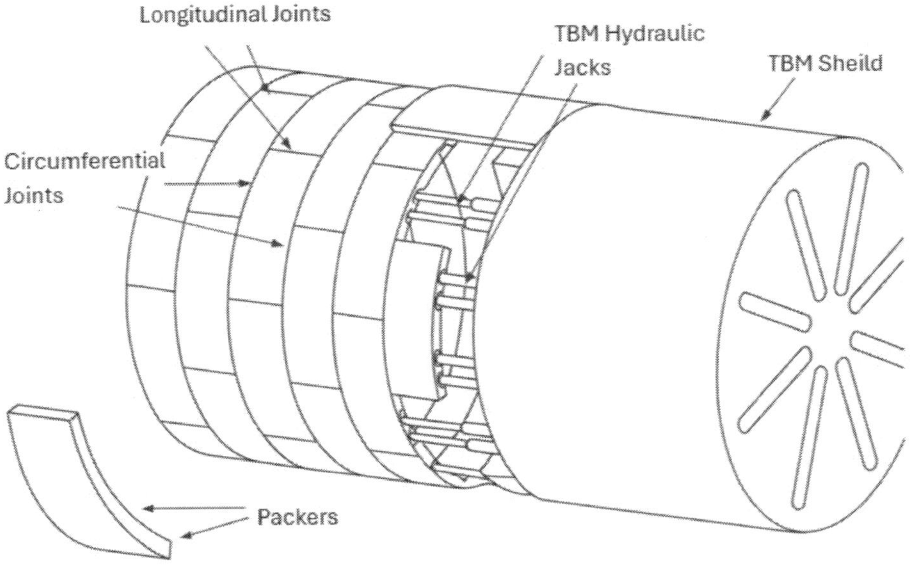

Figure 82: Usual construction process and configuration of segmental tunnel linings.

Segment Erection Mechanism

Segment erection occurs within the tail skin, just behind the TBM shield. This process typically involves a hydraulically actuated erector, which utilizes gripping devices to position segments accurately. Throughout this operation, tail seals maintain consistent contact with the previously installed segments to preserve pressure control and prevent material ingress [83, 548]. The segments are successively delivered from a backup system, such as conveyors or segment cars, and they must arrive in a specific order to facilitate efficient and safe installations [548].

The positioning and rotation of the first segment, often designated as the invert or reference segment, is crucial as it establishes the datum for the remaining segments. This initial placement necessitates utmost precision to avoid errors that could propagate through subsequent placements, leading to misalignment or improper gasket compression [83].

Sequential Installation of Standard Segments

Following the installation of the reference segment, standard segments are placed sequentially, with the erector rotating around the tunnel axis to align each segment with the previous one. As segments are fit together, their EPDM gaskets compress, and temporary bolts or dowels may be used to stabilize individual segments before subsequent segments are erected [83, 548]. The stringent control of segment orientation, gasket compression, and alignment during this phase is essential to mitigate risks such as edge damage or misalignment of bolts [83].

The final step in this installation cycle involves positioning a key segment that completes the ring. This component often requires careful manipulation due to its tapered design, ensuring a secure closure that facilitates the activation of circumferential gasket compression essential for effective waterproofing [83, 549].

Post-Installation Activities

With the ring installation complete, permanent bolts are fastened according to specified torque requirements, locking the geometry of the ring and maintaining compression of the gaskets, which is vital for long-term performance under ground and groundwater pressures [83, 548]. Following this, alignment checks ensure correct positioning and orientation, allowing for minor adjustments prior to advancing the TBM for the next cycle of installation.

Grouting and Ring Activation

Upon successful installation of the segments, annulus grouting occurs, whereby grout is injected into the space between the lining and the excavated ground. This process provides necessary ground support and load transfer to the lining, while meticulously controlling timing and pressure to prevent displacement or subsidence [83]. Subsequent TBM thrust is then gradually transferred to the newly installed ring, marking it as an active structural element within the tunnel ecosystem [83, 548].

Bolting and Doweling

Bolting and doweling systems are crucial connection methods used to join individual precast concrete segments in segmental tunnel linings, forming a cohesive structural ring. Segmental tunnel linings are primarily designed to operate under compression, and the role of bolts and dowels extends beyond mere assembly. They maintain structural integrity during erection, effectively control joint behaviour, and ensure reliable load transfer during various loading conditions.

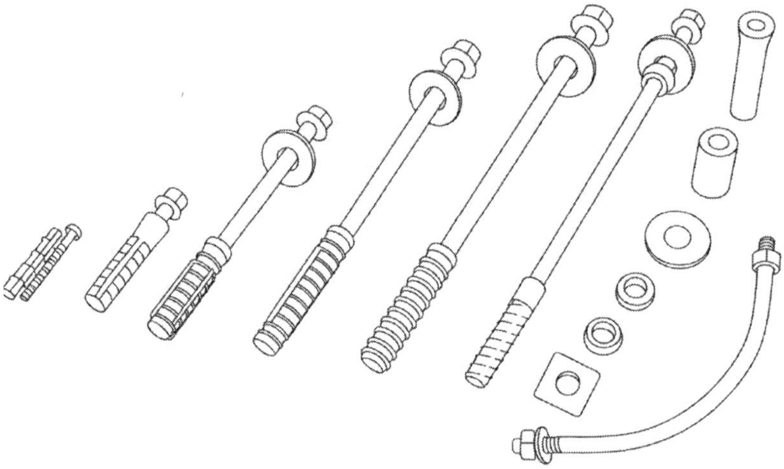

Figure 83: Fittings for Tunnel Segments.

The main function of these connection systems is to integrate individual segments into a unified structural entity rather than allowing them to act independently. This integration is vital during construction, as it helps maintain the geometry of the ring, supports the compression of gaskets for watertightness, and allows for the transfer of shear forces and resistance to bending effects across joints. Their role is particularly critical in the initial installation phase, prior to the ground support being fully engaged and before the annulus grouting hardens [550].

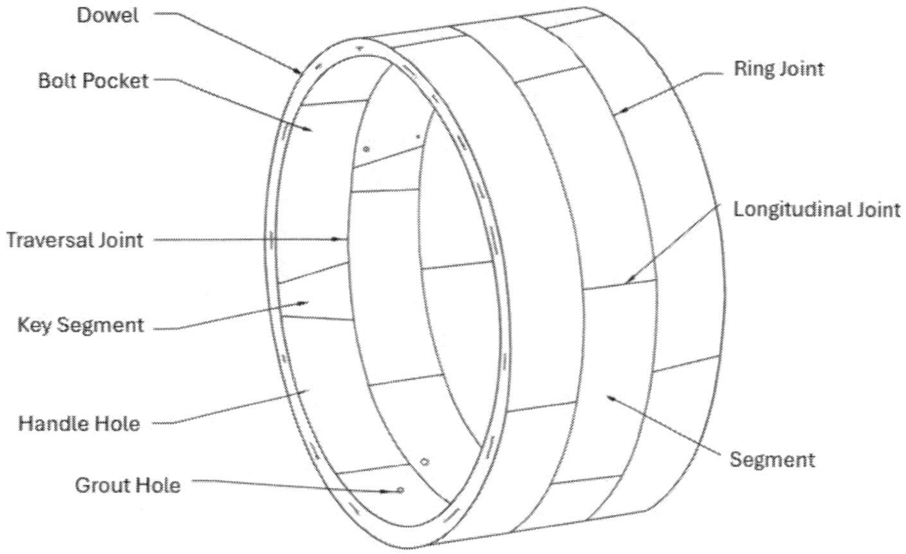

Figure 84: Segmented Tunnel arrangement.

Bolting systems are conventionally utilized across circumferential joints between segments and can also span longitudinal joints between adjacent rings. During the erection phase, bolts facilitate the alignment of segments by drawing them together and compressing the EPDM gaskets, which helps prevent rotation or displacement until the ring is fully closed [550, 551]. Additionally, the implementation of bolt pretension is significant as it generates a clamping force that increases frictional resistance, contributing to effective shear transfer between segments, controlling joint opening, and consequently reducing the risk of leakage and cracking at the joints [551, 552].

On the other hand, doweling systems involve short steel bars or pins, which provide direct mechanical interlock at the joints. Unlike bolts that depend on pre-tension, dowels resist shear through bearing action, making them particularly effective under high shear forces—commonly found in variable ground conditions or during tunnel boring machine (TBM) operations [552]. Dowels also assist in limiting relative movement between segments, thus preventing step formation at the joints and supporting the maintenance of circular geometry amid non-uniform loading [550, 552].

Both bolting and doweling are often used synergistically, where each system complements the other. While bolts primarily provide clamping force and frictional resistance, dowels guarantee shear transfer even when the capacity for friction diminishes due to long-term effects such as wear or relaxation [551, 552]. The integration of both systems forms semi-rigid joints that allow limited rotation, thus accommodating ground movement and construction-induced distortions without significant stress concentrations at the joints [553].

Proper installation of these systems is paramount to achieving the designed performance levels. Misalignment of bolt holes and dowel sockets can lead to installation challenges and unintended stress concentrations within the structure [550, 554]. Adhering to recommended tightening sequences and torque limits is critical to ensuring uniform compressive forces on gaskets, which can prevent cracking of segments and potential leakage paths over time [550].

In terms of long-term performance, while ring compression becomes the primary load-bearing mechanism post-installation, bolting and doweling continue to play significant roles by maintaining joint integrity throughout the service life of the tunnel. They control crack development, resist localized shear forces from ground movement, and alleviate joint opening that could compromise watertightness [555]. This redundancy in structural connections is essential for preserving the functional integrity of the lining, especially in unexpected loading scenarios [556].

Backfilling and Annular Grouting

Backfilling, or annular grouting, is a crucial process in tunnel boring, particularly when utilizing Tunnel Boring Machines (TBMs). This procedure involves filling the void between the excavated ground and the installed segmental lining, which is essential for maintaining structural integrity and minimizing ground movement.

During TBM advancement, a discrepancy in diameter between the TBM shield and the segmental lining creates an annular void. This void, formed as the cutterhead excavates a larger diameter than the lining, is significant for facilitating shield movement and segment installation. Without timely and effective backfilling, this void can lead to detrimental effects such as soil subsidence and surface settlement due to loss of ground support [381, 557, 558].

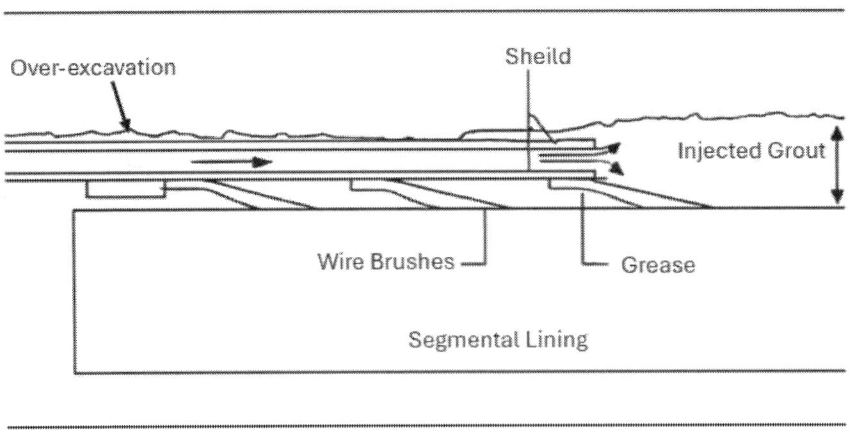

Figure 85: Annulus grouting behind tunnel segments.

Ground Stability and Support: The annular void presents a risk of soil relaxation and instability if not adequately filled. Studies demonstrate that immediate grouting post-segment installation effectively reduces soil movement and mitigates the risk of water ingress, which is vital in maintaining ground stability [30, 381, 558].

Timing of Grout Injection: Grout injection is typically initiated immediately after the installation of each ring of segments. Effective timing is critical; injection must occur promptly to stave off settlement and ensure ground support is restored before the TBM's next advancement. This is further emphasized through field observations indicating that grouting during active TBM motion helps in maintaining pressure and support [30, 381, 558].

Controlled Injection Techniques: The injection process involves the use of multiple ports in the TBM's tail skin, enabling uniform distribution of grout around the lining. Monitoring of injection pressures is essential to balance effective filling of the annular space while preventing undue stress on the newly installed segments [381, 558, 559]. Failure to manage injection pressure can result in either insufficient void filling leading to ground loss or excessive pressure causing structural damage to the lining [381, 559].

Grout Composition and Properties: In TBM applications, a two-component grout system is prevalent, comprised of a cement-based grout and an accelerator. This combination ensures that the grout remains fluid for effective placement yet develops early strength necessary for immediate ground support [558-560]. The formulation must meet several criteria, including fluidity for complete filling and adequate hardening properties to endure long-term structural loads [559, 560].

Structural Role of Annular Grout: Once the grout has set, it becomes integral to the tunnel's structural system. The hardened grout transfers radial loads to the lining and provides consistent confinement that reduces localized stresses and bending moments within the segments [557-559]. Additionally, the grout enhances the tunnel's watertightness, thus further ensuring its integrity over time through the reduction of potential leakage paths [558-560].

Settlement Control Settlement control is perhaps one of the most vital roles of annular grouting. Rapidly filling the annular void is essential to restraining ground volume loss and providing immediate support. Research illustrates that prompt grouting can prevent significant surface settlements, especially critical in shallow urban tunnels where infrastructure integrity is often jeopardized [30, 558].

Long-Term Performance: In the long term, the stability and durability of grouted annular voids are imperative under sustained loading and potential groundwater fluctuations. Rigorously ensuring the quality of the grout and the efficiency of installation practices directly influences the long-term performance of segmental linings [558, 559]. Structural failure due to grout shrinkage or degradation can lead to increased lining stress or create pathways for water infiltration, compromising tunnel safety and longevity [557, 558, 560].

Backfilling via annular grouting is a fundamental aspect of the TBM tunnelling process. By effectively filling the void created during excavation, annular grouting plays a pivotal role in stabilizing the surrounding ground, ensuring load transfer to the segmental lining, and controlling surface settlements. Timely execution, judiciously controlled pressures, and well-formulated grouts are essential to guarantee the successful operation and longevity of the tunnel structure.

Load Transfer and Long-Term Lining Behaviour

Load transfer describes how forces acting on the tunnel—such as ground pressure, groundwater pressure, and construction-induced loads—are progressively transmitted from the surrounding ground to the segmental lining and redistributed through the lining system. Unlike monolithic linings, segmental linings rely on the interaction between multiple components: the ground, annular grout, precast segments, joints, bolts, dowels, and gaskets. The effectiveness of this interaction governs both immediate structural response and long-term performance.

Immediately after ring erection, the segmental lining is not yet fully engaged with the surrounding ground. At this stage, loads are partly carried by the TBM shield, tail seals, and thrust system. The newly installed ring experiences localised forces from TBM thrust pads, temporary eccentric loading, and self-weight, while groundwater pressure may already be acting on the lining.

During this early phase, bolting, doweling, and gasket compression are critical for maintaining ring integrity. Load transfer is incomplete and uneven, making the lining more sensitive to misalignment, ovalisation, and joint opening. This construction-stage behaviour represents one of the most demanding conditions for the lining.

As annular grouting is injected and hardens, the load transfer mechanism changes fundamentally. The grout fills the tail void and establishes continuous contact between the lining and the surrounding ground. Once cured, the grout acts as a load transfer medium, transmitting radial ground pressures uniformly to the segmental lining.

Effective annular grout ensures that loads are distributed around the full circumference of the ring, reducing localised bending moments and stress concentrations in individual segments. Poor grout quality, incomplete filling, or delayed setting can result in uneven contact, causing increased bending, joint distress, and long-term serviceability issues.

With full ground–grout–lining interaction established, ring compression becomes the dominant load-carrying mechanism. Radial ground pressure and groundwater pressure generate circumferential compressive forces within the lining, allowing the ring to behave as a compression structure.

This hoop action is highly efficient, as concrete performs well in compression. Under uniform loading, compressive forces are transferred continuously across joints via concrete contact, bolt clamping forces, and compressed gaskets. In well-designed tunnels, long-term performance relies primarily on this compressive ring action rather than on bending or tensile resistance.

Despite the dominance of ring compression, long-term loading conditions are rarely perfectly uniform. Variations in ground stiffness, changes in groundwater level, consolidation of surrounding soils, and long-term creep can lead to differential loading around the tunnel perimeter. These effects cause gradual ovalisation of the lining, inducing bending moments in the segments.

Segment joints are designed to behave as semi-rigid connections, allowing limited rotation while maintaining force transfer. This controlled flexibility enables the lining to accommodate long-term deformations without brittle failure. Excessive ovalisation, however, can increase joint opening, reduce gasket compression, and raise the risk of cracking or leakage.

Shear forces continue to act on the lining throughout its service life due to ground movement, alignment curvature, and time-dependent deformation of the surrounding ground. These forces are transferred across joints through friction generated by ring compression, mechanical interlock provided by dowels or shear keys, and residual bolt forces.

Over time, some relaxation of bolt pre-tension may occur, increasing reliance on friction and dowel action. Proper joint design ensures that even with reduced bolt contribution, sufficient shear capacity remains to prevent relative displacement between segments and maintain ring geometry.

Groundwater pressure is a permanent load acting on the lining and significantly influences long-term behaviour. This pressure contributes to both radial loading and axial forces along the tunnel. EPDM gaskets and joint compression must remain effective over decades to maintain watertightness.

Changes in groundwater regime, such as seasonal fluctuations or long-term drawdown, can alter effective stresses in the ground and modify load transfer patterns. The lining must therefore be capable of accommodating these changes without loss of structural or hydraulic performance.

Long-term lining behaviour is influenced by time-dependent material properties, including concrete creep and shrinkage, grout stiffness evolution, and ground consolidation. Creep can lead to gradual redistribution of stresses within the lining, while shrinkage may slightly reduce contact pressure at joints if not adequately accounted for in design.

Design methodologies typically consider these effects through staged analysis, recognising that load paths and stress distributions evolve from the construction phase to the long-term equilibrium condition.

Long-term performance is governed not only by ultimate strength but also by serviceability criteria such as crack width control, joint opening limits, and leakage tolerance. Even small deformations can be critical in tunnels carrying water, utilities, or transport infrastructure.

Durability considerations include resistance to chemical attack, abrasion, and fatigue from repeated load cycles. Maintaining uniform load transfer through effective grout contact and joint integrity is essential to minimising deterioration mechanisms over the design life.

Chapter 9

Pressure Control and Face Stability

Pressure control at the tunnel face is one of the defining technical challenges of mechanised tunnelling and a primary determinant of safety, ground performance, and construction quality. As excavation advances, the removal of material at the face causes an immediate loss of confinement and a redistribution of in-situ stresses, creating conditions where instability, ground loss, or uncontrolled inflow can develop if support is not applied correctly. This chapter examines the principles, calculations, and operational strategies used to maintain face stability by balancing earth and groundwater pressures with mechanically applied support in EPB and slurry TBM operations.

The chapter builds from fundamental face stability concepts through Earth Pressure Balance theory and slurry pressure management, linking geotechnical design assumptions with real-time TBM control. It explains how support pressures are calculated, how pressure windows are defined, and how these theoretical limits are translated into practical operating parameters such as chamber pressure, extraction rate, conditioning, and slurry circulation. Particular attention is given to the consequences of operating outside the allowable pressure range, including over-excavation, settlement, face collapse, and blow-out.

By integrating pressure theory with construction practice, monitoring, and feedback control, this chapter demonstrates that face stability is not achieved through static design values alone, but through continuous, disciplined management of pressure throughout excavation. The concepts presented here underpin effective settlement control, safe machine operation, and the long-term performance of the tunnel lining, forming a critical foundation for successful mechanised tunnelling in both soft ground and mixed geological conditions

Face Stability Principles

Face stability in tunnelling is a critical aspect that ensures the integrity and safety of tunnel excavation. It pertains to the ability of the ground at the tunnel face to remain intact and adequately supported during the excavation process [561, 562]. As tunnelling advances, the excavation leads to a sudden reduction in confinement, which can trigger ground movement, collapse, or inflow of soil and water if not properly managed [563, 564]. Understanding and controlling face stability is, therefore, essential not only for maintaining machine safety but also for preventing settlement and ensuring the integrity of the tunnel lining.

The excavation process disrupts the in-situ stress equilibrium within the ground, prompting a redistribution of stresses around the tunnel opening. If the remaining ground strength is insufficient to resist these redistributed stresses, deformation or collapse can occur [565, 566]. Face stability principles are designed to ensure that the stresses acting at the tunnel face are counterbalanced, either by the inherent strength of the ground or by the artificial support applied during excavation [567, 568]. This concept is vital, particularly in environments where the surrounding rock or soil is weak and prone to instability.

In hard rock conditions, where geological features such as joints and bedding planes could dictate the stability of the tunnel face, instability can often be exacerbated by block fall-out, wedge failures, or stress-induced spalling [561, 569]. Various methods such as probe drilling, face bolting, and spiling may be employed to enhance stability and address these risks [570, 571]. By reinforcing or confining the rock mass ahead of the face, operators can effectively maintain face stability, ensuring that the rock behaves as an interlocked and stable system during excavation [565, 568].

Conversely, soft ground conditions—comprising clays, sands, and silts—pose unique challenges. The tunnel face in these scenarios typically cannot stand unsupported; therefore, face stability is primarily achieved through the application of support pressure [562, 572]. Earth Pressure Balance (EPB) and slurry TBMs utilize pressurized excavation chambers to provide the necessary support. In EPB machines, for instance, the excavated material acts as a support medium, allowing face pressure to counterbalance the effects of earth and groundwater pressure [573, 574]. Maintaining this balance is critical; if the applied face pressure becomes insufficient, it can lead to ground loss or surface settlement [278].

Groundwater significantly impacts the stability of tunnel faces, particularly in permeable soil types. The need to counteract both effective earth stresses and pore water pressure is paramount. Insufficient management of these hydraulic gradients can lead to risks such as piping or blow-in [575, 576]. Continuous monitoring of hydraulic conditions and adaptation in tunnelling strategies are essential to mitigate the risks associated with groundwater and maintain tunnel stability [561, 577].

The dynamic nature of ground conditions necessitates continuous monitoring to understand the evolution of face stability over time. Instruments such as pressure sensors and piezometers provide real-time feedback on ground responses, enabling operators to adjust excavation parameters, such as support pressures and excavation rates, accordingly [562,

576]. As face stability is not a static situation, effective management relies on monitoring signals of instability and responding proactively to changing geological or hydrological conditions [567, 578].

Earth Pressure Balance Theory

Earth Pressure Balance (EPB) theory is founded on the principle that stability of the tunnel face is achieved by balancing the in-situ ground stresses and groundwater pressures with an equal and opposing pressure applied directly at the tunnel face. In EPB tunnelling, this counter-pressure is not provided by slurry or compressed air, but by the excavated soil itself, which is retained and pressurised within a sealed excavation chamber immediately behind the cutterhead. The central theoretical objective of EPB operation is therefore to maintain stress equilibrium at the tunnel face so that excavation does not trigger ground collapse, excessive deformation, or uncontrolled inflow of soil and water.

Prior to excavation, the ground at tunnel depth is subjected to in-situ stresses arising from the weight of the overburden, horizontal earth pressures, and pore water pressure. Excavation removes material at the face and causes an abrupt loss of confinement, leading to stress redistribution within the surrounding ground. If this redistribution is not adequately counteracted, instability may develop. EPB theory dictates that the pressure within the excavation chamber should approximately equal the total stress acting at the face, including both effective earth pressure and groundwater pressure. When this balance is achieved, inward movement of the ground is prevented without inducing outward displacement or heave.

The excavation chamber is the core structural and functional element of an EPB TBM. It is a sealed volume located immediately behind the cutterhead and filled with excavated material maintained under controlled pressure. In theoretical terms, the soil within the chamber acts as a pressure-transmitting medium, behaving in a manner similar to a plastic or viscous material. For EPB principles to be effective, this soil must be sufficiently plastic to transmit pressure uniformly, have low permeability to restrict water flow, and remain homogeneous without segregation. When these conditions are satisfied, pressure applied within the chamber is transferred evenly to the tunnel face, maintaining stability.

Because natural soils rarely exhibit ideal EPB properties, soil conditioning is a fundamental component of EPB theory and practice. Conditioning agents such as foams, polymers, and bentonite are introduced at the cutterhead to modify soil behaviour. From a theoretical perspective, conditioning aims to reduce internal friction, increase cohesion and plasticity, control permeability, and improve flowability through the screw conveyor. Properly conditioned soil behaves as a controlled, deformable mass that can both support face pressure and be extracted at a regulated rate, which is essential for maintaining earth pressure balance.

In practice, EPB theory is implemented through coordinated control of several interacting operational parameters. The TBM advance rate governs the rate at which ground enters the excavation chamber, while the screw conveyor extraction rate controls the rate at which pressurised soil leaves the chamber. Screw torque and rotation speed influence internal pressure and resistance to soil flow, and the chamber volume and fill ratio must remain stable to avoid pressure fluctuations. Pressure equilibrium is achieved when the rate of soil inflow matches the rate of controlled outflow, with chamber pressure maintained within a defined operating range.

EPB theory defines a pressure window bounded by two critical stability limits. At the lower limit, known as the collapse limit, chamber pressure is insufficient to resist inward ground movement, leading to face instability, ground loss, or surface settlement. At the upper limit, excessive pressure can cause ground heave, hydraulic fracturing, or damage to surface structures. The acceptable operating pressure must lie within this window, which varies depending on tunnel depth, soil type, groundwater conditions, and the stiffness of the overburden.

Groundwater plays a significant role in EPB theory, particularly in water-bearing soils. Face support pressure must counteract both effective earth stresses and pore water pressure. EPB operation assumes that conditioned soil within the chamber has sufficiently low permeability to prevent rapid water inflow. If permeability is too high, water may enter the chamber, destabilising pressure balance and necessitating additional conditioning or operational adjustments. In highly permeable soils, the assumptions underpinning EPB theory become less reliable, and slurry TBMs may offer a more suitable alternative.

Surface settlement is closely linked to deviations from earth pressure balance. If chamber pressure drops below equilibrium, volume loss occurs at the face and settlement may develop at the surface. Conversely, excessive pressure can lead to uplift or heave. EPB theory therefore treats settlement control as a pressure management problem, in which maintaining stable face pressure minimises ground deformation and volume change around the tunnel.

EPB theory becomes more complex in transitional and mixed ground conditions, such as interfaces between clay and sand or between soil and rock. Different materials respond differently to applied pressure, complicating uniform stress transfer at the face. In these situations, EPB operation relies on enhanced soil conditioning strategies, reduced advance rates, and careful monitoring and adjustment of face pressure. Despite these complexities, the theoretical objective remains unchanged: to maintain equilibrium at the tunnel face despite heterogeneous ground behaviour.

Although EPB theory is robust and widely applied, it relies on several key assumptions, including the ability to condition soil to behave plastically, the capacity to control chamber pressure accurately, and a reasonably predictable ground response. In coarse gravels, highly permeable soils, or conditions with high groundwater inflow, these assumptions may not hold, limiting the effectiveness of EPB TBMs and requiring alternative tunnelling approaches.

The calculation of Earth Pressure Balance is based on determining the support pressure required at the tunnel face to maintain equilibrium between the forces acting on the ground and the pressure applied by the TBM. The required EPB face pressure is intended to counteract the total stress acting at the tunnel face, which consists of both soil stresses and groundwater pressure.

In simplified terms, EPB pressure is calculated so that:

Applied face pressure ≈ total ground pressure at tunnel depth

This ensures that the excavation neither causes inward collapse nor outward heave.

Components of Face Pressure

The total pressure acting at the tunnel face consists of three main components:

1. Vertical stress from overburden

This is caused by the weight of the soil above the tunnel and is calculated as:

$$\sigma_v = \gamma \times H$$

where γ is the unit weight of the ground and H is the depth to tunnel axis or face.

2. Horizontal earth pressure

The horizontal stress acting on the tunnel face is derived from the vertical stress using an earth pressure coefficient:

$$\sigma_h = K \times \sigma_v$$

where K is the earth pressure coefficient (K_o, K_a, or K_p depending on ground condition and stress state).

3. Pore water pressure

In saturated ground, groundwater pressure must be added:

$$u = \gamma w \times hw$$

where γw is the unit weight of water and hw is the hydraulic head above the tunnel face.

Total Required Face Support Pressure

The **required EPB face pressure** is calculated as the sum of effective horizontal stress and pore water pressure:

$$P_{ep\beta} \approx K \times \gamma \times H + \gamma w \times hw$$

This value represents the target chamber pressure that must be maintained inside the EPB excavation chamber to stabilise the face.

In cohesive soils, shear strength parameters (cohesion and internal friction angle) may allow a reduction in required pressure. In cohesionless or granular soils, little or no strength contribution is assumed, and the full pressure must be applied.

In practice, pressure at the tunnel face is not uniform across its full height. EPB calculations therefore typically assume a linear pressure distribution, with higher support pressure at the invert and lower pressure at the crown, reflecting the variation in overburden stress with depth. Rather than relying on a single value, design calculations define an average required face pressure together with a maximum allowable pressure at the invert and a minimum allowable pressure at the crown. The TBM is then operated so that chamber pressure remains within this calculated pressure envelope, ensuring stability across the entire face.

The selection of the earth pressure coefficient, K, has a strong influence on the calculated EPB support pressure and depends on the anticipated ground behaviour. The at-rest earth pressure coefficient, K_0, is most commonly adopted for EPB design because it reflects conditions of minimal ground deformation. The active earth pressure coefficient, K_a, is sometimes used for conservative lower-bound checks, while the passive earth pressure coefficient, K_p, is applied to define upper pressure limits to avoid heave or blow-out. Together, these coefficients are used to establish a pressure window bounded by collapse pressure at the lower limit and uplift or blow-out pressure at the upper limit.

Once the required face pressure has been calculated, it must be converted into practical TBM operating parameters. This includes setting the target chamber pressure, controlling screw conveyor speed and torque, adjusting the TBM advance rate, and regulating soil conditioning dosage. Earth pressure balance is maintained when the volume of soil entering the excavation chamber matches the volume extracted through the screw conveyor and when chamber pressure remains within the defined pressure window. Operators continuously adjust these interacting parameters to respond to changing ground conditions.

The calculated EPB pressure serves as a starting point rather than a fixed operating value. Actual face pressure is refined through observation and monitoring during tunnelling, using surface settlement measurements, face pressure sensor readings, chamber fill ratio, and trends in torque and thrust. If settlement is observed, chamber pressure is increased to reduce volume loss at the face. Conversely, if signs of heave or potential blow-out appear, pressure is reduced. EPB operation therefore functions as a feedback-controlled system rather than a static application of design pressure.

Design EPB pressures also incorporate safety margins and allowances to account for uncertainty and variability. These allowances typically address geological uncertainty, pressure losses within the chamber, limitations in instrumentation accuracy, and construction tolerances. Incorporating such margins ensures that short-term pressure fluctuations or minor deviations in operating conditions do not immediately lead to face instability or adverse ground response.

Slurry Pressure Management

The pressure distribution along the tunnel face varies with depth, a characteristic that must be carefully managed to maintain operational stability. Due to hydrostatic effects, pressure is typically greater at the tunnel invert compared to the crown, leading to a non-uniform pressure profile that must be accounted for during design and operation [319]. Accurate design calculations determine acceptable pressure limits that establish safe operational thresholds to mitigate risks of collapse or excessive deformation [579].

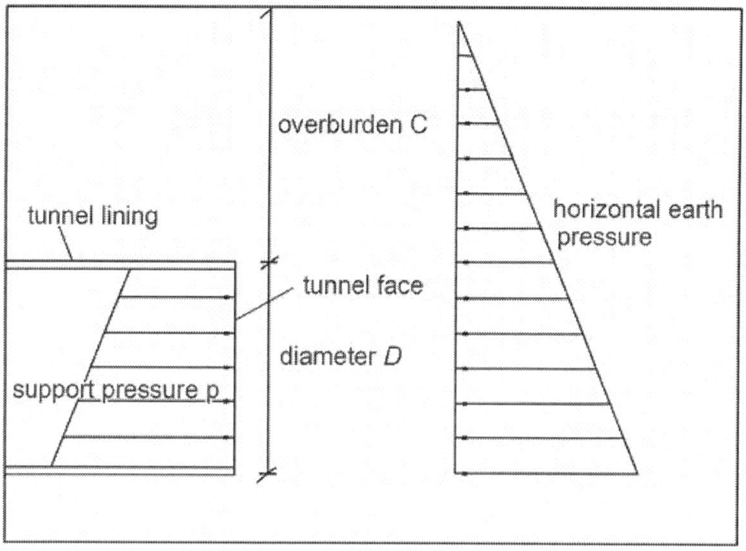

Figure 86: Stress condition at excavation face.

Figure 86 illustrates the principle of face support pressure balance in mechanised tunnelling, particularly as applied to EPB and slurry TBMs, by comparing the stresses acting on the tunnel face with the support pressure applied by the machine.

On the right-hand side, the triangular shaded area represents the horizontal earth pressure acting on the tunnel face. This pressure increases with depth because it is derived from the weight of the overburden above the tunnel. The vertical distance labelled overburden C represents the cover from ground surface to the tunnel axis (or crown reference), and the arrows indicate that horizontal stress acts laterally toward the excavation. The triangular shape reflects the fact that horizontal stress is lowest near the ground surface and highest at the tunnel invert.

At the centre of the diagram, the tunnel face is shown with a diameter D, representing the circular excavation profile created by the TBM. This face is the critical zone where in-situ stresses are released during excavation. Without support, the imbalance between the surrounding ground pressure and the void created by excavation would lead to face deformation or collapse.

On the left-hand side, the triangular shaded area labelled support pressure p represents the pressure applied by the TBM at the tunnel face. This pressure is generated either by conditioned soil in an EPB TBM or by pressurised slurry in a slurry TBM. Like the earth pressure, the support pressure is shown as varying with depth, increasing toward the invert to reflect the hydrostatic or pseudo-hydrostatic nature of the support medium.

The key concept illustrated by the diagram is that face stability is achieved when the support pressure distribution closely matches the horizontal earth pressure distribution. When the applied support pressure *p* balances the earth pressure acting across the face, the net force on the tunnel face is approximately zero. This balanced condition prevents inward movement of the ground (collapse and settlement) while also avoiding outward displacement (heave or blow-out).

The tunnel lining shown at the top left indicates that once excavation progresses and segments are installed, loads are transferred from the ground through the lining system. However, at the face itself, stability depends entirely on maintaining the correct support pressure profile before the lining is in place.

Role of the Filter Cake

A filter cake forms at the tunnel face as slurry infiltrates the surrounding soil, creating a low-permeability barrier that plays a vital role in slurry pressure management. The presence of this filter cake minimizes slurry loss while facilitating pressure transmission onto the soil structure [359]. However, the stability of this filter cake is essential; too thin a cake can lead to significant slurry loss, while an excessively thick cake can reduce excavation efficiency and create uneven pressure transmission dynamics [580]. Thus, careful management of the slurry's rheological properties and continuous monitoring is critical for maintaining an effective filter cake and, by extension, overall stability [581].

Pressure Control During Excavation

Real-time pressure control during excavation is achieved through the regulation of slurry circulation rates, pump pressures, and slurry levels within the excavation chamber. Leveraging pressure sensors allows for this immediate feedback, ensuring that operators can adapt their strategies to maintain optimal pressure levels and prevent destabilization of the tunnel face [30]. A rigid correlation between excavation rate and slurry circulation is crucial; an imbalance can trigger either insufficient face support or excessive pressures, both of which can lead to adverse outcomes [582].

Slurry Loss and Pressure Instability

One of the pivotal challenges in managing slurry pressure is the potential for slurry loss into permeable or fractured ground, which can drastically compromise face stability. To counteract these losses, various mitigation strategies may be employed, including altering slurry viscosity or density, implementing loss-control additives, and adjusting excavation rates [318]. Recognizing early signs of pressure fluctuations is vital for timely intervention—unexplained pressure drops, fluctuations in slurry consumption, or changes in flow can signal impending instability [359, 374].

Interaction with Groundwater

A comprehensive understanding of groundwater dynamics is paramount in slurry pressure management, particularly in saturated conditions. The interaction between earth pressures and pore water pressures must be continually monitored and managed; if groundwater pressure surpasses the applied slurry pressure, inflow and instability may occur [374, 583]. Effective adjustments to the slurry pressure must be responsive to changes in groundwater levels or permeability to ensure ongoing stability during excavation [319].

Relationship to Settlement Control

Maintaining optimal slurry pressure directly influences ground deformation and surface settlement. Inadequate pressure increases the likelihood of voids at the tunnel face, leading to settlement above the tunnel, while excessive pressure can result in ground heave and potential structural displacement [579]. Continuous monitoring, integrating face pressure and surface settlement data, allows operators to adjust pressures efficiently, ensuring ground movements remain within permissible limits throughout the tunnelling process [582].

Safety Margins and Operational Windows

The design of slurry pressures incorporates safety margins to account for geological variability and potential pressure losses within the system. These margins define operational windows, where the lower limit is the collapse pressure and the upper limit is the blow-out pressure [319]. The maintenance of these boundaries is a continuous operational concern, requiring seamless coordination among TBM operators, slurry management teams, and geotechnical monitoring personnel to navigate the complexities of safe tunnelling [374].

Over-Excavation and Settlement Control

Over-excavation refers to any excavation of ground beyond the theoretical tunnel profile required for the installed lining. In TBM tunnelling it most commonly appears as excess annular gap, local “overbreak” around the shield, or additional volume loss at the face and along the shield due to disturbance and extraction effects. Because tunnels are essentially a volume-balance problem, any avoidable extra excavation volume must be compensated by grout, confinement, and careful pressure control. If it is not, the ground relaxes toward the void, producing convergence around the tunnel and, in shallow ground, surface settlement.

A controlled annular void is unavoidable because the shield and cutterhead excavation diameter must exceed the outer diameter of the segmental lining to allow clearance for steering, shield passage, segment erection, and tail seals. However, over-excavation occurs when this clearance becomes larger than intended or when ground is disturbed beyond the excavation line. Common contributors include cutterhead overcut, excessive overcut on curves, shield taper/clearance effects, tool wear, uneven cutting in mixed ground, and face instability that causes ravelling or slumping into the excavation chamber. In granular or running ground, small losses at the face can rapidly propagate into larger voids. In cohesive ground, smearing and remoulding can still create a loosened zone that behaves as a volume loss source.

In mechanised tunnelling, over-excavation is also strongly influenced by operational balance. If extraction exceeds inflow (for EPB: screw conveyor removes too much; for slurry: spoil removal/pressure control leads to deficit), the chamber pressure can drop and the face can yield, creating additional volume loss. Similarly, if shield friction rises and the operator compensates with steering or thrust in a way that disturbs the ground, the excavation envelope can enlarge locally.

Settlement is driven by ground loss and stress relief. When additional void space is created around the tunnel—at the face, around the shield, or in the tail void—the surrounding ground moves inward to re-establish equilibrium. In shallow tunnels this inward movement migrates upward, forming a settlement trough at the surface. In cohesive soils, settlement may develop gradually as consolidation follows initial disturbance; in granular soils, settlement can occur quickly as particles rearrange and flow toward the void. The key point is that over-excavation increases the "missing volume" that must be supported or filled; if not managed, that missing volume becomes ground movement.

In practice, settlement control focuses on three linked zones:

1. **The face (front of the TBM):** Losses here are associated with inadequate support pressure, poor conditioning (EPB), slurry loss (slurry TBM), or unstable ground structures. Face losses typically produce the most direct and immediate settlement response.
2. **The shield (around the machine body):** As the shield advances, the annulus created by excavation and shield clearance can permit relaxation, particularly in soft ground or low confinement. Steering corrections and articulation can exacerbate this in curves.
3. **The tail void (behind the shield):** When a ring is erected, a continuous gap exists between the lining and ground. If annular grouting is delayed, incomplete, or shrinks

excessively, the ground can relax into the void, producing settlement even if the face was stable.

These zones are not independent. Face instability can create larger tail void demands; poor tail grouting can amplify movement initiated at the face.

Settlement control strategies for over-excavation

<u>Control excavation geometry and overcut</u>

The first settlement control measure is minimising the amount of unnecessary excavation. This includes careful design and operation of overcut tools, controlling shield clearance, and managing steering so the TBM does not "plough" or enlarge the profile. In curves, overcut must be sufficient for steering but not excessive; operators typically manage this through advance strategy (shorter strokes, controlled articulation) and careful monitoring of line and level to prevent repeated corrections that disturb the ground.

<u>Maintain face stability by managing support pressure</u>

Because over-excavation and settlement often initiate at the face, maintaining the correct support pressure profile is central.

- **EPB:** settlement control depends on stable chamber pressure, consistent chamber fill, controlled screw conveyor extraction, and effective soil conditioning to create a pressure-transmitting, low-permeability muck. Pressure fluctuations—especially under-pressure events—are strongly correlated with volume loss and settlement.
- **Slurry:** settlement control depends on maintaining slurry pressure within the design window, preventing slurry loss, and ensuring effective filter cake formation. Loss of slurry into permeable ground can cause pressure drops and void creation; excessive pressure can cause heave or hydraulic fracturing.

In both systems, the practical objective is to keep pressure stable within the allowable envelope (crown and invert limits), rather than chasing a single number.

<u>Optimise annular grouting to eliminate tail void loss</u>

Annular (tail void) grouting is the key control for the over-excavation that is unavoidable. Settlement is reduced when the annulus is filled quickly, completely, and uniformly with a grout that gains stiffness rapidly and maintains volume.

Key controls include grout volume tracking (placed volume vs theoretical annulus), grout pressure control (enough to support ground but not jack the lining), and mix design to reduce shrinkage and bleeding. Two-component grouts are widely used because they remain pumpable but set quickly, restoring confinement around the lining shortly after ring erection. Poor grouting practice—underfilling, delayed injection, blocked ports, low pressure, or high shrinkage—can turn the tail void into a persistent source of settlement.

Manage operational parameters to reduce disturbance

Even with correct pressures and grouting, poor operational discipline can generate over-excavation. Settlement control therefore includes managing advance rate, cutterhead torque, penetration per revolution, and stoppage frequency. Frequent stops in unstable ground can allow relaxation at the face; restarting can induce sudden changes in pressure and extraction. Excessive penetration can "tear" the ground rather than cut it cleanly. A stable, steady advance with controlled muck removal is generally the most settlement-friendly approach.

Monitoring and feedback control

In practice, settlement control is a closed-loop process. Operators correlate surface settlement data (levelling points, prisms, InSAR where applicable), subsurface instrumentation (extensometers, inclinometers), and TBM data (chamber pressure, screw speed, slurry pressure/flow, torque, thrust, grout volumes and pressures). When settlement trends increase, the typical response is to tighten the pressure control window, adjust conditioning, refine extraction, and increase focus on grout completeness and pressure consistency. The aim is to reduce volume loss at its source rather than attempting to compensate after settlement has developed.

Contingency measures when over-excavation is unavoidable

In difficult ground (voided strata, coarse gravels, high permeability, faults, or mixed face conditions), over-excavation may still occur despite best practice. Contingency settlement controls include pre-treatment (jet grouting, ground freezing, permeation grouting), forepoling/spiling in transitional ground, reduced advance rates, staged excavation, and, in extreme cases, changing the support method (e.g., shifting from EPB suitability to slurry where permeability and water pressures demand it). Compensation grouting from the surface or shafts may be used to correct settlement during construction in sensitive urban environments.

Over-excavation and settlement control in TBM tunnelling is fundamentally about minimising avoidable excavation volume and preventing ground loss at the face, along the shield, and in the tail void. Face pressure stability (EPB or slurry), careful steering and overcut management, and fast, complete annular grouting are the primary technical levers. Effective monitoring closes the loop, allowing the team to detect volume loss early and adjust pressures, conditioning, extraction, and grouting before settlement becomes unacceptable.

The following outlines a practical, numbers-based example that shows what over-excavation looks like in a TBM drive, where the "missing volume" comes from, and how it turns into settlement if not controlled.

Example: EPB TBM in shallow sandy silt under an urban road

Tunnel/lining geometry

- Excavated diameter (cutterhead): 6.30 m
- Segment outer diameter: 6.10 m
- Ring width (advance per ring): 1.50 m
- Normal annular gap (radial): $(6.30 - 6.10)/2 = 0.10\text{m}$ = 100 mm

1) The "normal" unavoidable annulus (what you expect)

Cross-sectional annulus area:

$$A = \frac{\pi}{4}(D_{exc}^2 - D_{seg}^2) = \frac{\pi}{4}(6.30^2 - 6.10^2)$$
$$= \frac{\pi}{4}(39.69 - 37.21) = \frac{\pi}{4}(2.48) \approx 1.95\ \text{m}^2$$

Volume per ring (1.5 m wide):

$$V_{annulus} = A \times 1.5 \approx 1.95 \times 1.5 = 2.93\ \text{m}^3$$

So the team expects to inject roughly ~2.9 m^3 of tail grout per ring (plus a small allowance for minor deformation).

2) Over-excavation event (what goes wrong)

A short reach of mixed ground causes:

- cutterhead overcut/uneven cutting,
- minor face ravelling (sand lenses),
- more steering corrections in a gentle curve.

The effective excavated diameter increases to 6.40 m over a few rings.

Recalculate annulus area:

$$A' = \frac{\pi}{4}(6.40^2 - 6.10^2) = \frac{\pi}{4}(40.96 - 37.21) = \frac{\pi}{4}(3.75) \approx 2.95\ \text{m}^2$$

Volume per ring:

$$V' = 2.95 \times 1.5 \approx 4.42\ \text{m}^3$$

Extra void created (over-excavation) per ring:

$$\Delta V = 4.42 - 2.93 = 1.49\ \text{m}^3/ring$$

If this occurs for 10 rings (~15 m of drive), the additional void volume is:

$$1.49 \times 10 \approx 14.9 \, \text{m}^3$$

That ~15 m^3 is the "missing volume" that must be filled and supported to prevent ground movement.

3) How it turns into settlement (the three zones)

Zone A: The face

During these rings, the screw conveyor is run slightly fast to maintain production. Extraction briefly exceeds inflow, so chamber pressure dips below target for short periods. That creates small face volume loss (ground moves into the chamber), which is often the fastest pathway to settlement, especially in granular layers.

Zone B: Around the shield

The slightly larger excavation plus steering corrections increases the clearance around the shield. The ground can relax into this enlarged envelope, especially if confinement is low (shallow cover) or the soil is loose.

Zone C: Tail void behind the shield

Even if the annulus should be filled with ~4.4 m^3/ring in this reach, the grout pump continues delivering the "normal" setpoint (~3.0 m^3/ring) because the change is not detected immediately. Result: underfilling by ~1.4–1.5 m^3 per ring, leaving voids that the ground will migrate into.

4) What you would observe on site

Within hours to days (often quickly in sandy silt), you might see:

- Surface settlement markers showing a developing trough above chainage ~15 m
- TBM data showing:
  - short pressure drops at the face,
  - increased torque/thrust variability,
  - grout volume placed below theoretical (placed vs required),
  - higher grout take later as voids gradually accept grout.

5) The settlement-control response (how you stop it)

A typical response sequence is:

1. Stabilise face pressure

- tighten the allowable pressure band,
- slow advance slightly,
- bring screw conveyor extraction back into balance,
- increase/optimise conditioning to reduce permeability and improve "plug flow".

2. Correct grout volume and pressure

- immediately update the theoretical grout take to ~4.4 m^3/ring,
- increase grout volume to match (plus allowance),
- maintain controlled grout pressure (enough to fill/support, not jack the lining).

3. Reduce further over-excavation

- reduce overcut in the curve if possible,
- minimise steering oscillations ("hunting"),
- check tool wear/overcutters and cutterhead condition.

4. If settlement continues

- implement contingency measures: pre-treatment ahead, slower advance, or compensation grouting (urban case).

This example shows that over-excavation isn't abstract: a seemingly small change (6.30 m → 6.40 m) can add ~1.5 m^3 of void per ring, and if grout/pressure control does not immediately compensate, that missing volume becomes ground movement, then settlement.

Blow-Outs and Face Collapse Mechanisms

Blow-outs and face collapse represent opposite failure modes of tunnel face instability, both arising from a loss of equilibrium between in-situ ground stresses and the support pressure applied at the tunnel face. A face collapse occurs when the applied support pressure is insufficient to resist inward ground movement, while a blow-out occurs when support pressure exceeds the ground's capacity to contain it, resulting in outward displacement or hydraulic failure. Both mechanisms can cause severe ground disturbance, sudden settlement or heave, damage to adjacent infrastructure, and significant safety risks.

Face collapse occurs when the face support pressure falls below the minimum level required to maintain stability. This under-pressure condition allows the ground ahead of the tunnel to deform and move inward toward the excavation. Collapse rarely occurs instantaneously; instead, it typically develops progressively. Early indicators include local loosening, ravelling, or running ground at the face. As confinement is lost, stresses redistribute into the surrounding ground, enlarging the failure zone. In shallow tunnels, this failure zone can propagate upward, forming settlement troughs or, in severe cases, sinkholes at the surface.

Common triggers for face collapse include insufficient EPB or slurry pressure, sudden pressure drops caused by excessive spoil extraction or slurry loss, and poor soil conditioning that results in high permeability or low cohesion. Highly permeable or running ground, such as sands and gravels, is particularly vulnerable, as are mixed-face conditions where unfavourably oriented discontinuities exist at soil–rock interfaces. Ground behaviour strongly influences the collapse mechanism: in granular soils, collapse can occur rapidly as particles flow toward the void, producing immediate settlement, while in cohesive soils collapse may develop more slowly as remoulded soil progressively loses strength and consolidates over time.

Blow-outs occur when face support pressure exceeds the resisting capacity of the ground. This over-pressure condition forces soil, slurry, or water outward from the face and can lead to surface uplift, hydraulic fracturing, or sudden release of pressurised material. When support pressure is too high, effective stress in the ground near the face is reduced. If effective stress approaches zero, the ground loses shear strength and the pressurised medium seeks preferential flow paths such as fractures, bedding planes, or loose zones. The result may be upward heave, slurry eruption, or sudden loss of confinement at the tunnel face.

Blow-outs are commonly associated with excessive EPB chamber pressure, over-dense or over-pressurised slurry, and rapid pressure increases that are not matched by controlled spoil extraction. They are particularly likely in areas of low overburden cover, such as beneath rivers or sensitive urban infrastructure, where the confining stress is limited. In highly permeable soils, pressure is transmitted rapidly through the ground, increasing the risk of hydraulic instability and outward failure.

In EPB TBMs, face collapse is most often linked to loss of chamber pressure due to excessive screw conveyor extraction or inadequate soil conditioning that allows water inflow and pressure dissipation. Blow-outs are less common but can occur if chamber pressure is increased too aggressively, especially in shallow ground, leading to surface heave or cracking. The plasticity and permeability of the conditioned muck are critical; if the muck cannot transmit pressure uniformly, localised zones of collapse or blow-out may develop even when average chamber pressure appears acceptable.

In slurry TBMs, face collapse typically results from slurry loss into permeable ground, fractures, or cavities, causing a rapid reduction in effective support pressure. Blow-outs occur when slurry pressure exceeds the ground's resistance, potentially leading to slurry escape to the surface or into adjacent strata. The formation and maintenance of a stable filter cake are central to preventing both failure modes. A weak or discontinuous filter cake increases the risk of collapse, while excessive pressure acting against a well-sealed face increases the risk of blow-out.

Low overburden significantly increases the likelihood of both face collapse and blow-out. With limited confining stress, the allowable pressure window becomes narrow, meaning that small deviations below the lower limit can trigger collapse, while small deviations above the upper limit can cause uplift or surface expression of pressure. This sensitivity makes shallow urban tunnelling particularly vulnerable and places high demands on precise pressure control and real-time monitoring.

Early recognition of instability is essential to preventing major failures. Warning signs include sudden changes in face or chamber pressure, unexpected increases in settlement or surface heave, changes in torque, thrust, or penetration rate, slurry losses or abnormal slurry returns, and irregular spoil characteristics such as running, watery, or segregated material. Rapid response to these indicators can prevent local instability from escalating into full face failure.

Prevention of both collapse and blow-out relies on maintaining support pressure within a defined pressure window bounded by the collapse pressure at the lower limit and the uplift or blow-out pressure at the upper limit. This requires stable advance rates, controlled spoil extraction, consistent conditioning, and avoidance of abrupt operational changes. Pressure adjustments should be gradual and coordinated with excavation and material removal. In high-risk ground conditions, additional measures such as ground improvement, reduced advance lengths, staged excavation, or changes in support method may be required. In sensitive environments, compensation grouting or pressure relief strategies may be used as contingency measures.

The following presents an example that demonstrates both face collapse and blow-out, tied directly to EPB and slurry TBM operation.

Example: Face Collapse and Blow-Out in Shallow Urban Tunnelling

A 6.5 m diameter tunnel is being driven beneath a dense urban area at a depth of approximately 12 m using an Earth Pressure Balance (EPB) TBM. The ground consists of interbedded silty clay and fine sand with a groundwater table close to the surface. Surface buildings and buried utilities impose strict settlement limits.

Example 1: Face Collapse Due to Under-Support

During advance beneath a road intersection, operators increase the screw conveyor speed to improve production. Although the average chamber pressure appears acceptable, the increased extraction rate causes a local pressure drop at the face, particularly in the sandy layers where permeability is higher.

Within minutes, operators observe:

- A drop in chamber pressure fluctuations
- Wet, running spoil at the screw conveyor outlet
- Increasing penetration rate with reduced cutterhead torque

At the face, the reduced support pressure allows sand to flow inward toward the excavation chamber. Initially, this manifests as local ravelling at the face. As excavation continues, the loss of confinement causes the failure zone to enlarge upward. Within hours, monitoring instruments detect surface settlement above the tunnel alignment, and a shallow depression forms in the pavement.

This event illustrates face collapse: the applied support pressure fell below the minimum required to resist inward ground movement, allowing progressive failure and volume loss that propagated to the surface.

Example 2: Blow-Out Due to Over-Support

Later in the drive, the TBM approaches a zone of very low cover beneath a creek crossing. To prevent settlement and water inflow, operators increase chamber pressure conservatively. However, the pressure increase is applied rapidly and exceeds the allowable upper limit for the low overburden conditions.

As chamber pressure rises:

- Screw extraction is not increased proportionally
- Effective stress in the ground ahead of the face is reduced
- Pore water pressure is exceeded locally

The over-pressurised conditioned soil forces its way into existing fractures and loose zones. Within minutes, muddy water and soil erupt at the creek bed, accompanied by surface heave and cracking of the surrounding ground. Chamber pressure drops suddenly as material escapes, and excavation must be halted.

This event represents a blow-out: face support pressure exceeded the ground's ability to contain it, leading to outward displacement and hydraulic failure.

Comparison of the Two Failure Modes in the Example

- In the face collapse, insufficient pressure allowed the ground to move *into* the tunnel, causing settlement.
- In the blow-out, excessive pressure forced material *out of* the tunnel, causing heave and surface expression.
- Both failures were caused by loss of equilibrium at the tunnel face, despite different operational errors.

- In both cases, shallow cover and mixed ground conditions narrowed the allowable pressure window, making the system highly sensitive to pressure deviations.

This example demonstrates that tunnel face stability is fundamentally a pressure balance problem rather than a simple matter of applying higher support pressure. Increasing pressure beyond what the ground can sustain does not inherently improve stability and can, in fact, introduce new failure risks.

When support pressure falls below the required level, the ground moves inward toward the excavation, leading to face collapse and surface settlement. Conversely, when support pressure exceeds the ground's capacity to contain it, outward displacement occurs in the form of blow-out or heave. Both outcomes are direct consequences of operating outside the allowable pressure range.

Stable TBM operation therefore depends on maintaining face support pressure within a narrow, well-defined operating window. Achieving this balance requires coordinated control of TBM advance rate, spoil extraction, and soil conditioning so that pressure at the face remains stable and evenly distributed.

Real-time monitoring and disciplined operational response are essential components of this process. Continuous interpretation of pressure data, spoil characteristics, machine performance, and ground response allows emerging instability to be detected early and corrected before local disturbances escalate into major ground failure.

Blow-outs and face collapse represent the two extremes of face instability in TBM tunnelling. Face collapse results from under-support and leads to inward ground movement and settlement, while blow-outs result from over-support and cause outward displacement, hydraulic failure, or heave. Both mechanisms are governed by the same fundamental principle: loss of equilibrium at the tunnel face. Successful tunnelling depends on maintaining support pressure within a narrow but well-defined operating window, supported by careful monitoring, responsive operation, and a sound understanding of ground behaviour across varying geological conditions.

Pressure Calculations and Control Strategies

Pressure calculations and control strategies are central to maintaining tunnel face stability in both EPB and slurry TBM operations. Their purpose is to ensure that the support pressure applied at the tunnel face balances the in-situ stresses from soil and groundwater closely enough to prevent collapse, excessive deformation, or blow-out, while also limiting surface settlement and heave. In practice, pressure control is not a single calculation but an ongoing, feedback-driven process that links design assumptions with real-time operational behaviour.

From a design perspective, required face support pressure is calculated by considering the total stresses acting at tunnel depth. These stresses include vertical overburden stress from the weight of soil above the tunnel, horizontal earth pressure derived using appropriate earth pressure coefficients (typically at-rest conditions for EPB design), and pore water pressure where groundwater is present. The calculated support pressure is therefore intended to counteract both effective earth stress and groundwater pressure at the face. Because stresses increase with depth, designers usually consider a pressure distribution across the tunnel face, with higher pressure at the invert and lower pressure at the crown, rather than assuming a uniform value.

These calculations define a pressure window rather than a single operating pressure. The lower bound of this window corresponds to the collapse pressure, below which inward ground movement and face instability may occur. The upper bound corresponds to the uplift or blow-out pressure, above which the ground may experience heave, hydraulic fracturing, or outward failure. The width of this window depends on ground type, permeability, strength, and overburden depth. In shallow tunnels or weak ground, the window can be very narrow, requiring especially precise control.

Once the design pressure window is established, it is translated into TBM operating parameters. In EPB machines, chamber pressure is controlled through the balance between excavation rate and spoil extraction via the screw conveyor. Advance rate governs how much ground enters the pressure chamber, while screw speed and torque regulate how much conditioned soil is removed. Soil conditioning plays a critical role by modifying the excavated material so it behaves as a pressure-transmitting, low-permeability mass. In slurry TBMs, pressure is controlled by regulating slurry density, slurry circulation rate, and the pressure applied within the excavation chamber, while ensuring that slurry losses are minimised and a stable filter cake forms at the face.

The following outlines a practical way to calculate a pressure window (lower = collapse, upper = uplift/blow-out) and then convert it into an axis setpoint window you can actually operate the TBM to.

1) Choose the calculation approach

For soft-ground EPB/slurry design checks, a common first-pass approach is to use effective stress earth pressure + pore water pressure:

$$p_{req}(z) = K\,\sigma_v'(z) + u(z)$$

Where:

- z= depth below ground surface at the point considered (crown, axis, invert)
- $\sigma_v'(z) = \gamma' z$= effective vertical stress
- $u(z) = \gamma_w z$= pore water pressure

- $\gamma' = \gamma_{sat} - \gamma_w$= submerged unit weight
- K= earth pressure coefficient

Then define:

- **Lower bound (collapse)**: use $K = K_a$(active) as a conservative "minimum support" proxy
- **Upper bound (uplift/blow-out)**: use $K = K_p$(passive) as an "upper limit" proxy

For a frictional soil with effective friction angle ϕ:

$$K_a = \frac{1 - \sin\phi}{1 + \sin\phi}, K_p = \frac{1 + \sin\phi}{1 - \sin\phi}$$

In cohesive/undrained clays you'd typically use an undrained stability formulation (e.g., s_u-based) for the *lower* bound—this example keeps it to the widely used ϕ'-based framework.

2) Calculate crown and invert bounds (this defines the "pressure envelope")

Let:

- Tunnel diameter D
- Cover to crown C
  So:
- Crown depth $z_c = C$
- Invert depth $z_i = C + D$

Compute:

$$p_{min}(z) = K_a\gamma' z + \gamma_w z$$
$$p_{max}(z) = K_p\gamma' z + \gamma_w z$$

Worked example (numbers)

Assume:

- $D = 6.0$ m
- $C = 10$ m→ $z_c = 10$ m, $z_i = 16$ m
- Saturated unit weight $\gamma_{sat} = 20$ kN/m^3
- Water unit weight $\gamma_w = 9.81$ kN/m^3

- $\gamma' = \gamma_{sat} - \gamma_w = 10.19$ kN/m^3
- $\phi' = 30° \rightarrow K_a = 0.333, K_p = 3.0$

Crown:

$$p_{min,c} = 0.333(10.19)(10) + 9.81(10) \approx 132 \text{ kPa}$$
$$p_{max,c} = 3.0(10.19)(10) + 9.81(10) \approx 404 \text{ kPa}$$

Invert:

$$p_{min,i} = 0.333(10.19)(16) + 9.81(16) \approx 211 \text{ kPa}$$
$$p_{max,i} = 3.0(10.19)(16) + 9.81(16) \approx 646 \text{ kPa}$$

So the pressure envelope across the face is roughly:

- Crown: 132 to 404 kPa
- Invert: 211 to 646 kPa

3) Convert the envelope into a TBM "axis pressure window"

TBM support pressure varies with height (hydrostatic-like). If you control pressure at the **axis**, then:

$$p_{crown} = p_{axis} - \gamma_m \frac{D}{2}$$
$$p_{invert} = p_{axis} + \gamma_m \frac{D}{2}$$

Where γ_m is the unit weight of the support medium (conditioned muck "equivalent" or slurry), e.g. ~11 kN/m^3.

To stay inside the envelope:

- $p_{crown} \geq p_{min,c}$ and $p_{crown} \leq p_{max,c}$
- $p_{invert} \geq p_{min,i}$ and $p_{invert} \leq p_{max,i}$

That gives:

$$p_{axis} \geq \max\left(p_{min,c} + \gamma_m \frac{D}{2}, p_{min,i} - \gamma_m \frac{D}{2}\right)$$
$$p_{axis} \leq \min\left(p_{max,c} + \gamma_m \frac{D}{2}, p_{max,i} - \gamma_m \frac{D}{2}\right)$$

Using $\gamma_m = 11$ kN/m^{3}and $D/2 = 3$ m $\Rightarrow \gamma_m D/2 = 33$ kPa:

Lower axis bound:

- $p_{min,c} + 33 = 132 + 33 = 165$ kPa
- $p_{min,i} - 33 = 211 - 33 = 178$ kPa
  → $p_{axis,min} = 178$ kPa

Upper axis bound:

- $p_{max,c} + 33 = 404 + 33 = 437$ kPa
- $p_{max,i} - 33 = 646 - 33 = 613$ kPa
  → $p_{axis,max} = 437$ kPa

Axis pressure window: ~178 to 437 kPa

That is the practical operating "window" for your axis setpoint, assuming the stated soil/water parameters.

4) What to do with it (control strategy)

- Start near the middle of the axis window (often slightly conservative toward the lower-middle to reduce heave risk in shallow cover).
- Keep short-term fluctuations inside the window (avoid sharp dips/spikes).
- If monitoring shows settlement → move upward within the window (and check extraction/conditioning/grout volumes).
- If heave/cracking/slurry loss signs → move downward within the window.

Pressure control strategies rely heavily on coordinated operation rather than independent parameter adjustment. Increasing chamber pressure without adjusting extraction or circulation can lead to over-support and blow-out, while increasing extraction without compensating pressure input can cause under-support and collapse. Operators therefore manage pressure through small, incremental changes, ensuring that inflow and outflow volumes remain balanced and that pressure fluctuations are minimised.

Real-time monitoring is fundamental to effective pressure control. Instrumentation typically includes face pressure sensors, chamber pressure gauges, screw torque and speed measurements, slurry pressure and flow sensors, and machine performance data such as thrust, torque, and penetration rate. These TBM parameters are interpreted alongside external monitoring, including surface settlement points, subsurface instrumentation, and observations of spoil characteristics. Changes in spoil consistency, unexpected torque variations, or abnormal pressure fluctuations often provide early warning of instability before visible ground movement occurs.

Pressure calculations are therefore used as a baseline, while actual operating pressures are continuously refined through feedback control. If settlement trends indicate volume loss, operators may slightly increase face pressure, adjust conditioning, or reduce extraction rates. If signs of heave or slurry loss appear, pressure is reduced or redistributed, and extraction or circulation is modified accordingly. This adaptive approach recognises that ground conditions are inherently variable and that theoretical calculations cannot capture all local effects.

To manage uncertainty, safety margins are built into both calculations and operational strategies. These margins account for geological variability, limitations in ground investigation data, sensor accuracy, and short-term operational disturbances such as stoppages or restarts. The goal is not to eliminate all fluctuation, which is unrealistic, but to ensure that fluctuations remain within tolerable limits that do not compromise stability.

In summary, pressure calculations define the required support conditions and establish safe operating limits, while pressure control strategies translate those calculations into coordinated TBM operation. Effective tunnelling depends on maintaining face pressure within a narrow operating window through balanced excavation and extraction, appropriate conditioning or slurry management, and continuous monitoring with responsive adjustment. Together, these elements ensure face stability, control settlement, and reduce the risk of collapse or blow-out throughout the tunnelling process.

Chapter 10

Spoil Handling and Logistics

Spoil handling and logistics form a critical interface between excavation at the tunnel face and surface-based construction, environmental management, and project delivery. In mechanised tunnelling, particularly with Tunnel Boring Machines, the removal, transport, treatment, and disposal or reuse of excavated material are not secondary activities but core operational systems that directly influence face stability, advance rate, equipment reliability, and regulatory compliance. Inefficient or poorly coordinated spoil logistics can rapidly constrain TBM performance, regardless of cutterhead design or ground support strategy.

This chapter examines spoil handling as an integrated engineering process rather than a simple material removal task. It explores how the physical, behavioural, and chemical characteristics of muck govern its movement through conveyors, pipelines, rail systems, and separation plants, and how these characteristics must be actively managed through conditioning, moisture control, and system selection. Particular attention is given to the contrasting requirements of Earth Pressure Balance and slurry TBM operations, where spoil itself becomes a functional component of face support and pressure control.

By linking underground excavation processes with surface infrastructure, waste treatment, and reuse pathways, this chapter highlights the systemic nature of spoil management in modern tunnelling projects. Understanding spoil handling and logistics is essential not only for maintaining continuous TBM operation, but also for achieving settlement control, minimising environmental impact, and meeting increasingly stringent regulatory and sustainability expectations across the full lifecycle of a tunnel project.

Muck Characteristics

The characteristics of muck, particularly in the context of Tunnel Boring Machines, play a critical role in determining operational efficiency and overall project performance. These characteristics encompass physical and behavioural properties of excavated material as it transitions through the muck handling chain, significantly influencing extraction reliability, transportation efficiency, and environmental compliance.

In tunnelling, muck is the excavated material produced by the Tunnel Boring Machine (TBM) as it cuts through soil or rock. It is the mixture of ground materials removed from the tunnel face and transported away from the TBM through the spoil handling system.

More specifically, muck can include:

- Soil (clays, silts, sands, gravels)
- Rock fragments or crushed rock
- Groundwater mixed with the excavated material
- Added conditioning agents (such as foam, polymers, or bentonite in EPB and slurry TBMs)

The term is deliberately broad because muck is not just "waste soil." Its physical behaviour—how it flows, sticks, drains water, or breaks apart—directly affects face stability, pressure control, conveyor performance, separation plant efficiency, and environmental handling.

In EPB TBM tunnelling, muck is expected to behave as a plastic, pressure-transmitting mass that supports the tunnel face while still flowing controllably through the screw conveyor and downstream logistics.

In slurry TBM tunnelling, muck refers to excavated particles suspended in slurry, which are hydraulically transported to a separation plant where solids are removed and the slurry is recycled.

In short, muck is the engineered spoil of tunnelling—material whose properties must be actively managed, not simply removed, to ensure safe, efficient, and stable TBM operation.

The physical properties of muck such as grain size distribution, moisture content, and density are pivotal in shaping its behaviour. For instance, grain size distribution affects particle compaction and flow behaviour; well-graded sands can create dense plugs in transfer systems, while uniform sands may segregate and run unconditionally if not adequately conditioned [584]. The moisture content plays a critical role in determining whether muck behaves as a solid, a plastic mass, or a slurry, with insufficient moisture leading to dust generation and excessive moisture resulting in instability and pressure loss during extraction [585, 586]. The density and bulk unit weight of muck influence the load on transportation systems, impacting energy requirements and wear rates on mechanical components [587, 588].

Additionally, muck consistency, characterized by plasticity and workability, is particularly essential for Earth Pressure Balance tunnelling operations. An ideal EPB muck exhibits a plastic consistency with moderate cohesion and limited permeability, which aids in transmitting pressure uniformly at the excavation face. This ensures continuous flow through conveyor systems and helps avoid issues such as blockages and excessive wear on equipment [589, 590].

The behavioural properties of muck under operational conditions focus on how it withstands mechanical handling and environmental exposure. For EPB tunnelling, effective muck conditioning ensures a controlled, deformable mass that supports chamber pressure [581, 585]. This is critical in preventing conditions that might lead to torque spikes or conveyor overloads, which can occur if the muck is either too stiff or excessively wet [588, 591]. The incorporation of conditioning agents such as foams and polymers can significantly improve muck workability, increasing the efficiency of the tunnelling process by reducing friction and enhancing flowability [293, 589].

In slurry tunnelling, muck behaviour is dictated by its interactions with the slurry medium. Key characteristics such as particle size, shape, and density affect slurry viscosity and the efficiency of hydraulic transport systems. Muck must be sufficiently suspended in slurry while also being separable at the treatment plant to prevent excessive wear on pumping systems and ensure consistent operational pressure [588, 591]. Fine content and clay mineral properties contributing to slurry stability are critical, as high proportions of coarse gravels can lead to wear and blockages in pipeline systems [335].

Chemical composition is an essential dimension of muck characteristics, influencing handling and disposal protocols. The presence of sulphides or heavy metals necessitates specific waste management practices to mitigate environmental impacts. This aspect is particularly vital in urban tunnelling projects where contamination concerns are paramount [592]. The interactions between muck and conditioning additives, such as bentonite or foams, can alter the chemical behaviour of excavated materials, potentially imposing constraints on treatment and disposal pathways [590, 593].

Example: Influence of Muck Characteristics on TBM Spoil Handling and Performance

A 7.2 m diameter tunnel is being excavated beneath an urban corridor using a Tunnel Boring Machine (TBM). The alignment passes through variable ground conditions consisting of silty clay, fine to medium sand, and occasional gravel lenses, with groundwater present near tunnel crown level. The project includes strict settlement limits and environmental controls on spoil handling and disposal.

As excavation progresses, the TBM generates muck, defined as the excavated material removed from the tunnel face and transported through the spoil handling system. In this project, the muck consists of a heterogeneous mixture of silty clay, sand, small gravel fragments, groundwater, and injected conditioning agents. Although this material might appear to be simple waste, its behaviour proves to be a controlling factor in tunnelling performance.

EPB TBM Muck Behaviour Example

In the silty clay sections, the natural ground has high plasticity but low permeability. When excavated without conditioning, the muck adheres to the screw conveyor and belt systems, causing buildup, high torque demand, and frequent stoppages. To address this, foam and polymer conditioning is introduced at the cutterhead. The conditioned muck becomes a plastic, pressure-transmitting mass with moderate cohesion and reduced internal friction. This allows the muck to support face pressure within the EPB chamber while flowing steadily through the screw conveyor and downstream conveyors.

In contrast, when the TBM enters a fine sand layer with elevated groundwater inflow, the muck characteristics change significantly. Without adequate conditioning, the muck behaves as running sand, releasing free water and flowing uncontrollably during extraction. This leads to pressure instability in the chamber, fluctuating screw loads, and an increased risk of settlement at the surface. Additional foam injection and polymer dosing are required to modify the sand into a cohesive, low-permeability muck that can be extracted predictably without pressure loss.

This example demonstrates how grain size distribution and moisture content directly influence muck behaviour and TBM operability. Well-graded sands compact and resist flow, while uniform sands run freely unless conditioned. Excess water transforms the muck from a controllable mass into a slurry-like material that undermines pressure control.

Slurry TBM Muck Behaviour Example

In a separate section of the same project, a slurry TBM is used to excavate beneath a river crossing where permeability is high. Here, muck consists of excavated sand and silt particles suspended in bentonite slurry. The characteristics of the muck—particularly particle size, shape, and clay content—affect slurry viscosity and filter cake formation at the tunnel face.

When coarse gravel content increases, abrasion in the slurry pumps and pipelines accelerates, and separation efficiency at the slurry treatment plant decreases. Conversely, high fines content increases slurry viscosity, making hydraulic transport more energy-intensive and reducing separation performance. Operators must continuously adjust slurry density and bentonite concentration to ensure that muck remains suspended during transport while still being efficiently separable at the plant.

Environmental and Disposal Considerations

Sampling of the muck reveals trace sulphides and hydrocarbons associated with historical land use along the alignment. This chemical composition affects disposal options and requires controlled handling to prevent acid generation and groundwater contamination. Conditioning additives further alter the chemical and physical properties of the muck, necessitating coordination between tunnelling operations, environmental management teams, and disposal facilities.

Key Lessons from the Example

This example illustrates that muck characteristics are not incidental by-products of excavation but engineered variables that must be actively managed. Grain size distribution, moisture content, density, plasticity, and chemical composition collectively determine whether muck can be extracted reliably, transported efficiently, and disposed of safely. In EPB tunnelling, the goal is a pressure-transmitting yet flowable muck. In slurry tunnelling, the objective is stable suspension with predictable separation behaviour.

Ultimately, effective management of muck characteristics enables stable face pressure, reliable spoil logistics, reduced equipment wear, regulatory compliance, and overall project efficiency—reinforcing the concept that muck is not merely waste, but a critical operational medium in mechanised tunnelling.

Screw Conveyors vs Slurry Transport

Screw conveyors and slurry transport systems are the two primary methods used to remove excavated material from the tunnel face in mechanised tunnelling. Each system reflects a fundamentally different approach to spoil handling and is closely tied to the type of TBM, ground conditions, and face support philosophy. Understanding the differences between these systems is essential for selecting appropriate tunnelling methods, managing face pressure, and ensuring reliable logistics.

Figure 87: Example of a Screw Conveyor. C. Löser, CC BY 3.0 DE, via Wikimedia Commons.

In Earth Pressure Balance (EPB) TBMs, spoil is removed using a screw conveyor. The screw conveyor extracts conditioned muck directly from the pressurised excavation chamber and discharges it into the tunnel backup system. In this configuration, the screw conveyor plays a dual role: it is both a material transport device and a critical pressure-control element. By regulating the screw rotation speed, torque, and fill ratio, operators control the rate at which material leaves the chamber, thereby maintaining the required face support pressure. For this reason, screw conveyors in EPB machines are designed to operate under pressure and are robustly sealed to prevent air or water ingress.

The effectiveness of screw conveyor transport depends heavily on muck characteristics. Ideally, EPB muck behaves as a plastic, pressure-transmitting mass that flows steadily through the screw without surging or clogging. If the muck is too dry or cohesive, it may adhere to the screw flights and casing, causing high torque and blockages. If it is too wet or granular, it may flow uncontrollably, leading to pressure loss at the face and instability. Soil conditioning using foams, polymers, or bentonite is therefore essential to ensure predictable screw conveyor performance.

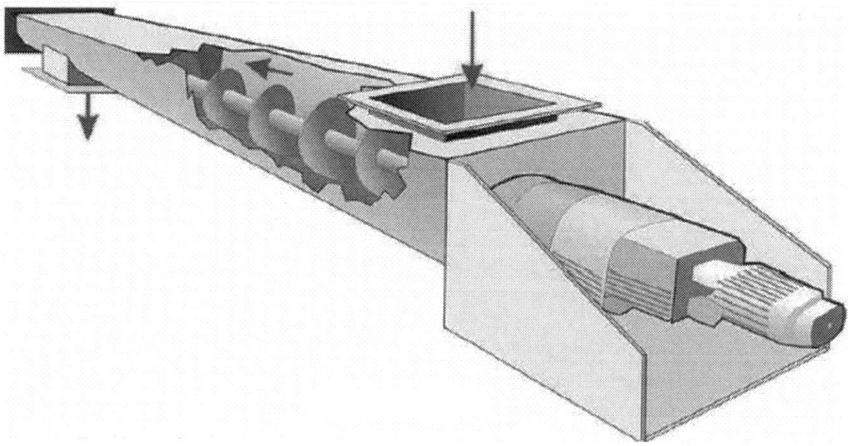

Figure 88: Depiction of a typical screw conveyor. Centers for Disease Control and Prevention, Public Domain, via Picryl.

Figure 88 illustrates a screw conveyor used in an Earth Pressure Balance (EPB) TBM, showing how excavated material (muck) is extracted from the pressurised excavation chamber while simultaneously controlling face pressure.

At the inlet , conditioned muck enters the screw conveyor directly from the excavation chamber behind the cutterhead. This zone is under face support pressure, and the screw opening is designed to admit material without allowing sudden pressure loss. The arrows at the inlet indicate downward and inward flow of muck into the conveyor trough.

Inside the screw casing, a rotating helical flight transports the muck axially toward the discharge. As the screw rotates, it performs two functions at once: it moves material and regulates pressure. Resistance to movement—created by muck consistency, casing confinement, and screw pitch—allows the chamber pressure to be maintained. By adjusting screw rotation speed and torque, operators control the rate of extraction and therefore the pressure balance at the tunnel face.

The diagram shows a progressive pressure reduction along the screw length. Near the inlet, the muck remains densely packed and pressurised. As it advances, the material gradually decompresses as confinement and resistance reduce. This controlled decompression is essential: it prevents abrupt pressure drops that could trigger face instability or ground loss.

At the outlet, the screw discharges muck into an unpressurised environment, such as a belt conveyor, skip, or muck car system. By the time the material reaches this point, pressure has been safely dissipated. The drive motor and gearbox at the discharge end provide the torque required to overcome internal friction, muck resistance, and pressure effects upstream.

In contrast, slurry TBMs remove spoil by hydraulic transport. Excavated material is mixed with slurry, typically a bentonite suspension, inside the excavation chamber and pumped as a slurry mixture through pipelines to a surface separation plant. In this system, face support pressure is maintained by the hydrostatic and pump-induced pressure of the slurry, rather than by controlling solid extraction through a mechanical device. Slurry transport decouples material removal from direct mechanical pressure control, allowing pressure to be maintained more independently of excavation rate.

Bentonite suspension is fundamentally a water-based mixture that incorporates finely dispersed bentonite clay, which upon mixing, produces a stable and viscous fluid. This suspension is particularly advantageous in various applications such as tunnelling, drilling, and ground support due to its ability to control pressure, stabilize excavations, and manage groundwater inflow in conditions where the ground is otherwise unstable [318, 594]. The primary mineral found in bentonite is montmorillonite, renowned for its significant water affinity and the tendency to swell when hydrated, leading to an effective engineering material [318, 595].

Properties and Functions of Bentonite Suspension

Hydration and Viscosity

When bentonite is mixed with water under controlled conditions, its particles separate into microscopic, plate-like structures, leading to an even suspension that demonstrates unique engineering characteristics [318, 596]. The hydration of clay particles enhances the viscosity of the resulting fluid, which is crucial for its functionality in slurry applications [594, 595]. The suspension demonstrates thixotropic behaviour, whereby it can form a weak gel when undisturbed but flows easily when agitated, allowing for effective pumping and circulation as well as sufficient support when stationary [318, 594].

Pressure and Fluid Management

One of the essential roles of bentonite suspension is its ability to transmit hydrostatic pressure evenly across excavation faces, limiting rapid fluid loss into surrounding permeable ground through the formation of an impermeable filter cake on soil surfaces [318, 579]. This property is critical in preventing collapses during excavation and optimizing drilling operations. Studies reveal that the suspension resists flow until a small shear force is applied, made possible due to its yield strength, a characteristic vital for maintaining stable conditions in water-bearing soils [579, 594].

Chemical Sensitivity and Water Quality

Bentonite suspensions are sensitive to chemical variations in the mixing water. The presence of dissolved salts, calcium ions, or contaminants can adversely affect critical properties such as swelling, viscosity, and sealing effectiveness [594, 595]. Consequently, meticulous control of water chemistry is paramount in setting up effective slurry systems for reliable performance in drilling and excavation applications [594].

Applications in Tunnelling and Ground Engineering

In slurry Tunnel Boring Machine (TBM) operations, bentonite suspension is employed to apply hydrostatic pressure at the tunnel face, stabilize the ground, and transport excavated materials hydraulically to the surface [318, 594]. The interaction of the slurry with the geotechnical environment leads to the formation of a robust filter cake that ensures minimal fluid loss and maintains pressure integrity, even in highly permeable conditions [579, 597]. Beyond tunnelling, bentonite suspensions are widely utilized in applications such as diaphragm wall excavation, borehole stabilization, and trench support, functioning as a temporary but highly effective support medium [594].

Slurry transport is particularly well suited to highly permeable, water-bearing soils such as sands and gravels, where maintaining pressure with conditioned muck would be difficult. The slurry medium stabilises the face, transports particles efficiently, and forms a filter cake that limits further slurry loss into the ground. However, the system requires extensive surface infrastructure, including pipelines, pumps, separation plants, and slurry storage, and is sensitive to slurry properties such as density, viscosity, and sand content.

Figure 89: Rolling Stock Muck Cars.

From a logistics perspective, screw conveyors offer a relatively compact and mechanically simple solution, with spoil handled as a bulk solid via belts, skips, or rail cars. This can be advantageous in confined urban sites with limited surface space. Slurry transport, by contrast, involves continuous hydraulic circulation between the TBM and surface facilities, demanding careful coordination between underground excavation and surface processing but enabling high production rates over long distances.

The two systems also differ in their sensitivity to material variability. Screw conveyors are highly sensitive to changes in muck consistency, requiring constant adjustment of conditioning and operating parameters. Slurry systems are more tolerant of variability in particle size but are sensitive to excessive coarse material that increases wear and to fine content that overloads separation plants. Each system therefore presents distinct operational risks and maintenance demands.

Conveyor Belts and Rail Haulage

In mechanized tunnelling, particularly with the use of Tunnel Boring Machines (TBM), the effective removal of excavated material—or muck—is critical for ensuring continuous operation and efficient project timelines. Conveyor belts and rail haulage systems are the two primary methods employed to transport this muck away from the TBM. Each method comes with its unique advantages and operates effectively under specific conditions dictated by the project parameters.

Conveyor Belts

Conveyor belt systems are particularly effective for long tunnels and high-production TBM drives. These systems allow for a continuous transport method, which is essential for maintaining a high rate of TBM advance. As muck is discharged directly from the TBM, it is quickly transferred onto a sequence of belt conveyors extending back to the portal or shaft [598, 599]. The primary advantage of this method is its ability to facilitate uninterrupted TBM operation. Because spoil removal is continuous, the TBM does not need to pause for wagon exchanges, thus enhancing overall productivity and reducing cycle time [598].

Operational challenges with conveyor belt systems include careful design considerations for belt speed, loading points, and spillage control. Issues such as sticky or moist muck can complicate operations, leading to increased maintenance needs [600]. Additionally, the durability of conveyor belts must be accounted for, as abrasive materials can accelerate wear, necessitating robust engineering and safety measures such as fire detection and dust suppression systems [598, 601].

Rail Haulage

In contrast, rail haulage systems utilize rail-mounted muck cars, which are loaded by the TBM and transported to the dumping site by locomotives [602]. This method is often preferred for shorter tunnels or those with complex alignments, where flexibility is a key operational requirement. Unlike conveyor systems, rail haulage is cyclical, necessitating that the TBM pauses for the exchange of muck cars, which can restrict production rates in high-efficiency tunnelling scenarios [602].

Rail systems offer advantages in projects involving intermittent excavation or multiple working faces. Muck cars can be designed for specific purposes, enabling a tailored approach to muck transport [602]. This adaptability allows rail systems to navigate changes in tunnel geometry and gradient; they can more easily manage logistical requirements as construction dynamics evolve [603].

Comparison and Selection Considerations

Selecting between conveyor belts and rail haulage systems depends on various project factors, including tunnel length, production rate, logistics strategy, and the geological characteristics of the excavation site. Conveyor belts are preferable for high muck volumes and long stretches of straight tunnels, where their efficiency in spoil removal significantly reduces operational costs per volume transported [598, 599]. Conversely, rail haulage is more advantageous where operational flexibility, ease of installation, and adaptability to changing conditions are necessary, especially in short or staged tunnels [602].

Many extensive tunnelling projects may also adopt hybrid approaches, such as using rail haulage during initial stages of excavation or in constrained environments, followed by conveyor systems once sufficient tunnel length has been achieved [602]. The combination of conveyor belts for muck removal and rail systems for supporting logistical tasks allows for an optimized solution that balances efficiency and operational flexibility [598, 602].

Overall, the decision between using conveyor belts and rail haulage systems in mechanized tunnelling is multifaceted. While conveyor belts enhance continuous muck transport and overall productivity, rail haulage provides operational flexibility suitable for more complex tunnelling environments. Understanding the specific project dynamics and constraints is essential for selecting the most effective muck transport method, ensuring that tunnelling operations remain safe, efficient, and productive.

Spoil Conditioning Agents

Natural soils and rocks often exhibit non-ideal characteristics for TBM operations. For instance, excavated material may be excessively dry, wet, permeable, abrasive, or stiff, which can jeopardize tunnel face stability, cause blockages in screw conveyors, and lead to increased wear on machinery. Research indicates that the incorporation of conditioning agents can mitigate these problems by bridging the gap between in-situ ground behaviour and the operational demands of the TBM [421, 604]. Specifically, conditioning techniques enhance muck's flowability and stability through optimized moisture content and the introduction of suitable agents [605].

Types of conditioning agents used:

Foams: Foams are among the most prevalent conditioning agents utilized in Earth Pressure Balance (EPB) TBMs. Composed of water, air, and surfactants, foams lower the internal friction of the muck, enhancing flowability and apparent cohesion [581, 606]. This adjustment allows the muck to behave as a cohesive, pressure-transmitting mass, crucial for supporting the tunnel face and ensuring seamless transportation through the screw conveyor. Furthermore, foams reduce adhesion to metal surfaces, thus decreasing wear on equipment [607].

Polymers: Polymers serve a complementary role, particularly effective in managing sands and silts. They bind particles together, reduce permeability, and limit segregation [608]. When dosed appropriately, polymers can improve muck cohesion without excessive water addition, which is vital under conditions where spoil disposal constraints are present [590]. The use of biodegradable polymers is also increasingly emphasized for environmental management purposes [306].

Bentonite: Bentonite and clay-based additives play a crucial role in both EPB and slurry tunnelling. Their inclusion enhances sealing and pressure transmission within the muck. The application of bentonite, even in small quantities, can stabilize chamber pressure and reduce permeability in granular soils, thereby minimizing risks associated with flow instability [605]. Research has also demonstrated that bentonite slurry can effectively form a filter cake when deployed in slurry TBMs, further ensuring hydrostatic pressure is maintained at the tunnel face [300].

Effective water management is integral to spoil conditioning. Maintaining optimal moisture levels directly influences muck consistency, density, and flow behaviour. Research shows that insufficient moisture can result in non-cohesive, dusty materials that compromise face pressure, while excessive moisture can lower shear strength and increase permeability [428]. Consequently, conditioning agents are injected at various points during tunnelling operations—at the cutterhead, within the excavation chamber, and into the mixing zone—to control these variables systematically [609].

Properly conditioned muck significantly enhances TBM operation, improving face stability through uniform pressure transmission, reducing permeability, and minimizing blockages [606, 607]. Studies have established that effective spoil conditioning leads to increased extraction efficiency and lower maintenance demands due to reduced wear and adhesion [428]. Conversely, inadequate conditioning is often linked with face instability and mechanical failures, underscoring its role as a critical operational lever in TBM tunnelling [610].

The choice of conditioning agents is also influenced by environmental and disposal considerations. Agents that adversely affect dewatering performance or landfill acceptance are being reevaluated in favour of low-toxicity, biodegradable options [611]. This trend balances high performance in tunnelling operations with environmental stewardship, highlighting the evolving nature of modern tunnelling projects.

Example: Application of Spoil Conditioning in an EPB TBM Drive Through Sandy Ground

An Earth Pressure Balance (EPB) Tunnel Boring Machine is deployed to construct a metro tunnel through predominantly dry to slightly moist sands with occasional silty layers. Site investigation identifies that the sands are poorly graded, highly permeable, and lack natural cohesion, making them unsuitable for EPB operation in their natural state.

During initial excavation without conditioning, the excavated material flows freely and fails to retain pressure within the excavation chamber. The TBM experiences:

- Rapid drops in chamber pressure during spoil extraction
- Running material entering the screw conveyor uncontrollably
- Increased wear on the screw conveyor flights due to abrasive contact
- Localised settlement observed at the surface

These issues demonstrate that the natural ground behaviour does not meet the operational requirements of EPB tunnelling.

To stabilise the excavation, a conditioning regime combining foam, polymers, and controlled water addition is introduced.

Foam injection is applied at the cutterhead using a defined foam injection ratio. The foam reduces friction between sand grains and introduces apparent cohesion, transforming the free-flowing sand into a plastic, deformable mass. This allows the muck to transmit pressure evenly to the tunnel face while flowing in a controlled manner through the screw conveyor. The lubricating effect of the foam also reduces adhesion and abrasion on metal surfaces, lowering wear rates on the screw conveyor and casing.

To further enhance performance, polymer additives are injected into the excavation chamber. The polymers bind fine particles within the sand matrix, reducing permeability and preventing segregation during pressure fluctuations. This enables stable chamber pressure to be maintained without excessive water addition, which is critical because spoil disposal regulations limit allowable moisture content. A biodegradable polymer is selected to ensure compliance with environmental requirements.

In short sections of particularly permeable sand, small quantities of bentonite are added to the conditioning mix. The bentonite reduces water migration through the muck and improves pressure retention within the chamber, stabilising face conditions during short stoppages and restart cycles [28]. This application aligns with research demonstrating the sealing and filter-forming capabilities of bentonite in granular soils.

Water content is carefully controlled to achieve the target muck consistency. Insufficient moisture is avoided to prevent dusty, non-cohesive behaviour and loss of face pressure, while over-wetting is prevented to avoid strength reduction and excessive permeability. Conditioning agents are injected at the cutterhead and within the chamber mixing zone, allowing continuous fine-tuning based on TBM performance data and spoil observations.

Following implementation of the conditioning strategy:

- Chamber pressure stabilises within the design pressure window
- Screw conveyor torque becomes steady and predictable
- Blockages and pressure losses are eliminated
- Surface settlement reduces to within allowable limits

The conditioned muck flows smoothly through the screw conveyor and downstream conveyor systems, improving extraction efficiency and significantly reducing mechanical wear and maintenance demand.

The selected foams and polymers are compatible with downstream dewatering and spoil reuse processes. The use of low-toxicity, biodegradable conditioning agents ensures landfill acceptance and regulatory compliance, reflecting current industry trends toward environmentally responsible tunnelling practices.

This example demonstrates how spoil conditioning is used in practice to convert non-ideal in-situ ground into an engineered muck suitable for EPB tunnelling. By applying foams, polymers, bentonite, and controlled water addition in response to ground behaviour, the TBM team maintains face stability, ensures reliable spoil extraction, reduces equipment wear, and controls settlement. Consistent with published research [26–36], the example highlights spoil conditioning as a critical operational control, rather than a supplementary measure, in modern TBM tunnelling.

Slurry Separation Plants

Slurry separation plants play a fundamental role in the process of slurry tunnel boring machine tunnelling, serving as a crucial link between underground excavation and surface operations. Their primary function is to separate excavated material from the supporting slurry to facilitate the reuse of the cleaned slurry while managing spoil effectively. This section elucidates how slurry separation systems operate, the impact on tunnel face stability, and the broader implications for environmental control.

In TBM tunnelling, excavation occurs under hydrostatic pressure provided by a bentonite slurry, which stabilizes the tunnel face and prevents inflow of groundwater. The mixture of excavated soil or rock with slurry is transported to the surface via pipelines, where a slurry separation plant is employed to remove the solid material. This is vital for maintaining the properties of the slurry, including density, viscosity, and sand content, which ensures its effective circulation back to the TBM for reuse. The degradation of slurry quality can lead to increased operational risks, including instability at the tunnel face and wear on essential equipment such as pumps and pipelines [579].

Without an effective separation system, the buildup of fine particles can hinder slurry effectiveness, leading to an increase in viscosity and density, and complicating pressure control mechanisms. Conversely, excessive cleaning can lead to a loss of filter cake formation capability at the face, increasing the risk of slurry loss and subsequent tunnel instability [579]. Thus, the separation plant is not merely auxiliary but a core component of the tunnelling operation [590].

Typically, a slurry separation plant employs a staged approach to filtering excavated material, starting with coarse screening to eliminate large fragments of gravel and rock. Primary screening utilizes vibrating or rotating scalping screens, which prevent damage to downstream machinery by reducing excessive loads [346]. Following this, hydrocyclones are employed, which leverage centrifugal force to separate materials by size and density. Specifically, desanders and desilters are used for sand and silt, respectively, with the underflow containing the solid waste and the overflow yielding cleaned slurry [579].

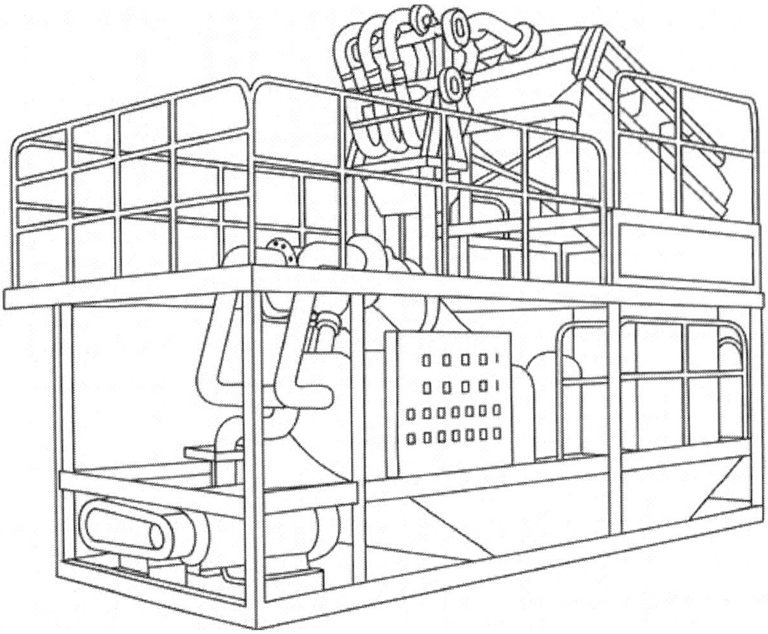

Figure 90: Diagram of a Shield, Pipe-jacking and TBM Slurry Separation System.

For projects with high clay or fine silt content, additional refinement stages involving centrifuges or filter presses are often necessary. These advanced processes address environmental discharge restrictions and ensure compliance with spoil management requirements, which are increasingly predicated on regulatory standards regarding waste treatment [297].

Handling and treating the separated solids is essential for compliance with environmental regulations, especially concerning water content and toxicity characteristics. Options for managing waste range from transportation to landfills to dewatering processes that prepare the material for reuse in construction or as fill material [346]. The influence of environmental regulations on spoil management highlights the intertwined relationship between operational efficiency and sustainability in TBM tunnelling practices [297].

Modern slurry separation plants are equipped with comprehensive instrumentation to monitor critical performance indicators, such as slurry density, viscosity, and flow rates, in real-time. This data allows for swift adjustments to slurry composition as conditions change, aligning TBM operations underground with surface-level separation efficiency [579]. The importance of maintaining a balance in slurry characteristics cannot be overstressed; both the accumulation of fine solid materials and over-expansion of the slurry can significantly affect tunneling performance and tunnel face stability [590].

Waste Management and Reuse

Waste management and reuse in mechanised tunnelling refer to the systematic handling, treatment, and beneficial use of excavated material (spoil or muck) and associated process by-products so that environmental impacts are minimised, regulatory requirements are met, and project efficiency is improved. In modern TBM projects, spoil is no longer treated simply as waste but as a material stream that can often be processed, reused, or valorised if properly managed.

The primary waste stream in TBM tunnelling is excavated ground, which may consist of soil, rock, or mixed materials, often combined with water and conditioning agents such as foams, polymers, or bentonite. Secondary waste streams include spent slurry, wash water from separation plants, worn cutting tools, filter cakes, and packaging or consumables from tunnelling operations. The physical, chemical, and moisture characteristics of these materials strongly influence how they must be managed.

The key objectives of waste management in tunnelling are to ensure safe handling, minimise environmental harm, comply with regulatory requirements, and reduce disposal volumes and costs. This involves controlling moisture content, managing fines and additives, preventing contamination of soil and water, and ensuring that waste streams are tracked and documented throughout the project lifecycle.

Before disposal or reuse, excavated material often requires treatment. In slurry TBM projects, separation plants remove solids from slurry, and further dewatering may be carried out using centrifuges, filter presses, or settling systems to reduce water content to acceptable levels. In EPB tunnelling, conditioned muck may need aeration, mixing, or drying to meet landfill or reuse specifications. Water recovered during treatment is commonly recycled back into the tunnelling process, reducing overall water demand.

Where ground conditions and regulatory frameworks permit, tunnelling spoil can be reused beneficially. Common reuse options include use as engineered fill, embankment material, landscaping soil, or raw material for construction applications such as bricks, aggregates, or cement feedstock. Clean, well-graded rock spoil may be crushed and reused as aggregate, while fine-grained soils may be suitable for land rehabilitation or noise bunds after treatment.

Reuse decisions depend on material properties, contamination levels, moisture content, and the presence of conditioning agents. Early planning and testing are essential to ensure that conditioning strategies are compatible with intended reuse pathways.

Environmental regulations play a central role in waste management and reuse. Authorities typically specify limits on moisture content, fines percentage, chemical additives, and contaminants such as heavy metals or sulphides. Spoil classification, waste tracking, and documentation are often mandatory, particularly in urban or sensitive environments. The selection of low-toxicity, biodegradable conditioning agents is increasingly driven by the need to simplify compliance and expand reuse options.

Effective waste management must be integrated with TBM logistics. Continuous spoil removal, stockpiling capacity, transport scheduling, and treatment plant performance all influence TBM productivity. Bottlenecks in waste handling can quickly constrain advance rates, making coordination between tunnelling operations and surface waste management systems essential.

From an engineering and project perspective, effective waste management and reuse reduce disposal volumes, lower environmental impact, and can significantly decrease project costs. They also support social licence to operate, particularly in urban projects where truck movements, noise, dust, and landfill use are highly scrutinised. Poor waste management, by contrast, can lead to environmental incidents, regulatory delays, and disruption to tunnelling operations.

Chapter 11

Tunnel Boring Calculations

Mechanised tunnelling relies on the precise interaction between ground behaviour, machine capability, and structural response. Tunnel Boring Machines do not operate by intuition or brute force alone; their performance is governed by a set of interrelated calculations that define how thrust, torque, penetration, face pressure, power consumption, and lining loads are balanced throughout excavation. These calculations form the quantitative backbone of TBM design, selection, and safe operation.

This chapter brings together the core engineering calculations that underpin mechanised tunnelling. It explains how forces and pressures are estimated, how machine limits are assessed, and how operational parameters are translated into predictable advance rates and structural demands. Rather than treating calculations as abstract design exercises, the chapter shows how they are used continuously—from preliminary design through day-to-day operation—to control risk, optimise productivity, and protect both the machine and the tunnel lining.

The focus is deliberately practical. Each calculation is framed in terms of what it represents physically at the tunnel face, within the TBM, and in the surrounding ground. Worked examples illustrate how simplified engineering relationships are applied, calibrated using precedent data, and refined using real-time monitoring during excavation. This approach reflects industry practice, where calculations provide a rational starting point but must be interpreted in light of observed ground behaviour and machine response.

By the end of this chapter, you should understand not only how key TBM calculations are performed, but why they matter. Thrust, torque, face pressure, penetration, power demand, and segment loading are not independent quantities—they form a tightly coupled system.

Successful mechanised tunnelling depends on managing this system within defined limits, using calculations as a guide and operational feedback as confirmation.

Thrust Force Calculations

Thrust force calculations are used to determine the axial force a Tunnel Boring Machine (TBM) must apply to advance the cutterhead into the ground while maintaining face stability, overcoming resistance at the cutterhead, and pushing the machine body forward without overstressing the lining or the machine.

In mechanised tunnelling, thrust is not a single resistance but the sum of several interacting components, each of which must be understood to size the thrust system and control TBM operation safely.

Purpose of thrust force calculations

Thrust calculations serve several critical engineering purposes throughout the life of a TBM project. From a design perspective, they are used to size the hydraulic thrust cylinders and thrust frame and to verify that the segmental lining and thrust-bearing arrangements can safely accommodate the applied loads. Accurate thrust estimation ensures that the machine has sufficient capacity to advance under the most demanding ground conditions without overstressing structural components.

From an operational standpoint, thrust calculations define acceptable penetration rates and inform advance strategies under varying ground conditions. By understanding the thrust required to overcome cutting resistance, face pressure, and friction, operators can apply thrust in a controlled manner that promotes steady excavation, efficient cutting, and predictable machine behaviour.

Thrust calculations are also fundamental to safety. Applying too little thrust can lead to poor penetration, loss of directional control, or unstable face conditions, while excessive thrust can crack lining segments, overload the tunnel face, or damage the TBM itself. Maintaining thrust within calculated limits is therefore essential to balancing productivity with structural integrity and face stability.

Main components of TBM thrust resistance

The total required thrust force is typically expressed as:

Total thrust = cutterhead resistance + face support effects + shield friction + machine resistance

Each component varies with ground type, depth, TBM type, and operational parameters.

1. Cutterhead cutting resistance

This is the force required to break and displace the ground at the tunnel face.

- **Soil TBMs (EPB / slurry)**
  Resistance depends on soil strength, conditioning effectiveness, cutterhead geometry, and penetration per revolution.

- **Rock TBMs**
  Resistance is governed by intact rock strength, jointing, cutter type (disc cutters), cutter spacing, and penetration depth.

In simplified form:

Cutting force ≈ unit cutting resistance × cutterhead area

Where unit cutting resistance is derived from laboratory tests, empirical correlations, or precedent projects.

Example 1: EPB TBM in Sand–Silt Ground (Soil TBM)

Project context
A 6.5 m diameter EPB TBM is excavating through medium-dense sand with silt. Laboratory testing and precedent projects indicate that, with effective foam conditioning, the unit cutting resistance of the conditioned soil is approximately:

- **Unit cutting resistance** = 80 kPa (80 kN/m^2)

Cutterhead area

$$A = \pi \left(\frac{D}{2}\right)^2 = \pi \left(\frac{6.5}{2}\right)^2 \approx 33.2 \text{ m}^2$$

Estimated cutting force

$$F_{\text{cut}} \approx 80 \text{ kN/m}^2 \times 33.2 \text{ m}^2 \approx 2\,660 \text{ kN}$$

Interpretation
This value represents the approximate thrust component required to overcome soil cutting resistance at the face. It informs:

- Minimum thrust demand during steady excavation
- Selection of thrust reserve for harder zones

- Assessment of whether segment bearing capacity can safely transmit the force

If conditioning deteriorates (e.g. foam breaks down or permeability increases), the unit cutting resistance may rise to 120–150 kPa, increasing thrust demand and risking pressure instability or screw conveyor overload.

Example 2: Hard Rock TBM with Disc Cutters (Rock TBM)

Project context
A 7.8 m diameter hard rock TBM is excavating intact granite with moderate joint spacing. Based on UCS testing and cutter performance models:

- **Uniaxial Compressive Strength (UCS)** ≈ 180 MPa
- Empirical correlations indicate an equivalent unit cutting resistance of approximately:
    - 1.2 MPa (1,200 kN/m^2) at the selected penetration depth

Cutterhead area

$$A = \pi \left(\frac{7.8}{2}\right)^2 \approx 47.8 \text{ m}^2$$

Estimated cutting force

$$F_{\text{cut}} \approx 1\,200 \text{ kN/m}^2 \times 47.8 \text{ m}^2 \approx 57\,400 \text{ kN}$$

Interpretation
This cutting resistance governs:

- Total thrust capacity required from hydraulic cylinders
- Number and size of disc cutters
- Penetration rate limits to avoid cutter overload

Unlike soil TBMs, conditioning plays no role here; instead, thrust must be matched with sufficient cutterhead torque to avoid stalling or excessive cutter wear.

Engineering Significance

These examples illustrate how the simplified relationship:

Cutting force ≈ unit cutting resistance × cutterhead area

is used as a first-order design and operational tool. While detailed TBM design relies on more refined cutter–ground interaction models, this approach provides:

- A practical estimate of thrust demand
- A basis for checking hydraulic capacity and lining limits
- A way to assess sensitivity to ground variability

In both soil and rock tunnelling, cutterhead cutting resistance is a fundamental input linking ground mechanics, machine capability, and safe operational control.

2. Face support pressure contribution (EPB and slurry TBMs)

In EPB and slurry machines, thrust must also overcome the reaction force created by face support pressure acting on the cutterhead.

This force is calculated as:

Face pressure force = face support pressure × cutterhead area

Although face pressure stabilises the ground, it creates a resisting force that the thrust system must overcome to advance the cutterhead.

Higher face pressure → higher required thrust.

Example: Face Support Pressure Contribution to Thrust (EPB / Slurry)

Given

A TBM has a **cutterhead diameter** $D = 6.5\ m$and is operating with a face support pressure setpoint of $p = 220\ kPa$(typical of shallow-to-moderate depth tunnelling with groundwater).

Step 1: Calculate cutterhead area

$$A = \pi\left(\frac{D}{2}\right)^2 = \pi\left(\frac{6.5}{2}\right)^2 = \pi(3.25)^2 \approx 33.2\ m^2$$

Step 2: Calculate face pressure reaction force

$$F_p = p \times A$$

Convert pressure units:

- $220\ kPa = 220\ kN/m^2$

Now compute:

$$F_p = 220\ kN/m^2 \times 33.2\ m^2 \approx 7{,}304\ kN$$

Result

Face support pressure contributes ~7,300 kN (≈ 7.3 MN) of resisting force that the thrust system must overcome just to advance the cutterhead.

What this means operationally

EPB TBM interpretation

In EPB, the pressurised conditioned muck in the chamber provides the support pressure. If the operator raises chamber pressure from, say, 220 kPa to 260 kPa to control settlement, the resisting force increases:

$$\Delta p = 40\ kPa = 40\ kN/m^2$$
$$\Delta F = 40 \times 33.2 \approx 1{,}328\ kN$$

So a modest pressure increase of 40 kPa adds **~1.3 MN** extra thrust demand.

Slurry TBM interpretation

In slurry TBMs, the same calculation applies, but the pressure is provided by slurry hydrostatic pressure in the excavation chamber. Increasing slurry pressure to prevent face instability directly increases the reaction force on the cutterhead—and therefore increases the thrust required for forward motion in exactly the same way.

Why this matters

This example shows why "more face pressure" is never free. Face pressure is essential for stability and settlement control, but it creates a substantial opposing force on the cutterhead. Therefore, as face pressure increases:

- Required thrust increases significantly
- Segment loading increases (for segmentally lined tunnels)
- Risk of stalling rises if torque and thrust are not balanced

3. Shield friction

Shield friction arises from contact between the TBM shield and the surrounding ground.

It depends on:

- Ground type and abrasiveness
- Overburden pressure
- Shield length and diameter
- Ground conditioning and lubrication (e.g. bentonite or polymer injection)

Shield friction can be approximated as:

Shield friction = friction coefficient × normal pressure × contact area

In soft ground, lubrication can significantly reduce this component. In squeezing or swelling ground, shield friction may dominate the thrust demand.

Example: Estimating shield friction and showing the effect of lubrication

A 6.5 m diameter EPB TBM is driving through soft ground with a steel shield length of 8 m. The shield is in contact with the ground over most of its length, so the shield–ground contact area can be approximated by the cylindrical surface area:

$$A \approx \pi D L = \pi \times 6.5 \times 8 \approx 163.36\ \mathrm{m}^2$$

Based on depth and ground conditions, the average normal pressure acting on the shield (radial contact pressure) is estimated at 250 kPa (i.e., 250 kN/m^2). For a "dry" contact condition (limited lubrication), an appropriate friction coefficient might be μ = 0.25.

Using the simplified relationship:

$$F_{fric} = \mu \times p_n \times A$$
$$F_{fric} = 0.25 \times 250 \times 163.36 \approx 10{,}210\ \mathrm{kN}$$

So, shield friction contributes about 10.2 MN of thrust demand in this case—often a major portion of total thrust in soft ground.

Same TBM, but with effective lubrication (bentonite/polymer injection)

Now assume the contractor injects bentonite/polymer lubrication along the shield, reducing the interface friction coefficient to $\mu = 0.10$ while the normal pressure remains similar:

$$F_{fric} = 0.10 \times 250 \times 163.36 \approx 4{,}084 \text{ kN}$$

With lubrication, shield friction drops to about 4.1 MN, a reduction of roughly 6.1 MN. This illustrates why lubrication is such a powerful control lever in soft ground: it can materially increase achievable advance rates and reduce the risk of stalling.

Contrast case: squeezing/swelling ground where friction can dominate

In squeezing clay or swelling ground, the normal pressure on the shield can rise substantially due to ground convergence and time-dependent closure. If the average normal pressure increases to 400 kPa and the interface becomes "stickier," giving $\mu = 0.35$, then:

$$F_{fric} = 0.35 \times 400 \times 163.36 \approx 22{,}870 \text{ kN}$$

Here, shield friction alone becomes ~22.9 MN, which can dominate the thrust demand and quickly approach the limits of the thrust system and/or the segment bearing capacity.

What the example demonstrates

This example shows how shield friction scales directly with (1) friction coefficient, (2) normal pressure, and (3) contact area. In soft ground, lubrication (bentonite/polymer) can cut shield friction dramatically. In squeezing or swelling ground, increasing normal pressure and higher friction can make shield friction the controlling factor for thrust demand and TBM advance.

4. Machine and system resistance

Additional thrust is required to overcome:

- Internal mechanical resistance

- Conveyor or slurry system drag during advance
- Misalignment or steering forces
- Articulation resistance in curved drives

These forces are usually estimated as a percentage allowance based on TBM type and experience.

Example: Allowance for machine and system resistance in thrust calculations

A 6.5 m diameter EPB TBM is excavating a mixed sand–silt deposit beneath an urban area. The primary thrust components have already been estimated as:

- Cutterhead cutting resistance: 4.5 MN
- Face support pressure reaction: 7.3 MN
- Shield friction (lubricated): 4.1 MN

This gives a base thrust demand of:

$$F_{base} = 4.5 + 7.3 + 4.1 = 15.9 \text{ MN}$$

However, this base value does not yet account for additional machine and system resistances.

Components of machine and system resistance

In this project, the following secondary resistances are identified:

- Internal mechanical losses in bearings, seals, and hydraulic systems
- Screw conveyor drag, as pressurised conditioned muck is extracted during advance
- Steering and alignment forces, particularly during small but frequent corrections in a shallow urban drive
- Articulation resistance, as the TBM negotiates a horizontal curve with a radius of 450 m

Based on manufacturer data and experience from similar EPB drives, these effects are conservatively estimated as 12% of the base thrust demand.

Calculation of additional resistance

$$F_{sys} = 0.12 \times 15.9 = 1.9 \text{ MN}$$

Total required thrust

$$F_{total} = F_{base} + F_{sys}$$
$$F_{total} = 15.9 + 1.9 = 17.8 \text{ MN}$$

Interpretation

In this example, machine and system resistance contributes nearly 2 MN of additional thrust, even though no single component is dominant on its own. This allowance ensures that:

- The TBM can maintain steady advance without operating at maximum hydraulic capacity
- Short-term increases in resistance (e.g. tighter curves or sticky muck) do not cause stalling
- Operational control remains stable during steering and articulation

Engineering insight

This example illustrates why thrust calculations always include a system resistance margin. Even in well-conditioned ground with controlled face pressure and shield friction, internal and operational losses can account for 10–20% of total thrust demand. Ignoring this component risks underestimating required thrust capacity and compromising operational robustness during real-world tunnelling conditions.

In segmentally lined tunnels, thrust is transmitted from the TBM to the ground through thrust shoes bearing directly on the installed lining. As a result, thrust calculations must account not only for the machine's hydraulic capacity but also for the structural limits of the lining. These checks include the maximum allowable compressive stress in the segments, the local bearing pressure beneath individual thrust shoes, and how the total thrust load is distributed across multiple shoes. In many cases, the governing constraint on allowable thrust is the strength and bearing capacity of the lining segments rather than the available thrust force of the TBM itself.

From an operational perspective, thrust is not applied as a constant or preset value. Instead, it is continuously adjusted in response to real-time excavation conditions. Operators monitor penetration rate, cutterhead torque, face pressure stability, shield friction, and the behaviour of the lining during thrusting. The objective is to apply the minimum thrust necessary to maintain a stable and continuous advance, thereby reducing mechanical wear, limiting stress on the lining, and minimising the risk of instability.

Thrust also interacts closely with cutterhead torque and penetration. Increasing thrust generally increases penetration, but only up to the limits imposed by ground strength and cutting efficiency. Applying excessive thrust without sufficient torque can cause cutter stalling, while high torque combined with low penetration indicates inefficient cutting and wasted energy. Effective excavation therefore relies on balanced control of thrust and torque to achieve optimal penetration.

Overall, thrust force calculations form a critical link between ground behaviour, TBM mechanical design, and lining performance. They ensure that the machine advances safely and efficiently while preserving face stability and protecting the structural integrity of the tunnel lining. In modern mechanised tunnelling, thrust control is not a matter of simply pushing harder to increase speed, but a carefully calculated and continuously managed parameter that responds to changing ground conditions and machine performance in real time.

Torque Requirements

Torque requirements define the rotational capacity needed at the cutterhead to cut the ground, overcome frictional resistance, and maintain stable excavation. While thrust controls *how hard* the TBM pushes forward, torque controls *whether the cutterhead can rotate and cut effectively*. Insufficient torque leads to cutterhead stalling and face instability; excessive torque increases wear, energy consumption, and mechanical stress.

1. Components of Required Cutterhead Torque

Total cutterhead torque is the sum of several resisting components:

$$T_{total} = T_{cut} + T_{face} + T_{fric} + T_{sys}$$

Where:

- T_{cut}= torque to cut and fragment the ground
- T_{face}= torque associated with face support pressure
- T_{fric}= torque from friction between cutterhead, muck, and ground
- T_{sys}= mechanical and system losses

Example: Breakdown of Cutterhead Torque Components

Project context

A 6.5 m diameter EPB TBM is excavating through conditioned sandy silt under moderate groundwater pressure in an urban environment.

- Cutterhead diameter, $D = 6.5\, m$
- Cutterhead radius, $R = 3.25\, m$
- Effective cutting radius, $r_{eff} \approx 0.65R = 2.1\, m$
- Cutterhead area,

$$A = \pi R^2 = \pi(3.25)^2 \approx 33.2\, m^2$$

1. Cutting Torque T_{cut}

Laboratory testing and precedent projects indicate a unit cutting resistance of:

$$\tau = 120\, kPa = 120\, kN/m^2$$

Cutting torque is approximated as:

$$T_{cut} = \tau \times A \times r_{eff}$$
$$T_{cut} = 120 \times 33.2 \times 2.1 \approx 8{,}370\, kN \cdot m$$
$$T_{cut} \approx 8.4\, MN \cdot m$$

This represents the torque required to shear, fracture, and displace the ground at the face.

2. Face Support Pressure Torque T_{face}

The EPB chamber is operated at a **face support pressure** of:

$$p = 220\, kPa$$

The friction coefficient between conditioned muck and cutterhead steel is estimated as:

$$\mu = 0.25$$

Face pressure–related torque is:

$$T_{face} = \mu \times p \times A \times r_{eff}$$
$$T_{face} = 0.25 \times 220 \times 33.2 \times 2.1 \approx 3{,}840\, kN \cdot m$$
$$T_{face} \approx 3.8\, MN \cdot m$$

This torque arises because pressurised muck resists cutterhead rotation while stabilising the face.

3. Frictional Torque T_{fric}

Additional friction occurs due to:

- Adhesion of muck to cutterhead openings
- Interaction between excavated material and cutterhead structure

Based on experience in similar ground, frictional torque is taken as 15% of cutting torque:

$$T_{fric} = 0.15 \times T_{cut} = 0.15 \times 8.4 \approx 1.3\ MN \cdot m$$

4. Mechanical and System Losses T_{sys}

Mechanical losses include:

- Main bearing resistance
- Gearbox inefficiency
- Seal and drive-train losses

These are conservatively estimated as 7% of the subtotal torque:

$$T_{sys} = 0.07 \times (T_{cut} + T_{face} + T_{fric})$$
$$T_{sys} = 0.07 \times (8.4 + 3.8 + 1.3) \approx 0.95\ MN \cdot m$$

5. Total Required Cutterhead Torque

$$T_{total} = T_{cut} + T_{face} + T_{fric} + T_{sys}$$
$$T_{total} = 8.4 + 3.8 + 1.3 + 1.0 \approx 14.5\ MN \cdot m$$

Interpretation

In this example:

- Cutting torque is the dominant component
- Face pressure contributes a significant secondary resistance
- Friction and system losses together add nearly 15–20% to the total demand

This means the TBM must be equipped with a cutterhead drive system capable of delivering at least 14.5 MN·m of continuous torque, with additional margin for transient conditions such as harder lenses, pressure fluctuations, or poor conditioning.

Key engineering insight

This example shows why cutterhead torque is not just a function of ground strength. Even in well-conditioned soils, face pressure, friction, and mechanical losses can add several meganewton-metres to the torque demand. Reliable TBM performance depends on correctly estimating and balancing all four torque components, not just the cutting resistance alone.

2. Cutting Torque (Primary Component)

Soil TBMs (EPB / Slurry)

Cutting torque depends on:

- Soil shear strength and density
- Conditioning effectiveness
- Cutterhead opening ratio and tool type
- Penetration per revolution

A simplified expression is:

$$T_{cut} \approx \tau \times A \times r_{eff}$$

Where:

- τ= unit cutting resistance (kPa)
- A= cutterhead area (m^2)
- r_{eff}= effective radius of cutting action (≈ 0.6–0.7 × cutterhead radius)

Example
For a 6.5 m TBM:

- $A = 33.2\ m^2$
- $r_{eff} = 2.2\ m$
- $\tau = 120\ kPa$

$$T_{cut} \approx 120 \times 33.2 \times 2.2 \approx 8.8\ MN \cdot m$$

Rock TBMs

For hard rock TBMs, torque is governed by:

- Disc cutter forces
- Number and spacing of cutters
- Penetration per cutter

$$T_{cut} = \sum(F_{disc} \times r_{disc})$$

Where disc force is derived from laboratory indentation tests and empirical models (e.g. CSM, NTNU).

Example 1: Soil TBM (EPB / Slurry) – Cutting Torque

Project context

A 6.5 m diameter EPB TBM is excavating through conditioned sandy silt with moderate shear strength. Laboratory testing and precedent data indicate a unit cutting resistance of 120 kPa, reflecting effective soil conditioning and moderate penetration per revolution.

Given

- Cutterhead diameter, $D = 6.5\ m$
- Cutterhead radius, $R = 3.25\ m$
- Cutterhead area,

$$A = \pi R^2 = \pi(3.25)^2 \approx 33.2\ m^2$$

- Effective cutting radius,

$$r_{eff} \approx 0.65R \approx 2.2\ m$$

- Unit cutting resistance,

$$\tau = 120\ kPa = 120\ kN/m^2$$

Calculation

Using the simplified expression:

$$T_{cut} \approx \tau \times A \times r_{eff}$$
$$T_{cut} \approx 120 \times 33.2 \times 2.2 \approx 8{,}770\ kN \cdot m$$
$$T_{cut} \approx 8.8\ MN \cdot m$$

Interpretation

This value represents the torque required to shear and displace the conditioned soil at the face. Improved conditioning (lower τ) would reduce this torque, while poorer conditioning or higher penetration per revolution would increase it.

Example 2: Hard Rock TBM – Cutting Torque

Project context

A hard rock TBM with disc cutters is driving through moderately strong jointed rock. Cutter forces are derived from indentation testing and NTNU/CSM-based estimates.

Given

- Number of disc cutters: 40
- Average normal force per cutter:

$$F_{disc} = 300\ kN$$

- Average cutter radius from centre:

$$r_{disc} = 2.5\ m$$

Calculation

Cutting torque is the sum of moments from all disc cutters:

$$T_{cut} = \sum(F_{disc} \times r_{disc})$$
$$T_{cut} = 40 \times 300 \times 2.5 = 30{,}000\ kN \cdot m$$
$$T_{cut} = 30\ MN \cdot m$$

Interpretation

In hard rock TBMs, cutting torque is dominated by disc cutter forces, which are strongly influenced by:

- Rock strength and abrasivity
- Joint spacing and orientation
- Penetration per cutter
- Cutter spacing and layout

Unlike soil TBMs, face pressure plays little role in cutting torque; instead, torque scales directly with disc cutter loading.

Engineering takeaway

These examples show the fundamental difference between soil and rock TBMs:

- **Soil TBMs:** cutting torque is governed by unit cutting resistance, cutterhead size, and conditioning effectiveness.
- **Rock TBMs:** cutting torque is governed by disc cutter forces and geometry.

In both cases, cutting torque is the primary component of total cutterhead torque and sets the baseline for drive system sizing and operational limits.

3. Torque from Face Support Pressure

In EPB and slurry TBMs, pressurised soil or slurry acts on the cutterhead and generates **shear resistance to rotation**.

$$T_{face} = \mu \times p \times A \times r_{eff}$$

Where:

- μ= friction coefficient between muck/slurry and cutterhead
- p= face pressure (kPa)

Example

- $p = 220\ kPa$
- $\mu = 0.25$

$$T_{face} \approx 0.25 \times 220 \times 33.2 \times 2.2 \approx 4.0\ MN \cdot m$$

Example: Torque from Face Support Pressure (EPB / Slurry TBM)

Project context

A 6.5 m diameter EPB TBM is operating in shallow urban ground with groundwater present. To control face stability and settlement, the excavation chamber is maintained at a face support pressure of 220 kPa. The pressurised, conditioned muck in contact with the cutterhead resists rotation through friction.

Given

- Cutterhead diameter, $D = 6.5\,m$
- Cutterhead radius, $R = 3.25\,m$
- Cutterhead area:

$$A = \pi R^2 = \pi(3.25)^2 \approx 33.2\,m^2$$

- Effective cutting radius:

$$r_{eff} \approx 0.65R \approx 2.2\,m$$

- Face support pressure:

$$p = 220\,kPa = 220\,kN/m^2$$

- Friction coefficient between conditioned muck and steel cutterhead:

$$\mu = 0.25$$

Calculation

The torque arising from face support pressure is estimated using:

$$T_{face} = \mu \times p \times A \times r_{eff}$$

Substituting values:

$$T_{face} = 0.25 \times 220 \times 33.2 \times 2.2$$
$$T_{face} \approx 4{,}000\,kN \cdot m$$
$$T_{face} \approx 4.0\,MN \cdot m$$

Interpretation

This result means that approximately 4 MN·m of cutterhead torque is required solely to overcome the shear resistance created by the pressurised muck or slurry acting on the cutterhead. This torque does not contribute to cutting the ground; it is a direct consequence of maintaining face stability.

If the operator increases face pressure—for example, from 220 kPa to 260 kPa to control settlement—the torque contribution increases proportionally. A 40 kPa increase would add roughly:

$$\Delta T \approx 0.25 \times 40 \times 33.2 \times 2.2 \approx 0.7\,MN \cdot m$$

Engineering significance

This example highlights an important operational trade-off in EPB and slurry tunnelling:

- Higher face pressure improves stability and settlement control, but
- It directly increases cutterhead torque demand, raising the risk of stalling and accelerating wear.

As a result, face pressure, thrust, and torque must always be managed together within the TBM's mechanical limits and the ground's allowable pressure window.

4. Frictional Torque

Additional torque is required to overcome:

- Muck adhesion to cutterhead and openings
- Friction in the excavation chamber
- Resistance at the cutterhead periphery

This component is strongly influenced by spoil conditioning.

Well-conditioned EPB muck → low friction → lower torque
Poor conditioning → sticky or abrasive muck → torque spikes

Typically estimated as 10–30% of cutting torque, depending on conditions.

Example: Estimating frictional torque and showing the impact of conditioning

Project context
A 6.5 m EPB TBM is driving through clayey silt. Cutting torque has been estimated (from unit cutting resistance and cutterhead geometry) as:

- Cutting torque, $T_{cut} = 9.0\ MN \cdot m$

The TBM team is assessing the additional frictional torque required due to muck interaction with the cutterhead and excavation chamber.

Case A: Well-conditioned muck (low friction)

Foam conditioning is performing well, and the muck behaves as a plastic, lubricated mass with limited adhesion. In this case, frictional torque is conservatively taken as 12% of cutting torque:

$$T_{fric} = 0.12 \times T_{cut}$$
$$T_{fric} = 0.12 \times 9.0 = 1.08\ MN \cdot m$$

Interpretation
With good conditioning, the additional frictional torque is about 1.1 MN·m. Cutterhead

rotation remains stable, torque signals are smooth, and the TBM maintains steady penetration without frequent stalls.

Case B: Poor conditioning (sticky clay → torque spikes)

Foam quality drops due to water chemistry changes and insufficient injection rate. The muck becomes sticky and adheres to cutterhead openings, and the excavation chamber exhibits increased friction and clogging. Frictional torque is now estimated at 28% of cutting torque:

$$T_{fric} = 0.28 \times 9.0 = 2.52\ MN \cdot m$$

Interpretation
With poor conditioning, frictional torque rises to ~2.5 MN·m, adding more than 1.4 MN·m compared with the well-conditioned case. Operationally, the TBM may show:

- Sudden torque spikes during rotation
- Reduced penetration at the same thrust
- Increased cutterhead power draw
- Higher wear and clogging around openings

Engineering takeaway

This example shows why frictional torque is commonly estimated as 10–30% of T_{cut} and why it is highly sensitive to spoil conditioning:

- Well-conditioned muck → low adhesion and lower friction → T_{fric}≈ 10–15% of T_{cut}
- Poorly conditioned muck → sticky or abrasive behaviour → higher resistance and torque spikes → T_{fric}≈ 20–30% of T_{cut}

In practice, a rise in measured torque without a corresponding increase in penetration is a strong indicator that frictional torque is increasing—often signalling the need to adjust conditioning dosage, water content, or operating parameters before stalling or face instability develops.

5. Mechanical and System Losses

System losses include:

- Main bearing resistance

- Gearbox and seal losses
- Drive train inefficiencies

These are usually estimated as 5–10% of total calculated torque based on TBM design.

Example: Estimating mechanical and system losses in cutterhead torque

Project context
A 6.5 m EPB TBM is operating in conditioned sandy silt. The primary torque components have already been estimated as:

- Cutting torque, $T_{cut} = 8.8\ MN \cdot m$
- Face pressure torque, $T_{face} = 4.0\ MN \cdot m$
- Frictional torque, $T_{fric} = 1.4\ MN \cdot m$

Step 1: Calculate subtotal torque

$$T_{sub} = T_{cut} + T_{face} + T_{fric}$$
$$T_{sub} = 8.8 + 4.0 + 1.4 = 14.2\ MN \cdot m$$

This subtotal represents the torque required at the cutterhead excluding internal mechanical losses.

Step 2: Apply mechanical and system loss allowance

Based on TBM manufacturer data and experience with similar EPB machines, mechanical and system losses (main bearing friction, gearbox inefficiency, seal drag, and drive train losses) are estimated at 7% of the subtotal torque.

$$T_{sys} = 0.07 \times T_{sub}$$
$$T_{sys} = 0.07 \times 14.2 \approx 1.0\ MN \cdot m$$

Step 3: Total torque including system losses

$$T_{total} = T_{sub} + T_{sys}$$
$$T_{total} = 14.2 + 1.0 = 15.2\ MN \cdot m$$

Interpretation

In this example, mechanical and system losses contribute about 1 MN·m, or roughly 7% of the total cutterhead torque requirement. Although this component does not contribute directly to cutting or face stability, it is unavoidable and must be included to ensure the drive system can operate reliably under real conditions.

Engineering significance

This example shows why mechanical and system losses cannot be ignored in torque calculations. Even in well-conditioned ground, these losses can account for 5–10% of total torque demand, and underestimating them risks undersizing the cutterhead drive, overheating gearboxes, or operating continuously at the limit of installed capacity.

6. Total Torque Requirement (Worked Example)

Using the earlier values:

- Cutting torque: 8.8 MN·m
- Face pressure torque: 4.0 MN·m
- Frictional allowance (15% of cutting): 1.3 MN·m
- Mechanical losses (7%): 1.0 MN·m

$$T_{total} \approx 15.1\ MN \cdot m$$

This value is then checked against:

- Installed motor and gearbox capacity
- Thermal limits
- Cutterhead structural capacity

Example: Total Torque Requirement and Design Checks

A 6.5 m EPB TBM is excavating in conditioned sandy silt under a face pressure setpoint of 220 kPa. Based on ground data and prior calculations, the main cutterhead torque components are estimated as:

- Cutting torque, T_{cut}= 8.8 MN·m
- Face pressure torque, T_{face}= 4.0 MN·m
- Frictional torque, T_{fric}= 15% of T_{cut}= $0.15 \times 8.8 = 1.3$MN·m
- Mechanical/system losses, T_{sys}= 7% of $\left(T_{cut}+T_{face}+T_{fric}\right)$

$$T_{sys} = 0.07 \times (8.8 + 4.0 + 1.3) = 0.07 \times 14.1 \approx 1.0 \text{ MN\cdotpm}$$

Total required torque

$$T_{total} = T_{cut} + T_{face} + T_{fric} + T_{sys} = 8.8 + 4.0 + 1.3 + 1.0 = 15.1 \text{ MN\cdotpm}$$

So, the TBM should be capable of delivering ~15.1 MN·m of cutterhead torque for steady operation in these conditions (with additional margin for transients such as harder lenses, poor conditioning, or pressure spikes).

Checks against TBM capability and limits

1) Installed motor and gearbox capacity

Assume the TBM has a rated cutterhead drive capacity of:

- Continuous torque capacity: 16.5 MN·m
- Peak (short-duration) torque capacity: 20 MN·m

Comparison:

- Continuous margin = $16.5 - 15.1 = 1.4$MN·m (~8.5% margin)
  This is acceptable for steady-state excavation, but it indicates the machine is operating relatively close to continuous capacity—so conditioning quality and pressure stability matter.

2) Thermal limits (power check at operating RPM)

Thermal limits are usually checked via power demand, because motors and gearboxes are heat-limited under sustained load.

If the cutterhead operates at 2.0 rpm, angular velocity is:

$$\omega = 2.0 \times \frac{2\pi}{60} \approx 0.209 \text{ rad/s}$$

Power required:

$$P = T \times \omega = 15.1 \times 10^6 \times 0.209 \approx 3.16 \times 10^6 \text{ W} \approx 3.2 \text{ MW}$$

If the installed drive power is 3.5 MW continuous, the system is within thermal capacity, but again with limited headroom—meaning sustained increases in torque (e.g., sticky muck) could push temperatures up and require operational changes (reduce RPM, improve conditioning, adjust pressure).

3) Cutterhead structural capacity

A structural check confirms the cutterhead can safely transmit the required torque without overstressing arms, hubs, welds, and tool mounts.

Assume the cutterhead structure is certified for:

- Allowable continuous torque: 18 MN·m

Comparison:

- Structural margin = $18 - 15.1 = 2.9$MN·m (~16% margin)
  This suggests the structure is adequate for the calculated torque demand.

What this example demonstrates

A calculated $T_{total} \approx 15.1$MN·m is not just a number—it becomes meaningful only after it is checked against:

- Drive capacity (continuous and peak torque)
- Thermal capacity (power and heat rejection at intended RPM)
- Structural capacity (cutterhead strength and fatigue robustness)

If any check is tight, the typical mitigations are operational and ground-interface controls: improve conditioning (reduce T_{fric}), optimise penetration/RPM, stabilise face pressure, and avoid sustained operation near peak torque.

7. Operational Control of Torque

Torque is not a fixed design value. Operators manage it by:

- Adjusting thrust and penetration
- Modifying cutterhead RPM
- Changing conditioning dosage
- Managing face pressure stability

High torque with low penetration → poor cutting efficiency
Low torque with high penetration → risk of cutter damage or stall

Example: Operational control of torque during an EPB TBM drive

A 6.5 m EPB TBM is tunnelling through sandy silt with occasional clay lenses. The control room trends show real-time values for torque, thrust, penetration rate, RPM, chamber pressure, and screw conveyor parameters. The operator's goal is to keep torque within a stable band while maintaining production and face stability.

Situation 1: High torque with low penetration (poor cutting efficiency)

What the operator sees

Over a 10–15 minute interval, the TBM trends shift to:

- **Cutterhead torque:** rises from 12 MN·m to 17 MN·m
- **Penetration rate:** drops from 45 mm/min to 15 mm/min
- **Thrust:** remains high and steady
- **Chamber pressure:** stable, no major drop
- **Muck appearance:** sticky, lumpy, poor flow through screw conveyor

Interpretation

This is a classic sign of increasing frictional resistance at the cutterhead and in the chamber—often caused by insufficient conditioning or a transition into a more cohesive clay-rich lens. The machine is "working hard" (high torque) but not advancing efficiently (low penetration).

Operational response

Operators manage torque by adjusting multiple levers:

1. **Increase conditioning dosage**
    - Increase foam injection ratio (FIR) and/or polymer dosage to reduce adhesion and internal friction.
2. **Reduce thrust slightly**
    - Avoid forcing the cutterhead into a sticky face, which can spike torque further and overload cutters.
3. **Lower RPM** (temporarily)
    - Reduces instantaneous power demand and can stabilise rotation while conditioning takes effect.
4. **Stabilise face pressure**
    - Avoid sudden chamber pressure oscillations while muck properties are changing.

Outcome

Within 20 minutes, torque drops back to 13–14 MN·m and penetration recovers to 35–40 mm/min, indicating improved cutting efficiency and reduced clogging.

Situation 2: Low torque with high penetration (risk of cutter damage or stall)

What the operator sees

Later in the shift, the TBM enters a sandier band with lower resistance:

- **Cutterhead torque:** drops to 7 MN·m
- **Penetration rate:** increases to 80–90 mm/min
- **Thrust:** remains high
- **Vibration trends:** increase slightly
- **Torque fluctuations:** begin to spike intermittently

Interpretation

At first glance, low torque looks good—but high penetration at high thrust can indicate the cutterhead is biting too aggressively. In soils this can lead to unstable cutting and pressure fluctuations; in mixed ground it can cause sudden torque spikes, tool shock loading, or even stalling if the cutterhead hits a harder lens unexpectedly.

Operational response

To keep torque stable and protect cutters:

1. **Reduce thrust to control penetration per revolution**
   - Aim for a smoother, controlled bite rather than "over-penetrating."
2. **Increase RPM slightly** (if within design limits)
   - Maintains production while reducing penetration per revolution.
3. **Keep chamber pressure stable**
   - High penetration can destabilise the chamber balance if inflow/outflow rates aren't matched.
4. **Fine-tune conditioning**
   - Not necessarily more conditioning—sometimes less water/foam is needed to avoid overly fluid muck and pressure instability.

Outcome

Penetration stabilises around 55–60 mm/min, torque rises modestly to 9–10 MN·m, vibration reduces, and torque spikes become less frequent—indicating smoother cutting and improved machine stability.

What this example demonstrates

- High torque + low penetration usually means the cutterhead is fighting friction/adhesion, so operators respond by improving conditioning, moderating thrust, and stabilising RPM and pressure.
- Low torque + high penetration can indicate overly aggressive cutting and instability risk, so operators reduce thrust and manage penetration per revolution while keeping pressure and muck behaviour consistent.

In practice, good torque control is not a single adjustment—it is a coordinated operational strategy using thrust, RPM, conditioning, and pressure stability to keep cutting efficient and the TBM operating safely within its mechanical limits.

Penetration Rate Estimation

The penetration rate describes how far a TBM advances per unit time and is one of the key indicators of excavation efficiency. It links thrust, torque, cutterhead speed, and ground resistance, and is therefore both a design parameter and an operational control variable. Penetration rate is usually expressed as mm/min or m/day, and is estimated differently for soil TBMs (EPB/slurry) and hard-rock TBMs, although the underlying mechanics are similar.

1. Fundamental relationships

At its simplest, penetration rate $v$is the product of:

$v =$ penetration per revolution × cutterhead rotational speed

$$v = p_{rev} \times n$$

Where:

- p_{rev}= penetration per revolution (mm/rev)
- n= cutterhead speed (rev/min)

Estimating penetration rate therefore requires estimating penetration per revolution, which is governed by the balance between applied thrust and ground cutting resistance.

Example: Fundamental penetration rate relationship in practice

A 6.5 m EPB TBM is operating in conditioned sandy silt. Based on thrust, torque, and ground resistance assessments, engineers estimate that the cutterhead is achieving a penetration per revolution of:

$$p_{rev} = 2.0 \text{ mm/rev}$$

The operator sets the cutterhead rotational speed to:

$$n = 3.0 \text{ rev/min}$$

Using the fundamental relationship:

$$v = p_{rev} \times n$$
$$v = 2.0 \times 3.0 = 6.0 \text{ mm/min}$$

Interpretation

In this example, the TBM advances at a penetration rate of 6 mm/min, which corresponds to approximately 8.6 m/day under continuous operation.

If ground resistance increases—due to poorer conditioning or a stiffer soil layer—the penetration per revolution may drop to, say, 1.2 mm/rev at the same RPM. The penetration rate would then reduce to:

$$v = 1.2 \times 3.0 = 3.6 \text{ mm/min}$$

Conversely, if thrust and conditioning are improved so that penetration per revolution increases to 2.5 mm/rev, the penetration rate rises to:

$$v = 2.5 \times 3.0 = 7.5 \text{ mm/min}$$

Engineering insight

This example illustrates that penetration rate is fundamentally controlled by penetration per revolution, which itself depends on the balance between applied thrust and ground cutting resistance. Increasing RPM alone will not increase advance if penetration per revolution is limited by torque capacity, tool performance, or face stability. Effective penetration control therefore focuses first on achieving an efficient and stable penetration per revolution, then selecting an appropriate cutterhead speed to convert that penetration into steady advance.

2. Penetration per revolution – soil TBMs (EPB / slurry)

For EPB and slurry TBMs, penetration per revolution depends on:

- Net thrust acting on the cutterhead
- Soil shear strength and density
- Effectiveness of conditioning
- Cutterhead opening ratio and tool geometry
- Face support pressure (which opposes penetration)

A simplified engineering expression is:

$$p_{rev} \approx \frac{F_{net}}{A \times \tau}$$

Where:

- F_{net}= net thrust available for cutting (kN), after subtracting face pressure and friction
- A= cutterhead area (m^2)
- τ= unit cutting resistance of the soil (kPa)

This expression does not represent an exact physical law but captures the proportional relationship used in preliminary estimation and operational benchmarking.

Example – EPB TBM in sandy silt

Given:

- Cutterhead diameter = 6.5 m → $A = 33.2\ m^2$
- Applied thrust = 18 MN
- Face pressure reaction = 7 MN
- Shield friction and losses = 4 MN

$$F_{net} = 18 - (7 + 4) = 7\ MN$$

Assume:

- Unit cutting resistance $\tau = 120\ kPa = 120\ kN/m^2$

$$p_{rev} \approx \frac{7{,}000}{33.2 \times 120} \approx 1.75\ mm/rev$$

If cutterhead speed:

$$n = 2.5\, rpm$$

Then penetration rate:

$$v = 1.75 \times 2.5 \approx 4.4\, mm/min$$

(≈ 6.3 m/day at continuous operation)

3. Penetration per revolution – hard rock TBMs

For hard-rock TBMs, penetration is governed by **disc cutter indentation** into the rock. Penetration per revolution depends on:

- Normal force per disc
- Rock strength (UCS, tensile strength)
- Disc diameter and edge geometry
- Cutter spacing

A simplified relationship is:

$$p_{rev} \approx \frac{F_{disc}}{k \times \text{UCS}}$$

Where:

- F_{disc}= normal force per disc
- k= empirical constant derived from indentation tests (CSM/NTNU models)

Example – hard rock TBM

Assume:

- Average disc force = 300 kN
- UCS = 100 MPa
- Empirical factor $k = 20$

$$p_{rev} \approx \frac{300}{20 \times 100} = 0.15\, mm/rev$$

At:

$$n = 6\,rpm$$
$$v = 0.15 \times 6 = 0.9\,mm/min$$

(≈ 1.3 m/day)

Penetration rate cannot be increased indefinitely by simply raising thrust or cutterhead rotational speed. Practical limits are imposed by the mechanical and geotechnical response of the system. As ground resistance increases, torque demand rises, and the cutterhead may approach or exceed its continuous torque capacity, restricting further increases in RPM. Excessive penetration per revolution can overload cutting tools, leading to vibration, shock loading, accelerated wear, or cutter damage. Similarly, high RPM increases power demand and thermal loading on motors, gearboxes, and bearings, as well as increasing abrasive wear on tools and muck-handling components. For these reasons, operators aim to identify an optimum combination of penetration per revolution and RPM in which torque remains within continuous capacity, cutting efficiency is high but controlled, and face stability and conditioning performance are maintained.

In practice, penetration rate is managed through continuous adjustment and feedback control rather than fixed settings. Operators rely on real-time monitoring of thrust and torque trends, cutterhead RPM, chamber pressure stability (for EPB and slurry machines), and observed muck behaviour and extraction performance. These inputs allow the tunnelling team to diagnose inefficiencies and respond before instability or mechanical problems develop. Typical responses include improving conditioning, reducing thrust, or stabilising face pressure when low penetration is accompanied by high torque, indicating excessive friction or poor cutting efficiency. Conversely, when penetration is high but torque is low, thrust may be reduced or RPM increased to avoid overly aggressive cutting and tool shock. Penetration fluctuations often signal imbalance at the face or in material extraction, prompting adjustments to face pressure or spoil removal rates.

From an engineering perspective, penetration rate estimation is not a single calculation but a linked performance assessment that integrates multiple aspects of the tunnelling system. It reflects the interaction between ground strength and variability, TBM thrust and torque capacity, cutterhead design and tool layout, and the chosen face support and conditioning strategy. Accurate penetration rate estimation allows engineers to predict advance rates and construction durations, size thrust and torque systems appropriately, and manage wear, energy consumption, and face stability risks.

In modern TBM tunnelling, penetration rate is therefore best understood not as a target to be maximised, but as the outcome of a balanced and adaptive operating strategy. Effective control requires coordinated management of thrust, torque, RPM, conditioning, and face pressure in response to real ground behaviour, ensuring safe, efficient, and sustainable excavation rather than simply higher instantaneous advance rates.

Cutter Wear Prediction

Cutter wear prediction is the process of estimating how quickly cutting tools (disc cutters, scrapers, picks) degrade during TBM excavation so that tool life, maintenance intervals, costs, and advance rates can be planned realistically. While exact wear cannot be predicted deterministically, engineering calculations combine rock/soil properties, tool loading, and operational parameters with empirical models calibrated from precedent projects.

Cutter wear is governed by the combined interaction of ground properties, cutter loading, machine kinematics, operating duration, and the effectiveness of conditioning and cooling systems. These factors do not act independently; instead, they reinforce or mitigate each other, determining the rate at which cutting tools degrade during excavation.

Ground abrasivity and strength are primary drivers of wear. In rock tunnelling, parameters such as uniaxial compressive strength (UCS), tensile strength, quartz content, and the Cerchar Abrasivity Index (CAI) are commonly used to characterise how aggressively the rock will wear cutting tools. In soft-ground tunnelling, wear is influenced by sand content, particle angularity, material hardness, and the proportion of fines, with coarse, angular, quartz-rich soils producing significantly higher abrasion than cohesive or fine-grained materials.

Cutter loading is another critical factor. The normal force applied to each cutter—whether disc cutters in hard rock or scrapers and picks in soil—directly affects contact stresses and wear intensity. Higher penetration per cutter or per revolution generally increases productivity but also raises wear rates by increasing contact energy and tool loading at the cutting interface.

Cutterhead kinematics strongly influence how wear develops across the cutterhead. Cutter radius plays a key role, as cutters located near the periphery of the cutterhead travel a longer distance per revolution and therefore experience higher sliding velocities and wear rates than inner cutters. Cutterhead rotational speed (RPM) and the resulting peripheral speed further control the rate of abrasive interaction between tools and ground, with higher speeds leading to accelerated wear.

Operational time and distance provide the cumulative exposure that ultimately determines total wear. The number of cutterhead revolutions, total operating hours, and metres excavated all contribute to the progressive loss of cutter material. Even under moderate ground conditions, extended operation at high RPM or high penetration can result in significant cumulative wear.

Finally, conditioning and cooling play an important moderating role, particularly in EPB and slurry TBMs. Effective soil conditioning in EPB tunnelling reduces friction, adhesion, and abrasive contact, thereby lowering wear on cutters. In slurry TBMs, slurry properties such as density, viscosity, and temperature influence both abrasion and cooling of cutting tools. Poor conditioning or inadequate cooling can significantly accelerate wear, even in ground that would otherwise be considered manageable.

Together, these five interacting factors define the wear environment at the cutterhead and form the basis for cutter wear prediction, tool selection, and maintenance planning in mechanised tunnelling.

1. Basic wear rate concept

At its simplest, cutter wear is expressed as a wear rate per unit excavation:

$$\text{Wear rate} = \frac{\text{wear depth or mass loss}}{\text{distance excavated}}$$

Typical units:

- mm of cutter ring loss per 100 m
- kg of steel loss per 1,000 m
- cutters replaced per 100 m or per ring

Example: Cutter Wear Rate in a Hard-Rock TBM Drive

A 7.2 m diameter hard-rock TBM is excavating a tunnel through moderately abrasive granite. Disc cutters have an initial ring thickness of 40 mm. Cutter inspections are carried out during planned interventions every 500 m.

After advancing 500 m, inspection records show that an average of 6 mm of cutter ring thickness has been lost on the outer gauge cutters.

Using the basic wear rate concept:

$$\text{Wear rate} = \frac{\text{Wear depth}}{\text{Distance excavated}}$$

$$\text{Wear rate} = \frac{6 \text{ mm}}{500 \text{ m}} = 0.012 \text{ mm/m}$$

This can be expressed in a more practical tunnelling format as:

- 1.2 mm of cutter ring wear per 100 m, or

- 12 mm of cutter ring wear per 1,000 m

Given a usable cutter ring thickness of 30 mm (allowing a safety margin), the estimated cutter life becomes:

$$\frac{30\ \text{mm}}{12\ \text{mm per } 1{,}000\ \text{m}} \approx 2{,}500\ \text{m}$$

This estimate allows engineers to plan cutter changes approximately every 2.0–2.5 km, depending on safety margins and observed variability.

Alternative Expression: Cutter Consumption Rate

On the same drive, records show that 8 disc cutters are replaced over a 1,000 m section due to reaching wear limits.

The cutter consumption rate can be expressed as:

- 0.8 cutters per 100 m, or
- 8 cutters per kilometre

This form is particularly useful for cost estimation, spares planning, and logistics.

2. Rock TBMs – Disc cutter wear prediction

Cerchar-based approach (widely used)

For hard rock TBMs, cutter wear is commonly correlated with the Cerchar Abrasivity Index (CAI).

A simplified empirical form is:

$$W = k \times \text{CAI} \times F_{disc} \times v$$

Where:

- W= wear rate indicator
- k= empirical constant (from CSM/NTNU databases)
- F_{disc}= normal force per disc
- v= peripheral velocity of the cutter

Higher CAI, higher disc force, and higher RPM all increase wear.

Example (hard rock TBM)

- CAI = 3.5 (abrasive quartz-rich rock)
- Average disc force = 280 kN
- Cutter radius = 3.0 m
- RPM = 6 → peripheral speed ≈ 1.9 m/s

From precedent data, this corresponds to:

- ~1 disc ring replacement every 150–200 m
- Higher wear on outer cutters due to higher velocity

Example: Disc cutter wear prediction using a Cerchar-based approach (hard rock TBM)

A hard rock TBM is driving through quartz-rich granite and the site investigation reports a Cerchar Abrasivity Index (CAI) of 3.5, indicating a highly abrasive rock mass. The TBM operates with an average normal force per disc cutter of 280 kN, and the cutters of interest are on the outer part of the cutterhead at a radius of 3.0 m. The cutterhead speed is 6 rpm.

Step 1: Calculate peripheral velocity at the cutter radius

Peripheral velocity is:

$$v = 2\pi r \times \frac{\text{RPM}}{60}$$

$$v = 2\pi(3.0) \times \frac{6}{60} = 18.85 \times 0.1 \approx 1.89 \text{ m/s}$$

So the cutter at 3.0 m radius runs at ~1.9 m/s.

Step 2: Use the simplified empirical wear indicator form

$$W = k \times CAI \times F_{disc} \times v$$

Using a representative calibration constant for this example (from project/precedent calibration),

$$k = 2.0 \times 10^{-4}$$

(with F_{disc} in kN and v in m/s; W is a relative wear index used for comparison and planning).

Now substitute:

$$W = (2.0 \times 10^{-4}) \times 3.5 \times 280 \times 1.89$$
$$W \approx (2.0 \times 10^{-4}) \times 1852 \approx 0.37$$

Interpretation: $W \approx 0.37$ indicates high wear severity, and it is mainly driven by the combination of high CAI, high disc force, and high peripheral velocity.

Step 3: Translate the wear indicator into a practical replacement interval (using precedent calibration)

On this project, precedent data and early intervention records show that for granite with CAI ≈ 3.5, operating at ~280 kN per disc and ~1.9 m/s peripheral speed, the TBM typically requires:

- ~1 disc ring replacement every 150–200 m (outer/gauge cutters)
- Longer life for inner cutters, because their radius (and therefore v) is lower

Why outer cutters wear faster (simple comparison)

If an inner cutter is at 1.5 m radius at the same RPM:

$$v_{inner} = 2\pi(1.5) \times \frac{6}{60} \approx 0.94 \text{ m/s}$$

That is about half the velocity, so (all else equal) the wear indicator W would also be roughly half, explaining why gauge/outer cutters typically control intervention frequency.

3. Soil TBMs (EPB / slurry) – Wear prediction

In soft ground, wear is driven more by abrasion and adhesion than crushing strength.

Key parameters include:

- Sand content and grain angularity
- Quartz percentage
- Conditioning quality (foam, polymers, bentonite)
- Presence of gravels or cobbles

A simplified engineering expression is:

$$W \propto A_s \times v \times t$$

Where:

- A_s= abrasive fraction of soil
- v= cutter peripheral speed
- t= operating time

Example (EPB TBM in sandy silt)

- Sand content: 55%
- Quartz content: high
- RPM = 3
- Conditioning: moderate foam, limited polymer

Observed wear rate from similar projects:

- Scraper/pick replacement every 250–400 m
- Accelerated wear when conditioning deteriorates

Improved conditioning (foam + polymer) can reduce wear by 30–50%.

Example: Wear prediction for a soil TBM (EPB) in sandy silt

An Earth Pressure Balance (EPB) TBM with a 6.5 m cutterhead diameter is excavating through a sandy silt formation in an urban tunnel. Ground investigation indicates a sand content of approximately 55%, with a high quartz fraction in the sand particles, giving the soil a relatively high abrasivity despite its low strength. Occasional fine gravel lenses are also present.

The cutterhead operates at an average speed of 3 rpm, and soil conditioning is initially limited to moderate foam injection, with little polymer addition.

Step 1: Identify dominant wear drivers

In this ground, wear is governed primarily by abrasion and adhesion, rather than crushing strength. The key contributors are:

- High abrasive fraction (A_s) due to quartz-rich sand
- Cutter peripheral speed (v), which increases sliding abrasion
- Operating time (t), which reflects cumulative exposure

- Conditioning quality, which controls friction and particle interaction

Step 2: Apply the simplified wear relationship

The qualitative engineering relationship is:

$$W \propto A_s \times v \times t$$

For this case:

- $A_s \approx 0.55$(55% sand, high quartz)
- Cutterhead radius ≈ 3.25 m
- Peripheral speed at outer cutters:

$$v = 2\pi r \times \frac{n}{60} = 2\pi(3.25) \times \frac{3}{60} \approx 1.02 \text{ m/s}$$

With continuous operation, wear accumulates steadily with time at this velocity.

Step 3: Translate into observed wear rates (from precedent projects)

Based on data from comparable EPB drives in sandy silt:

- Scraper or pick tools typically require replacement every 250–400 m
- Outer cutters and scrapers wear faster than inner tools due to higher peripheral speed
- Localised gravel lenses cause short-term spikes in wear rate

Operators also observe that when foam quality deteriorates or chamber conditioning becomes inconsistent, torque rises and wear accelerates noticeably.

Step 4: Effect of improved conditioning

Midway through the drive, the conditioning strategy is upgraded to include:

- Higher-quality foam (improved bubble stability)
- Addition of a polymer to bind fine particles and reduce abrasion

Following these changes:

- Muck becomes more plastic and less abrasive at the cutterhead
- Adhesion to tools decreases

- Torque fluctuations reduce

Measured outcomes show:

- Tool replacement intervals extending to 400–600 m
- An overall 30–50% reduction in wear rate, consistent with published experience

What this example demonstrates

- In soil TBMs, wear is controlled by abrasive fraction, cutter speed, and time, not compressive strength.
- Quartz-rich sands are a dominant wear driver even in "soft" ground.
- Poor or marginal conditioning leads to rapid wear and frequent tool changes.
- Improved foam and polymer conditioning can significantly extend tool life—often by 30–50%—making conditioning a primary lever for wear management in EPB tunnelling.

4. Peripheral speed correction

Wear is strongly dependent on cutter radius:

$$v_{peripheral} = 2\pi r \times \text{RPM}$$

Outer cutters experience:

- Higher sliding distance per revolution
- Higher contact energy
- Faster wear

Design and maintenance planning therefore assumes:

- Outer cutters wear 1.5–3× faster than inner cutters

Example: Peripheral speed correction and why outer cutters wear faster

A TBM cutterhead is operating at a constant 6 rpm. Engineers want to compare expected wear on an inner cutter versus an outer (gauge) cutter based on peripheral speed (and therefore sliding distance and contact energy).

Given

- Cutterhead speed: RPM = 6
- Inner cutter radius: $r_{ini} = 1.5\ m$
- Outer cutter radius: $r_{out} = 3.0\ m$

Peripheral speed is calculated as:

$$v = 2\pi r \times \frac{RPM}{60}$$

Step 1: Calculate peripheral speeds

Inner cutter

$$v_{in} = 2\pi(1.5) \times \frac{6}{60} = 9.42 \times 0.1 \approx 0.94 \text{ m/s}$$

Outer cutter

$$v_{out} = 2\pi(3.0) \times \frac{6}{60} = 18.85 \times 0.1 \approx 1.89 \text{ m/s}$$

Step 2: Compare speeds (proxy for wear severity)

$$\frac{v_{out}}{v_{in}} = \frac{1.89}{0.94} \approx 2.0$$

So the outer cutter has about twice the peripheral speed, meaning it travels twice the sliding distance per second and experiences higher contact energy at the cutting interface.

Step 3: Translate to a practical wear planning assumption

If an inner cutter ring typically lasts 900 m before reaching its wear limit, a reasonable planning assumption using the "outer cutters wear 1.5–3× faster" rule is:

- 2× faster wear (this example) → outer cutter life ≈ 900 / 2 = 450 m

So planners would expect:

- Inner cutters: ~900 m life

- Outer cutters: ~450 m life

This is why gauge cutters often control intervention frequency—they wear out sooner because they operate at much higher peripheral speed, even if the ground and cutter forces are similar.

What this example shows

- Peripheral speed scales directly with radius, so outer cutters accumulate far more sliding distance.
- Higher sliding distance leads to higher abrasion work and faster wear.
- Maintenance planning therefore assumes outer cutters typically wear 1.5–3× faster than inner cutters, and intervention strategy is usually designed around outer/gauge cutter life.

5. Life prediction for a cutter (worked concept)

If a disc cutter ring has:

- Allowable wear depth = 20 mm
- Predicted wear rate = 0.12 mm per 10 m

Then estimated life:

$$\text{Life} = \frac{20}{0.12} \times 10 \approx 1{,}670\ m$$

This value is adjusted for:

- Local geology changes
- Torque spikes and vibration
- Maintenance access strategy

Example: Cutter life prediction using a wear-rate approach

A hard-rock TBM is equipped with disc cutters that have a defined allowable wear limit before replacement is required. Engineers estimate cutter life based on predicted wear rates derived from laboratory abrasivity tests and precedent project data.

Given

- Allowable wear depth of disc cutter ring: 20 mm
- Predicted average wear rate: 0.12 mm per 10 m of excavation

Step 1: Calculate basic cutter life

First, determine how many 10 m intervals are required to reach the allowable wear depth:

$$\text{Number of intervals} = \frac{20}{0.12} \approx 166.7$$

Convert this to excavation distance:

$$\text{Estimated cutter life} = 166.7 \times 10 \approx 1{,}670 \text{ m}$$

Base prediction: the cutter ring is expected to reach its wear limit after approximately 1.67 km of tunnelling under steady conditions.

Step 2: Apply practical adjustment factors

In real TBM operations, this base value is rarely used directly. Engineers apply correction factors to account for operational and geological variability.

1. Local geology changes

If the alignment includes short sections of more abrasive rock (e.g. higher quartz content), engineers may reduce predicted life by 10–30%.

- Adjusted life (20% reduction):

$$1{,}670 \times 0.8 \approx 1{,}340 \text{ m}$$

2. Torque spikes and vibration

Frequent torque peaks or cutterhead vibration increase micro-chipping and thermal damage. Monitoring data may justify a further **10–15%** reduction.

- Further adjusted life (10% reduction):

$$1{,}340 \times 0.9 \approx 1{,}200 \text{ m}$$

3. Maintenance and access strategy

If cutter changes are only possible at planned interventions (e.g. every 1,000 m), engineers may conservatively schedule replacement earlier to avoid unplanned stoppages.

Step 3: Final operational planning value

Based on the above, the project team may adopt:

- **Planned cutter life:** ~1,200 m
- **Inspection trigger:** ~900–1,000 m
- **Replacement window:** aligned with planned intervention points

This example illustrates that cutter life prediction is not a single deterministic number. A simple wear-rate calculation provides a baseline, but final planning values are refined using geological knowledge, real-time TBM performance data, and maintenance constraints. This approach allows engineers to balance tool utilisation, safety, and programme reliability while avoiding unexpected cutter failures.

Cutter wear prediction is an iterative process rather than a fixed, upfront calculation. As tunnelling progresses, initial predictions are continuously refined using operational feedback from the machine and the construction process. Key inputs include records of actual cutter changes, trends in cutterhead torque and vibration, changes in penetration efficiency, and visual inspection of cutting tools during planned or unplanned interventions. These data provide direct evidence of how the ground and operating conditions are affecting cutter performance in practice, allowing predictions to be updated and made more reliable as the drive advances.

Sudden increases in cutter wear rate are often linked to identifiable changes in conditions. In EPB tunnelling, loss or degradation of soil conditioning commonly leads to higher friction and abrasion at the cutterhead. Entry into localised abrasive lenses, such as sand or quartz-rich zones, can also cause wear rates to increase sharply compared with surrounding ground. Excessive penetration per revolution, whether due to high thrust or low RPM, may overload cutters and accelerate wear through increased contact stresses and vibration. Recognising these correlations allows operators and engineers to adjust operating parameters or conditioning strategies before excessive damage occurs.

From an engineering perspective, accurate cutter wear prediction is essential for effective project planning and risk management. Reliable estimates allow engineers to quantify cutter consumption and associated costs, plan cutter change intervals and maintenance interventions, and optimise thrust, torque, and RPM to balance productivity against tool life. By anticipating wear-related stoppages, projects can avoid unplanned shutdowns that disrupt schedules and increase costs.

In mechanised tunnelling, cutter wear is therefore a production constraint as critical as thrust or torque capacity. Effective wear prediction integrates laboratory-derived indices such as CAI, empirical models based on precedent projects, and real-time operational feedback. This combined approach enables tunnelling teams to manage advance rates, tool life, and safety in a balanced and controlled manner throughout the drive.

Face Pressure Calculations

Face pressure calculations determine the support pressure that must be applied at the tunnel face to maintain stability during excavation. The objective is to keep the ground at the face in a state of equilibrium by counteracting earth and groundwater pressures without inducing collapse (under-pressure) or blow-out/heave (over-pressure). These calculations underpin both Earth Pressure Balance (EPB) and slurry TBM operation and define the allowable pressure window for safe tunnelling.

1. Components of Required Face Support Pressure

The required face pressure is typically calculated as the sum of:

1. Effective earth pressure
2. Pore water pressure
3. Allowances for uncertainty and operational control

In simplified form:

$$p_{face} = \sigma_h' + u$$

Where:

- σ_h'= effective horizontal earth stress at tunnel axis
- u= pore water pressure at tunnel axis

Example: Calculation of required face support pressure at tunnel axis

Consider a 6.5 m diameter EPB TBM driving a shallow urban tunnel through saturated sandy silt. The tunnel axis is located 15 m below ground level, and the groundwater table is at the surface.

Step 1: Calculate vertical effective stress

Total vertical stress at tunnel axis:

$$\sigma_v = \gamma \times z$$

Assume:

- Total unit weight of soil, $\gamma = 20$ kN/m^3
- Depth to tunnel axis, $z = 15$ m

$$\sigma_v = 20 \times 15 = 300 \text{ kPa}$$

Pore water pressure at tunnel axis:

$$u = \gamma_w \times z = 9.81 \times 15 \approx 147 \text{ kPa}$$

Effective vertical stress:

$$\sigma_v' = \sigma_v - u = 300 - 147 = 153 \text{ kPa}$$

Step 2: Calculate effective horizontal earth stress

Assume at-rest earth pressure conditions, appropriate for EPB tunnelling with limited ground deformation:

$$\sigma_h' = K_0 \times \sigma_v'$$

Let:

- $K_0 = 0.6$

$$\sigma_h' = 0.6 \times 153 \approx 92 \text{ kPa}$$

Step 3: Determine required face support pressure

Using the simplified face pressure relationship:

$$p_{\text{face}} = \sigma_h' + u$$
$$p_{\text{face}} = 92 + 147 = 239 \text{ kPa}$$

Step 4: Engineering interpretation

- Effective earth pressure component (σ'_h): 92 kPa

- Pore water pressure component (u): 147 kPa
- Base calculated face support pressure: ≈ 240 kPa

In practice, this value would be treated as the central design pressure, around which a pressure window is defined. Allowances would then be added for geological uncertainty, pressure losses, and operational fluctuations, resulting in a lower collapse limit and an upper blow-out or uplift limit.

Engineering significance

This example illustrates that face support pressure is not an arbitrary operational setting but the direct sum of effective ground stress and groundwater pressure at tunnel depth. The calculation provides a rational starting point for EPB or slurry TBM operation, ensuring that the applied face pressure balances in-situ stresses and maintains tunnel face stability.

2. Earth Pressure Component

Vertical stress

The total vertical stress at tunnel depth is:

$$\sigma_v = \gamma H$$

Where:

- γ= bulk unit weight of ground (kN/m^3)
- H= depth to tunnel axis (m)

Horizontal effective stress

Horizontal stress is estimated using an earth pressure coefficient:

$$\sigma_h' = K \sigma_v'$$

Where:

- K= earth pressure coefficient
  - K_0(at-rest) → commonly used for EPB design
  - K_a(active) → lower-bound collapse check
  - K_p(passive) → upper-bound blow-out check
- $\sigma_v' = \sigma_v - u$

For most EPB and slurry designs, $K_0$is adopted because tunnelling aims to minimise ground deformation.

Example: Earth pressure component for face pressure calculation

Consider a slurry or EPB TBM driving a tunnel with its axis located 18 m below ground level in saturated silty sand. The groundwater table is at ground surface.

Step 1: Calculate total vertical stress

The total vertical stress at tunnel axis depth is calculated from the bulk unit weight of the ground:

$$\sigma_v = \gamma \times H$$

Assume:

- Bulk unit weight of soil, $\gamma = 19$ kN/m^3
- Depth to tunnel axis, $H = 18$ m

$$\sigma_v = 19 \times 18 = 342 \text{ kPa}$$

Step 2: Calculate pore water pressure

$$u = \gamma_w \times H$$

Where:

- Unit weight of water, $\gamma_w = 9.81$ kN/m^3

$$u = 9.81 \times 18 \approx 177 \text{ kPa}$$

Step 3: Calculate effective vertical stress

$$\sigma_v' = \sigma_v - u$$
$$\sigma_v' = 342 - 177 = 165 \text{ kPa}$$

Step 4: Calculate effective horizontal stress using earth pressure coefficient

For EPB and slurry TBM design, at-rest earth pressure is typically adopted because tunnelling aims to limit ground deformation.

Assume:

- At-rest earth pressure coefficient, $K_0 = 0.55$

$$\sigma'_h = K_0 \times \sigma'_v$$
$$\sigma'_h = 0.55 \times 165 \approx 91 \text{ kPa}$$

Step 5: Interpretation and pressure window checks

- **Effective horizontal earth pressure (σ'_h):** ~91 kPa
- This value represents the **earth pressure component** of the required face support pressure.
- To define the full operating window:
  - **Lower-bound collapse check:** use K_a(e.g. 0.3–0.35)
  - **Upper-bound blow-out check:** use K_p(often >2.0 in sands)

For normal EPB and slurry operation, the design face pressure is centred around the K_0-based value, with upper and lower limits defined using $K_a$and $K_p$to prevent collapse and uplift respectively.

Engineering significance

This example shows how the earth pressure component of face support pressure is derived directly from vertical stress, groundwater conditions, and the selected earth pressure coefficient. Using K_0reflects the objective of mechanised tunnelling: maintaining near in-situ stress conditions while controlling deformation, settlement, and face stability.

3. Pore Water Pressure Component

Pore water pressure depends on groundwater level:

$$u = \gamma_w h_w$$

Where:

- γ_w= unit weight of water (~9.81 kN/m^3)
- h_w= height of water column above tunnel axis (m)

In water-bearing soils, this term can be a dominant component of face pressure, especially in shallow tunnels below the water table.

Example: Pore water pressure component in face pressure calculation

Consider a TBM tunnel with its axis located 12 m below the groundwater table in saturated sand. The groundwater table is assumed to be at or near ground surface, which is typical in shallow urban tunnelling or river crossings.

Step 1: Identify water column height

The height of the water column acting at the tunnel axis is:

$$h_w = 12 \text{ m}$$

Step 2: Calculate pore water pressure

Pore water pressure is calculated as:

$$u = \gamma_w \times h_w$$

Where:

- Unit weight of water, $\gamma_w = 9.81 \text{ kN/m}^3$

$$u = 9.81 \times 12 \approx 118 \text{ kPa}$$

Step 3: Engineering interpretation

- Pore water pressure at tunnel axis: ~118 kPa
- This pressure acts equally in all directions and must be fully counteracted by the TBM face support system (EPB muck pressure or slurry pressure).
- In shallow tunnels, this component can represent more than half of the total required face pressure, particularly where effective earth stresses are relatively low.

In water-bearing soils, especially below rivers, canals, or floodplains, pore water pressure often dominates the face pressure calculation. Even if the soil itself is relatively weak, failure to counteract groundwater pressure can lead to:

- Sudden inflow of water and soil
- Face instability or collapse
- Hydraulic piping or blow-outs

This example highlights why accurate groundwater profiling and real-time pressure control are critical in EPB and slurry TBM operations. Face pressure must always be sufficient to balance both effective earth pressure and pore water pressure, not just the weight of the soil.

4. Pressure Distribution Across the Face

Face pressure is not uniform. Because overburden increases with depth, pressure is typically assumed to vary linearly:

- Maximum pressure at the invert
- Minimum pressure at the crown

Design therefore defines:

- Average face pressure
- Minimum crown pressure (collapse limit)
- Maximum invert pressure (uplift or blow-out limit)

The TBM is operated to keep chamber pressure within this pressure envelope, not at a single fixed value.

Example: Pressure distribution across the tunnel face

Consider a 6.5 m diameter EPB TBM excavating a tunnel with its axis at 15 m below ground level in saturated sandy silt. The groundwater table is at ground surface. From prior calculations, the average required face pressure at tunnel axis has been determined as 240 kPa.

Step 1: Determine vertical pressure gradient

Pressure increases with depth due to the combined unit weight of soil and water. For simplicity, assume an average pressure gradient of:

$$\Delta p = \gamma_{\text{soil}} \approx 20 \text{ kPa/m}$$

(The water pressure component is already included in the average face pressure.)

The tunnel radius is:

$$R = 6.5/2 = 3.25 \text{ m}$$

Step 2: Calculate crown and invert pressures

Crown (top of face)

The crown is 3.25 m above the tunnel axis:

$$p_{\text{crown}} = p_{\text{avg}} - (\gamma \times R)$$
$$p_{\text{crown}} = 240 - (20 \times 3.25) = 240 - 65 \approx 175 \text{ kPa}$$

This value represents the minimum allowable pressure. If pressure at the crown drops below this level, face collapse and ground loss may initiate.

Invert (bottom of face)

The invert is **3.25 m below** the tunnel axis:

$$p_{\text{invert}} = p_{\text{avg}} + (\gamma \times R)$$
$$p_{\text{invert}} = 240 + 65 \approx 305 \text{ kPa}$$

This value represents the maximum allowable pressure. Exceeding it may cause uplift, hydraulic fracturing, or blow-outs.

Step 3: Define the operating pressure envelope

From the above calculations, the pressure envelope is:

- **Minimum crown pressure (collapse limit):** ~175 kPa
- **Average face pressure (design target):** ~240 kPa
- **Maximum invert pressure (uplift/blow-out limit):** ~305 kPa

The **allowable operating window** is therefore approximately:

$$175 \text{ kPa} \leq p_{\text{face}} \leq 305 \text{ kPa}$$

Step 4: Operational interpretation

The TBM is not operated at a single pressure value. Instead, operators aim to keep the average chamber pressure such that:

- Crown pressure never falls below the collapse limit
- Invert pressure never exceeds the uplift or blow-out limit

Small short-term fluctuations are acceptable, but sustained operation outside this envelope significantly increases risk.

This example demonstrates why face pressure control is framed as a pressure envelope, not a single design number. Linear pressure distribution across the face means that safe operation depends on simultaneously protecting the crown from under-pressure and the invert from over-pressure. Effective EPB and slurry TBM operation therefore requires continuous monitoring and adjustment to keep chamber pressure within this defined range as ground conditions change.

5. Collapse and Blow-Out Limits (Pressure Window)

Lower limit – collapse pressure

If face pressure drops below this value:

- Ground yields inward
- Volume loss occurs
- Settlement or sinkholes may develop

Often checked using **active earth pressure** assumptions.

Upper limit – blow-out or uplift pressure

If face pressure exceeds this value:

- Effective stress approaches zero
- Hydraulic fracturing or uplift may occur
- Slurry or soil may escape to the surface

Often checked using passive earth pressure or uplift criteria.

The allowable operating pressure window lies between these two limits.

Example: Defining collapse and blow-out limits (pressure window)

A 6.5 m diameter EPB TBM is driving a shallow tunnel in saturated silty sand. The tunnel axis is 15 m below ground level, and the groundwater table is at the surface. Assume:

- Bulk unit weight of ground: $\gamma = 20$ kN/m^3
- Unit weight of water: $\gamma_w = 9.81$ kN/m^3
- Depth to axis: $H = 15$ m

Step 1: Calculate total vertical stress and pore pressure at tunnel axis

$$\sigma_v = \gamma H = 20 \times 15 = 300 \text{ kPa}$$
$$u = \gamma_w H = 9.81 \times 15 \approx 147 \text{ kPa}$$

Effective vertical stress:

$$\sigma_v' = \sigma_v - u = 300 - 147 = 153 \text{ kPa}$$

Step 2: Lower limit (collapse pressure) using active earth pressure K_a

Assume a conservative active coefficient for sandy silt:

$$K_a = 0.33$$

Effective horizontal stress at collapse limit:

$$\sigma_h'(\text{collapse}) = K_a \sigma_v' = 0.33 \times 153 \approx 50 \text{ kPa}$$

Collapse-limit face support pressure:

$$p_{\text{collapse}} = \sigma_h'(\text{collapse}) + u = 50 + 147 = 197 \text{ kPa}$$

Interpretation: If chamber pressure drops below ~200 kPa, the face may begin to yield inward, producing volume loss and settlement.

Step 3: Upper limit (blow-out / uplift pressure) using passive earth pressure K_p

Assume passive coefficient for the same soil:

$$K_p = 3.0$$

Effective horizontal stress at blow-out limit:

$$\sigma'_h(\text{blowout}) = K_p \sigma'_v = 3.0 \times 153 = 459 \text{ kPa}$$

Blow-out-limit face support pressure:

$$p_{\text{blowout}} = \sigma'_h(\text{blowout}) + u = 459 + 147 = 606 \text{ kPa}$$

Interpretation: Pressures approaching this magnitude risk reducing effective stress to very low values locally and can initiate hydraulic fracturing, uplift, or slurry/soil escape—especially in shallow ground where the *true* blow-out limit may be lower due to cover sensitivity.

Step 4: Define the allowable operating pressure window

The pressure window lies between the two limits:

- **Collapse limit:** $p_{\text{collapse}} \approx 197$ kPa
- **Blow-out/uplift limit:** $p_{\text{blowout}} \approx 606$ kPa

So the theoretical pressure window at tunnel axis is:

$$197 \text{ kPa} \leq p_{\text{face}} \leq 606 \text{ kPa}$$

In practice, engineers would narrow this window further by:

- checking crown and invert pressures (linear distribution),
- applying uplift criteria for shallow cover,
- adding safety margins for uncertainty and sensor error.

If face pressure falls below the collapse limit, the ground can move inward, causing ground loss and settlement. If face pressure approaches or exceeds the blow-out/uplift limit, effective stress can be driven toward zero locally and failure may occur outward, producing uplift, hydraulic fracturing, or material escape. Safe TBM operation therefore depends on maintaining pressure inside a defined operating window, refined with distribution checks and real-time monitoring.

In Earth Pressure Balance (EPB) TBMs, face support is provided by pressurised, conditioned excavated material retained within the excavation chamber. The calculated face pressure is achieved and controlled through coordinated adjustment of the screw conveyor extraction rate, the TBM advance rate, and the chamber fill ratio. For this system to function effectively, the excavated soil must be conditioned to behave as a plastic, low-permeability mass capable of transmitting pressure uniformly to the tunnel face. If the muck is too permeable or lacks cohesion, pressure dissipation can occur, undermining face stability.

In slurry TBMs, face pressure is provided hydrostatically by a bentonite slurry that fills the excavation chamber. The applied pressure is governed primarily by the slurry density and the slurry level within the chamber, which together define the hydrostatic head acting on the tunnel face. In permeable ground, the formation of a stable filter cake on the face is critical. This filter cake limits slurry loss into the ground and ensures that the calculated pressure can be maintained consistently across the face.

Design face pressure values incorporate safety margins to account for uncertainties and operational realities. These allowances typically include geological variability, pressure losses across cutterhead openings, instrumentation accuracy limits, and short-term fluctuations during excavation. As a result, the calculated pressure represents an initial target rather than a fixed operating value.

During tunnelling, face pressure is continuously refined using real-time data and observational feedback. Operators rely on face pressure sensors, surface settlement monitoring, TBM thrust and torque trends, and observed spoil behaviour and extraction performance to fine-tune pressure control. This feedback-based adjustment is essential to responding to changing ground conditions and maintaining stability throughout the drive.

From an engineering perspective, face pressure calculations form a critical link between ground mechanics, TBM operational control, and settlement risk management. They highlight that face stability is fundamentally a balance problem rather than a question of maximising support pressure. Insufficient pressure leads to collapse and settlement, while excessive pressure can cause uplift, blow-outs, or damage to surrounding infrastructure. Successful tunnelling depends on maintaining face pressure within a narrow, well-defined operating window.

In modern TBM operations, face pressure is therefore not treated as a single design number but as a continuously managed control parameter. Its value evolves as excavation progresses, reflecting real ground behaviour and the dynamic interaction between the TBM, the ground, and the support system.

Advance Rate and Productivity Modelling

Advance rate and productivity modelling are used to predict how fast a TBM will excavate and how long a tunnel drive will take. These calculations integrate penetration mechanics, machine limits, and operational availability, translating geotechnical and mechanical performance into programme duration, cost, and risk.

1. Fundamental Advance Rate Relationship

At the most basic level, the instantaneous advance rate is governed by cutterhead penetration:

$$v = p_{\text{rev}} \times n$$

Where:

- v= instantaneous advance rate (mm/min or m/h)
- p_{rev}= penetration per revolution (mm/rev)
- n= cutterhead speed (rev/min)

This relationship defines the best-case mechanical advance, assuming continuous excavation with no interruptions.

Example: Fundamental advance rate relationship (instantaneous advance)

Consider an EPB TBM operating in conditioned sandy silt with the following parameters measured during steady excavation:

- Penetration per revolution, $p_{\text{rev}} = 18\text{mm/rev}$
- Cutterhead speed, $n = 2.5\text{rev/min}$

Using the fundamental relationship:

$$v = p_{\text{rev}} \times n$$
$$v = 18 \times 2.5 = 45 \text{ mm/min}$$

Convert to practical units

$$45 \text{ mm/min} = 2.7 \text{ m/h}$$

Engineering interpretation

- The instantaneous mechanical advance rate is 2.7 m/h.
- This represents the best-case advance during active boring only, assuming:

- continuous rotation,
- stable face pressure,
- no interruptions for ring build, maintenance, or stoppages.

In practice, the net advance rate over a shift or day will be lower once ring erection, thrust transfer, grouting, inspections, and downtime are included. Nonetheless, this simple calculation provides the baseline against which operational efficiency and productivity losses are assessed.

This example shows how advance rate is fundamentally controlled by penetration per revolution and RPM, both of which are constrained by thrust, torque capacity, cutter condition, and ground resistance. Productivity improvements therefore often come from improving penetration efficiency and operational continuity—not merely increasing cutterhead speed.

2. Penetration per Revolution Estimation

Penetration per revolution is not arbitrary; it results from a balance between:

- Applied thrust
- Available cutterhead torque
- Ground cutting resistance
- Cutter condition and spacing
- Conditioning effectiveness (EPB/slurry)

Simplified conceptual form:

$$p_{\text{rev}} \propto \frac{F_{\text{thrust}}}{R_{\text{cut}}}$$

Where R_{cut}represents ground cutting resistance.

In practice, penetration is capped by:

- Maximum allowable cutter load
- Torque limits
- Vibration and tool life considerations

Example: Estimating penetration per revolution from thrust vs cutting resistance

A 6.5 m EPB TBM is excavating through a dense sandy silt. The crew wants to estimate a realistic penetration per revolution based on the balance between available thrust and ground cutting resistance, then check whether torque and tool limits cap the result.

Step 1: Define a simple thrust-to-resistance estimate

Assume the TBM is applying an average total thrust of:

- $F_{\text{thrust}} = 18$ MN(18,000 kN)

From site performance correlations (or precedent data), the cutting resistance for this ground at the current conditioning state is estimated as:

- $R_{\text{cut}} = 1{,}200$ MN/m

This value represents the "push required per unit penetration" (i.e., how much thrust is needed to achieve 1 m of penetration under the current cutting conditions).

Using the simplified conceptual form:

$$p_{\text{rev}} \propto \frac{F_{\text{thrust}}}{R_{\text{cut}}}$$

A practical estimate is:

$$p_{\text{rev}} \approx \frac{F_{\text{thrust}}}{R_{\text{cut}}}$$

$$p_{\text{rev}} \approx \frac{18}{1200} = 0.015 \text{ m/rev} = 15 \text{ mm/rev}$$

Initial estimate: $p_{\text{rev}} \approx 15$ mm/rev

Step 2: Check if cutter load caps penetration

Suppose the cutterhead has **30 primary cutting tools**, and for a quick check we assume thrust is reasonably distributed:

$$F_{\text{tool}} \approx \frac{18{,}000}{30} = 600 \text{ kN/tool}$$

But the allowable working load per tool (based on tool design and wear limits) is:

- $F_{\text{allow}} = 500$ kN/tool

This indicates the applied thrust is **too high for tool limits**, so penetration must be reduced by lowering thrust:

Revised total thrust:

$$F_{\text{thrust, capped}} = 500 \times 30 = 15{,}000\ \text{kN} = 15\ \text{MN}$$

Recalculate penetration:

$$p_{\text{rev}} \approx \frac{15}{1200} = 0.0125\ \text{m/rev} = 12.5\ \text{mm/rev}$$

Cutter load–capped penetration: $p_{\text{rev}} \approx 12.5$ mm/rev

Step 3: Check torque limits (secondary cap)

Assume that at 12.5 mm/rev, the predicted cutting torque demand rises to 14 MN·m, but the TBM continuous torque capacity is:

- T_{cont} = 12 MN·m

This means torque will become limiting. Operators must either:

- reduce penetration (reduce thrust), or
- reduce RPM / improve conditioning to lower resistance.

If improved foam + polymer conditioning reduces cutting resistance by ~15% (common order-of-magnitude effect), then:

$$R_{\text{cut,new}} = 0.85 \times 1200 = 1020\ \text{MN/m}$$

With thrust capped at 15 MN:

$$p_{\text{rev,new}} \approx \frac{15}{1020} = 0.0147\ \text{m/rev} \approx 14.7\ \text{mm/rev}$$

But because torque is still limiting, operators might choose to hold penetration at a safer:

- $p_{\text{rev}} \approx 10\text{–}12$ mm/rev

to keep torque within continuous capacity and avoid vibration/tool shock.

What this example demonstrates

- Penetration per revolution emerges from the thrust vs cutting resistance balance (conceptually $p_{\text{rev}} \propto F_{\text{thrust}}/R_{\text{cut}}$).
- Even if the thrust-based estimate is high, penetration is often capped by:
  - maximum allowable cutter load,
  - torque limits, and
  - vibration and tool life considerations.
- Conditioning improvements reduce effective cutting resistance, allowing higher penetration at the same thrust or the same penetration at lower thrust/torque, improving stability and tool life.

3. Mechanical Advance Rate (Gross Boring Rate)

The gross boring rate (GBR) represents advance during active excavation only:

$$\text{GBR} = p_{\text{rev}} \times n \times 60$$

Typical values:

- Soft ground EPB TBM: 30–80 mm/min (1.8–4.8 m/h)
- Slurry TBM: 20–60 mm/min
- Hard rock TBM: 10–40 mm/min

GBR reflects machine–ground interaction, not overall productivity.

Example: Mechanical Advance Rate (Gross Boring Rate)

Consider an EPB TBM operating in well-conditioned sandy clay with the following steady-state excavation parameters:

- Penetration per revolution, p_rev = 15 mm/rev
- Cutterhead speed, n = 3.0 rev/min

The gross boring rate (GBR) is calculated as:

GBR = p_rev × n × 60

GBR = 15 × 3.0 × 60 = 2,700 mm/h

Converting to metres per hour:

GBR = 2.7 m/h

Engineering interpretation

- The TBM advances 2.7 m per hour while actively boring.
- This value lies comfortably within the typical range for soft-ground EPB TBMs (1.8–4.8 m/h).
- The calculated GBR reflects pure excavation performance, governed by the interaction between cutterhead penetration, rotation speed, and ground resistance.

Importantly, this is not the overall production rate. Ring erection, grouting, inspections, maintenance, and operational pauses are excluded. Once these non-boring activities are included, the net advance rate over a shift or day will be significantly lower.

GBR is used to:

- compare TBM performance between different ground conditions,
- assess the effect of conditioning and cutting efficiency,
- define the upper bound of achievable advance during excavation.

However, overall productivity depends far more on cycle time and utilisation than on GBR alone, which is why GBR should always be interpreted as a mechanical performance indicator rather than a programme rate.

4. Cycle Time and Ring Build Time

Actual progress is governed by the excavation cycle, which includes:

- Excavation (boring)
- Ring build (segment erection)
- Thrust transfer and grouting
- Short operational pauses

For segmentally lined tunnels, average advance per cycle:

$$\text{Advance per cycle} = \text{ring width}$$

Average cycle time:

$$t_{\text{cycle}} = t_{\text{boring}} + t_{\text{erection}} + t_{\text{grouting}} + t_{\text{reset}}$$

Example: Cycle time and ring build time in a segmentally lined tunnel

Consider a 6.5 m diameter EPB TBM installing a 1.5 m wide segmental lining ring in soft ground.

Step 1: Define advance per cycle

For a segmentally lined tunnel, one full excavation and lining cycle advances the tunnel by exactly one ring width:

Advance per cycle = ring width = **1.5 m**

Step 2: Define typical cycle components

Based on observed site performance, the average times for each activity are:

- **Boring time (t_boring):** 18 minutes
  (time to excavate 1.5 m at the current penetration rate)
- **Ring erection time (t_erection):** 15 minutes
  (lifting, placing, bolting segments, and closing the ring)
- **Grouting and thrust transfer (t_grouting):** 7 minutes
  (annulus grouting, pressure stabilisation, transfer of thrust to the new ring)
- **Reset / short operational pauses (t_reset):** 5 minutes
  (checks, alignment confirmation, system stabilisation)

Step 3: Calculate total cycle time

t_cycle = t_boring + t_erection + t_grouting + t_reset

t_cycle = 18 + 15 + 7 + 5 = 45 minutes

Step 4: Calculate net advance rate from cycle time

Net advance rate:

Net advance rate = advance per cycle / cycle time

Net advance rate = 1.5 m / 0.75 h = **2.0 m/h**

Engineering interpretation

- Although the gross boring rate during excavation may be higher, the actual tunnel advance is governed by the full cycle time.

- In this example, ring erection and associated activities account for over 60% of the cycle, meaning improvements in lining logistics or erection efficiency would have a greater impact on productivity than increasing penetration alone.
- Cycle-based calculations are therefore central to realistic programme planning and productivity modelling.

This example shows that TBM productivity is controlled by the complete excavation–lining cycle, not just cutting performance. Reducing ring build time by even a few minutes per cycle can translate into significant gains in daily and monthly advance.

5. Net Advance Rate (NAR)

The **net advance rate** accounts for all activities within the normal excavation cycle:

$$\text{NAR} = \frac{\text{ring width}}{t_{\text{cycle}}}$$

This is the most relevant metric for day-to-day production planning.

Example:

- Ring width = 1.5 m
- Average cycle time = 45 min

$$\text{NAR} = \frac{1.5}{0.75} = 2.0\ \text{m/h}$$

Example: Net Advance Rate (NAR) for day-to-day production planning

Consider a segmentally lined EPB TBM tunnel with the following operating parameters:

- **Ring width:** 1.5 m
- **Average excavation cycle time:** 45 minutes

Step 1: Convert cycle time to hours

45 minutes = 0.75 hours

Step 2: Calculate the net advance rate

Net Advance Rate (NAR) is calculated as:

NAR = ring width / cycle time

NAR = 1.5 m / 0.75 h = **2.0 m/h**

Step 3: Engineering interpretation

- The TBM advances an average of 2.0 metres per hour when all routine activities within the excavation cycle are included.
- This value already accounts for:
    - active boring,
    - segment erection,
    - annulus grouting,
    - thrust transfer and short operational pauses.
- NAR is therefore far more representative of actual daily performance than gross boring rate.

Why NAR matters in practice

Net advance rate is the key metric used for:

- shift and daily production targets,
- manpower and logistics planning,
- short-term programme tracking.

While penetration rate and cutterhead performance define how fast the TBM *can* cut, NAR defines how fast the tunnel is actually built.

6. Utilisation and Availability Factors

Real projects are affected by interruptions such as:

- Cutter changes
- Maintenance
- Survey checks
- Ground treatment
- Logistics delays

Productivity modelling therefore introduces a utilisation factor (U):

$$\text{Effective advance rate} = \text{NAR} \times U$$

Typical utilisation:

- Well-run urban EPB TBM: 40–55%
- Complex ground / long logistics: 30–40%

- Hard rock TBM with frequent cutter changes: 25–35%

Example: Utilisation and availability factors in productivity modelling

Consider an urban EPB TBM with a calculated net advance rate (NAR) of 2.0 m/h, based on ring width and average cycle time. While this represents the rate during normal excavation cycles, the TBM does not operate continuously due to unavoidable interruptions.

Step 1: Identify typical interruptions

Over a typical working day, time is lost to:

- cutter inspections and replacements,
- routine maintenance and servicing,
- survey and guidance checks,
- occasional ground treatment or probe drilling,
- delays in segment or muck logistics.

As a result, the TBM may only be productively excavating for a portion of the available time.

Step 2: Apply utilisation factor

For a well-run urban EPB TBM, a realistic utilisation factor is:

U = 0.50 (50%)

Step 3: Calculate effective advance rate

Effective advance rate is calculated as:

Effective advance rate = NAR × U

Effective advance rate = 2.0 m/h × 0.50 = 1.0 m/h

Step 4: Daily production estimate

Assuming a 24-hour operation:

Daily advance = 1.0 m/h × 24 h = **24 m/day**

Step 5: Comparison with other conditions

For the same NAR of 2.0 m/h:

- **Complex ground or long logistics (U = 0.35):**
  Effective advance rate = 0.7 m/h → ~17 m/day
- **Hard rock TBM with frequent cutter changes (U = 0.30):**
  Effective advance rate = 0.6 m/h → ~14 m/day

Engineering interpretation

This example shows that utilisation has a first-order influence on project duration. Even with a strong mechanical advance rate, poor availability or frequent interruptions can halve overall productivity. Consequently, improvements in maintenance planning, cutter change strategy, and logistics often deliver greater schedule benefits than marginal increases in penetration or RPM.

In productivity modelling, utilisation factors convert theoretical performance into realistic programme outcomes, making them essential for reliable planning and risk assessment.

7. Daily and Monthly Production

Daily advance:

$$\text{Daily production} = \text{Effective advance rate} \times 24$$

Monthly advance:

$$\text{Monthly production} = \text{Daily production} \times \text{working days}$$

These values feed directly into:

- Programme schedules
- Cost forecasts
- Resource planning

Example: Daily and Monthly Production Estimation

Consider an EPB TBM operating in an urban tunnel with the following productivity parameters already established:

- **Net advance rate (NAR):** 2.0 m/h
- **Utilisation factor (U):** 0.50

From this, the **effective advance rate** is:

Effective advance rate = NAR × U
Effective advance rate = 2.0 × 0.50 = **1.0 m/h**

Step 1: Calculate daily production

Daily production is calculated as:

Daily production = Effective advance rate × 24

Daily production = 1.0 × 24 = 24 m/day

Step 2: Calculate monthly production

Assume:

- **Working days per month:** 26 days

Monthly production is:

Monthly production = Daily production × working days

Monthly production = 24 × 26 = 624 m/month

Engineering interpretation

- The TBM is expected to advance approximately 24 metres per day under normal operating conditions.
- Over a typical month, this equates to around 620–630 metres of tunnel.
- These values represent realistic production, incorporating excavation, lining installation, maintenance, and operational interruptions.

Daily and monthly production estimates feed directly into:

- programme and milestone schedules,
- cost and cash-flow forecasts, and
- resource planning, including labour, segment supply, spoil handling, and maintenance logistics.

Small changes in utilisation or cycle efficiency can therefore translate into large differences in total project duration and cost, making these calculations central to TBM planning and control.

Advance rate is highly sensitive to a range of interacting parameters rather than a single controlling factor. Key influences include penetration per revolution, cutterhead RPM limits, the efficiency of ring build and lining installation, cutter wear rates, face pressure stability, and the reliability of spoil removal and logistics systems. Experience shows that relatively small improvements in ring build time, segment handling, or overall utilisation often deliver greater productivity gains than attempts to increase raw penetration alone. This reflects the cyclic nature of TBM operations, where non-cutting activities can dominate total advance time.

From an engineering perspective, advance rate and productivity modelling provide a framework that links ground behaviour with TBM mechanical limits, lining and logistics cycles, and maintenance strategy. They demonstrate that productivity is fundamentally system-controlled rather than driven solely by cutterhead speed or thrust. In modern TBM projects, the highest and most reliable production is achieved through balanced optimisation of thrust, torque, penetration, ring build efficiency, spoil conditioning, and downtime management, rather than by maximising any single parameter in isolation.

In practice, advance rate should be understood as a probabilistic outcome rather than a fixed design value. It is informed by analytical calculations, precedent project data, and continuous operational feedback as excavation progresses. This approach allows engineers and operators to respond to changing ground conditions and machine performance, maintaining productivity while managing risk, wear, and face stability.

Power Consumption Estimates

Power consumption estimates quantify the electrical power required to operate a TBM under given ground and operational conditions. They are used to size electrical infrastructure, assess energy costs, manage thermal limits, and evaluate operational efficiency. In practice, TBM power demand is the sum of several interacting components, dominated by cutterhead rotation and thrust-related systems.

1. Core Power Relationship

At its simplest, mechanical power is the product of torque and rotational speed (for rotating systems) or force and velocity (for linear systems):

- **Rotational power (cutterhead):**
  Power = Torque × Angular speed
- **Linear power (thrust):**
  Power = Thrust force × advance velocity

These mechanical demands are then adjusted for drive efficiency to estimate electrical input power.

Example: Core power relationship for a TBM

A. Rotational power – cutterhead

Assume an EPB TBM operating at steady excavation with:

- Cutterhead torque, T = 10 MN·m
- Cutterhead speed = 3.0 RPM
- Drivetrain efficiency, $\eta = 0.92$

First convert RPM to angular speed:

Angular speed, $\omega = 2\pi \times \text{RPM} / 60$
$\omega = 2\pi \times 3.0 / 60 \approx 0.314$ rad/s

Mechanical cutterhead power:

Power_mech = Torque × Angular speed
Power_mech = 10 × 0.314 = 3.14 MW

Electrical input power (allowing for efficiency):

Power_elec = Power_mech / η
Power_elec = 3.14 / 0.92 ≈ 3.41 MW

Interpretation:
About 3.4 MW of electrical power is required to drive the cutterhead under these conditions.

B. Linear power – thrust system

Assume at the same time the TBM applies:

- Total thrust force, F = 15 MN
- Advance rate, v = 2.0 m/h

Convert advance rate to m/s:

$v = 2.0 / 3600 \approx 0.00056$ m/s

Mechanical thrust power:

Power_thrust = Thrust force × advance velocity
Power_thrust = 15 × 0.00056 ≈ 0.0084 MW (8.4 kW)

Allowing for hydraulic efficiency, say $\eta = 0.80$:

Electrical power ≈ 0.0084 / 0.80 ≈ 0.0105 MW (10.5 kW)

Interpretation:
Compared with the cutterhead, the thrust system consumes very little power during steady advance; its significance lies more in force capacity than energy demand.

What this example shows

- Rotational power at the cutterhead dominates TBM power consumption.
- Linear thrust power is comparatively small, even at high thrust levels.
- Electrical power demand is always higher than mechanical demand due to system inefficiencies.

This is why TBM power calculations focus primarily on torque, RPM, and cutterhead efficiency, with thrust power treated as a secondary but necessary component.

2. Cutterhead Power Requirement

The cutterhead is typically the largest single power consumer on a TBM.

Mechanical cutterhead power is calculated as:

$Power\_cutterhead = T \times \omega$

Where:

- T = cutterhead torque (MN·m)
- ω = angular speed (rad/s)

Angular speed is related to RPM by:

$\omega = 2\pi \times RPM / 60$

Electrical power demand is then:

$Electrical\ power = Power\_cutterhead / \eta$

Where:

- η = overall drivetrain efficiency (typically 0.90–0.95)

This calculation captures the power needed to overcome ground cutting resistance, face pressure effects, and friction at the cutterhead.

Example: Cutterhead power requirement calculation

Consider a 6.5 m diameter EPB TBM operating in dense, well-conditioned sandy silt under steady excavation conditions. The following operating parameters are measured:

- **Cutterhead torque (T):** 11 MN·m
- **Cutterhead speed:** 3.5 RPM
- **Overall drivetrain efficiency (η):** 0.92

Step 1: Convert RPM to angular speed

Angular speed is calculated as:

$\omega = 2\pi \times RPM / 60$

$\omega = 2\pi \times 3.5 / 60 \approx 0.366$ rad/s

Step 2: Calculate mechanical cutterhead power

Mechanical power required at the cutterhead is:

$Power\_cutterhead = T \times \omega$

$Power\_cutterhead = 11 \times 0.366 \approx 4.03$ MW

This is the mechanical power needed to overcome:

- ground cutting resistance,
- face pressure effects (EPB chamber pressure acting on the cutterhead),
- friction between the cutterhead, muck, and surrounding ground.

Step 3: Calculate electrical input power

Allowing for drivetrain efficiency:

$Electrical\ power = Power\_cutterhead / \eta$

$Electrical\ power = 4.03 / 0.92 \approx 4.38$ MW

Engineering interpretation

- The cutterhead alone requires approximately 4.4 MW of electrical power under these conditions.
- This confirms why the cutterhead drive is typically the dominant power consumer on a TBM.

- Any increase in torque (harder ground, poor conditioning, higher face pressure) or RPM will immediately increase power demand.

This example shows how cutterhead power consumption directly reflects ground conditions and operational control. Rising electrical power at constant RPM is often an early indicator of increasing cutting resistance, deteriorating conditioning, or cutter wear, making power monitoring a key tool for real-time TBM performance management.

3. Thrust System Power

The thrust system consumes power to push the TBM forward against face resistance and shield friction.

Mechanical thrust power is:

$Power_{thrust} = F_{thrust} \times v$

Where:

- F_{thrust} = total thrust force (MN)
- v = advance velocity (m/s)

Although thrust power is usually smaller than cutterhead power, it becomes significant in:

- squeezing or high-friction ground,
- long shields,
- high advance rates.

Hydraulic inefficiencies are accounted for by applying a system efficiency factor.

Example: Thrust system power calculation

Consider a soft-ground EPB TBM advancing under steady conditions with the following parameters:

- **Total thrust force (F_{thrust}):** 20 MN
- **Net advance rate:** 2.4 m/h
- **Hydraulic system efficiency (η):** 0.80

Step 1: Convert advance rate to m/s

$v = 2.4 \text{ m/h} \div 3600$
$v \approx 0.00067 \text{ m/s}$

Step 2: Calculate mechanical thrust power

Mechanical thrust power is:

Power_thrust = F_thrust × v

Power_thrust = 20 × 0.00067
Power_thrust ≈ 0.0134 MW (13.4 kW)

This is the mechanical power required to push the TBM forward against:

- face support reaction,
- shield–ground friction,
- internal resistance during advance.

Step 3: Account for hydraulic inefficiencies

Electrical input power is higher due to hydraulic losses:

Electrical thrust power = Power_thrust / η

Electrical thrust power = 0.0134 / 0.80
Electrical thrust power ≈ 0.0168 MW (16.8 kW)

Engineering interpretation

- The thrust system requires less than 20 kW of electrical power under these conditions.
- Compared with the cutterhead (typically several megawatts), thrust power is minor during normal advance.
- However, thrust power demand increases significantly in:
    - squeezing ground where shield friction dominates,
    - long TBM shields with high contact area,
    - high advance rates where velocity increases.

This example highlights that thrust systems are force-critical rather than power-critical. Their design is governed by maximum thrust capacity and load transfer to the lining, while their contribution to overall power consumption is usually secondary—except in difficult ground where friction and resistance escalate rapidly.

4. Auxiliary System Power

In addition to cutting and thrusting, TBMs require power for multiple auxiliary systems, including:

- Screw conveyors (EPB) or slurry pumps (slurry TBMs)
- Segment erectors and hydraulic systems
- Guidance, control, lighting, ventilation
- Slurry separation plant (for slurry TBMs)

These loads are typically estimated from manufacturer data or precedent projects and added as fixed or variable allowances.

Example: Auxiliary system power allowance in a TBM power budget

Assume a 6.5 m EPB TBM operating on a 24/7 urban drive. The cutterhead and thrust power have already been estimated separately. The project team now builds an auxiliary power budget using typical manufacturer allowances and precedent values.

Auxiliary loads (EPB TBM)

- **Screw conveyor drive:** 250 kW
  (varies with muck density, extraction rate, and torque)
- **Hydraulic power pack (thrust, articulation, erector hydraulics):** 400 kW
  (higher during ring build and steering corrections)
- **Segment erector (peak during erection):** 120 kW
  (often intermittent; not always running)
- **Guidance, control, instrumentation, comms:** 30 kW
  (mostly steady)
- **Conveyors on backup / belt transfer systems:** 180 kW
  (depends on belt length, gradient, and throughput)
- **Ventilation fans, cooling, lighting (tunnel services):** 300 kW
  (often treated as a relatively steady load)

Total auxiliary power (EPB)

Total auxiliary power = 250 + 400 + 120 + 30 + 180 + 300
Total auxiliary power = 1,280 kW ≈ 1.28 MW

How it's used

- During steady boring, the erector load may be near zero, so auxiliaries might average closer to 1.1 MW.
- During ring build, hydraulic and erector loads increase, and auxiliaries may peak near 1.3–1.5 MW.

This auxiliary total is then added to the primary consumers (cutterhead + thrust) to estimate:

- typical operating power, and
- peak installed power requirement (with margin).

Slurry TBM note (contrast example)

If the same project used a slurry TBM, the auxiliary budget would typically include:

- slurry feed/return pumps: often 1–3 MW combined (major auxiliary load), and
- slurry separation plant: commonly 0.5–2 MW depending on flow rate and soil type,

which can make auxiliaries a much larger share of total site power demand than in EPB tunnelling.

5. Total Installed Power vs Operating Power

Two different power values are important:

- **Installed power:**
  The maximum available electrical capacity, sized for worst-case conditions (hard ground, peak torque, simultaneous operation of systems).
- **Operating (average) power:**
  The typical power draw during steady excavation, usually 60–80% of installed capacity.

Design calculations ensure installed power exceeds peak demand with margin, while productivity and cost modelling focus on average power consumption.

Example: Total installed power vs operating (average) power for a TBM

A project is planning a 6.5 m EPB TBM for an urban tunnel. The electrical team needs to size the site power supply (installed power), while the planning team needs a realistic average power for energy cost and productivity modelling.

1) Installed power (worst-case capacity)

The TBM supplier provides the following maximum / peak ratings for worst-case conditions (dense ground, high torque, simultaneous operation during transitions and ring-build overlap):

- **Cutterhead drive (peak):** 6.0 MW
- **Screw conveyor (peak):** 0.35 MW
- **Hydraulic power packs (peak):** 0.60 MW
- **Backup conveyors + auxiliaries (peak):** 0.80 MW

- **Ventilation/cooling/lighting allowance (peak):** 0.40 MW

Peak simultaneous demand:

Total peak = 6.0 + 0.35 + 0.60 + 0.80 + 0.40
Total peak = 8.15 MW

The designer then adds a margin (e.g., 15%) for uncertainty, start-up surges, and contingencies:

Installed power = 8.15 × 1.15 = 9.37 MW

Installed electrical capacity (rounded): ~9.5 MW

This is the value used to size:

- substations and transformers,
- switchgear and cabling,
- generator backup (if required).

2) Operating (average) power during steady excavation

During typical steady-state boring in the dominant ground type, measured/expected loads are lower:

- **Cutterhead (typical):** 4.2 MW
- **Screw conveyor (typical):** 0.25 MW
- **Hydraulics (typical):** 0.40 MW
- **Backup conveyors + auxiliaries (typical):** 0.65 MW
- **Ventilation/cooling/lighting:** 0.40 MW

Average operating demand:

Operating power = 4.2 + 0.25 + 0.40 + 0.65 + 0.40
Operating power = 5.90 MW

Now compare operating to installed:

Operating / Installed = 5.90 / 9.37 ≈ 0.63 (63%)

This aligns with the typical rule that average operating power is about 60–80% of installed capacity.

Why this matters

- Installed power (~9.5 MW) ensures the TBM can handle worst-case peaks without tripping or stalling, even when several systems run simultaneously.
- Operating power (~5.9 MW) is the realistic value used for:
  - energy cost forecasts (kWh),
  - carbon footprint estimates,
  - heat load and cooling assessments,
  - day-to-day power monitoring benchmarks.

In short, installed power is a "can we handle peaks?" design number, while operating power is the "what do we actually draw?" planning number.

6. Energy Consumption per Metre

For cost and sustainability analysis, power is often converted into energy per metre of tunnel:

Energy per metre = Average power / Advance rate

Units:

- kWh per metre

Higher penetration at the same power reduces energy per metre, while inefficient cutting, poor conditioning, or excessive torque increase it.

Example: Energy consumption per metre of tunnel

Assume a 6.5 m EPB TBM operating in mixed sand–silt ground during steady excavation. From power monitoring and advance records, the following typical values are observed:

- **Average electrical power draw:** 5.8 MW
- **Net advance rate (NAR):** 2.3 m/h

Step 1: Convert power to energy per hour

Power is already an energy rate:

5.8 MW = 5,800 kWh per hour

Step 2: Calculate energy per metre

$$\text{Energy per metre} = \frac{\text{Average power}}{\text{Advance rate}}$$
$$= \frac{5{,}800\ \text{kWh/h}}{2.3\ \text{m/h}} \approx 2{,}520\ \text{kWh/m}$$

Interpretation

This means the TBM consumes approximately 2.5 MWh of electrical energy for every metre of tunnel constructed under these conditions.

Sensitivity illustration

Improved performance

After optimising spoil conditioning and reducing cutterhead torque, the advance rate increases while power remains similar:

- Average power: 5.7 MW
- Advance rate: 3.0 m/h

$$\text{Energy per metre} = \frac{5{,}700}{3.0} \approx 1{,}900 \text{ kWh/m}$$

→ Energy per metre drops by ~25%, even though installed power is unchanged.

Poor cutting conditions

If conditioning deteriorates and torque rises:

- Average power: 6.2 MW
- Advance rate: 1.8 m/h

$$\text{Energy per metre} = \frac{6{,}200}{1.8} \approx 3{,}440 \text{ kWh/m}$$

→ Energy per metre increases sharply, signalling inefficient excavation.

Engineering significance

Energy consumption per metre is a powerful performance indicator because it combines mechanical efficiency, ground response, and operational quality into a single metric. In practice:

- **Lower kWh/m** → efficient cutting, good conditioning, balanced thrust and torque
- **Rising kWh/m** → increased resistance, poor conditioning, cutter wear, or instability

For modern TBM projects, tracking energy per metre is central to cost control, carbon reporting, and early detection of deteriorating excavation conditions.

Power consumption in TBM operations is highly sensitive to a number of interrelated parameters. Ground strength and abrasivity have a direct influence on the torque required for cutting, while penetration per revolution and cutterhead RPM determine how much work is being done per unit time. Face pressure in EPB and slurry TBMs also affects power demand by increasing resistance at the cutterhead when pressures rise. The effectiveness of spoil conditioning plays a critical role, as well-conditioned material reduces friction and cutting resistance, whereas poor conditioning increases torque and power draw. Cutter wear state further amplifies these effects, with worn or damaged tools causing sudden increases in resistance. As a result, abrupt changes in torque or face pressure are immediately reflected as power spikes in the electrical system.

From an engineering perspective, power consumption estimates provide an essential link between ground behaviour, TBM mechanical design, electrical infrastructure requirements, and operating cost and carbon footprint. They illustrate that efficient TBM operation is not achieved by maximising power input, but by using available power effectively. Well-conditioned ground, balanced application of thrust and torque, and stable face pressure reduce cutting resistance and therefore lower overall power demand.

In modern tunnelling practice, power consumption is treated as a performance indicator as important as advance rate. A rising power demand at constant penetration is often an early warning sign of deteriorating cutting conditions, inadequate conditioning, or increasing cutter wear. Continuous monitoring of power therefore supports proactive operational control and predictive maintenance, helping to maintain productivity while managing risk and energy use.

Segment Load Calculations

Segment load calculations determine the forces and stresses acting on precast concrete tunnel lining segments during construction and throughout the service life of the tunnel. These calculations are essential to ensure that segments are structurally adequate during installation, grouting, and long-term ground loading, and that they meet both safety and durability requirements.

Segment load calculations serve several critical engineering objectives:

- **Structural design:** sizing segment thickness, reinforcement, and concrete strength
- **Construction safety:** preventing cracking, crushing, or joint damage during thrusting and ring build
- **Serviceability:** limiting long-term deformation, leakage, and joint opening
- **Durability:** ensuring adequate performance under sustained loads and environmental exposure

Because tunnel segments experience very different loading conditions over time, calculations are usually divided into construction-stage loads and long-term in-ground loads.

Construction-stage loading

Thrust forces during TBM advance

During excavation, thrust cylinders apply axial force to the installed lining via thrust shoes. Segment load checks must verify:

- Axial compressive stress in segments
- Local bearing stress beneath thrust shoes
- Load distribution across multiple shoes

The average axial stress in a ring can be approximated as:

$$\sigma_{axial} = \frac{F_{thrust}}{A_{ring}}$$

Where:

- F_{thrust}= total applied thrust force
- A_{ring}= net concrete cross-sectional area of the ring

Local checks ensure that bearing stresses do not exceed allowable concrete compressive limits at shoe contact points.

Worked example: Thrust forces during TBM advance

Consider a 6.8 m internal diameter EPB TBM advancing through soft ground with a segmental concrete lining.

Given data

- **Total thrust force applied:**

$$F_{thrust} = 55 \text{ MN}$$

- **Segment ring geometry:**
  - Internal diameter = 6.8 m
  - Segment thickness = 0.35 m
- **Concrete strength:**
  - Characteristic compressive strength $f_{ck} = 50$ MPa
- **Number of thrust shoes:**
  - 12 thrust shoes bearing on the ring

Step 1: Calculate ring cross-sectional area

The concrete cross-sectional area of the ring is approximated as:

$$A_{ring} = \pi\left(R_o^2 - R_i^2\right)$$

Where:

- $R_i = 3.40$ m
- $R_o = 3.40 + 0.35 = 3.75$ m

$$A_{ring} = \pi(3.75^2 - 3.40^2) = \pi(14.06 - 11.56) = \pi \times 2.50 \approx 7.85\ \text{m}^2$$

Step 2: Average axial compressive stress in the ring

$$\sigma_{axial} = \frac{F_{thrust}}{A_{ring}}$$

$$\sigma_{axial} = \frac{55}{7.85} \approx 7.0\ \text{MPa}$$

Interpretation:
An average axial stress of 7 MPa is well below the concrete's compressive capacity, even after applying partial safety factors. This confirms that the global ring compression is acceptable.

Step 3: Local bearing stress beneath thrust shoes

Although the average stress is low, thrust is transferred through discrete thrust shoes, producing higher local bearing stresses.

Assume:

- Effective bearing area per shoe = 0.18 m^2

Total bearing area:

$$A_{bear,total} = 12 \times 0.18 = 2.16\ \text{m}^2$$

Average bearing stress:

$$\sigma_{bear} = \frac{55}{2.16} \approx 25.5\ \text{MPa}$$

Step 4: Check against allowable bearing stress

For construction-stage loading, allowable concrete bearing stress is often limited to:

- 0.6–0.7 × f_{ck}

$$\sigma_{allow} \approx 0.6 \times 50 = 30 \text{ MPa}$$

Result:

$$25.5 \text{ MPa} < 30 \text{ MPa}✓$$

The local bearing stress is acceptable, with margin.

Step 5: Load distribution considerations

Design also checks that:

- Thrust shoes are evenly loaded (no excessive eccentricity)
- Adjacent segments share load effectively
- Joint rotation does not concentrate stress on individual segments

In practice, misalignment, steering corrections, or incomplete ring contact can increase local stresses, so operational thrust limits are often set below theoretical maxima.

Engineering takeaway

This example shows that:

- Average axial stress checks ensure global ring safety
- Local bearing checks often govern allowable thrust
- Segment capacity, not hydraulic thrust capacity, frequently controls TBM operation

As a result, thrust force during TBM advance is managed not simply to achieve penetration, but to protect the newly installed lining from crushing, cracking, or joint damage.

<u>Handling, erection, and temporary conditions</u>

Segments are also checked for:

- Lifting stresses during handling and erection
- Eccentric loading during partial ring assembly
- Temporary unsupported spans before ring closure

These conditions often govern reinforcement detailing even though they are short-term.

Worked example: Handling, erection, and temporary conditions for segmental linings

Consider a precast concrete segment used in a 6.8 m diameter EPB TBM tunnel.

Segment properties

- Segment width (axial): 1.5 m
- Segment thickness: 0.35 m
- Segment arc length: ≈ 2.2 m
- Segment mass: 7.5 t (≈ 74 kN)
- Concrete strength: 50 MPa
- Reinforcement designed for both permanent and temporary loads

1. Lifting stresses during handling and erection

Each segment is lifted from storage, transported, and erected using two lifting anchors located near the intrados.

Loading condition

- Self-weight = 74 kN
- Lifted at two points → each anchor carries ~37 kN
- Segment behaves as a simply supported beam between lifting points

The critical bending moment during lifting is approximated as:

$$M = \frac{W \cdot L}{4}$$

Where:

- $W = 74$ kN
- $L = 2.0$ m(distance between anchors)

$$M = \frac{74 \times 2.0}{4} = 37 \text{ kNm}$$

Engineering implication

This bending moment may exceed the permanent-service bending moment in the completed ring. As a result, temporary lifting stresses often govern minimum reinforcement, particularly near anchor zones and intrados faces.

2. Eccentric loading during partial ring assembly

During ring erection, segments are installed sequentially and temporarily bolted together. Before ring closure:

- The partially assembled ring is not self-supporting
- Thrust shoes may bear on an incomplete ring
- Loads are eccentric and unevenly distributed

Example condition

- 4 segments installed out of a 6-segment ring
- Thrust force applied: 45 MN
- Load shared by only part of the ring → uneven compression

This creates:

- Local bending at segment joints
- Tensile stresses at the intrados of unsupported segments
- High shear forces at temporary bolts and dowels

Engineering implication

Segments are detailed with:

- Extra reinforcement near circumferential joints
- Ductile detailing to tolerate short-term eccentric compression
- Temporary bolt capacity checked independently of permanent design

3. Temporary unsupported spans before ring closure

Before the key segment is installed, the last-open segment may act as a cantilever or short-span beam, especially if the erector releases support prematurely.

Example condition

- Unsupported span = 1.2 m
- Segment self-weight = 74 kN
- Resulting mid-span bending moment:

$$M = \frac{wL^2}{8}$$
$$M = \frac{74 \times 1.2^2}{8} \approx 13.3 \text{ kNm}$$

Although short-lived, this moment:

- Can cause cracking at the intrados
- Often governs crack-control reinforcement
- Is checked under construction-stage load combinations

Why these checks matter

Although handling, erection, and temporary conditions exist only for minutes or hours, they often:

- Control reinforcement quantity and placement
- Govern lifting anchor design
- Dictate minimum concrete strength at demoulding
- Drive construction tolerances and erection procedures

In many tunnel projects, segments that are perfectly adequate for permanent ground loads would fail during handling if these temporary conditions were ignored.

Engineering takeaway

This example shows that segment design is not governed solely by long-term ground loading. Lifting, partial assembly, and temporary unsupported conditions frequently produce the highest bending stresses, and therefore play a decisive role in reinforcement detailing, crack control, and construction safety—even though they are short-term states.

Grouting and early-age loading

Once a ring is erected, annular grouting introduces radial pressure on the lining before the grout has hardened and before full ground contact is achieved.

Key checks include:

- Uniform radial pressure from grout injection
- Asymmetric pressure due to uneven grout fill
- Early-age concrete strength limitations

At this stage, loads may be higher than long-term values because grout pressure can temporarily exceed in-situ ground stress.

Example: Grouting and early-age loading of a segmental lining ring

Consider a 6.5 m internal diameter EPB TBM tunnel constructed with precast concrete segments. Immediately after a ring is erected inside the tail shield, two-component annular grout is injected to fill the gap between the lining and the surrounding ground.

Grouting scenario

- **Annular gap thickness:** 120 mm
- **Grout injection pressure at ports:** 350 kPa
- **In-situ horizontal ground stress at tunnel axis:** 220 kPa
- **Concrete age at grouting:** 6–8 hours
- **Early-age concrete compressive strength:** ~12 MPa (≈ 35–40% of 28-day strength)

At this stage, the lining is not yet fully supported by the ground. The primary radial load acting on the ring is therefore grout pressure, not ground pressure.

Uniform radial loading check

Assuming uniform grout distribution around the ring, the lining experiences a temporary external pressure of approximately:

$$p_{grout} = 350 \text{ kPa}$$

This value exceeds the long-term ground pressure (220 kPa), meaning the lining is subjected to a short-term load case that is more severe than its final design condition.

Segment checks at this stage verify that:

- Compressive stresses in the concrete remain below early-age strength limits

- Hoop compression does not cause cracking or joint opening
- Gasket compression remains within allowable limits

Asymmetric loading scenario

During injection, grout distribution is rarely perfectly uniform. In this example:

- Invert grout fills rapidly and reaches 350 kPa
- Crown grout lags due to air entrapment, reaching only 200 kPa

This creates an asymmetric radial pressure distribution, inducing:

- Local bending in the ring
- Higher compressive stress at the invert
- Tensile stress at the crown intrados

Even though this condition is temporary, it can govern:

- Minimum reinforcement requirements
- Crack control criteria
- Joint detailing at segment interfaces

Early-age concrete limitation

Because the concrete has not yet reached full strength, allowable stresses are reduced. Designers typically apply:

- Early-age strength reduction factors
- Crack control limits stricter than long-term cases

In this example, checks confirm that the combination of grout pressure and early-age strength does not exceed allowable compressive or tensile limits, even under asymmetric loading.

Engineering significance

This example illustrates why grouting and early-age loading often govern segment design, even though they act for only a short duration. Grout pressures can temporarily exceed in-situ ground stresses, and uneven grout fill can introduce bending effects not present in long-term conditions.

As a result, segment load calculations must explicitly consider:

- Peak grout injection pressures
- Potential asymmetry in grout distribution

- Reduced early-age concrete capacity

Ignoring these effects can lead to cracking, joint damage, or leakage that compromises lining performance long before long-term ground loads become critical.

Long-term ground and water loads

Ground pressure

After the grout hardens and ground–lining interaction is established, the lining is subjected to earth pressures from the surrounding ground.

These are typically estimated using:

- At-rest earth pressure (K_0) for most bored tunnels
- Non-uniform pressure distribution (higher at invert, lower at crown)

The vertical stress at tunnel depth is:

$$\sigma_v = \gamma \cdot H$$

And horizontal stress:

$$\sigma_h = K \cdot \sigma_v$$

Where:

- γ= unit weight of ground
- H= depth to tunnel axis
- K= earth pressure coefficient

Example: Ground pressure acting on a segmental tunnel lining

Consider a segmentally lined bored tunnel constructed at moderate depth in urban soil, where the annular grout has fully hardened and effective ground–lining interaction has been established. At this stage, the temporary grout pressure has dissipated and the lining is primarily loaded by in-situ earth pressures.

Ground and tunnel conditions

- **Tunnel axis depth (H):** 25 m
- **Ground type:** stiff clay with some silt
- **Unit weight of ground (γ):** 19 kN/m^3

- **Earth pressure coefficient (K_0):** 0.6 (at-rest condition, appropriate for bored tunnels)

Because mechanised tunnelling aims to minimise ground deformation, at-rest earth pressure is typically adopted for long-term lining design.

Vertical stress at tunnel depth

The total vertical stress acting at the tunnel axis is calculated as:

$$\sigma_v = \gamma \cdot H$$
$$\sigma_v = 19 \times 25 = 475 \text{ kPa}$$

This represents the overburden load from the ground above the tunnel.

Horizontal ground pressure

The horizontal stress acting on the lining is estimated using the at-rest earth pressure coefficient:

$$\sigma_h = K_0 \cdot \sigma_v$$
$$\sigma_h = 0.6 \times 475 = 285 \text{ kPa}$$

This horizontal pressure forms the basis of the long-term radial loading on the lining.

Non-uniform pressure distribution around the ring

In practice, ground pressure around the tunnel circumference is **not uniform**:

- **Invert:** higher pressure due to greater overburden and stress concentration
- **Crown:** lower pressure due to reduced vertical confinement

For design, engineers typically assume a **linear pressure distribution** from crown to invert, with:

- **Average radial pressure:** ≈ 285 kPa
- **Maximum pressure at invert:** slightly higher than average
- **Minimum pressure at crown:** slightly lower than average

This non-uniform distribution induces **hoop compression and bending** in the segmental lining, which are checked against concrete strength, reinforcement capacity, and joint performance.

Engineering significance

This example shows how long-term ground pressure on a tunnel lining is derived directly from basic geotechnical parameters—unit weight, depth, and earth pressure coefficient. Once grout hardens, these pressures replace temporary construction loads and govern:

- Long-term structural capacity of segments
- Crack control and watertightness
- Joint and gasket performance

Accurate estimation of ground pressure is therefore fundamental to ensuring that the lining performs safely and durably throughout the tunnel's design life.

Pore water pressure

In water-bearing ground, hydrostatic pressure acts on the lining:

$$u = \gamma_w \cdot h_w$$

This pressure contributes directly to radial loading and often governs design in soft ground or subaqueous tunnels.

Example: Pore water pressure acting on a tunnel lining

Consider a bored tunnel constructed below the groundwater table in saturated sandy clay, where groundwater contributes significantly to the loading on the segmental lining.

Groundwater conditions

- **Tunnel axis depth below groundwater table (h_w):** 18 m
- **Unit weight of water (γ_w):** 9.81 kN/m^3

Because the ground is fully saturated, hydrostatic pressure acts directly on the lining over the full circumference.

Calculation of pore water pressure

Pore water pressure at the tunnel axis is calculated as:

$$u = \gamma_w \cdot h_w$$
$$u = 9.81 \times 18 \approx 177 \text{ kPa}$$

This represents the hydrostatic pressure acting radially inward on the lining.

Contribution to total lining load

This pore water pressure acts in addition to effective earth pressure. For example, if the effective horizontal earth pressure at tunnel depth is 250 kPa, the total radial pressure acting on the lining becomes:

$$p_{total} = \sigma_h' + u = 250 + 177 = 427\ \text{kPa}$$

In many soft-ground or subaqueous tunnels, the pore water pressure component can equal or exceed the effective stress component, making it the governing load case for lining design.

Pressure distribution around the tunnel

Hydrostatic pressure varies with elevation:

- Maximum pressure at invert (greatest water head)
- Minimum pressure at crown (lowest water head)

Design therefore assumes a non-uniform pressure distribution, often approximated as linear from crown to invert, contributing to bending as well as hoop compression in the lining.

Engineering significance

This example illustrates why pore water pressure is often critical in tunnel lining design. In water-bearing ground:

- It directly increases radial loading on the lining
- It influences gasket compression and watertightness
- It can govern segment thickness and reinforcement

Accurate assessment of pore water pressure is therefore essential for safe, durable tunnel performance, particularly in shallow tunnels below the water table and in river or seabed crossings.

Structural response of the segmental lining

Ring action and bending

Segmental linings behave as compressed rings with bending due to non-uniform loading. Structural analysis typically considers:

- Axial force (N)

- Bending moment (M)
- Shear force (V)

Analysis methods include:

- Closed-form elastic ring solutions
- Beam–spring models (ground–lining interaction)
- Finite element models for complex conditions

Combined stresses are checked using interaction equations:

$$\frac{N}{N_{allow}} + \frac{M}{M_{allow}} \leq 1.0$$

Worked example: Ring action with bending under non-uniform loading

A 6.5 m ID segmental lining ring is analysed for long-term loading. Ground pressure is higher at the invert than the crown, so the ring is in overall compression (ring action) but also develops bending due to ovalisation.

Given (from structural analysis)

Using a beam–spring model (ring on radial ground springs), the designer obtains peak internal actions at a critical section:

- **Axial force:** $N = 18$ MN(compression)
- **Bending moment:** $M = 3.0$ MN-m
- **Shear force:** $V = 1.2$ MN(reported for joint and local checks)

Segment/ring capacities (from section design checks including reinforcement, joint behaviour, and code factors):

- **Allowable axial capacity:** $N_{allow} = 30$ MN
- **Allowable bending capacity:** $M_{allow} = 5.0$ MN-m

Combined action interaction check

A simple interaction form is applied:

$$\frac{N}{N_{allow}} + \frac{M}{M_{allow}} \leq 1.0$$

Substitute values:

$$\frac{18}{30} + \frac{3.0}{5.0} = 0.60 + 0.60 = 1.20$$

Result:
$1.20 > 1.0 \rightarrow$ not acceptable for that load case and capacity set.

What the engineer would do next (typical design response)

Because the ring is compression-dominated but bending is pushing the interaction over the limit, common remedies include:

- increasing segment thickness or section modulus (reduces bending stress and increases M_{allow})
- increasing longitudinal reinforcement (increases M_{allow}, improves crack control)
- improving joint stiffness / adding dowels / optimising bolt layout (reduces peak moment concentrations)
- updating ground–lining spring assumptions if justified by better geotechnical data (may reduce predicted M)

Why shear V still matters

Even if the interaction equation is satisfied, the reported shear (here $V = 1.2$MN) is used for:

- joint shear checks (dowels, shear keys, friction capacity)
- step/displacement control between segments
- local concrete shear/bearing near joints and pockets

Engineering takeaway

This example shows the practical meaning of "compressed ring with bending":

- Ring action produces large compressive axial force N.
- Non-uniform loading causes ovalisation and bending M.
- Designers check combined capacity (and separately check shear for joints).
- If the interaction exceeds 1.0, the lining needs redesign or the load model needs refinement.

Joint behaviour

Segment joints are critical weak points and must be checked for:

- Joint compression capacity
- Rotation and opening
- Bolt forces and dowel shear

Gaskets must remain compressed under service loads to maintain watertightness.

Worked example: Segment joint behaviour under service loading

A precast concrete segmental lining is analysed at a circumferential joint located near the springline, where combined axial force and bending cause joint rotation and non-uniform compression. The objective is to verify that the joint remains closed, bolts/dowels are adequate, and gaskets stay compressed to maintain watertightness.

Given (from ring analysis)

- **Axial force at joint:** $N = 12$ MN(compression)
- **Bending moment at joint:** $M = 1.8$ MN\cdotpm
- **Joint width (segment thickness):** $t = 0.35$ m
- **Joint lever arm (effective):** $z = 0.25$ m
- **Concrete compressive strength:** $f_{ck} = 50$ MPa

Joint hardware and gasket:

- **Bolts:** 6 × M36, grade 8.8
- **Dowels:** 2 × Ø40 mm shear dowels
- **EPDM gasket:** required minimum compression = 3 mm
- **Initial gasket compression after erection:** 6 mm

1. Joint compression and opening check

The bending moment causes non-uniform contact pressure across the joint. The peak compressive force on the compressed edge is approximated as:

$$N_{max} = \frac{N}{2} + \frac{M}{z}$$
$$N_{max} = \frac{12}{2} + \frac{1.8}{0.25} = 6 + 7.2 = 13.2 \text{ MN}$$

Average compressive stress at the joint face:

$$\sigma_c = \frac{N_{max}}{A_{joint}}$$

Assuming joint contact area $A_{joint} = 0.35 \times 1.5 = 0.525 \text{ m}^2$:

$$\sigma_c \approx \frac{13.2}{0.525} \approx 25 \text{ MPa}$$

Check:
25 MPa < allowable joint compression (≈ 0.6 f_{ck}≈ 30 MPa) → OK

On the opposite edge, compression is reduced but remains positive, so no joint opening occurs under service load.

2. Joint rotation and gasket performance

Calculated joint rotation from the ring model:

- **Rotation:** $\theta = 0.0015$ rad

Resulting reduction in gasket compression at the least-compressed edge:

$$\Delta g = \theta \times t = 0.0015 \times 350 \approx 0.5 \text{ mm}$$

Remaining gasket compression:

$$6.0 - 0.5 = 5.5 \text{ mm}$$

Check:
5.5 mm > 3 mm minimum → gasket remains compressed and watertight

3. Bolt force check (joint restraint)

Bending induces tensile force in bolts at the opening side. Approximate bolt force:

$$F_{bolt} = \frac{M}{n \cdot z}$$

$$F_{bolt} = \frac{1.8}{6 \times 0.25} = 1.2 \text{ MN (total)} \Rightarrow 200 \text{ kN per bolt}$$

Allowable tensile capacity of M36 grade 8.8 bolt ≈ 450 kN

Check:
200 kN < 450 kN → OK

4. Dowel shear check

Shear force transferred across the joint:

- **Shear demand:** $V = 0.6$ MN

Shear capacity per Ø40 dowel ≈ 400 kN

Total dowel capacity:

$$2 \times 400 = 800 \text{ kN}$$

Check:
600 kN < 800 kN → OK

Dowels also limit step formation and improve ring stiffness.

Engineering significance

This example shows how joint behaviour governs lining performance, even when the ring itself is adequate:

- Axial force and bending concentrate stresses at joints
- Joint rotation must remain small enough to keep gaskets compressed
- Bolts control opening and alignment
- Dowels provide positive shear transfer and stiffness

In mechanised tunnelling, joint checks often control segment detailing, gasket selection, and bolt/dowel layout, because joints are the most vulnerable points for cracking and leakage—even when the concrete section itself is strong enough.

Load combinations and design cases

Typical design load cases include:

- Construction thrust only
- Grout pressure + self-weight
- Ground pressure + water pressure
- Combined axial force and bending
- Extreme cases (partial loss of ground support, seismic loading, fire where relevant)

Safety factors and partial load factors are applied in accordance with applicable design codes.

Example: Typical segmental lining load combinations and design cases

Assume a 6.5 m ID segmental lining with 350 mm thickness and 1.5 m ring width in a shallow urban tunnel. The designer sets up a suite of load cases to cover construction, early-age, long-term, and extreme events. (Exact factors vary by code; the example below shows a typical *engineering* approach using partial factors.)

Load Case 1 — Construction thrust only (TBM advance)

Purpose: check that thrust shoes do not crush or crack segments during advance.

- Total TBM thrust (characteristic): $F_k = 55\text{MN}$
- Partial factor on construction action: $\gamma_F = 1.2$

Design thrust:

$$F_d = 1.2 \times 55 = 66 \text{ MN}$$

Checks:

- average ring compressive stress
- local bearing stress under each shoe
- joint compression and bolt pocket cracking risk

Load Case 2 — Grout pressure + self-weight (early-age)

Purpose: check the ring when grout pressure is high and concrete is still gaining strength.

- Grout pressure at ports (characteristic): $p_{g,k} = 350\text{kPa}$
- Segment self-weight (ring, characteristic): $G_k = 0.55\text{MN}$ per ring
- Partial factors: $\gamma_p = 1.3, \gamma_G = 1.0$(self-weight often treated as well-known)

Design loads:

$$p_{g,d} = 1.3 \times 350 = 455 \text{ kPa}$$
$$G_d = 1.0 \times 0.55 = 0.55 \text{ MN}$$

Checks:

- hoop compression and bending from non-uniform grout fill
- joint rotation/opening and gasket compression
- early-age concrete stress limits (reduced f_c)

Load Case 3 — Long-term ground pressure + water pressure (service)

Purpose: check permanent lining capacity and watertightness.

- Vertical stress at axis (characteristic): $\sigma_{v,k} = 480\text{kPa}$
- At-rest coefficient: $K_0 = 0.6 \rightarrow \sigma'_{h,k} = 0.6 \times \sigma'_{v,k}$
- Water head above axis: $h_w = 18\text{m} \rightarrow u_k = 9.81 \times 18 \approx 177\text{kPa}$
- Partial factors (illustrative): $\gamma_{soil} = 1.2, \gamma_{water} = 1.2$

Design pressures (simplified):

$$\sigma'_{h,d} = 1.2 \times \sigma'_{h,k}$$
$$u_d = 1.2 \times 177 \approx 212 \text{ kPa}$$

Total design radial pressure at axis (concept):

$$p_d \approx \sigma'_{h,d} + u_d$$

Checks:

- ring compression + bending under non-uniform crown/invert loading
- crack width and serviceability
- joint closure and gasket compression (no leakage)

Load Case 4 — Combined axial force and bending (critical N–M interaction)

Purpose: capture non-uniform loading/ovalisation where both N and M are significant.

From a beam–spring model, assume at a critical section:

- $N_k = 18\text{MN}$

- $M_k = 3.2$MN·m
- Partial factor for combined effects: $\gamma = 1.2$

Design actions:

$$N_d = 1.2 \times 18 = 21.6 \text{ MN}$$
$$M_d = 1.2 \times 3.2 = 3.84 \text{ MN\cdotpm}$$

Checks:

- section capacity using N–M interaction
- joint rotation/opening at peak moment locations
- bolt tension and dowel shear

Load Case 5 — Extreme case: partial loss of ground support (void / washout)

Purpose: ensure robustness if local support is lost (e.g., void behind lining due to grout loss or ground erosion).

Assume a **120° sector** of the ring loses external support, increasing bending and joint demand.

Model output (characteristic):

- $M_k$increases to 6.0 MN·m
- $N_k$reduces locally to 10 MN (less uniform compression)

Apply a higher accidental/extreme factor (illustrative): $\gamma = 1.0$to 1.1 depending on code philosophy.

Checks:

- cracking/ultimate capacity under high bending
- joint opening (watertightness may be compromised)
- need for local strengthening or contingency grouting strategy

In addition to the primary construction and service load cases, segmental tunnel linings are often checked against project-specific extreme conditions. Seismic loading, where relevant, is typically represented as racking deformation imposed on the tunnel cross-section rather than as a simple force. This deformation induces additional bending and joint rotation in the lining, and design checks focus on the lining's ability to accommodate these imposed displacements without excessive cracking, joint opening, or loss of structural integrity.

Fire loading may also be considered in tunnels where fire risk is credible, such as transport or utility tunnels. Elevated temperatures reduce concrete strength and stiffness and can lead to spalling of the segment surface. Design checks therefore assess the residual load-bearing capacity of the lining after thermal degradation and determine whether passive fire protection, increased cover, or special concrete mixes are required to maintain safety during and after a fire event.

Together, these cases illustrate that a segmental lining is not designed for a single governing load. Instead, engineers develop a comprehensive suite of load combinations that address the full life cycle of the tunnel. These typically include TBM thrust during construction, grout pressure at early age, long-term ground and water pressures in service, combined axial force and bending effects, and extreme or accidental scenarios that test the robustness of the system.

Appropriate partial safety factors are then applied in accordance with the governing design codes to each load combination. This approach ensures that the lining achieves adequate strength, serviceability, and watertightness from installation through long-term operation, while retaining sufficient reserve capacity to cope with unusual or adverse events over the tunnel's design life.

Segment load calculations form a critical link between TBM operation, ground mechanics, and structural concrete design. They provide the analytical framework that ensures forces generated during excavation and lining installation are safely transferred through the tunnel lining without causing damage or distress.

Through these calculations, engineers verify that thrust forces applied by the TBM do not overstress or crack newly installed rings, particularly during thrust transfer and early-age concrete conditions. They also ensure that annular grouting pressures are sufficient to support the ground and fill the tail void without exceeding the lining's structural capacity or inducing localised damage.

Segment load analysis further confirms that long-term ground and groundwater loads can be resisted throughout the design life of the tunnel without excessive cracking, deformation, or loss of watertightness. This includes checking axial compression, bending, shear, and joint behaviour under both uniform and non-uniform loading conditions.

Equally important, these calculations underpin the performance of segment joints and gaskets. Accurate load modelling ensures that joints remain closed, gaskets are adequately compressed for watertightness, and differential movements are controlled within acceptable limits.

In mechanised tunnelling, the segmental lining is therefore not a passive shell but an active, load-bearing structural system. Its performance depends directly on realistic load assumptions, sound structural design, and disciplined construction control throughout excavation and lining installation.

Chapter 12

Surveying, Guidance, and Control

Accurate surveying, guidance, and control form the backbone of successful mechanised tunnelling. While a Tunnel Boring Machine provides the mechanical means to excavate the ground, it is the surveying and guidance systems that ensure the tunnel is constructed to its intended line, level, and geometry, safely and predictably. In modern TBM operations, alignment control is not a single task performed at the start or end of a drive, but a continuous, integrated process that links design intent, real-time machine data, ground response, and construction practice.

This chapter examines how tunnel alignment is established, monitored, and controlled throughout TBM excavation. It explores the role of underground survey control networks, laser guidance systems, gyroscopic navigation, and real-time data acquisition in maintaining positional accuracy. The chapter also addresses how deviations are detected and corrected, how settlement monitoring informs operational decisions, and how feedback from ground behaviour is used to refine control strategies. Together, these systems transform tunnelling from a purely mechanical activity into a tightly managed engineering process, capable of delivering millimetre-level accuracy in complex ground and constrained urban environments.

By understanding the principles and practical application of surveying, guidance, and control, engineers and operators can minimise alignment drift, reduce ground disturbance, protect segmental linings, and achieve reliable breakthroughs. This chapter therefore provides essential insight into how modern TBM projects maintain control over geometry, risk, and performance from launch to completion.

Tunnel Alignment Control

Tunnel alignment control encompasses a set of surveying, guidance, and steering practices essential for maintaining a Tunnel Boring Machine on its designed horizontal and vertical alignment while ensuring the required tunnel geometry. This control mechanism is not a one-off task but a continuous loop that integrates a reference survey network, a TBM guidance system, strategic steering decisions at the cutterhead and shield, and verification checks during and after tunnel ring installation [612, 613]. Ensuring precise alignment is essential for successful tunnelling operations and minimizing construction risks.

The foundation of effective tunnel alignment begins with a robust underground control network. This network transfers surface coordinates to the tunnel via shafts or portals, using a series of permanent or semi-permanent survey stations, known as control points, to define the project datum [614]. These control points must be safeguarded from construction disturbances and regularly checked for closure errors, as a weak control network can lead to detectable drift and systematic steering errors later during tunnelling operations [612, 615].

The dynamics of the TBM position are primarily assessed against this control network, which is referred to as the “truth” for the machine’s location. If the control network is weak, characterized by instability, poor redundancy, or overly long, unverified traverses, subsequent errors in the machine's position can emerge as unexplained deviations during the tunnelling process [616]. Continuous verification, such as total station observations to TBM-mounted prisms and regular calibration checks of control points, plays a pivotal role in ensuring alignment accuracy [617].

Modern TBM guidance systems rely on real-time measurements of the machine’s position and orientation concerning the control network. Typically, these systems incorporate laser guidance: a laser is positioned on a surveyed control point behind the TBM, projecting a beam along the tunnel axis to facilitate precise tracking [612, 613]. Sensors within the TBM detect the laser beam, allowing the guidance computer to compute any lateral offsets, vertical offsets, and deviations in yaw or roll relative to the design line. In instances where line-of-sight is compromised, such as during curves or over long distances, alternative technologies like total stations, prisms, gyroscopes, and Inertial Measurement Units (IMUs) are often employed in hybrid systems [614, 616].

The alignment control process is typically expressed in three correlated components: position (x-y), elevation (z), and attitude (yaw, pitch, roll). Even slight deviations in yaw or pitch can result in significant cumulative errors over time, necessitating vigilant monitoring of these parameters [618, 619]. As the TBM manoeuvres through varying ground conditions, real-time data from these systems inform corrective actions required to maintain alignment.

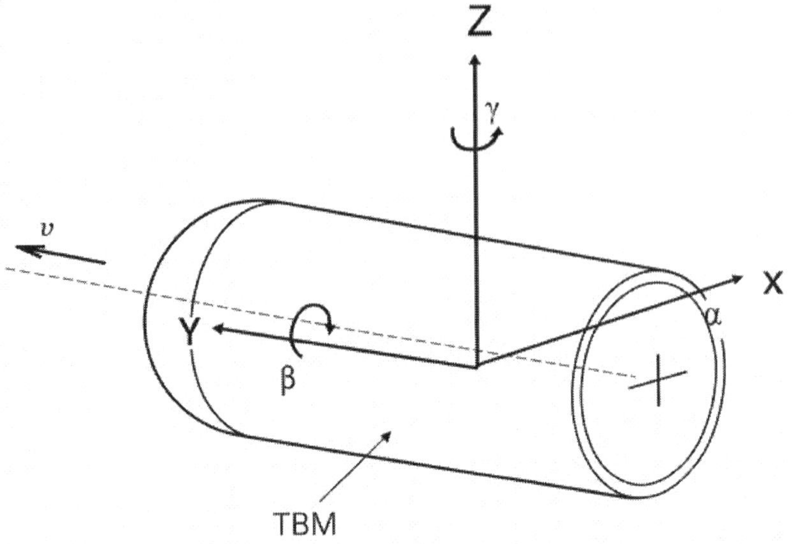

Figure 91: The yaw, pitch, and roll angles of a tunnel boring machine (TBM).Adapted from He et al. (2020) [620]

Figure 91 illustrates the three rotational degrees of freedom used to describe and control a Tunnel Boring Machine's orientation in space: yaw, pitch, and roll. Together with position (X, Y, Z), these angles define the TBM's full spatial state for surveying, guidance, and steering.

Reference axes shown

- **X-axis**: Longitudinal axis of the TBM (along the tunnel direction).
- **Y-axis**: Horizontal transverse axis (left–right across the tunnel).
- **Z-axis**: Vertical axis (up–down).

These axes form a right-handed coordinate system fixed to the tunnel design datum.

Yaw (α) — horizontal steering

- Rotation about the Z-axis.
- Describes whether the TBM's nose points left or right relative to the design centreline.
- **Effect**: A yaw error causes lateral deviation that grows with distance if not corrected.
- **Control**: Differential thrust, articulation, and gradual steering corrections.
- **Monitoring**: Laser/total-station guidance reports lateral offset and heading (yaw) error.

Pitch (β) — vertical steering

- Rotation about the Y-axis.
- Describes whether the TBM points up or down relative to the design gradient.
- **Effect**: Pitch errors accumulate into vertical deviation (too high/too low).
- **Control**: Vertical thrust balance, articulation, and controlled penetration.
- **Monitoring**: Vertical offset and gradient trend from guidance sensors.

Roll (γ) — rotation about the tunnel axis

- Rotation about the X-axis.
- Describes twist of the TBM around its own axis.
- Effect: Roll does not change centreline position directly, but it strongly affects:
    - Segment orientation and ring build accuracy
    - Use of tapered rings in curves
    - Gasket compression uniformity and joint behaviour
- **Control**: Erector alignment, thrust shoe balance, and consistent ring build practice.
- **Monitoring**: Roll sensors and checks on ring rotation/cant.

Why all three matter together

- A TBM can be on line but have a yaw error that predicts future drift.
- A small pitch error can push the machine off grade over several strokes.
- Uncontrolled roll can introduce systematic lining bias even if guidance shows acceptable offsets.

Effective alignment control therefore manages yaw, pitch, and roll simultaneously, using small, continuous corrections rather than large, late adjustments.

Steering within the TBM is accomplished through precise control of thrust application and the articulation of the shield. In soft ground conditions, operators employ selective thrust cylinder control (differential thrust) paired with adjustable articulation joints to facilitate movement towards the intended direction [616, 618]. Conversely, in harder rock formations, steering is largely influenced by the cutterhead's interaction with the surrounding geology, the behaviour of the thrust/gripper system, and meticulous management of penetration and torque [612, 618].

A significant operational challenge confronted in TBM steering is the inherent lag in steering corrections. As TBMs are long, rigid structures moving through a deformable medium, over-corrections can lead to unintended paths and greater excavation disturbance [612, 616]. Hence, effective alignment control necessitates implementing small, incremental corrective actions rather than aggressive adjustments, particularly when navigating curves where overcut and shield clearance requirements complicate alignment maintenance [613, 618].

The process of segment installation is an integral component of alignment control rather than a standalone operation. The final tunnel centreline is determined by the accuracy of each ring's construction, including factors like ring rotation, joint closure, and handling of the segments [615]. Variability during ring assembly can introduce geometric drift, even with TBM guidance indicating acceptable offsets. Consequently, it is essential to monitor ring geometry closely for parameters such as ring position, roundness, segment stepping, and cumulative orientation [614, 616].

Closing the loop of alignment control through diligent survey verification is essential. Continuous estimates produced by guidance systems must be validated against independent survey checks to confirm consistency with the control network. This verification may involve traverse extensions, gyroscopic checks, and as-built inspections of the tunnel lining, particularly in high-risk zones where tolerances are stringent [614, 617]. Setting "alert" and "action" thresholds for tolerances and control limits empowers the team to respond proactively before deviations escalate into significant alignment issues [615].

Laser Guidance Systems

Laser guidance systems play a critical role in the precise control of tunnel alignment during TBM excavation. By continuously providing high-precision information regarding the TBM's position and orientation relative to the design centreline, these systems enable operators to implement small, controlled corrections in real-time, which is crucial for maintaining alignment accuracy over long drives and through variable ground conditions [612, 621].

At the heart of a laser guidance system is a fixed reference line established along the designed tunnel axis. A laser beam projected along this axis is detected by a target mounted on the TBM. Correct alignment is indicated when the laser beam strikes a known point on this target. Any deviation of the laser spot signals misalignment, while the orientation of the target can reveal angular discrepancies. This system relies on geometric relationships to translate laser measurements into actionable alignment data [19].

The system comprises several essential components, including:

- **Laser Emitter**: Typically mounted on a surveyed control point, it projects a narrow beam aligned to the tunnel design. This laser must be periodically checked to ensure its alignment with the project control network [19].
- **Laser Target or Sensor**: Located on the TBM, this component captures the projected laser beam. Traditional methods utilized graduated target plates, whereas modern approaches adopt CCD or CMOS sensors capable of detecting the beam's position with sub-millimetre accuracy [612, 621].
- **Guidance Computer**: This processes data from the target, converting it into essential metrics such as horizontal and vertical offsets and angular errors.

Continuous feedback is provided to the operator, who can then take corrective actions [19, 622].

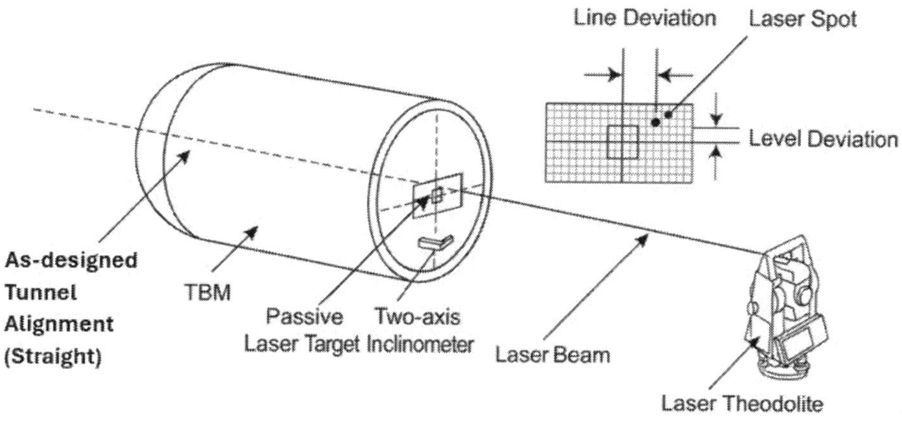

Figure 92: Tunnel Boring Machine positioning.

Figure 92 illustrates how a TBM laser guidance system measures and controls tunnel alignment by continuously comparing the machine's actual position and orientation with the as-designed tunnel centreline. It shows how a fixed reference established by surveying is transferred to the TBM and used in real time to guide steering decisions during excavation.

The dashed line represents the as-designed tunnel alignment. This centreline is the fundamental reference for the drive, and all guidance measurements are made relative to it. Any deviation from this line must be identified early and corrected gradually to avoid cumulative alignment errors and excessive steering forces.

At the rear of the TBM, a laser theodolite or fixed laser station is set up on surveyed control points. This instrument projects a narrow laser beam precisely along the design centreline and grade. Once correctly aligned and stabilised, the laser beam forms the fixed spatial reference against which the TBM's position is continuously assessed.

The laser beam travels along the tunnel and intersects a passive laser target mounted inside the TBM. This target is typically a grid or graduated plate. Where the beam strikes the target, it creates a laser spot whose position directly indicates how the TBM is positioned relative to the design alignment.

When the laser spot is centred on the target, the TBM is correctly aligned in both line and level. Any movement of the spot away from the centre indicates deviation from the design alignment. The magnitude and direction of this movement quantify the error.

The horizontal displacement of the laser spot on the target represents line deviation, or left–right error. This corresponds to yaw error in the TBM and is corrected by operators using differential thrust, articulation systems, or steering cylinders. The vertical displacement of the spot represents level deviation, or up–down error, which reflects pitch error relative to the design gradient. Vertical alignment is corrected by adjusting thrust balance and penetration control.

In addition to the laser target, a two-axis inclinometer is mounted on the TBM. This instrument measures the machine's pitch and roll independently of the laser beam. Its data allows the guidance system to distinguish between true positional deviation and angular rotation of the TBM body. Roll, which is rotation about the tunnel axis, does not directly move the laser spot but affects segment orientation and ring build accuracy, so it is monitored alongside laser guidance.

Together, these components form an integrated guidance system. The laser provides a fixed position reference for line and level, the target converts the laser beam into measurable offsets, and the inclinometer supplies angular context. The guidance computer combines these inputs to present the operator with real-time information on offset, heading, and grade. Steering corrections are then applied incrementally over several thrust strokes, keeping deviations within tolerance while minimising ground disturbance and lining stress.

Through continuous comparison of the TBM's position and orientation against the design alignment, operators can apply incremental corrections when necessary. The system supports the philosophy of making early and gradual corrections rather than large, late adjustments, which can exacerbate ground disturbances and internal stresses within the TBM and its linings [612].

Laser guidance is particularly effective in managing yaw (horizontal direction) and pitch (vertical direction). The system's feedback allows for real-time adjustments, which are critical during excavation in challenging geological conditions [621]. The monitoring of roll, though indirectly managed through geometric measures and sensors, integrates seamlessly into the overall control framework provided by the laser system [612].

These laser guidance systems can achieve millimetre-level precision over extensive distances; however, this accuracy is influenced by factors such as the stability of the laser emitter, tunnel air quality, line-of-sight distances, and frequency of recalibration [19, 612]. In long drives, especially when dealing with curved alignments or obstructions, the laser must be repositioned frequently. Supplemental technologies, including total stations and inertial navigation systems, are often employed for comprehensive accuracy maintenance [19, 622].

While laser guidance does not replace traditional surveying methods, it is heavily reliant on them. Surveying teams establish and maintain the control network, periodically validating the position of the TBM independently of the laser measures. Modern systems are beginning to integrate laser feedback with operational parameters such as thrust and torque to enable proactive alignment management before critical tolerance limits are reached [19].

Laser guidance systems are pivotal in ensuring the safety and accuracy of TBM operation. They facilitate tight alignment tolerances, minimize the need for aggressive steering manoeuvres, and ultimately allow for precision during breakthroughs at key construction junctions. Consequently, laser guidance has transformed tunnel alignment from a retrospective verification task into a proactive element of real-time tunnelling operations [612, 621].

Setting Up and Using a TBM Laser Guidance System

The implementation of a Tunnel Boring Machine (TBM) laser guidance system is essential for precise excavation in tunnel construction. This process involves multiple phases that ensure accuracy and efficiency throughout the tunnelling operation. Below, we detail the setup and utilization of such systems in a disciplined manner, supported by relevant literature.

1. Pre-setup Requirements

Survey Control Network

Prior to operating the laser guidance system, establishing a survey control network is critical. This involves setting primary and secondary control points from the tunnel portal or shaft using total stations or gyroscopic instruments for accuracy verification [612]. The control points must be transferred into the tunnel with traverses later closed to confirm positional accuracy before commencing TBM operations [441]. Furthermore, the design centreline and grade must be well-defined within the tunnel's coordinate system to ensure that the TBM follows the intended alignment throughout its path [612].

Environmental Checks

Ensuring environmental conditions are conducive to laser operation is essential. A clear line of sight must be maintained to minimize disruptions caused by dust or inadequate ventilation [296]. The selection of a stable mounting location for the laser is equally important; ideally, it should be secure from vibrations and accidental contact [612].

2. Laser Station Setup

Mounting

The laser should be mounted securely on a rigid structure. This could either be a bracket affixed to the tunnel lining or a robust tripod positioned securely behind the TBM [441]. To facilitate optimal performance, the laser should be set at a distance that provides a long sightline yet remains maintainable—typically between 50 to 150 meters [612].

Alignment

Precise alignment of the laser to the design centreline and grade is crucial. This process is conducted using survey instruments to lock in all adjustment settings, with the setup coordinates duly recorded for future reference [296]. Verification against independent survey marks is also necessary to eliminate any bias [441].

Verification

Documenting the tolerances and acceptance criteria—often sub-millimetre at the target over the specified distance—ensures the readiness of the laser system for operational tasks [441].

3. TBM Target / Sensor Setup

Target Positioning

The installation of a laser target or electronic sensor on the TBM is designed to provide feedback on its position relative to the tunnel design. The positioning must ensure that the target plane is squared with the TBM's reference frame to avoid measurement errors [296].

Calibration

Calibrating the system is critical. This involves zeroing the system when the TBM is on the design line and ensuring that any offsets are accurately applied [441]. Calibration extends to mapping the laser spot data to horizontal and vertical offsets effectively [612].

System Check

A comprehensive system check is performed to confirm live readouts displaying lateral and vertical offsets, yaw, and pitch. This aspect is crucial for ongoing adjustments throughout the tunnelling process [612].

4. Using the System During Excavation

Continuous Monitoring

During excavation, the guidance display provides real-time feedback about the TBM's position, allowing operators to observe deviations and adjust operations as necessary

[441]. Continuous monitoring is fundamental to observing both absolute deviation and the rate at which changes occur.

Steering Corrections

Operators can apply small, incremental corrections based on this feedback, employing mechanisms such as differential thrust or adjusting articulation. Correcting early, while deviations remain minimal, aids in maintaining alignment and preventing unnecessary stress on the lining system [612].

5. Managing Yaw, Pitch, and Roll

Effective management of the TBM's yaw, pitch, and roll is essential for maintaining trajectory. Yaw can be adjusted using lateral thrust, while pitch is managed through thrust balance and controlled penetration. For roll, sensors monitor the machine's status, ensuring that corrections are applied efficiently [612].

6. Leapfrogging the Laser

As the TBM progresses, the leapfrogging of the laser is necessary to maintain measurement accuracy. Stopping at predetermined intervals allows for repositioning the laser, realigning it to the design parameters, and verifying it against pre-set checks before proceeding [441].

7. Survey Validation and QA

Maintaining stringent quality assurance measures is vital. Regular independent checks should verify TBM position against laser readouts, and any discrepancies must be investigated immediately to ensure ongoing accuracy [612].

8. Dealing with Limitations

Addressing operational limitations such as air quality, vibrations, and obstacles is key to maintaining laser accuracy. Dust suppression and environmental control strategies help mitigate visual disruptions, while vibration checks post-excavation ensure that mounting systems remain stable [441].

9. Integration with TBM Controls

Modern laser guidance systems often integrate with TBM controls, combining data from thrust and torque settings with predictive deviations to adjust operations proactively, enhancing overall efficiency [612].

10. Practical Tolerances

Typical practical tolerances for laser guidance systems in TBM operations usually hover at project-specific levels, ensuring narrow margins for horizontal and vertical offsets to facilitate precise breakthroughs [441].

Key Takeaway

A TBM laser guidance system operates effectively when approached as a comprehensive control loop rather than solely a measurement tool. This methodology encompasses precise setup, continuous monitoring, gradual corrections, and frequent validation, contributing to successful tunnelling operations with millimetre-level precision.

Gyroscopic Navigation

Gyroscopic navigation is a crucial technique employed in tunnelling to accurately determine and control the direction of tunnel boring machines without relying on line-of-sight methods. This advanced guidance method is especially valuable in long drives, curved alignments, and shafts where conventional laser guidance systems may prove impractical. By offering an independent directional reference, gyroscopic systems help maintain precision in tunnel alignment even over extended distances where optical methods face significant challenges [623, 624].

The fundamental operation of gyroscopic navigation hinges on the gyroscope's capability to detect true north by leveraging Earth's rotation. Once initialized at a known surveyed location, the gyroscope provides a stable heading reference without the need for external cues [625]. The system measures the TBM's heading relative to true north, incorporating positional data to generate a reliable directional reference crucial for maintaining alignment. This capability is particularly beneficial in environments where GNSS is ineffective, such as deep tunnels [624, 626].

Despite their high accuracy, laser guidance systems exhibit inherent limitations that gyroscopic navigation effectively overcomes. These limitations include the requirement for a clear line of sight, degradation of accuracy over long distances, and the need for frequent realignment in curved tunnel sections. In contrast, gyroscopic navigation systems maintain consistent accuracy, providing orientation control where laser operation may be intermittent or impossible [624, 626]. This capacity to sustain an accurate directional reference reduces the risk of cumulative alignment errors, thereby enhancing the reliability of tunnelling projects.

There are several types of gyroscopic systems employed in tunnelling operations. North-seeking gyroscopes are widely recognized for their role in establishing the initial tunnel orientation and verifying heading at intervals during excavation [624]. Additionally, Inertial Measurement Units (IMUs) are increasingly integrated into TBM guidance systems. These units combine the functions of gyroscopes, which detect angular rotation, with accelerometers, which track linear motion, allowing for comprehensive continuous monitoring of the TBM's position and orientation [623].

The gyroscopic navigation process typically commences with the initialization of the gyro at a known point, where it stabilizes to ascertain the true north direction. This initial heading then forms a reference for the TBM concerning the planned alignment [624]. As tunnelling progresses, gyroscopic data may be continuously monitored or collected via portable gyro surveys to ensure alignment accuracy. The data from the gyroscope is often integrated with other sources, such as TBM advancement distance, laser offsets, and inclinometer readings, to ascertain the three-dimensional position and orientation of the TBM [624, 627].

Although gyroscopic systems provide critical navigational data, they are not immune to drift, particularly in IMUs over extended periods [628, 629]. To mitigate this issue, periodic checks against survey control and integration with laser or total-station measurements are implemented to correct accumulated errors. The combination of gyroscopic and optical methods forms hybrid systems that enhance both long-term stability and positional accuracy, ensuring cm-level breakthrough precision is achieved for modern mechanized tunnelling [623, 624].

From an engineering perspective, gyroscopic navigation is indispensable for conducting long TBM drives, especially in challenging environments characterized by limited survey access, intricate curves, and the need for high precision during breakthroughs in urban or sensitive areas [614, 624]. The independent reference to true north provided by gyroscopic systems significantly curtails alignment errors, verifying their integral role in achieving optimal alignment in tunnelling operations. These systems not only complement laser guidance but also ensure directional integrity over distances that pose challenges to conventional methods [623].

Settlement Monitoring

Settlement monitoring is the systematic measurement and interpretation of ground movements caused by tunnelling. Its purpose is to detect, quantify, and control both surface and subsurface deformation so that risks to buildings, utilities, and infrastructure are identified early and mitigated through timely operational or engineering responses.

Mechanised tunnelling inevitably disturbs the ground. Even where face support and annular grouting are well controlled, some degree of volume loss occurs at the tunnel face, around the shield, or within the tail void. If not managed, this loss can propagate upward and outward through the ground, resulting in surface settlement or deformation at depth. Settlement monitoring provides the essential feedback loop that links TBM operation, such as face pressure, advance rate, conditioning, and grouting, to ground response, enabling proactive control rather than reactive repair.

Settlement monitoring addresses both surface and subsurface movements. At the surface, monitoring focuses on protecting buildings, pavements, roads, railways, and buried services. Measurements typically capture vertical settlement, occasional heave caused by over-pressure, and the shape and width of the settlement trough along and across the tunnel alignment. These data are critical for understanding how tunnelling-induced movements are distributed spatially and how they evolve as excavation progresses.

Below ground, subsurface monitoring instruments measure how deformation develops with depth. This information helps engineers distinguish between different mechanisms, such as face-related volume loss, shield friction effects, tail-void collapse, or long-term consolidation. Typical parameters include vertical displacement at multiple depths, horizontal movement toward the tunnel axis, and changes in groundwater levels, which can significantly influence settlement behaviour.

A range of instruments is used to capture these movements. Surface levelling points consist of precise markers installed along the tunnel alignment and on adjacent structures, providing high-accuracy vertical displacement data and often serving as the primary trigger for action levels. Prisms mounted on buildings or infrastructure are observed automatically by total stations, allowing near-real-time monitoring of both vertical and horizontal movements.

Figure 93: Robotic total station on tripod. Pierre5018, CC BY-SA 4.0, via Wikimedia Commons.

Inclinometers installed in boreholes measure lateral ground movement with depth and are particularly useful for assessing shear deformation and potential impacts on deep foundations. Extensometers measure vertical displacement at multiple depths within a single borehole, helping identify whether settlement is originating near the tunnel crown, within the overburden, or at ground surface. Piezometers monitor groundwater pressures, as changes in pore pressure can precede or accompany settlement and are essential for diagnosing hydraulic contributions to ground movement. In some projects, remote sensing techniques such as InSAR are also used to supplement ground-based systems and identify broader settlement trends over large urban areas.

Surface levelling points

Surface levelling points are fixed reference markers installed on the ground surface or on structures such as buildings, pavements, and retaining walls. They are typically metal studs or bolts set into stable material so that their vertical position can be measured repeatedly with high precision.

Measurements are taken using precise optical or digital levels referenced to a stable benchmark outside the zone of influence. By comparing successive readings, engineers can detect very small vertical movements—often to an accuracy of ±0.5 mm or better.

Because they directly measure vertical displacement at the surface, levelling points are the most reliable indicator of settlement affecting third-party assets. For this reason, they are commonly used as the primary trigger for alert, action, and alarm levels during tunnelling.

Prisms and total stations

Prisms are reflective targets mounted on buildings, infrastructure, or ground fixtures. Automated total stations are set up at fixed locations and programmed to observe these prisms at regular intervals, often every few minutes.

The total station measures angles and distances to each prism and calculates its three-dimensional position (horizontal and vertical). Changes in these coordinates over time indicate movement of the monitored object.

This system allows near-real-time monitoring of both settlement and lateral displacement, making it especially valuable in dense urban environments where structures may experience complex movement patterns. Automated prism monitoring also reduces the need for manual surveying and enables rapid response if movements accelerate.

Inclinometers

Inclinometers are installed in vertical or near-vertical boreholes drilled into the ground. A grooved casing is placed in the borehole and grouted in position so that it moves with the surrounding soil.

A probe containing tilt sensors is lowered down the casing and measures inclination at regular depth intervals. By comparing readings over time, engineers can calculate lateral ground movement profiles with depth.

Inclinometers are particularly useful for identifying shear zones, ground movement toward the tunnel, and potential impacts on deep foundations or retaining structures. They help distinguish surface settlement caused by shallow volume loss from deeper ground deformation mechanisms.

Extensometers

Extensometers measure vertical displacement at multiple depths within a single borehole. Anchors are fixed at selected depths in the ground, and each anchor is connected to a reference head at the surface via rods, wires, or vibrating-wire sensors.

As the ground moves, the relative movement between anchors and the reference head is recorded. This allows engineers to determine where vertical deformation is occurring within the soil profile.

Extensometers are especially useful for identifying whether settlement originates near the tunnel crown, within compressible soil layers, or at the surface. This information helps diagnose whether movement is associated with face loss, shield effects, tail-void closure, or long-term consolidation.

Piezometers

Piezometers measure groundwater pressure (pore water pressure) in the ground. They are installed in boreholes at specific depths and may be standpipe, pneumatic, or vibrating-wire types.

Changes in pore pressure can significantly affect effective stress in the soil and therefore influence settlement behaviour. For example, a drop in pore pressure can increase effective stress and trigger consolidation settlement, while over-pressurisation can lead to heave or hydraulic instability.

Monitoring pore pressure alongside settlement allows engineers to distinguish between mechanically induced settlement and hydraulically driven ground movement, which is critical for diagnosing problems and selecting appropriate mitigation measures.

Remote sensing (InSAR)

Interferometric Synthetic Aperture Radar (InSAR) is a satellite-based remote sensing technique that detects ground movement by comparing radar images of the same area taken at different times.

By analysing phase differences between images, InSAR can measure ground displacement over wide areas with millimetre-scale sensitivity. It is particularly useful for identifying long-term or regional settlement trends in urban environments.

While InSAR does not replace ground-based instruments—due to limitations in temporal resolution, depth sensitivity, and line-of-sight—it provides valuable contextual data and can highlight areas requiring closer on-site monitoring.

Monitoring layouts and measurement frequency are designed based on tunnel depth and diameter, ground conditions, and the sensitivity of nearby assets. Monitoring points are typically arranged in transverse sections across the tunnel to capture the settlement trough, longitudinally along the alignment to track changes with TBM advance, and directly on or within critical structures. Measurement frequency increases as the TBM approaches, passes beneath, and moves away from a monitored section, often shifting from weekly readings to daily or continuous monitoring.

Settlement monitoring operates within a framework of predefined trigger levels. An alert level indicates that movement exceeds expected trends and requires increased observation and review. An action level triggers operational changes, such as adjusting face pressure, conditioning dosage, or grouting parameters. An alarm level requires immediate intervention, which may include slowing or stopping excavation, implementing ground treatment, or activating compensation grouting. These thresholds are project-specific and are directly linked to allowable movements for affected assets.

Interpreting settlement data in isolation is insufficient; it must be integrated with TBM operational parameters. Engineers assess settlement trends alongside face pressure stability, advance rate and stoppages, screw conveyor or slurry balance, grout volume and pressure, and conditioning effectiveness. Increasing settlement ahead of the face may indicate under-pressure or inadequate conditioning, while settlement behind the shield often points to insufficient or delayed annular grouting. Conversely, heave can signal excessive face pressure or overly aggressive grouting. This integrated interpretation allows engineers to identify the underlying cause of movement and apply targeted corrective measures.

Settlement monitoring transforms tunnelling from a blind excavation into a controlled ground–structure interaction process. It protects third-party assets and public safety, validates design assumptions and pressure windows, enables early intervention before damage occurs, and provides a defensible record for compliance and risk management. In modern mechanised tunnelling, settlement monitoring is not merely an observational requirement but a core control system that underpins safe, predictable, and accountable tunnel construction in sensitive ground and urban environments.

Real-Time Data Acquisition

Real-time data acquisition in tunnelling refers to the continuous collection, transmission, processing, and interpretation of monitoring and operational data from Tunnel Boring Machines (TBMs) as excavation proceeds. This capability allows operators and engineers to gain actionable insights from ground behaviour and machine performance quickly, enabling timely adjustments to various operational parameters to prevent issues like settlement, instability, or damage [618, 630].

In the context of tunnelling, the notion of "real-time" does not always imply millisecond control intervals. Instead, it signifies that the data latency must be sufficiently low to influence immediate decision-making. Specifically, it allows for adjustments to TBM parameters such as pressure, torque, and thrust within seconds to minutes, while automated surveying and structural monitoring may operate with latencies ranging from minutes to hours [463, 631].

Requirements for Real-Time Data

1. **Continuous Monitoring**: TBM operational data is captured through various onboard sensors that record parameters such as face pressure, thrust force, cutterhead torque, and conditioning injection rates. These metrics help describe the interaction of the TBM with the ground in real time [618, 630] .
2. **Real-Time Decision Support**: The acquired data not only informs operators but also can signal necessary changes in operations. For instance, if an increasing torque is noted while penetration remains constant, this may indicate the need for enhanced conditioning or reduced thrust [630, 631].
3. **Data Source Quality**: Reliable real-time data acquisition hinges on the quality of the data gathered from sensors, including aspects like calibration and redundancy, as poor-quality data risks leading to detrimental decisions [632].

Data Acquisition and Transmission Processes

Sensor and Instrumentation

Dedicated sensors—such as pressure transducers, load cells, and flow meters—are used to measure specific parameters. These sensors convert physical phenomena into electrical signals, which are then collected and logged by Programmable Logic Controllers (PLCs) or data loggers [370, 630].

Data Logging and Transmission

The captured data undergoes time-stamping and synchronization before being transmitted to surface servers over communication lines such as fibre optic cables or wireless networks. This architecture supports the integration of data into a coherent backbone for analysis [463, 618].

Data Processing and Visualization

Processing Techniques

Upon transmission, raw data typically requires filtering to eliminate noise and calculate derived metrics (e.g., penetration per revolution). The processed data is then compared with design limits and historical trends to provide context for ongoing operations [463, 633].

Visualization Methods

Visual representations, such as dashboards, provide operators with real-time information and historical trends, allowing for immediate situational awareness. This includes displaying rolling averages, tolerance bands, and alerts for parameters exceeding predefined thresholds [631, 633, 634].

Integration with TBM Control Systems

In modern, advanced systems, real-time data directly influences TBM control. Operators can adjust key parameters based on live feedback, improving decision-making efficiency. This interconnectivity enhances responses to real-time situations rather than relying on reactive measures based on historical data [463, 631].

Reliability and Data Quality

The effectiveness of real-time data acquisition is contingent upon trustworthy data. Thus, a framework ensuring sensor calibration, robust communications, and clear data ownership is essential for effective monitoring and interpretive accuracy [632, 633].

Deviations and Correction Strategies

Mechanized tunnelling utilizing Tunnel Boring Machines is a sophisticated process that necessitates meticulous management of deviations from predefined alignment, geometry, and operational parameters. These deviations can compromise the structural integrity of the tunnel, increase construction costs, and extend project timelines if not effectively addressed. Understanding these deviations and their correction mechanisms is crucial for maintaining safe and efficient tunnelling operations.

Types and Causes of Deviations

In TBM operations, deviations are classified into three primary categories: alignment deviations, operational deviations, and ground response deviations.

Alignment Deviations:

- **Horizontal deviation (yaw)** occurs when the TBM veers sideways from its designated centreline.
- **Vertical deviation (pitch)** happens when there is a deviation from the designed gradient, causing the TBM to operate above or below the intended elevation.
- **Roll deviations** relate to the TBM's misalignment around its longitudinal axis, which impacts the accuracy of segment assembly [441].

Operational Deviations:

- These include abnormalities in thrust, torque, and penetration rates, which may arise from unexpected geological conditions or machinery performance issues [65, 635]. Unstable face pressures and irregular muck behaviour are also indicative of operational deviations [5].

Ground Response Deviations:

- These are characterized by unexpected settlement or heave of the ground, shifts in pore water pressure, and deformations that may endanger surrounding structures [636, 637]. A thorough understanding of these factors is necessary to interpret and mitigate their risks effectively.

Detection and Monitoring of Deviations

Integrated monitoring systems are fundamental in identifying these deviations promptly. Continuous feedback mechanisms play a vital role, including:

- **Laser guidance systems** provide real-time data on horizontal and vertical offsets, enabling operators to maintain alignment [638].
- **Inclinometers** measure roll deviations and overall heading stability [5].
- **TBM onboard sensors** track changes in thrust, torque, penetration rates, and face pressure, offering insight into the TBM's operational state [35].

Trend analysis is of utmost importance, as persistent minor deviations can signal larger potential issues if left unaddressed. Early detection allows the implementation of small, manageable corrections rather than waiting until larger deviations intensify [5].

Correction Strategies for Deviations

Correcting deviations effectively requires a clear understanding of their root causes. Common contributors include ground variability, face resistance from geological anomalies, mechanical wear, and pressure imbalances within the TBM systems [65, 635, 637]. Therefore, rectifying only the observable symptoms without addressing the underlying causes often exacerbates the issue.

Strategies for Different Types of Deviations:

1. **Horizontal Alignment Corrections**:
    - Adjustments are primarily executed through differential thrust management, where thrust is increased on one side of the TBM while decreased on the opposite side. This method ensures a gradual return to the intended alignment, minimizing ground disturbance [65, 441].
2. **Vertical Alignment Corrections**:
    - Control is achieved via changes in thrust balance between the TBM's upper and lower cylinders and slight adjustments to the penetration rate. This must be coordinated with face pressure management to prevent instability, such as crown settlement or invert heave [635, 637].
3. **Roll Alignment Corrections**:
    - Roll deviations are addressed through lining control strategies, including modifications to the segment erection sequence and maintaining careful consistency in the construction practices [5].

If deviations like excessive thrust or torque are observed, practices such as improving the conditioning of muck or adjusting cutter operation parameters may be employed [13, 636]. Moreover, timely rebalancing of extraction rates and slurry density is vital to stabilize the TBM environment [5, 637].

Establishing tolerances for horizontal and vertical alignment, roll, and ground response is critical. Adjustments should trigger at predefined alert levels rather than reactively correcting after exceeding final tolerances, thus ensuring proactive management throughout the tunnelling process [639, 640].

Effective deviation management in TBM excavation is paramount to ensuring project success, compliance with design tolerances, and minimizing excessive ground disturbance. By utilizing modern monitoring technologies and embracing an understanding of the multifaceted nature of deviations, tunnelling operations can remain within safe operational envelopes. The focus should be on early detection, thorough understanding of causes, and disciplined incremental corrections to enhance the predictability and accuracy of tunnel construction.

Chapter 13

Ground Conditioning and Improvement

Ground conditioning and improvement are fundamental to the safe, efficient, and predictable execution of mechanised tunnelling. In modern TBM operations, ground is no longer treated as a fixed constraint to be overcome, but as an engineering system whose behaviour can be actively modified to meet the demands of excavation, face support, spoil transport, and settlement control. This chapter introduces the principles, techniques, and control parameters used to transform variable in-situ ground into a stable, manageable medium that works in harmony with the TBM.

The chapter focuses on the practical engineering methods used to control soil and groundwater behaviour ahead of, at, and behind the tunnel face. It explains how conditioning agents such as foam, polymers, and bentonite are used to adjust rheological properties, permeability, and cohesion in Earth Pressure Balance (EPB) and slurry TBMs. It also examines how quantitative soil conditioning ratios translate real-time ground response into controllable operational inputs, linking geotechnical behaviour directly to TBM performance.

Beyond conditioning, the chapter explores ground improvement techniques applied where conditioning alone is insufficient or where additional safety margins are required. These include grouting methods, ground freezing, and jet grouting, each selected to address specific risks such as high groundwater pressure, unstable soils, shallow cover, or sensitive surface structures. Emphasis is placed on understanding not only how each technique works, but why it is chosen, how it interacts with TBM operation, and what controls govern its effectiveness.

Taken together, the topics in this chapter demonstrate that successful mechanised tunnelling depends on an integrated approach in which excavation, conditioning, improvement, and monitoring function as a single system. Ground conditioning and improvement are therefore not auxiliary measures, but core engineering controls that underpin face stability, alignment accuracy, settlement management, and long-term tunnel performance throughout the construction lifecycle.

Foam Injection

Foam injection has become a critical ground conditioning technique used in mechanised tunnelling, specifically with Earth Pressure Balance Tunnel Boring Machines. The primary purpose of foam injection is to alter the mechanical and hydraulic properties of the excavated soil, facilitating both the support of the tunnel face and the predictable flow of muck through the TBM and the accompanying spoil handling system. This is particularly necessary as many naturally occurring soils, particularly sands, silts, and mixed-face materials, do not inherently exhibit the plastic, cohesive, and low-permeability characteristics required for optimal EPB operation [306, 641].

The effectiveness of EPB tunnelling relies heavily on the behaviour of the excavated soil. When properly conditioned with foam, the soil can achieve the necessary characteristics to behave as a cohesive mass, thus providing stability at the tunnel face. Previous studies have demonstrated that the integration of foam allows for improved soil flow and face stability, which is critical for tunnelling in groundwater-rich conditions [642, 643]. Specifically, foam injection reduces internal friction and shear strength, thereby improving flowability and controlling groundwater inflows [306, 644].

Foam injection employs a controlled mixture of water, compressed air, and surfactants to create a stable foam that modifies the properties of the excavated muck. When introduced to the soil mass, foam bubbles disperse and occupy void spaces, which enhances the apparent cohesion among soil particles and lowers permeability [585]. This dual action serves to maintain chamber pressure effectively and reduces the risk of water ingress during tunnelling operations, significantly aiding in face control and preventing collapse [645].

The conditioning of the soil with foam alters its rheological properties, enhancing its plasticity and facilitating a paste-like flow essential for smooth progression through the excavation chamber and screw conveyor. Laboratory tests indicate that foam helps reduce torque fluctuations and mechanical wear in the machinery, crucial for efficient operation [195, 308]. Controlled introduction of foam at strategic locations, such as the cutterhead or the screw conveyor, optimises its benefits based on real-time soil conditions [301, 427].

Rheological properties describe how a material flows and deforms when a force is applied. In simple terms, rheology explains whether a material behaves like a solid, a liquid, or something in between, and how that behaviour changes under stress, time, or movement. Rather than focusing on what a material is made of, rheology focuses on how it responds when it is pushed, sheared, squeezed, or stirred.

At its core, rheology links three fundamental aspects of material behaviour: stress, which is the applied force such as pressure or shear; strain or deformation, which is how much the material changes shape; and time, which determines whether the response is immediate, delayed, or permanent. Many materials encountered in tunnelling—such as soils, slurries, conditioned muck, grout, and concrete—do not behave as ideal solids or ideal fluids. Rheology provides the concepts and models needed to describe and predict this complex, non-ideal behaviour.

One of the most important rheological properties is viscosity, which describes a material's resistance to flow. Materials with low viscosity flow easily, like water, while those with high viscosity flow slowly, like honey or thick slurry. In tunnelling, slurry viscosity directly affects face pressure transmission and the efficiency of slurry separation plants, while the viscosity of conditioned EPB muck influences screw conveyor performance and torque demand.

Another critical property is yield stress, defined as the minimum stress required before a material begins to flow. Below the yield stress, the material behaves like a solid; above it, the material flows like a fluid. Common examples include toothpaste, which remains in place until squeezed, and EPB-conditioned soil, which must be stiff enough to support the tunnel face yet capable of flowing during extraction. Yield stress is one of the most important rheological parameters in EPB tunnelling because it governs both face stability and spoil flow.

The relationship between shear stress and shear rate further characterises how materials respond to applied movement. Newtonian materials, such as water, have a constant viscosity regardless of shear rate. In contrast, non-Newtonian materials change viscosity when shear is applied. Most tunnelling materials—foams, slurries, and conditioned soils—are non-Newtonian, meaning their flow resistance varies with operating conditions.

Plasticity describes a material's ability to deform without cracking and to retain its shape after deformation. Materials with high plasticity deform smoothly, while those with low plasticity tend to behave in a brittle or collapsible manner. In EPB tunnelling, sufficient plasticity is essential for uniform transmission of face pressure and for smooth, controlled flow through the screw conveyor.

Thixotropy is another important rheological behaviour, referring to materials whose viscosity decreases with continued shear and recovers when the material is at rest. Such materials are stiff when stationary, flow easily when agitated, and then stiffen again once movement stops. This behaviour is particularly desirable in tunnelling, as EPB muck should remain stable in the excavation chamber but flow during extraction, while bentonite slurries should support the face yet remain pumpable in pipelines.

Many tunnelling materials also exhibit elastic or viscoelastic behaviour. An elastic response means deformation is recoverable when the load is removed, while a viscous response results in permanent deformation. Viscoelastic materials display a combination of both behaviours. Conditioned soils, grout, and slurries commonly show viscoelastic characteristics, influencing how they transmit loads and respond to changes in stress.

To analyse and design around these behaviours, engineers use simplified rheological models. The Bingham plastic model, which incorporates a yield stress and a linear viscosity, is commonly applied to EPB muck and slurry. More flexible models, such as the Herschel–Bulkley model, can capture shear-thinning or shear-thickening behaviour. These models enable calculation of pressure losses, pumping power requirements, face support performance, and overall flow stability.

Rheological properties are central to mechanised tunnelling because they govern face stability, excavation efficiency, torque and thrust demand, settlement control, and slurry transport and separation performance. Poorly controlled rheology can lead to pressure loss, excessive machine loads, blockages, or increased settlement, even in ground that appears favourable from a purely strength-based perspective.

Simple examples illustrate this clearly. Dry sand, with low cohesion and no yield stress, cannot support a tunnel face. Foam-conditioned sand develops apparent cohesion and controlled viscosity, allowing stable EPB operation. Bentonite slurry, with thixotropic behaviour and a defined yield stress, can support the face while remaining pumpable through long pipelines.

The operational effectiveness of foam injection relies on meticulous control over several key parameters, including the Foam Injection Ratio (FIR) and Foam Expansion Ratio (FER) [373]. FIR represents the volume of foam relative to the excavated soil, which must be balanced to prevent over-conditioning or under-conditioning the soil. For example, typical FIR values range between 20% to 80%, depending on specific soil types and conditions [646]. Meanwhile, FER governs the balance of air to liquid in the foam; variations in these ratios can significantly affect the viscosity and flow characteristics of the conditioned muck.

1. Foam Injection Ratio (FIR)

What FIR represents

The Foam Injection Ratio (FIR) expresses how much foam is injected relative to the volume of ground excavated. It controls the *intensity of conditioning*.

- Low FIR → weak conditioning, poor pressure retention
- High FIR → strong conditioning, risk of over-softening or pressure loss

Definition

$$\mathrm{FIR} = \frac{V_{\mathrm{foam}}}{V_{\mathrm{soil}}} \times 100\%$$

Where:

- V_{foam}= volume of foam injected (m^3)
- V_{soil}= volume of excavated soil (m^3)

FIR is usually expressed as a percentage.

How to calculate excavated soil volume

For one TBM advance step or over a time interval:

$$V_{\text{soil}} = A_{\text{tunnel}} \times L$$

Where:

- A_{tunnel}= tunnel cross-sectional area (m^2)
- L= advance length (m)

Worked example (FIR)

- TBM diameter = 6.5 m
- Tunnel area:

$$A = \frac{\pi D^2}{4} = \frac{\pi \times 6.5^2}{4} \approx 33.2 \text{ m}^2$$

- Advance length = 1.5 m (one ring)

$$V_{\text{soil}} = 33.2 \times 1.5 \approx 49.8 \text{ m}^3$$

Assume:

- Foam injected during this ring = 20 m^3

$$\text{FIR} = \frac{20}{49.8} \times 100 \approx 40\%$$

Interpretation:
An FIR of ~40% is typical for sandy silt or mixed ground, providing sufficient conditioning without excessive softening.

2. Foam Expansion Ratio (FER)

What FER represents

The Foam Expansion Ratio (FER) describes foam quality, i.e. how much air is introduced relative to liquid.

- Low FER → wet foam, higher water content
- High FER → dry foam, higher air content

FER affects plasticity, cohesion, and moisture control, but not the total foam volume (that's FIR).

Definition

$$\text{FER} = \frac{V_{\text{foam}}}{V_{\text{liquid}}}$$

Where:

- V_{foam} = total foam volume (air + liquid)
- V_{liquid} = volume of liquid used (water + surfactant)

Worked example (FER)

Assume:

- Liquid injected = 1.0 m^3
- Generated foam volume = 15 m^3

$$\text{FER} = \frac{15}{1.0} = 15$$

Interpretation:

- FER ≈ 10–15 → balanced foam (common EPB range)
- FER < 8 → very wet foam (risk of over-watering)
- FER > 20 → very dry foam (risk of poor coating / uneven conditioning)

3. Relationship between FIR and FER

FIR and FER control different aspects of conditioning:

Parameter	Controls	Practical effect
FIR	Quantity of foam	Strength of conditioning
FER	Foam texture	Moisture vs air balance

You can have:

- **High FIR + low FER** → lots of wet foam (risk of slurry-like muck)
- **Low FIR + high FER** → dry, insufficient conditioning
- **Moderate FIR + moderate FER** → optimal EPB behaviour

4. Combined operational example

Target conditions:

- Sandy silt with groundwater
- Desired behaviour: plastic, low-permeability muck

Selected parameters:

- FIR = 45%
- FER = 12

If:

- Soil excavated per ring = 50 m^3
- Foam required:

$$V_{\text{foam}} = 0.45 \times 50 = 22.5\ \text{m}^3$$

- Liquid required:

$$V_{\text{liquid}} = \frac{22.5}{12} \approx 1.9\ \text{m}^3$$

This defines:

- Water flow rate

- Surfactant dosage
- Air flow rate

These values are then set in the foam generation system and adjusted in real time.

Moreover, proper injection pressure and distribution are critical; uneven distribution can lead to localized soil conditioning, which may compromise stability and flow characteristics [311]. Effective management of foam injection is therefore imperative not only for maintaining tunnel face stability and minimizing settlements but also for ensuring efficient machine performance and reducing operational risks, such as clogging and excessive wear on equipment [293].

Beyond technical performance, environmental considerations are increasingly pertinent in the application of foam. Modern tunnelling projects prioritize the use of low-toxicity, biodegradable surfactants that adhere to environmental regulations regarding spoil disposal and site rehabilitation [379, 645]. This has prompted further research into the composition of foam materials employed in EPB tunnelling, enhancing both operational efficiency and compliance with ecological standards.

Foam injection serves as a foundational technique in EPB tunnelling, underpinning its success in maintaining face stability, controlling pressure, and facilitating muck extraction in various ground conditions. The approach necessitates precise tailoring of foam properties and dosages to underlying geological and operational circumstances. Thus, foam injection should not be viewed as an auxiliary measure; rather, it must be regarded as a core component of modern EPB tunnelling strategies.

Polymer and Bentonite Use

Natural soils often do not exhibit the stability required for TBM operation, with excessive permeability or instability complicating tunnel face support and spoil flow predictability [323]. Bentonite and polymers are specifically introduced to engineer soil behaviour, bridging the gap between in-situ soil properties and the operational demands of TBMs. Their respective roles include reducing soil permeability, enhancing stability, controlling groundwater inflow, and facilitating smoother spoil flow [195, 318, 647]. The synergistic use of these materials allows for fine-tuned control of soil mechanics, thereby optimizing the tunnelling process [321].

Polymers

Polymers used in tunnelling, such as polyacrylamides and cellulose-based compounds, are water-soluble organic materials that, when mixed with water, yield viscous solutions capable of robustly interacting with fine soil particles [195, 318]. This interaction fosters the aggregation of soil particles into larger structures, thereby reducing segregation and permeability, which is vital in maintaining face stability under high groundwater conditions [302, 381]. When polymers are injected into the excavation chamber or cutterhead, they bind soil particles and provide significant improvements in cohesive strength and flowability without drastically increasing moisture content—an important consideration for spoil disposal [648, 649].

Polymers are particularly effective in sand and silty conditions, where they can enhance cohesion and flowability while also managing moisture limits [318, 650] . Their integration into conditioning mixes often targets control of hydrostatic pressure, contributing to face stability during breaks in the tunnelling process [651].

Bentonite

Bentonite, especially sodium montmorillonite, hydrates to create a thixotropic slurry with exceptionally low permeability characteristics, crucial for sealing and pressure maintenance during tunnelling operations [321, 581]. The slurry forms a filter cake on the tunnel face, significantly reducing groundwater inflow and stabilizing the tunnel environment against destabilization [318, 579]. Proper management of bentonite application is essential; excessive use can lead to operational challenges such as clogging [649, 652].

In slurry TBMs, bentonite is indispensable, serving as a sealant and a transport medium for excavated soils. Adjustments to bentonite concentration allow for the management of slurry properties—density, viscosity, and yield stress, all critical for effective tunnelling system performance [647, 653].

Combined Application of Polymers and Bentonite

Combining polymers and bentonite is particularly advantageous for addressing the complexities of high-permeability sandy soils or mixed-face conditions. While bentonite provides the necessary sealing and pressure retention, polymers enhance cohesion and mitigate spoil segregation, leading to more predictable flow characteristics during excavation [648, 654]. This cooperative approach allows for more precise adjustments based on real-time feedback regarding soil and tunnelling conditions [318, 580].

The application of these materials requires continuous operational control. The balancing act in dosing is critical: insufficient or excessive amounts of either agent can lead to substantial operational difficulties—such as uncontrolled groundwater flows or material sticking [648, 650]. Therefore, understanding local soil conditions and effectively managing agent concentrations is fundamental to maintaining tunnel stability and operational efficiency.

Environmental Considerations

Modern tunnelling practices emphasize environmental sustainability. The choice of polymers often revolves around biodegradability and low toxicity to adhere to environmental regulations [323, 655]. The interaction between ground conditioning agents and groundwater necessitates careful consideration of both material choice and dosage to ensure minimal environmental impact.

Polymers and bentonite are pivotal elements in engineering ground behaviour in mechanised tunnelling. Their role extends beyond mere additives; these materials are integral to the performance and safety of tunnelling operations, facilitating the management of permeability, groundwater, face stability, and spoil extraction. The effective integration of these agents underscores the modern philosophy in tunnelling that actively engineers ground behaviour rather than simply responding to existing conditions.

Soil Conditioning Ratios

Soil conditioning ratios play a crucial role in the operation of Earth Pressure Balance (EPB) tunnel boring machines (TBMs). These ratios are quantitative metrics employed to control the amount of conditioning agents—such as foam, polymer, water, and bentonite—added to the excavated soil. Their significance lies in transforming the in-situ soil characteristics into engineered muck with specific rheological properties necessary for effective tunnelling.

In EPB tunnelling, the excavated soil serves as the structural support for the tunnel face. For optimal face stability during excavation, the conditioned soil must achieve several critical conditions:

1. **Uniform Face Pressure:** The support pressure transmitted to the tunnel face must be uniformly distributed to prevent face instability [307, 328].
2. **Low Permeability:** Achieving low permeability is essential to prevent water inflow, which can destabilize the tunnel face and lead to complications during excavation [581, 656].
3. **Plastic Flow Through the Screw Conveyor:** The soil must flow easily into and through the TBM's screw conveyor without excessive resistance [586].
4. **Avoid Segregation:** The conditioned muck should maintain homogeneity to avoid issues such as segregation or clogging within the TBM [657, 658] .

These requirements are often unmet by natural soils, necessitating the use of soil conditioning ratios to dictate how much modification is required to ensure stability and flowability.

Soil conditioning fundamentally alters the properties of in-situ materials. Various additives play distinct roles in modifying soil characteristics:

- **Foams and Polymers:** These agents can significantly reduce soil permeability and enhance its plasticity, which is crucial for maintaining the integrity of the tunnel face under pressure [586, 658, 659].
- **Bentonite:** When utilized correctly, bentonite can interact with other conditioning agents to provide improved stability and reduce the risk of clogging during tunnelling processes [307, 659].
- **Water:** Moisture helps achieve the desired consistency of the muck, facilitating its flowability in the TBM systems [328, 581, 590].

The choice and ratio of these agents determine how effectively the soil can be transformed from its natural state into suitable muck for EPB excavation.

Conditioning ratios are not static; they must be monitored and adjusted in real-time based on the specific conditions encountered during tunnelling. This dynamic management allows for:

- **Better Control of Rheology:** Engineers can manipulate the properties of the muck to optimize flow characteristics and support capacity continually [307, 328, 656].
- **Enhanced Operational Safety:** By ensuring that the soil conditions remain favourable, risks associated with unexpected geological changes can be minimized, enhancing both safety and efficiency during the tunnelling process [431, 657].

The calibrated ratios foster a deep understanding of muck behaviour, enabling operators to predict changes and adjust conditions on-the-fly, thereby improving overall project timelines and reducing the likelihood of operational failures.

Main soil conditioning ratios

1. Foam Injection Ratio (FIR)

Definition

The Foam Injection Ratio expresses how much foam is added relative to the volume of excavated soil.

$$\text{FIR} = \frac{\text{Volume of foam injected}}{\text{Volume of excavated soil}} \times 100\%$$

What it controls

- Apparent cohesion
- Plasticity
- Flowability

- Pressure transmission

Typical ranges

- Fine sands / silts: 30–80%
- Mixed-face ground: 20–60%
- Clayey soils: 10–30% (or none)

Engineering meaning

- Low FIR → insufficient conditioning, running ground
- Excessive FIR → overly soft muck, pressure instability

FIR is one of the most critical control parameters in EPB tunnelling.

2. Foam Expansion Ratio (FER)

Definition
FER describes how "dry" or "wet" the foam is by comparing foam volume to liquid volume.

$$\text{FER} = \frac{\text{Foam volume}}{\text{Liquid volume}}$$

Typical values

- FER = 5–10 → wet foam (more water)
- FER = 10–20 → balanced foam
- FER = 20–30 → dry foam (more air)

What it controls

- Water addition to the soil
- Bubble stability
- Lubrication vs sealing balance

Engineering meaning

- Low FER → increases moisture, improves cohesion but risks wet spoil
- High FER → improves lubrication with minimal water, but weaker sealing

FER is adjusted together with FIR—they must always be interpreted as a pair.

3. Polymer Dosage Ratio

Definition
Polymer dosage is usually expressed as mass or volume of polymer per volume of soil or per volume of water used to prepare the solution.

Examples:

- kg polymer / m^3 of soil
- g polymer / m^3 of injected water

Typical order of magnitude

- 0.1–1.0 kg/m^3 of soil (project-dependent)

What it controls

- Permeability reduction
- Particle binding
- Segregation control
- Pressure retention

Engineering meaning

- Low dosage → limited sealing effect
- Excessive dosage → sticky muck, high torque, disposal problems

Polymers are often used to supplement foam, not replace it.

4. Bentonite Content Ratio

Definition
Bentonite addition is typically expressed as:

- kg bentonite / m^3 of soil (EPB), or
- % bentonite by weight in slurry (slurry TBM)

Typical values

- EPB: very low, targeted additions only
- Slurry TBM: 3–7% bentonite by weight (varies by ground)

What it controls

- Sealing and filter cake formation

- Yield stress and pressure stability
- Groundwater control

Engineering meaning

- Insufficient bentonite → pressure loss in permeable ground
- Excessive bentonite → sticky muck, separation difficulties

5. Water Addition Ratio

Definition
The volume of water added relative to excavated soil volume.

Role

- Adjusts consistency and density
- Supports foam and polymer effectiveness

Engineering meaning

- Too little water → dry, non-cohesive muck
- Too much water → low shear strength, high permeability

Water is rarely used alone; it is managed through FIR and FER.

Soil conditioning ratios do not operate in isolation; they are interdependent parameters that must be managed as a coordinated system. The Foam Injection Ratio (FIR) controls how much foam is introduced into the excavated soil, while the Foam Expansion Ratio (FER) governs the character of that foam by defining the balance between air and liquid. Polymer dosage primarily influences sealing and apparent cohesion, whereas bentonite content controls permeability reduction and pressure retention. Together, these ratios determine the overall rheological behaviour of the conditioned muck.

Different combinations of ratios are used to achieve specific conditioning outcomes. A high FIR combined with a high FER produces a relatively dry, well-lubricated soil with minimal additional water, suitable where flowability is required without increasing moisture content. A moderate FIR supplemented with polymer injection provides strong cohesion and sealing while keeping water content under control, which is particularly useful where spoil disposal limits apply. In highly permeable sandy soils, the combined use of foam and bentonite delivers both improved flowability and effective sealing, enabling stable face pressure to be maintained. Effective conditioning therefore relies on balancing these ratios rather than maximising any single parameter.

The following provides practical examples showing how each main soil conditioning ratio is applied on a real EPB or slurry TBM. The intent is to show *how the numbers are used*, not just what they mean.

1. Foam Injection Ratio (FIR) — worked example
Definition

$$\text{FIR} = \frac{\text{Foam volume injected}}{\text{Excavated soil volume}} \times 100\%$$

Example
A 6.5 m EPB TBM excavates 10 m^3 of soil per ring.
The conditioning system injects 5 m^3 of foam into the cutterhead and chamber.

$$\text{FIR} = \frac{5}{10} \times 100 = 50\%$$

Interpretation

- FIR = **50%**, suitable for **fine sand / silty sand**
- Expect:
  - Good apparent cohesion
  - Stable pressure transmission
  - Controlled screw conveyor flow

If FIR dropped to 15%, the same sand would likely become running and unstable.
If FIR increased to 90%, the muck could become too soft and compressible, risking pressure instability.

2. Foam Expansion Ratio (FER) — worked example
Definition

$$\text{FER} = \frac{\text{Foam volume}}{\text{Liquid volume}}$$

Example
A foam generator produces 6 m^3 of foam using 0.4 m^3 of water + surfactant solution.

$$\text{FER} = \frac{6}{0.4} = 15$$

Interpretation

- FER = **15** → balanced foam
- Good compromise between:
  - Lubrication
  - Sealing
  - Moisture control

If FER were 6, the foam would be wet → higher water content, stronger cohesion but disposal risk.
If FER were 25, the foam would be very dry → excellent lubrication but weaker sealing.
FER must always be read together with FIR.

3. Polymer Dosage Ratio — worked example

Example (per soil volume)
An EPB TBM excavates 10 m^3 of sandy silt per ring.
Polymer injection rate = 4 kg per ring

$$\text{Polymer dosage} = \frac{4}{10} = 0.4 \text{ kg/m}^3$$

Interpretation

- 0.4 kg/m^3 → typical for permeable sands/silts
- Effects:
  - Reduced permeability
  - Improved cohesion
  - Less water segregation

If dosage dropped to 0.1 kg/m^3, sealing would be weak.
If increased above 1.0 kg/m^3, muck could become sticky, increasing torque and clogging risk.

4. Bentonite Content Ratio — worked examples
EPB TBM (targeted use)
An EPB TBM encounters a highly permeable sand lens.

- Excavated soil: 10 m^3 per ring
- Bentonite added: 30 kg

$$\text{Bentonite content} = \frac{30}{10} = 3 \text{ kg/m}^3$$

Interpretation

- Very low but effective sealing addition
- Helps retain pressure during stoppages
- Excess beyond this could cause stickiness and screw overload

Slurry TBM (slurry concentration)
A slurry plant prepares 100 m^3 of slurry with 5 tonnes of bentonite.

$$\text{Bentonite concentration} = \frac{5{,}000}{100{,}000} = 5\%$$

Interpretation

- 5% bentonite → typical operating range
- Ensures:
  - Stable filter cake
  - Adequate yield stress
  - Reliable face pressure

Below 3% → risk of slurry loss
Above 7% → separation and pumping problems

5. Water Addition Ratio — worked example
An EPB TBM excavates 10 m^3 of soil per ring.
Total water added (via foam + polymer solution) = 1.5 m^3

$$\text{Water ratio} = \frac{1.5}{10} = 15\%$$

Interpretation

- 15% water addition → controlled moisture
- Enough to activate foam and polymers without flooding the muck

If water ratio fell below 5%, muck could be dusty and non-cohesive.
If it exceeded 30%, shear strength and pressure retention would drop.

Combined conditioning example (realistic EPB case)
Ground: saturated silty sand
Target behaviour: plastic, low permeability, stable pressure

Parameter	Value
FIR	45%
FER	18
Polymer dosage	0.35 kg/m^3
Bentonite	2 kg/m^3
Water ratio	~12%

Result

- Stable face pressure
- Smooth screw conveyor torque
- Minimal settlement
- Acceptable spoil disposal characteristics

In practice, soil conditioning ratios are adjusted continuously in response to real-time feedback from the TBM and the ground. Operators monitor indicators such as face pressure stability, screw conveyor torque, penetration consistency, and the visual and mechanical behaviour of the muck. Settlement monitoring results are also integrated into this feedback loop, linking underground performance directly to surface response.

Typical operational responses follow well-established patterns. Running or unstable ground conditions often prompt an increase in FIR or the introduction of polymer to improve cohesion and pressure transmission. Torque spikes or signs of clogging may require a reduction in bentonite content or improved foam distribution to reduce stickiness and friction. Loss of face pressure or increased water inflow generally indicates the need for enhanced sealing, achieved through additional polymer or bentonite dosing.

From an engineering perspective, soil conditioning ratios transform complex and variable ground behaviour into controllable operational parameters. They enable stable face support in EPB tunnelling, control permeability and groundwater inflow, reduce wear and clogging in mechanical systems, and minimise settlement and surface risk. Most importantly, they provide a direct link between ground mechanics and TBM operation. In modern mechanised tunnelling, success is therefore defined not by the quantity of conditioning agents used, but by how precisely their ratios are selected, monitored, and adjusted to match evolving ground conditions in real time.

Grouting Techniques

Grouting techniques are critical to the success of mechanized tunnelling and ground improvement. They serve several essential functions, including void filling, load transfer, groundwater control, and settlement management. The application of these techniques must be strategic as grouting is not a uniform operation; rather, it comprises various approaches tailored to specific engineering challenges encountered during tunnelling operations [660, 661].

In the context of TBM (Tunnel Boring Machine) tunnelling, grouting plays a vital role in maintaining structural integrity and stability. Grouting is primarily employed to fill voids created due to excavation activities, notably the annular gap between the TBM shield and surrounding ground, facilitating effective load transfer and minimizing settlement risks. These techniques directly contribute to ground stability by preventing volume loss and managing groundwater ingress, thus enhancing the overall robustness of the tunnel structure [660, 661].

Grouting also serves a key function in reducing soil permeability, thereby controlling groundwater flow and protecting sensitive urban infrastructure situated above or adjacent to tunnelling projects. For instance, compensation grouting techniques are used to inject low-viscosity grout proactively or reactively to mitigate the impacts of tunnelling on above-ground structures [660, 661]. Furthermore, it has been emphasized that the effectiveness of annular grouting is crucial in preventing post-excavation settlement and ensuring uniform load transfer to the lining [660].

Specific Grouting Techniques in Tunnelling

Several specialized grouting techniques are employed during the different stages of tunnelling:

1. Annular Grouting

This technique involves filling the void between the TBM shield and the tunnel lining with grout immediately after ring erection. Effective management of injection pressure helps ensure complete contact with the surrounding ground and prevents overstressing the segments [660]. The control mechanisms governing the injection process, including pressure and volume, are pivotal to achieving desired outcomes like minimizing water migration and stabilizing the tunnel structure during the early stages of construction [660, 661].

2. Compensation Grouting

Used primarily in urban areas where tunnelling might affect sensitive structures, compensation grouting involves the installation of pipes such as tube-à-manchette (TAM) to monitor and manage surface movements effectively. This technique allows for localized uplift and conveys ground conditioning, thus making tunnelling beneath critical assets feasible [661].

Figure 94: CGV Shaft SA 2 Härtelstraße, City Tunnel Leipzig – Compensation Grouting Method. Frank Eritt, CC BY 3.0, via Wikimedia Commons.

3. Permeation Grouting

This approach targets soil's pore spaces without disturbing soil structure, effectively reducing permeability and controlling groundwater inflow. It is particularly useful in granular soils and water-bearing strata, enhancing the stability of the tunnel face and mitigating pressure losses during TBM operations [660].

Figure 95: Installing TAM (Tube-A-Manchette) Pipes for Permeation Grouting. MTA Construction & Development Mega Projects, CC BY 2.0, via Flickr.

4. Fracture Grouting

In contrast with permeation grouting, fracture grouting employs higher pressure to intentionally fracture and displace the ground. This technique is particularly beneficial for densifying loose fills and providing additional support. However, it necessitates careful control to avoid unintended consequences like damage or excessive ground uplift [661].

5. Jet Grouting

Jet grouting involves employing high-pressure jets to mix grout with in-situ soil, enhancing its strength and impermeability. This technique is frequently applied in pre-support situations, stabilizing tunnels, shafts, or cut-off blocks as part of comprehensive groundwater management strategies [661].

6. Contact Grouting

Applied post-construction to fill residual voids or improve load transfer, contact grouting becomes critical when initial grouting efforts are insufficient. This method aims to rectify unexpected settlement and fortify the tunnel lining for long-term durability [661].

Figure 96: Crossover Contact Grouting. MTA Construction & Development Mega Projects, CC BY 2.0, via Flickr.

Integrative Approach to Grouting Techniques

Successful implementation of grouting techniques demands careful control and monitoring of numerous parameters including grout rheology, injection pressure, and sequencing [660, 661]. These factors directly affect the effectiveness of the grouting process, ensuring optimized interactions between the tunnel and surrounding ground. Notably, grouting not only serves as a method of managing immediate tunnelling impacts but also acts as a strategic component of a broader engineering system that encompasses face pressure management and soil conditioning [660].

Grouting techniques are integral to the mechanized tunnelling process, enhancing ground-structure interaction while managing the risks associated with tunnelling operations. Each specific grouting approach serves distinct purposes tailored to the challenges posed by different geological and structural contexts, collectively ensuring safe and predictable tunnel construction.

Ground Freezing

Ground freezing, often referred to as artificial ground freezing (AGF), is a ground-improvement and temporary support technique used in tunnelling and underground construction to stabilise soil and control groundwater by converting the ground itself into a frozen, load-bearing mass. By deliberately freezing the pore water within the ground, engineers transform otherwise unstable or permeable soils into a material that behaves similarly to weak rock. Although highly reliable, ground freezing is energy-intensive and is therefore typically reserved for situations where other ground improvement or support methods are ineffective or carry unacceptable risk.

By freezing the pore water in soil or rock, ground freezing produces several fundamental changes in ground behaviour. Strength and stiffness increase significantly, permeability is reduced to near zero, and a continuous, self-supporting frozen block is formed. Even soils such as sands, silts, or gravels—which would normally be unstable or prone to running—can behave as a coherent, predictable mass once frozen. This transformation underpins the method's effectiveness in high-risk ground conditions.

Ground freezing works through the installation of freeze pipes drilled into the ground around or ahead of the excavation. These steel pipes are typically spaced between 0.8 and 1.5 metres apart, with spacing determined by soil type, groundwater conditions, and the required thickness of the frozen body. Once installed, a coolant is circulated through the pipes to extract heat from the ground.

Two main freezing systems are used. In brine freezing, which is the most common approach, a calcium chloride brine cooled to approximately –25 to –35°C is circulated continuously through the freeze pipes. Freezing develops gradually, usually over several weeks, allowing controlled formation of the frozen ground. Liquid nitrogen freezing is a faster but more expensive alternative, involving the direct injection of liquid nitrogen at –196°C. This method can freeze ground within hours or days and is typically reserved for emergencies or very short-duration applications due to its high cost.

As cooling continues, individual frozen bulbs form around each freeze pipe. These bulbs progressively expand and overlap, eventually merging into a continuous frozen wall, block, or arch. Excavation is only permitted once this frozen body is confirmed to be continuous and has reached the required design temperature and strength.

Several freezing layouts are used depending on the application. In shaft freezing, freeze pipes are arranged in a circular pattern to create a frozen cylinder around a shaft, a configuration commonly used for deep shafts in water-bearing ground. For tunnel faces or cross passages, freeze pipes are drilled horizontally or radially to create a frozen plug ahead of the excavation, a method widely used for cross passages between parallel TBM tunnels. Block freezing involves freezing an entire volume of ground and is employed where maximum control and robustness are required.

One of the key advantages of ground freezing is its applicability across a wide range of ground conditions. It is effective in sands and silts, coarse gravels, mixed-face ground, high groundwater pressure environments, and running or highly unstable soils. Its primary limitation is heat flow rather than soil type, meaning freezing performance depends more on thermal conditions than geotechnical variability.

The engineering properties of frozen ground are highly favourable for excavation. Typical compressive strengths range from 1 to 5 MPa depending on soil type and temperature, permeability is effectively eliminated, and cohesion is greatly increased. Frozen ground also exhibits excellent stand-up time, allowing safe excavation and support installation. Its homogeneity and predictability are key reasons why ground freezing is trusted in high-risk or sensitive projects.

Because of its complexity and risk profile, ground freezing requires continuous monitoring. Temperature sensors are installed in the ground to verify freezing progress and uniformity, while groundwater pressure, surface settlement, and subsurface deformation are also monitored. Energy input and cooling system performance are tracked throughout the freezing period. Excavation is only allowed once target temperatures—often –10°C or colder—are achieved throughout the frozen zone.

Excavation in frozen ground proceeds in a manner similar to excavation in weak rock. The frozen material can be excavated using roadheaders, hand mining, or small mechanical excavators, depending on geometry and access. Permanent or temporary support systems, such as shotcrete or segmental linings, are installed before thawing begins to ensure load transfer is controlled.

After permanent support is in place, the refrigeration system is shut down and the ground is allowed to thaw gradually. Thawing is typically natural and slow, allowing loads to transfer progressively to the lining. Controlled thawing is essential to avoid sudden settlement or loss of ground support.

Ground freezing offers several distinct advantages. It works in virtually any soil type, provides excellent groundwater cut-off, delivers highly predictable ground behaviour, avoids chemical contamination, and is fully reversible as a temporary method. However, it also has notable disadvantages, including high energy consumption, long lead times for brine systems, complex monitoring requirements, risk of frost heave if poorly controlled, and higher cost compared to most grouting methods. As a result, it is usually considered a last-resort or high-risk solution rather than a routine construction technique.

Typical tunnelling applications include cross passages between TBM tunnels, deep shafts in water-bearing ground, tunnels beneath rivers or sensitive structures, emergency stabilisation following ground loss, and complex interfaces between stations and TBM drives.

From an engineering perspective, ground freezing is often described as the "gold standard" of temporary ground support. It replaces uncertain and variable soil behaviour with an engineered frozen mass, decouples excavation safety from groundwater pressure, and enables construction in conditions that would otherwise be impractical or unsafe. In mechanised tunnelling, ground freezing is not about speed or productivity; it is about certainty, safety, and control where the consequences of failure are unacceptable.

Jet Grouting and Soil Stabilisation

Jet grouting is a ground improvement and soil stabilisation technique used in tunnelling and underground construction to create strong, low-permeability soil–cement bodies in situ. Unlike permeation or compensation grouting, jet grouting deliberately breaks down the natural soil structure and replaces it with an engineered soil–cement composite formed by mixing the in-situ material with cementitious grout. This fundamental difference allows jet grouting to produce ground with predictable strength and hydraulic properties, largely independent of the original soil fabric.

Because of this capability, jet grouting is one of the most versatile stabilisation methods available. It is widely applied where precise geometry, high strength, or reliable groundwater cut-off is required, particularly in complex urban tunnelling environments and at critical interfaces such as portals, shafts, and cross passages.

In functional terms, jet grouting improves the ground by increasing its strength and stiffness, reducing permeability, and creating defined structural elements such as columns, panels, or blocks. These treated zones can provide either temporary or permanent support, control groundwater inflow, and limit settlement. The resulting material behaves as a soil–cement composite whose properties can be tailored through grout design, jet energy, and execution parameters.

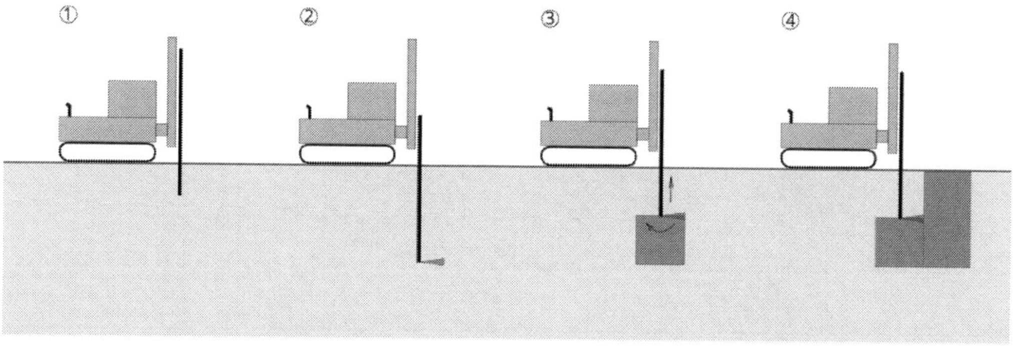

Figure 97: Jet Grouting - 1) Boring in the drill pipe to the requested depth. 2) Starting the soil injection. 3) Pulling up the drill pipe while rotating and continuously injecting cement suspension. 4) Repeating the process and adding the columns. 5gon12eder, CC BY-SA 4.0, via Wikimedia Commons.

The core principle of jet grouting is the use of extremely high-pressure grout injection, typically in the range of 300–600 bar, through small nozzles at the end of a rotating drill rod. The high-energy jet erodes and disaggregates the in-situ soil, mixes it thoroughly with cement grout, and forms a cylindrical or panel-shaped body as the rod is rotated and withdrawn. Once the grout hardens, the treated zone exhibits strength and permeability characteristics far superior to those of the original ground.

Several jet grouting systems are used in practice, selected according to soil conditions and performance requirements. Single-fluid systems (Jet 1) inject cement grout only and are the simplest configuration, generally suited to softer soils such as clays and silts, producing relatively small column diameters. Double-fluid systems (Jet 2) combine cement grout with compressed air, which increases jet energy and erosion radius, resulting in larger and more uniform columns; this configuration is widely used in tunnelling. Triple-fluid systems (Jet 3) use a high-pressure water jet for cutting, an air shroud to boost jet energy, and a separate cement grout for mixing, allowing the largest column diameters and making them suitable for dense sands and gravels, albeit at higher complexity and cost.

Jet grouting offers exceptional flexibility in geometry. It can be used to form individual columns, typically 0.6–2.0 m in diameter, overlapping columns that create walls or blocks, panels for groundwater cut-off, arch or umbrella supports above tunnels, bottom plugs for shafts, and transition zones at portals or station interfaces. The final geometry is controlled by drilling position, jet pressure, rotation speed, and withdrawal rate, allowing engineers to shape the improved ground to suit specific design needs.

Figure 98: Jet grouting. U.S. Army Corps of Engineers Sacramento District, Public domain, via Wikimedia Commons.

The technique is applicable to a wide range of soil conditions, including clays and silts, sands and silty sands, mixed ground, and fill or made ground. However, limitations arise in very coarse gravels or cobble layers, where uniform mixing is difficult, and in ground containing obstructions such as boulders or debris, which can cause variability in column quality. As a result, the performance of jet grouting depends heavily on execution control and quality assurance.

The engineering properties of jet-grouted soil vary with soil type and grout formulation, but typical unconfined compressive strengths range from about 1 to 10 MPa, with higher values achievable through optimisation. Permeability is usually reduced to the order of 10^{-7} to 10^{-9}

m/s, making the treated ground effectively impermeable. Stiffness is comparable to weak concrete, and durability is sufficient for both temporary and permanent works. Because of inherent variability, trial columns, coring, and laboratory testing are essential parts of design verification.

In tunnelling and underground construction, jet grouting is commonly used for pre-support of shallow tunnels, stabilisation of portals and shafts, improvement of ground for cross passages, groundwater cut-off beneath rivers or water-bearing strata, replacement or supplementation of ground freezing, stabilisation of mixed-face or faulted ground, and reduction of settlement risk beneath sensitive structures. In TBM projects, it is typically applied ahead of the face or around critical interfaces rather than continuously along the alignment.

Jet grouting is highly sensitive to execution, and strict control is required over injection pressure and flow rate, rotation speed, withdrawal rate, grout mix design, and column overlap and spacing. Quality assurance relies on coring and strength testing, permeability testing, geophysical verification, and continuous monitoring of grout take and injection pressure to confirm that the intended ground improvement has been achieved.

The technique offers many advantages, including applicability to a wide range of soils, the ability to create structural ground elements with defined geometry, excellent groundwater control, flexibility of layout, and suitability for both temporary and permanent works without the need for large excavations. However, it also has limitations, such as high spoil generation from grout–soil returns, variability linked to execution quality, sensitivity to operator skill, reduced effectiveness in coarse gravels, and higher cost compared with simpler grouting methods. For these reasons, jet grouting is typically applied in a targeted manner rather than indiscriminately.

Compared with other stabilisation methods, jet grouting replaces the soil structure rather than merely modifying it, as in permeation grouting. Unlike compensation grouting, which adjusts ground volume to manage movement, jet grouting fundamentally improves ground strength and stiffness. Compared with ground freezing, jet grouting is permanent and less energy-intensive, though freezing can offer greater predictability in extreme conditions.

From an engineering perspective, jet grouting is a powerful tool that allows engineers to engineer the ground itself rather than simply support it. In mechanised tunnelling, it is particularly valuable where face stability margins are tight, groundwater pressures are high, settlement tolerance is low, or precise geometry control is essential. When properly designed and executed, jet grouting delivers robust and predictable ground behaviour, bridging the gap between natural soil variability and the controlled demands of modern tunnelling.

Chapter 14

Health, Safety, and Risk Management

Health, safety, and risk management form the foundation of responsible mechanised tunnelling. As illustrated throughout this chapter, Tunnel Boring Machine operations combine high-energy mechanical systems, pressurised excavation environments, complex geotechnical conditions, and confined underground workplaces. These interacting systems create a risk profile that is dynamic rather than static, requiring continuous monitoring, disciplined operational control, and integrated engineering safeguards.

Unlike many construction environments where hazards are visible and discrete, TBM risks often develop through gradual system imbalance, pressure drift, torque escalation, groundwater migration, gas accumulation, or human-factor degradation. Effective risk management therefore depends not only on compliance procedures, but on real-time parameter control, predictive monitoring, and strong safety culture embedded at every operational level.

This chapter examines the principal hazard domains in mechanised tunnelling, including face instability, flooding, fire, gas exposure, confined space risks, high-pressure systems, hyperbaric intervention, and the critical influence of human factors and fatigue. It further explores how structured Safety Management Systems (SMS), emergency response planning, and data-driven operational control convert complex underground risks into manageable engineering processes.

Ultimately, safe TBM operation is not achieved by reacting to incidents—it is achieved by maintaining balance across the interconnected systems of machine, ground, lining, and people. The content of this chapter demonstrates that technical performance and safety performance are inseparable: the safest tunnel drives are those where engineering discipline, monitoring, and risk management operate continuously and cohesively throughout excavation.

TBM Operational Hazards

Tunnel Boring Machine (TBM) operational hazards arise from the interaction between heavy mechanical systems, pressurised excavation environments, variable ground conditions, and confined underground workplaces. Although mechanised tunnelling has significantly improved safety compared to conventional excavation methods, it introduces its own complex risk profile. These risks must be actively managed through robust engineering design, continuous monitoring, and disciplined operational control.

TBM hazards are rarely isolated events; they are system-based and interdependent. A change in ground behaviour may increase cutterhead torque, which may alter face pressure balance, which in turn may influence settlement or lining loads. Understanding operational hazards therefore requires a holistic view that integrates mechanical, geotechnical, hydraulic, electrical, and human factors.

One of the most critical hazards in TBM tunnelling is face instability and potential ground collapse. In EPB and slurry TBMs, the excavation face is stabilised by pressurised soil or slurry. If face pressure drops below the required stability threshold, ground may collapse inward, sudden inflow of soil or water may occur, surface settlement or sinkholes may develop, and worker safety at the cutterhead may be compromised. Common causes include incorrect face pressure settings, inadequate spoil conditioning, over-extraction through the screw conveyor, or highly permeable and variable ground conditions. Conversely, excessive face pressure can result in blow-outs, hydraulic fracturing, or ground heave. Maintaining pressure within a defined operational window is therefore a primary safety control.

In water-bearing ground, uncontrolled groundwater inflow presents a major hazard. Flooding of the excavation chamber, erosion of surrounding soil, loss of ground support, and contamination of slurry systems can occur if hydrostatic balance is not maintained. Slurry TBMs rely on controlled slurry density and filter cake formation, while EPB machines depend on low-permeability conditioned muck. Failure of sealing systems or pressure imbalance can rapidly escalate into an inrush event, particularly under rivers or in high groundwater environments.

Mechanical overload and equipment failure represent another significant hazard category. TBMs operate under extreme thrust and torque forces. Excessive thrust can crack newly installed segments; torque spikes can overload the drive system; main bearings can overheat; screw conveyors may block; and slurry pumps or pipelines can rupture. Mechanical failures underground are difficult and time-consuming to repair and may require hyperbaric intervention, increasing exposure to additional risks. Sudden changes in torque, vibration, or temperature are often early indicators of developing problems and must be interpreted promptly.

Hyperbaric interventions are among the highest-risk activities in mechanised tunnelling. When cutterhead maintenance is required under pressure, workers may enter compressed air or slurry environments. Hazards include decompression sickness, barotrauma, nitrogen narcosis in deep interventions, and limited emergency response capability. Strict medical screening, decompression protocols, and specialised procedures are mandatory to manage these risks.

Hyperbaric interventions are specialised maintenance operations carried out inside the pressurised excavation chamber of a Tunnel Boring Machine (TBM) when the machine is operating below groundwater level or in unstable ground. During these interventions, workers enter an environment where the pressure is intentionally increased above atmospheric levels to maintain face stability and prevent water or soil inflow.

These operations are among the highest-risk activities in mechanised tunnelling and require strict engineering, medical, and procedural controls.

Types of Hyperbaric Systems

Compressed air intervention is most commonly associated with Earth Pressure Balance (EPB) TBMs operating in soft ground. In this method, the excavation chamber is pressurised using compressed air, typically at pressures up to about 3–4 bar, although higher pressures may be used under specialised procedures and regulatory controls. Workers access the chamber through a manlock (airlock), where they are gradually compressed to match chamber pressure before entry. Because slurry is not normally present in EPB chambers, the surrounding soil must be sufficiently cohesive and impermeable to remain stable when air replaces the pressurised muck at the face. This technique is widely used in soft-ground EPB drives but requires careful control to prevent blow-out or loss of face support.

Slurry-supported hyperbaric intervention is used in slurry TBMs, where the excavation chamber is normally filled with bentonite slurry under hydrostatic pressure. For intervention, the slurry may be partially displaced, and the chamber pressurised with compressed air or, in some cases, mixed gas. Because slurry systems are often used in high groundwater environments, the required support pressures may be higher than in EPB operations. As a result, slurry TBMs can accommodate higher intervention pressures, but this also increases physiological risk and procedural complexity. The transition from slurry-supported conditions to an air-supported working environment must be carefully managed to maintain face stability and prevent uncontrolled inflow.

In extreme conditions—such as deep tunnels beneath rivers or offshore environments—mixed-gas or saturation systems may be required. These systems typically use gas mixtures such as heliox or trimix to reduce the risk of nitrogen narcosis at elevated pressures. Mixed-gas operations demand specialised medical supervision, diving-level safety protocols, and dedicated life-support systems. In some cases, saturation techniques are employed, whereby workers remain under pressure in dedicated chambers between shifts rather than decompressing after each intervention. Although rare and costly, these systems are used where very high pressures make conventional compressed air procedures unsafe or impractical.

Key Risks of Hyperbaric Intervention

Decompression sickness (DCS) is one of the primary physiological hazards associated with hyperbaric intervention. When workers are exposed to elevated pressure, inert gases—primarily nitrogen—dissolve into the bloodstream and body tissues. If decompression occurs too quickly, these dissolved gases form bubbles within the body, which can obstruct blood flow or damage tissues. Symptoms may range from joint and muscle pain ("the bends") to serious neurological impairment, paralysis, or even death. For this reason, controlled decompression schedules based on approved decompression tables are mandatory, and compliance is strictly supervised.

Barotrauma results from pressure changes affecting air-filled spaces within the body. During compression or decompression, inadequate equalisation can cause injury to the ears, sinuses, or lungs. Middle ear barotrauma is particularly common if pressure is increased too rapidly or if a worker is unable to equalise properly. Lung barotrauma, although less common, can be severe if breath-holding or improper breathing occurs during decompression. Rigorous medical screening and training are required to ensure workers are fit for hyperbaric exposure and understand proper equalisation techniques.

Nitrogen narcosis is a cognitive impairment that can occur at elevated pressures when nitrogen acts as a narcotic agent within the central nervous system. It can reduce alertness, impair judgement, and slow reaction times—effects that are especially dangerous in confined, high-risk environments such as a TBM excavation chamber. The likelihood and severity increase with depth and pressure. In higher-pressure interventions, mixed-gas systems are sometimes used to reduce nitrogen concentration and mitigate this risk.

Oxygen toxicity is another serious hazard in hyperbaric environments. While oxygen is essential for life, excessive partial pressure of oxygen at elevated pressures can cause neurological symptoms such as seizures or pulmonary damage with prolonged exposure. Hyperbaric systems therefore carefully control oxygen concentration and exposure time within breathing mixtures. Gas management protocols are strictly regulated to maintain safe oxygen partial pressures.

Emergency response in a pressurised environment presents additional challenges. Evacuation cannot occur instantly, as workers must undergo controlled decompression before returning to atmospheric pressure. This delay complicates emergency medical treatment and rescue operations. Furthermore, fire risk is elevated in compressed air environments due to increased oxygen partial pressure and confined conditions. As a result, hyperbaric interventions require detailed emergency planning, fire-resistant materials, specialised rescue procedures, and highly trained personnel to manage potential incidents safely.

Hyperbaric Procedure Overview

Step 1 – Preparation

- Ground and pressure conditions verified.
- Cutterhead rotation stopped and secured.
- Pressure stabilised.
- Medical clearance confirmed.
- Decompression schedule prepared.

Step 2 – Compression

Workers enter the manlock.

Pressure is gradually increased to match chamber pressure.

Equalisation takes several minutes to over an hour depending on depth.

Step 3 – Intervention Work

Typical duration:

- 30–90 minutes depending on pressure and medical limits.

Work must be:

- Highly planned
- Tool-prepared
- Time-efficient

Because allowable exposure time is limited.

Step 4 – Decompression

Workers return to the manlock.

Pressure is reduced in controlled stages.

Total decompression time can exceed working time.

Medical observation follows.

Operational Controls

Hyperbaric interventions require:

- Specialist hyperbaric supervisors
- Medical support and emergency chamber facilities
- Continuous pressure monitoring
- Communication systems
- Rescue procedures
- Certified pressure vessels

In many jurisdictions, hyperbaric tunnelling is regulated under diving or pressure vessel legislation.

Confined space and environmental hazards are inherent in underground construction. TBM tunnels have limited ventilation and escape routes, and hazards may include reduced oxygen levels, accumulation of hazardous gases such as methane or hydrogen sulphide, dust and airborne contaminants, heat stress from machinery and geology, and restricted visibility. Continuous atmospheric monitoring, ventilation management, and emergency preparedness are essential components of safe operation.

Segment handling and lining installation introduce further risks. Segment erection involves heavy lifting in confined spaces, pinch points during ring closure, temporary instability before full ring completion, and thrust transfer onto newly installed segments. Errors in ring build sequence or use of damaged segments can compromise structural integrity and pose risks to workers and long-term lining performance.

Fire and electrical hazards are also significant in TBM operations. High-voltage electrical systems, pressurised hydraulic circuits, and flammable materials such as lubricants and conveyor belts create potential ignition sources. Underground fire presents severe evacuation challenges. Overheated motors, cable faults, hydraulic oil leaks, and friction-generated heat are common contributors. Effective fire detection, suppression systems, compartmentalisation, and emergency procedures are critical.

Spoil handling and logistics systems—such as screw conveyors, belt conveyors, slurry pipelines, and rail haulage—introduce mechanical and pressure-related hazards. Risks include entanglement, belt rupture, pipeline overpressure, and sudden release due to blockages. These systems must be properly interlocked with TBM controls to prevent overload and unintended restart.

Settlement and surface impact hazards, though not immediately visible underground, are critical from a public safety perspective. Excessive settlement can damage buildings, utilities, and transport infrastructure. Settlement often reflects upstream operational imbalance, such as insufficient face pressure, inadequate tail void grouting, or excessive advance rate. Surface and subsurface monitoring systems therefore function as integral safety controls, not merely compliance tools.

Alignment deviations and breakthrough risks also present operational challenges. Significant deviation from the designed alignment can result in missed shaft connections, excessive structural stress, or clashes with stations and cross passages. Corrections must be gradual and controlled, as aggressive steering increases ground disturbance and settlement risk.

Human factors amplify many of these hazards. Inadequate interpretation of monitoring data, delayed response to warning signs, poor communication between shifts, and inconsistent parameter control can escalate manageable deviations into serious incidents. Modern TBM safety depends heavily on real-time data acquisition and disciplined, informed operational response rather than reactive correction.

Effective hazard management integrates face pressure control, spoil conditioning optimisation, torque and thrust monitoring, settlement surveillance, mechanical condition monitoring, redundant guidance systems, and comprehensive emergency planning. Risk is controlled by maintaining balanced system behaviour rather than maximising production speed.

Ultimately, TBM operational hazards illustrate that mechanised tunnelling is a dynamic control process. The machine, ground, lining, and monitoring systems operate as a single interconnected system. Most serious incidents arise not from a single failure but from cumulative imbalance—pressure drift, conditioning degradation, unnoticed torque trends, or gradual settlement increase. Safe TBM operation depends on maintaining parameters within defined envelopes, detecting trends early, applying small incremental corrections, and integrating geotechnical, mechanical, and structural controls. In modern tunnelling, operational safety is inseparable from technical performance; the safest drives are those where the system remains balanced, predictable, and continuously monitored throughout excavation.

Figure 99: San Francisco Muni Central Subway Tunnel Boring Machine tour, May 30, 2013. Eric Fischer, CC BY 2.0, via Wikimedia Commons.

Confined Space Risks

Tunnel boring operations, particularly those utilizing Tunnel Boring Machines (TBMs), are beset by a unique configuration of risks attributable to their confined operational environment. The inherent characteristics of TBM tunnelling, such as restricted access, limited ventilation, heavy mechanical systems, and dynamic ground conditions, introduce specific safety hazards that necessitate meticulous risk management frameworks.

Definition of Confined Spaces in TBM Operations

A confined space within the context of tunnelling is commonly defined as any enclosed or partly enclosed area that is not designed for continuous occupancy and that can give rise to hazardous conditions. Within a TBM operation, such spaces include various critical areas like the excavation chamber during interventions, the cutterhead housing, screw conveyor casings, and other enclosed areas which pose multiple atmospheric, mechanical, and operational risks [395, 662]. For instance, these areas are often characterized by the accumulation of hazardous gases, mechanical entrapments, and restricted escape routes, all of which amplify the risks associated with confined spaces in tunnelling operations [450, 663].

Atmospheric Hazards

One of the primary concerns in confined space operations is atmospheric hazards. There is a significant risk of oxygen deficiency, which can occur from the displacement of air by various gases generated during excavation processes or from chemical reactions in the surrounding geology [434, 450]. Specific hazardous gases—including methane, carbon dioxide, hydrogen sulphide, and carbon monoxide—can be released during tunnelling, some of which present serious explosion or toxicity risks [392, 664]. Methane, for example, can lead to explosive atmospheres in the absence of effective ventilation, while hydrogen sulphide poses lethal health risks even at low concentrations [665, 666]. To mitigate these hazards, continuous gas monitoring and forced ventilation systems are critical, ensuring that oxygen levels are maintained and harmful gases diluted to within safe limits [667, 668].

Mechanical Hazards

Mechanical risks are also substantial in the confined spaces of a TBM. Workers are frequently in close proximity to rotating cutterheads, high-torque drives, conveyors, and other heavy machinery, where restricted access can exacerbate the consequences of incidents such as entanglement, crushing, or pinching injuries [326, 669]. Specific hazards such as the jamming of TBM cutterheads, often linked to geological unpredictability, require rigorous risk evaluations and proactive management strategies [497, 670]. Moreover, the operation of mechanical equipment in such confined spaces demands stringent adherence to lockout–tagout procedures, interlocked guarding systems, and permit-to-work controls to safeguard worker safety during maintenance and operational activities [662, 671].

Fire and Heat Stress Risks

Fire risks in confined environments are particularly acute due to the accumulation of flammable materials and the presence of high-voltage electrical systems [663, 672]. The design of TBMs often includes integrated fire detection and suppression systems, alongside emergency refuge chambers to manage potential fire hazards effectively [670, 672]. Alongside fire hazards, heat stress poses a significant risk, especially in deeper tunnels or warm geological settings. The operation of machinery and the ambient ground conditions can lead to excessive heat accumulation, necessitating effective ventilation, hydration protocols, and monitored work-rest cycles to prevent heat-related illnesses among workers [320, 434].

Human Factors

Human factors also play a critical role in the context of confined space risks. Conditions such as fatigue, situational unawareness, and communication breakdown between shifts can significantly compromise safety efforts [664, 672]. The reliance on real-time monitoring systems and enhanced communication protocols is fundamental in maintaining operational safety and fostering a responsive work environment that can adapt to the changing conditions typical of TBM tunnelling [662, 669].

High-Pressure Systems

High-pressure systems in Tunnel Boring Machine (TBM) operations are fundamental to excavation control but they are also among the most serious sources of risk. These systems include face support pressure (EPB or slurry), hydraulic thrust and articulation circuits, slurry transport pipelines, compressed air systems, grout injection systems, and high-pressure lubrication lines. Because TBMs operate by balancing and controlling large forces underground, any loss of containment, imbalance, or uncontrolled release of pressure can escalate rapidly into mechanical failure, ground instability, or personnel injury.

High-pressure systems in TBMs generally operate in three main domains: excavation support pressure, hydraulic/mechanical pressure, and fluid transport pressure. Each carries distinct hazards, but they are interlinked. A failure in one domain often propagates into others.

Face support pressure hazards are central in EPB and slurry TBMs. In EPB machines, the excavation chamber is filled with pressurised conditioned soil. In slurry machines, hydrostatic slurry pressure balances earth and water pressures. If chamber pressure drops below the required stability threshold, ground collapse or water inflow may occur. Conversely, excessive pressure can cause blow-outs, hydraulic fracturing, or surface heave. The hazard is therefore not simply “high pressure” but imbalance—pressure outside the defined operational window. Sudden pressure loss may result from over-extraction through the screw conveyor, slurry loss into permeable ground, filter cake breakdown, or mechanical failure of seals. Sudden over-pressure may arise from aggressive adjustments, blocked extraction systems, or incorrect slurry density control.

Slurry transport systems present additional high-pressure risks. Slurry pipelines may operate at several bar of pressure to transport excavated material to the separation plant. Pipeline rupture, flange failure, or valve malfunction can result in high-velocity slurry discharge, posing both injury and flooding risks. Cavitation in slurry pumps can cause mechanical damage and unpredictable pressure fluctuations. High-permeability ground conditions increase the risk of slurry loss and pressure instability at the face, potentially leading to surface expression of slurry.

Hydraulic systems on TBMs operate at extremely high pressures, often in the range of 250–350 bar or more. These systems power thrust cylinders, articulation rams, erectors, and cutterhead drives. Hydraulic hose failure or fitting rupture can produce high-velocity fluid jets capable of penetrating skin (hydraulic injection injury), which is a severe medical emergency. Hydraulic leaks also increase fire risk, particularly in confined underground spaces with hot surfaces and electrical equipment.

Compressed air systems introduce further hazards. In EPB machines, compressed air may be used to support the face during interventions. In hyperbaric maintenance, workers enter pressurised environments. Sudden decompression, seal failure, or manlock malfunction can create dangerous pressure transitions. Compressed air also increases fire intensity due to elevated oxygen availability and airflow.

Grouting systems, particularly two-component tail void grouting, operate under controlled injection pressure. Excessive grout pressure may overstress newly erected segments or induce ground heave. Insufficient pressure leads to incomplete void filling and settlement. Grout lines themselves are pressurised systems and may rupture if blocked or improperly maintained.

High-pressure lubrication systems are often used at the shield skin to reduce friction in squeezing ground. Failure of lubrication delivery can increase shield friction dramatically, raising thrust demand and potentially overloading segments or cylinders.

Mechanical overpressure and overload are indirect high-pressure hazards. Excessive thrust applied through segmental linings can cause localised crushing or cracking. Excessive cutterhead torque may overload gearboxes or bearings. These conditions often arise when pressure imbalance at the face increases resistance, demonstrating the systemic nature of TBM high-pressure risks.

Personnel exposure risks are amplified by the confined underground environment. High-pressure leaks, slurry releases, or hydraulic bursts may not have safe discharge paths. Emergency response is complicated by limited access and evacuation constraints. In hyperbaric environments, pressure-related medical risks such as decompression sickness or barotrauma are significant.

Risk management of high-pressure systems in TBMs relies on layered engineering controls. These include pressure sensors with redundancy, automatic interlocks, relief valves, overpressure alarms, real-time monitoring of thrust and torque, and defined operational pressure envelopes. Pressure changes are applied gradually, and extraction rates are coordinated with face support levels to prevent imbalance. Hydraulic lines are regularly inspected, protected from mechanical damage, and fitted with burst protection sleeves. Slurry systems include pressure monitoring at multiple points along pipelines to detect abnormal losses.

Procedural controls are equally important. Operators are trained to interpret pressure trends rather than single readings. Sudden pressure changes are treated as warning signals. Permit systems govern maintenance on pressurised systems, including lockout–tagout and depressurisation protocols.

Engineering significance lies in understanding that TBM tunnelling is fundamentally a controlled pressure balance system. The machine advances only because forces are applied and resisted in equilibrium. High-pressure systems make this possible—but they also concentrate energy in confined spaces. Safe TBM operation depends on maintaining those pressures within defined limits, detecting deviations early, and ensuring that no single failure can cascade into uncontrolled ground movement or mechanical damage.

In essence, high-pressure systems in TBMs are both the enabling mechanism of mechanised tunnelling and one of its primary hazard domains. Managing them effectively requires continuous monitoring, disciplined operation, robust mechanical design, and integration between geotechnical understanding and mechanical control.

Fire, Flooding, and Gas Hazards

Fire, flooding, and gas hazards represent some of the most serious life-safety risks in Tunnel Boring Machine (TBM) operations because they occur within confined underground environments where escape routes are limited, travel distances are long, and personnel depend heavily on engineered systems such as ventilation, power supply, and communications. Unlike many surface construction incidents, underground events can escalate rapidly due to restricted access, reduced visibility, and the interaction between mechanical systems and ground conditions. In TBM tunnelling, these hazards must be understood as system-based rather than isolated risks.

Real-world TBM incidents involving fire, flooding, and gas demonstrate how underground hazards develop from system imbalance, equipment failure, or unexpected ground conditions. While mechanised tunnelling is generally very safe compared to historical methods, these events show why modern TBM operations rely heavily on monitoring, redundancy, and disciplined response.

Fire Incidents

<u>Seattle SR 99 Tunnel Fire (2016)</u>

In 2016, while constructing the Alaskan Way Viaduct Replacement Tunnel in Seattle, a significant fire broke out due to the ignition of a conveyor belt and associated components. Contributing factors included friction and mechanical overheating of equipment, the presence of combustible materials, and confined ventilation conditions, which posed serious smoke propagation issues. Although workers were evacuated safely, the incident led to a suspension of operations and highlighted the need for enhanced safety measures, including the adoption of fire-resistant materials, improved automatic suppression systems, and stringent housekeeping practices underground [376, 673].

Channel Tunnel TBM Fire (1987)

A fire incident during excavation works in the Channel Tunnel (UK–France) showcased how quickly underground fires could escalate due to air flow driven by ventilation systems. Although the fire did not occur directly in the TBM chamber, its consequences prompted significant changes in fire safety standards, including stricter compartmentalization measures, enhanced fire-resistance ratings for cables, and improved smoke control protocols during evacuation planning [636, 674].

Flooding Incidents

Guangzhou Metro TBM Inrush (2005)

During the construction of the Guangzhou Metro in 2005, a TBM encountered an unexpected water-bearing sand layer, leading to sudden inflows of water and sand. This incident caused localized ground collapse and surface settlement, resulting in notable project delays. The failure underscored the importance of accurate hydrogeological modelling and the need for proactive measures such as grouting and pressure maintenance to mitigate risks associated with high-permeability zones [36, 675].

Kuala Lumpur SMART Tunnel Flooding

In Malaysia, the SMART tunnel construction highlighted critical challenges in groundwater management during excavation phases. Although designed as a flood bypass system, maintaining hydrostatic balance in mixed-face and karstic conditions proved complex. This incident underlined the necessity of real-time pressure monitoring systems to respond effectively to potential flooding and the importance of redundant pumping capacity in construction practices [69, 676].

Gas Incidents

Los Angeles Metro Methane Explosion (1985)

The accumulation of methane in underground works posed a catastrophic risk during the Los Angeles Metro tunneling operations in 1985, culminating in an explosion upon encountering an ignition source. Insufficient detection systems and the absence of explosion-proof equipment were identified as primary causes, prompting California to enforce stricter regulations for methane zones, which subsequently led to the adoption of explosion-proof TBM systems and mandatory continuous gas monitoring [677, 678].

Cologne Subway Collapse and Gas Interaction (2009)

In Cologne, during the construction of the North-South Stadtbahn, a structural collapse occurred, exacerbated by interactions between groundwater migration and gas migration. Although primarily a structural failure linked to slurry wall issues, the incident demonstrated how gas migration pathways could escalate risks in urban tunnelling, emphasizing the necessity of comprehensive risk assessments in gassy and water-laden environments [64, 679].

Hyperbaric and Pressure-Related Incidents

Brightwater Conveyance TBM Hyperbaric Fatality (2006)

A tragic incident occurred in 2006 during a hyperbaric intervention on a slurry TBM in the Brightwater Conveyance Tunnel, resulting in a worker fatality. Contributing factors included complex compressed air procedures, elevated working pressures, and medical complications during decompression. This incident profoundly impacted hyperbaric protocols, promoting greater oversight and improvements in medical procedures associated with TBM operations [680, 681].

Patterns Observed Across Incidents

Across these various incidents involving fire, flooding, and gas, several recurring themes can be observed:

1. **Early Warning Signs**: Many incidents exhibited early warning signs, such as pressure drift, torque spikes, and abnormal gas readings, indicating the importance of continuous monitoring [682, 683].
2. **System Interactions**: The interplay of multiple systems, where water inflow can damage electrical systems posing fire risks, was evident across incidents, stressing the need for integrated risk management [684, 685].
3. **Human Response**: The human element, including adherence to protocols and procedural discipline, significantly influenced the severity of the incidents. Effective communication and training among workers can mitigate potential hazards significantly [481, 686].

Fire is one of the most critical underground threats because smoke and toxic gases can spread quickly through tunnels. Common ignition sources include high-voltage trailing cables, transformers, overheated motors, damaged insulation, and hydraulic oil leaks under pressure. Mechanical friction from misaligned conveyor belts, overheated bearings, or cutterhead gearboxes can also generate sufficient heat for ignition. The presence of lubricants, greases, conveyor belts, timber packing materials, and, where applicable, diesel fuel increases the fire load. Underground, heat and smoke accumulate rapidly, ventilation systems may unintentionally feed oxygen to the fire, and visibility can fall to zero within minutes. In many cases, smoke inhalation presents a greater danger than flame exposure.

Fire risk is controlled through a combination of engineering and procedural measures. Engineering controls include heat and smoke detection systems, automatic fire suppression systems such as water mist or foam, flame-retardant hydraulic fluids, fire-resistant cables, and physical separation of electrical and hydraulic systems. Compartmentalisation within the TBM backup further limits fire spread. Procedural controls involve strict housekeeping, hot-work permitting, routine inspection of belts and hydraulic systems, and regular emergency evacuation drills. Effective fire management depends on early detection and rapid isolation.

Flooding hazards arise from groundwater inrush, slurry system failures, pipe ruptures, or surface water intrusion through shafts. Face instability caused by under-pressure in EPB machines, slurry loss in permeable soils, or breakdown of the filter cake in slurry TBMs can allow uncontrolled water inflow. High hydrostatic pressures beneath rivers or coastal areas increase the risk significantly. System failures such as pump breakdown, pipeline rupture, or seal failure at the tail shield can also lead to rapid water ingress. In confined underground conditions, flooding can result in loss of face support, ground collapse, electrical damage, equipment loss, and potential entrapment of personnel. In high-pressure environments, water inrush may occur suddenly and with considerable force.

Flood control relies on maintaining face pressure within the calculated stability window, continuous groundwater monitoring, redundant pumping systems, automated pressure and flow alarms, and, where designed, emergency bulkheads or closure gates. Controlled depressurisation procedures are essential in slurry systems. In high-risk zones, pre-treatment methods such as grouting or ground freezing may be implemented to reduce permeability and strengthen the ground before excavation. As with fire, early detection and system balance are fundamental to preventing escalation.

Gas hazards originate from both geological formations and operational processes. Methane, hydrogen sulphide, carbon monoxide, carbon dioxide, and oxygen-deficient atmospheres are among the most significant threats. Methane is flammable and explosive within certain concentration ranges and may accumulate in high points of tunnels. Hydrogen sulphide is toxic even at low concentrations, particularly in organic-rich soils. Carbon monoxide can result from incomplete combustion, while carbon dioxide may displace oxygen in poorly ventilated areas. Oxygen deficiency itself can occur where other gases displace breathable air. These conditions can lead to explosion, toxic exposure, impaired judgement, or asphyxiation.

Gas management depends on continuous monitoring through fixed detection systems and portable personal gas detectors, robust forced ventilation designed for worst-case release scenarios, and explosion-proof electrical equipment where required. In gassy ground, methane drainage systems may be installed. Strict no-ignition policies apply in hazardous zones. Ventilation systems are engineered to control airflow direction and maintain gas concentrations below defined safety thresholds.

These hazards are closely interrelated. Flooding may damage electrical systems and increase fire risk. Gas accumulation combined with electrical faults can lead to explosion. Fire can disable pumps or ventilation systems, worsening flooding or gas build-up. Slurry loss may destabilise the ground, increasing both water inflow and gas migration pathways. Because TBM operations are integrated systems, hazard management must also be integrated, combining geotechnical control, mechanical reliability, electrical safety, and environmental monitoring.

Effective hazard management relies on real-time environmental monitoring, integrated control systems, redundant life-support infrastructure, defined evacuation routes and refuge chambers, reliable communication systems, and regular emergency drills. Projects establish trigger levels for gas concentration, water level, temperature rise, smoke detection, and pressure deviation. Early detection, disciplined response, and maintenance of system balance are essential to preventing minor deviations from developing into major underground emergencies.

Emergency Response Planning

Emergency Response Planning (ERP) in tunnel boring machine (TBM) operations is essential for navigating the unique challenges presented by confined underground environments. These operations are marked by low-frequency yet high-consequence events, necessitating well-structured systems for effective risk management. Unlike surface construction, emergencies in tunnelling present complications such as long travel distances, limited access points, and the interdependencies of mechanical, electrical, hydraulic, and geotechnical systems. Therefore, effective ERP must integrate into project design from the outset rather than being merely procedural.

The primary objectives of ERP in TBM operations include the protection of life, hazard stabilization, the maintenance of critical systems, and the coordination of safe evacuations or rescues. TBM operations must prepare for possible incidents such as fires, flooding, gas releases, or structural failures. Effective ERP aims to establish immediate responses based on predefined protocols to avert improvisation under stress, thus ensuring rapid and coordinated action during crises [687, 688].

The ERP process begins with rigorous risk identification through hazard identification (HAZID) and risk assessment (RA). This process involves evaluating credible worst-case scenarios, including fire occurrences, slurry pipeline ruptures, methane ignition, sudden groundwater inrushes, air pressure incidents, or loss of ventilation [689, 690]. For each scenario, specific protocols are laid out regarding detection systems, evacuation routes, communication strategies, and coordination with external emergency services. This proactive approach minimizes reaction times and maximizes safety [20, 691].

Components of TBM Emergency Planning

Life-Support Systems: A key aspect of ERP is ensuring the reliability of life-support systems, including ventilation, power supply, lighting, and pumping systems. These systems require redundancy to operate effectively under emergency conditions. Strategically located refuge chambers equipped with compressed air or oxygen supplies, emergency rations, and communication links serve as temporary safe havens when immediate evacuation is not feasible [692, 693].

Fire Response Planning: Fire response protocols are particularly critical in underground settings where smoke is a primary hazard. Effective ERP includes the installation of early detection systems and automatic suppression systems. Measures such as electrical zone isolation and controlled shutdown of non-essential systems can mitigate the risk posed by fires [694, 695]. Moreover, evacuation strategies must consider ventilation control to direct smoke away from escape routes, enhancing safety for workers and emergency responders [696, 697].

Flooding and Gas Emergency Management: Flood response strategies focus on rapid stabilization through trigger-level guidelines for unexpected water influx or pump failures. The incorporation of emergency bulkheads and backup pumps are common strategic responses to flooding events [690, 698]. In the context of gas emergencies, the continuous monitoring of atmospheric conditions is vital. Defined alarm thresholds for various gases inform necessary actions such as power isolation and ignition control, followed by staging evacuations and reconfiguring ventilation systems for safety [20, 699].

Hyperbaric Scenarios: Hyperbaric emergencies require highly specialized responses, including immediate access to medical protocols and trained personnel. The development of controlled decompression procedures is paramount in case of equipment failure during mixed-gas interventions; coordination with surface medical facilities must also be rehearsed to ensure timely, effective responses [690, 700].

Communication and Drills

Communication is pivotal in any ERP, particularly in the complex underground environment. Ensuring reliable, continuous communication through underground radio systems and backup powers helps maintain critical information flow during crises [697, 701]. Additionally, regular training and unannounced emergency simulations are essential to test and improve the efficacy of response protocols, analysing weaknesses and updating procedures based on drill outcomes [20, 688].

Integration of Modern Technologies

The incorporation of real-time monitoring and data acquisition systems into ERP protocols allows for automated alerts when predefined thresholds are exceeded. This technology reduces reliance on human recognition, enabling faster responses to emergent situations [690, 692].

Safety Management Systems

A Safety Management System (SMS) for Tunnel Boring Machines (TBM) represents a crucial framework designed to enhance safety in high-hazard environments typical of underground construction. The SMS is systematic, documented, and continuously monitored, integrating engineering controls, operational procedures, human factors, and compliance with regulatory requirements. In mechanized tunnelling, safety is not attained through isolated rules; instead, it is achieved through disciplined system control across various interconnected components and processes.

An effective TBM SMS focuses on ensuring that components such as face pressure control, ground conditioning, electrical integrity, confined space management, and emergency planning operate as a coordinated safety architecture rather than as unconnected systems. This comprehensive approach aligns with international safety frameworks, such as ISO 45001, while being tailored to tackle the unique challenges of pressurized excavation, variable ground conditions, and confined underground workplaces [702, 703].

While standardized frameworks provide a baseline, TBM SMS requires significant customization. The inherent risks, such as those posed by fluctuating ground conditions and the complex dynamics within a tunnelling machine, necessitate a SMS that can adeptly handle small deviations in machine or ground behaviour, which, if unnoticed, can lead to serious incidents [704, 705].

The cornerstone of any SMS is structured hazard identification, typically conducted through HAZID (Hazard Identification) and HAZOP (Hazard and Operability study) workshops. These processes enable project teams to anticipate risks associated with various operational scenarios, such as the possibility of face pressure loss or slurry inrush when operating an Earth Pressure Balance (EPB) TBM [706]. For example, prior to launching a TBM beneath a river, teams identify critical risks and develop strategies, including pressure window analyses that define operational limits [707, 708].

By integrating safety thresholds into TBM control systems, theoretical geotechnical risks become operational imperatives. This approach allows for immediate alarm notifications and actions if pressure deviations occur, transforming statistical risk data into enforceable operational limits [709, 710]. Such a proactive stance on risk management ensures that the SMS can respond dynamically to real-time data—treating ongoing monitoring and intervention as critical components of safety.

Modern TBM SMS are heavily data-driven, focusing on real-time monitoring of various parameters, including face pressure, thrust, torque, and gas levels. This continuous data stream allows for early recognition of potentially dangerous trends and patterns, which is essential for preventing catastrophic failures [711, 712]. For instance, simultaneous spikes in torque with reduced screw speed can indicate escalating risk, prompting immediate escalation procedures before an incident occurs [713].

Defining operational envelopes—or parameter windows—is a vital aspect of maintaining safety in TBM operations. For example, a project might establish safe face pressure limits within specified bounds to prevent ground collapse or blow-out events [714]. The SMS enforces operational parameters by mandating technical reviews for deviations outside defined limits and implementing controls to halt operations if necessary [715].

Permit-to-work systems are essential within TBM projects, particularly when conducting high-risk tasks like hyperbaric interventions or maintaining hydraulic systems. Each designated task requires documented risk assessments, isolation verifications, and supervisory approvals to ensure that safety evaluations have been meticulously completed [716, 717]. This prevents informal decision-making and enforces a culture of safety awareness among all operational personnel.

Particularly regarding hyperbaric conditions, a specialized subsystem within the SMS addresses unique hazards and physiological risks. Strict protocols regarding medical clearance, approved decompression schedules, and standby rescue team obligations are critical for ensuring worker safety during high-pressure interventions [708, 718].

A TBM Safety Management System stands apart from general construction safety systems due to its parameter-controlled, geotechnically integrated nature and sensitivity to early signs of operational imbalance. By comprehensively addressing risks systematically and through real-time monitoring, these systems can effectively prevent serious incidents, ensuring the safety of teams working within mechanized tunnelling environments. The effective management of TBM operations hinges on recognizing and correcting minor deviations promptly, maintaining the balance within the complex machine–ground–lining dynamics.

Human Factors and Fatigue

Human operators are tasked with interpreting complex data streams that include face pressure, torque, thrust, screw speed, slurry density, settlement readings, and gas levels. Mastery over these parameters necessitates not just monitoring but skilled pattern recognition and judgment [719, 720]. Research shows that fatigue can severely impair cognitive performance, leading to decreased situational awareness, slower reaction times, and suboptimal decision-making (e.g., delayed responses to pressure changes or misinterpretation of torque spikes) which are critical in a TBM setting [721, 722].

The socio-technical nature of TBM operations means that decisions made under fatigue frequently escalate operational risks. Team communication, a key human factor, can break down, particularly during shift handovers. Poorly documented changes in parameters or anomalies can lead to context loss, as observed in various industries, underscoring the importance of systematic communication protocols [723, 724].

Fatigue emerges as a primary risk factor in TBM operations due to the continuous nature of these projects, which often require 24/7 work with rotating shifts, leading to sleep debt accumulation and reduced alertness, especially during night shifts [724, 725]. The environmental conditions of tunnel operations, characterized by limited natural light, high noise levels, and pressure, further exacerbate fatigue [721, 722]. It is not merely physical fatigue that poses threats—cognitive fatigue plays a substantial role in impairing the operator's ability to detect subtle signs that could signal potential danger. Studies demonstrate that fatigue can lead to inaccurate responses to critical alarms and a normalization of deviant behaviour concerning operational parameters, often resulting in accidents [724, 726, 727].

The handling of complex data inputs in TBM operations creates significant cognitive load, particularly when operators must attend to multiple simultaneous signals while anticipating the effects of adjustments made in real-time [728, 729]. Poorly designed human-machine interfaces (HMI) can cloud data clarity, increasing the risk of errors as operators may miss crucial signals during operational shifts. Effective HMI designs should focus on clarity, prioritizing critical data and presenting it in an intuitive manner to mitigate cognitive overload, thereby enhancing situational awareness [723, 730].

The culture within which TBM operations occur plays a crucial role in managing human factors and mitigating risks associated with fatigue [731, 732]. A robust safety culture that encourages the reporting of fatigue, anomalies, and near-misses without stigma can significantly enhance safety outcomes. Supervisors must cultivate an environment where operational pauses can be authorized when parameters deviate from safe operating boundaries [724, 733]. Optimally, organizations should implement formalized protocols and training systems aimed at enhancing trend recognition, operational clarity, and communication efficiency, which are vital for effective responses to potential risks [729, 731, 734].

To address the multifaceted challenges posed by human factors in TBM operations, organizations can adopt systematic approaches to risk management. This includes protocols for fatigue risk management, structured rest breaks, and monitoring working hours to ensure that workers are fit for work [721, 726]. Training programs should extend beyond procedural compliance to emphasize critical thinking, situational awareness, and data interpretation, thereby ensuring that personnel are equipped to manage the cognitive demands of their environment effectively.

Chapter 15

Environmental and Community Impacts

Chapter 15, *Environmental and Community Impacts*, expands the focus of tunnelling from purely technical performance to the broader interface between underground construction and the urban environment. While previous chapters examine machine control, ground behaviour, and structural response, this chapter addresses how those same engineering processes manifest at the surface—through settlement, vibration, groundwater change, noise, spoil handling, and community perception.

Mechanised tunnelling does not occur in isolation. Every adjustment to face pressure, conditioning ratio, advance rate, or grouting regime has a corresponding environmental consequence. Surface movement reflects subsurface volume loss. Groundwater pressure changes mirror excavation balance. Noise and vibration reveal mechanical interaction between machine and ground. In dense urban settings, these effects are measured not only in millimetres and kilopascals, but in regulatory thresholds, public safety margins, and community confidence.

This chapter therefore frames environmental management as an extension of system control. Settlement prediction using Gaussian models links geotechnical theory to real-world risk. Differential movement analysis connects millimetre-scale deformation to structural damage mechanisms. Groundwater monitoring integrates hydrogeology with operational pressure control. Noise, vibration, and spoil logistics demonstrate that environmental performance is inseparable from mechanical efficiency.

The chapter emphasises that mitigation is proactive rather than reactive. Baseline monitoring, trigger–action–response frameworks, hydrogeological modelling, and transparent reporting transform environmental protection into a structured engineering process. GNSS, InSAR, piezometers, geophones, and digital dashboards are not supplementary tools—they are core components of risk management architecture.

In modern urban tunnelling, technical success is measured not only by breakthrough accuracy or advance rate, but by how effectively the excavation remains in equilibrium with the surrounding city. Chapter 15 explores that equilibrium—showing how environmental stewardship, community engagement, and precise TBM control converge to deliver infrastructure safely beneath complex urban environments.

Surface Settlement and Vibration

Surface settlement and vibration are two of the most significant ground-response effects associated with mechanised tunnelling. Although Tunnel Boring Machines (TBMs) are designed to minimise disturbance compared with conventional excavation methods, the creation of an underground void inevitably alters the in-situ stress state of the surrounding soil or rock. These stress redistributions may propagate upward to the surface or laterally to adjacent structures, utilities, and sensitive infrastructure. For this reason, the control of settlement and vibration is a central objective of geotechnical design, TBM operation, and integrated monitoring systems.

Surface settlement occurs when ground moves downward as a result of volume loss around the tunnel. In TBM operations, volume loss may arise from insufficient face support pressure allowing ground to move inward (face loss), over-excavation or inadequate lubrication around the shield (shield loss), incomplete or delayed annular grouting behind the lining (tail void loss), long-term consolidation of cohesive soils, or groundwater drawdown that reduces pore pressure and induces compression. Even relatively small volume losses at tunnel depth can translate into measurable settlement at the surface, particularly in soft or compressible soils.

Settlement from a single-bore tunnel typically forms a Gaussian-shaped trough at the surface. The trough geometry depends on tunnel depth, diameter, and ground stiffness. Shallow tunnels tend to produce narrower and deeper settlement troughs, while deeper tunnels generate wider and flatter profiles. Maximum settlement generally occurs above the tunnel axis, with settlement decreasing laterally away from the centreline. In practice, empirical relationships are used to estimate settlement based on tunnel diameter, depth to axis, volume loss percentage, and soil stiffness. Among these variables, volume loss is often the primary operational control parameter.

A common way to *calculate* the “Gaussian settlement trough” above a single-bore TBM tunnel is the Peck formulation:

$$S(x) = S_{\max} \exp\left(-\frac{x^2}{2i^2}\right)$$

Where:

- $S(x)$= settlement at horizontal offset $x$from tunnel centreline
- $S_{\max}$= maximum settlement (at $x = 0$)
- i= trough width parameter (controls how "wide" the trough is)

A widely used empirical link for trough width is:

$$i = K\, z_0$$

Where:

- z_0= depth to tunnel axis (m)
- K= empirical constant (depends on ground; often ~0.25–0.35 in sands/gravels, ~0.4–0.6 in clays—project dependent)

Step 1 — Relate settlement trough area to volume loss

Volume loss VL(as a fraction) is the key driver. For a tunnel of diameter D:

$$A = \frac{\pi D^2}{4} \Rightarrow V_s = VL \cdot A$$

Where:

- A= excavated area (m^2)
- V_s= volume loss per metre of tunnel ($m^3/m = m^2$)

For a Gaussian trough, the "area under the settlement curve" per metre of tunnel is:

$$V_s = \sqrt{2\pi}\, S_{\max}\, i$$

So:

$$S_{\max} = \frac{VL \cdot A}{\sqrt{2\pi}\, i}$$

Worked example (single bore)

Given

- TBM diameter: $D = 6.5$ m
- Depth to axis: $z_0 = 20$ m(shallow urban-ish)

- Ground constant: $K = 0.5$(representative "clayey/stiffer" behaviour)
- Volume loss: $VL = 1.0\% = 0.01$

1) Excavated area

$$A = \frac{\pi(6.5)^2}{4} = 33.18 \text{ m}^2$$

2) Volume loss per metre

$$V_s = 0.01 \times 33.18 = 0.3318 \text{ m}^2$$

3) Trough width parameter

$$i = Kz_0 = 0.5 \times 20 = 10 \text{ m}$$

4) Maximum settlement

$$S_{\max} = \frac{0.3318}{\sqrt{2\pi} \times 10} = \frac{0.3318}{2.5066 \times 10} = 0.01324 \text{ m} \approx 13.2 \text{ mm}$$

5) Settlement at offsets from centreline

Using

$$S(x) = S_{\max} \exp\left(-\frac{x^2}{2i^2}\right) \text{ with } S_{\max} = 13.2 \text{ mm and } i = 10 \text{ m:}$$

- $x = 0\text{m}$: $S(0) = 13.2\text{mm}$
- $x = 5\text{m}$: $S(5) = 13.2\, e^{-25/200} \approx 11.7\text{mm}$
- $x = 10\text{m}$: $S(10) = 13.2\, e^{-100/200} \approx 8.0\text{mm}$
- $x = 15\text{m}$: $S(15) = 13.2\, e^{-225/200} \approx 4.3\text{mm}$
- $x = 20\text{m}$: $S(20) = 13.2\, e^{-400/200} \approx 1.8\text{mm}$

So you get a peak settlement of ~13 mm over the tunnel axis, tapering off laterally in a bell-shaped (Gaussian) curve.

Showing "shallow = narrow/deep" vs "deep = wide/flat" (same tunnel, same VL)

Keep $D = 6.5\text{m}$ and $VL = 1\%$and $K = 0.5$, change only depth z_0:

- **Shallower** $z_0 = 12\text{m} \rightarrow i = 6\text{m} \rightarrow S_{\max} \approx 22.1\text{mm}$ (**narrower, deeper trough**)
- **Moderate** $z_0 = 20\text{m} \rightarrow i = 10\text{m} \rightarrow S_{\max} \approx 13.2\text{mm}$
- **Deeper** $z_0 = 30\text{m} \rightarrow i = 15\text{m} \rightarrow S_{\max} \approx 8.8\text{mm}$ (**wider, flatter trough**)

That's the key behaviour you described: greater depth increases i(wider trough) and therefore reduces $S_{\max}$for the same volume loss.

Why volume loss is "the" control knob

From $S_{\max} = \frac{VL \cdot A}{\sqrt{2\pi}\, i}$, you can see:

- $S_{\max}$increases linearly with VL(double $VL \rightarrow$ double $S_{\max}$)
- $S_{\max}$increases with tunnel size (bigger A)
- $S_{\max}$decreases as the trough widens (bigger i, often deeper tunnel / different soil response)

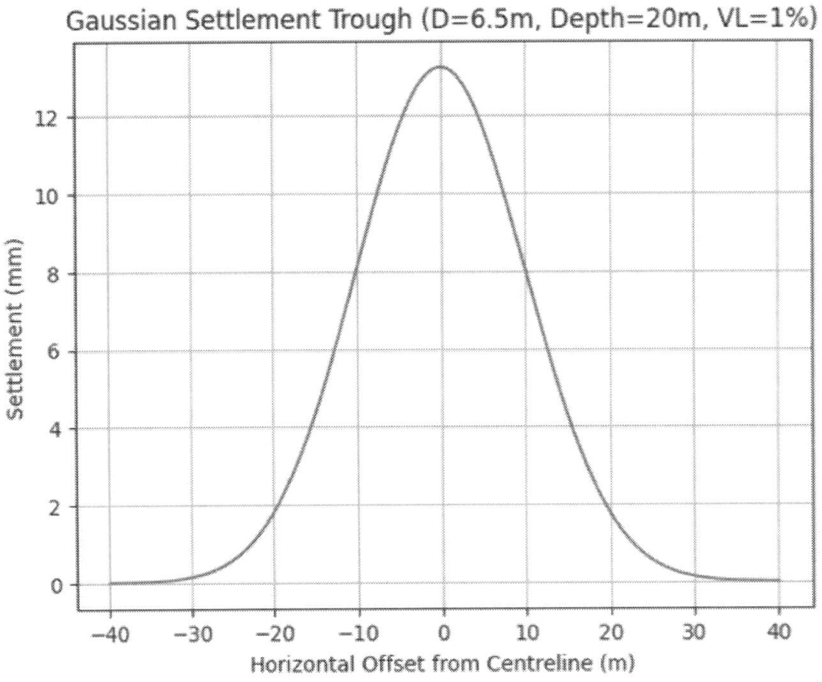

Figure 100: Gaussian Settlement Trough (D=6.5m, Depth=20m, VL=1%),

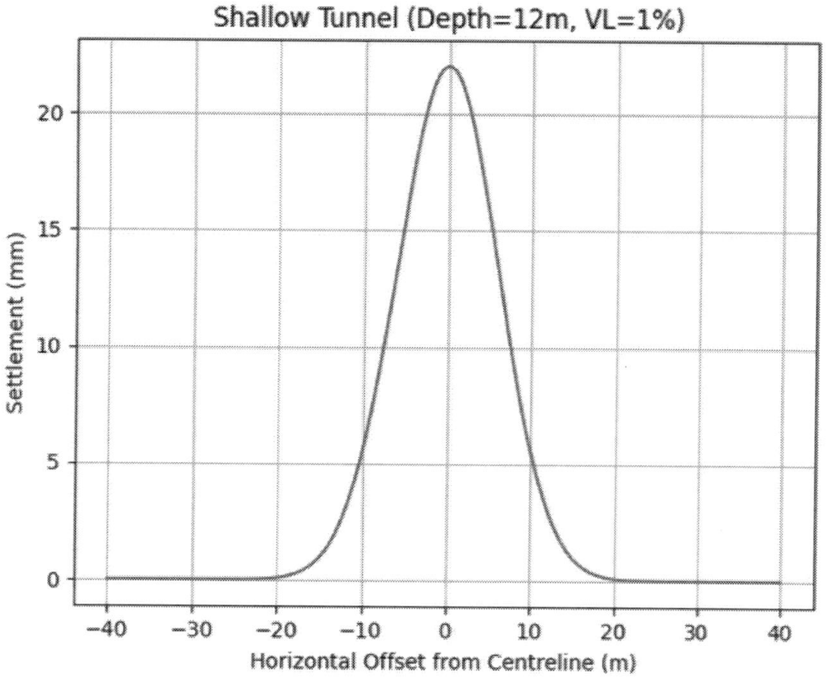

Figure 101: Shallow Tunnel (Depth=12m, VL=1%).

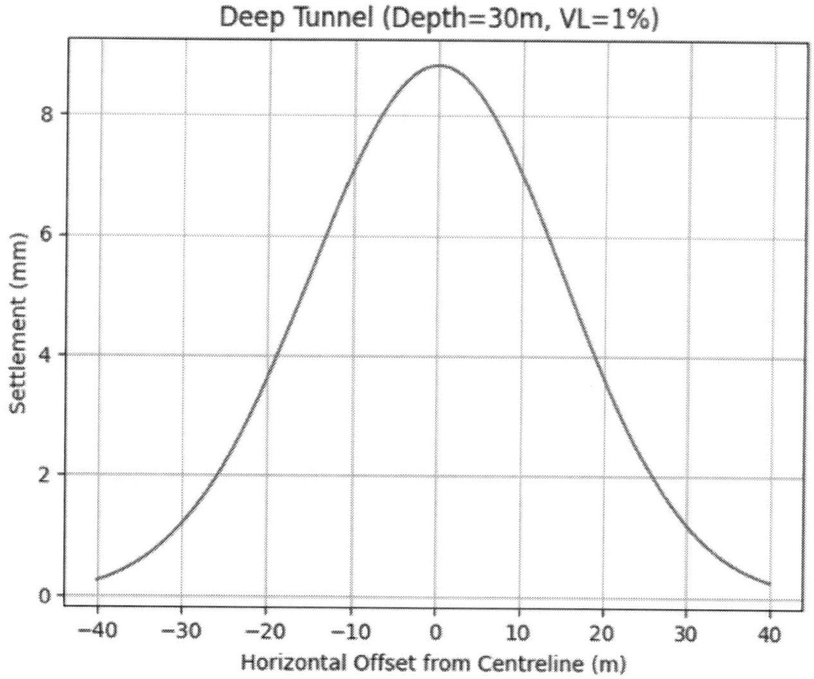

Figure 102: Deep Tunnel (Depth=30m, VL=1%).

The three diagrams above illustrate how the Peck Gaussian settlement formulation behaves for a 6.5 m diameter TBM with 1% volume loss in clayey ground (K = 0.5).

<u>Base Case (Depth = 20 m)</u>

The first diagram shows the classic Gaussian settlement trough at 20 m depth. The maximum settlement occurs directly above the tunnel axis (x = 0) and is approximately 13 mm. Settlement reduces smoothly and symmetrically away from the centreline, forming the characteristic bell-shaped curve.

This curve reflects the equation:

$$S(x) = Smax \cdot \exp(-x^2 / 2i^2)$$

where the trough width parameter $i = Kz_0$ controls how wide the settlement distribution spreads.

<u>Shallow Tunnel (Depth = 12 m)</u>

The second diagram shows the same tunnel and volume loss, but at a shallower depth of 12 m. Because:

$$i = Kz_0$$

the trough width reduces to 6 m. Since the same volume loss must be distributed over a narrower area, the maximum settlement increases to approximately 22 mm.

This produces:

- A narrower trough
- A steeper curvature
- A higher peak settlement

This is why shallow tunnels tend to create sharper and larger peak settlements for the same volume loss.

<u>Deep Tunnel (Depth = 30 m)</u>

The third diagram shows the same tunnel at 30 m depth. Now:

$$i = 15 \text{ m}$$

The settlement spreads over a much wider zone, and the peak settlement reduces to about 9 mm.

This produces:

- A wider trough
- A flatter surface profile

- Lower peak settlement

This demonstrates a fundamental tunnelling principle:

For the same tunnel diameter and volume loss, increasing depth widens the settlement trough and reduces maximum settlement.

The engineering interpretation of the Peck settlement formulation highlights three fundamental relationships that govern surface response above a single-bore TBM tunnel.

First, volume loss is the primary control variable. Maximum settlement increases directly and linearly with the percentage of volume loss (VL). If volume loss doubles, peak settlement also doubles. This linear relationship makes volume loss the most critical operational parameter in urban tunnelling. Regardless of depth or ground type, excessive volume loss will translate directly into increased surface movement.

Second, tunnel diameter influences settlement magnitude. A larger tunnel diameter increases the excavated cross-sectional area, which increases the potential volume of ground loss for the same percentage VL. Because maximum settlement is proportional to excavated area, larger diameter tunnels inherently carry higher settlement risk if volume loss is not tightly controlled.

Third, tunnel depth governs how settlement is distributed. Increasing depth increases the trough width parameter, i, which spreads settlement laterally across a wider area. As a result, deeper tunnels generally produce wider but flatter settlement troughs with lower peak values for the same volume loss. Conversely, shallow tunnels concentrate deformation into a narrower zone, resulting in higher peak settlement.

In practical TBM engineering, these relationships reinforce a clear operational objective: control volume loss. This is achieved through stable face pressure control, effective spoil conditioning, proper and timely annular grouting, and disciplined advance rate management. When these systems are balanced, volume loss remains within predicted limits and settlement remains manageable.

The Gaussian method provides a clear and intuitive link between TBM operational parameters and surface risk. It translates geotechnical behaviour into measurable outcomes and offers a practical framework for design prediction, monitoring interpretation, and real-time operational adjustment. For this reason, it remains one of the most widely applied settlement prediction tools in urban mechanised tunnelling.

In mechanised tunnelling, settlement is closely linked to TBM control parameters. In EPB and slurry machines, face pressure must be maintained within a defined pressure window: under-pressure allows ground to yield inward and increases settlement risk, while over-pressure may cause uplift or blow-out. Annular grouting must match the theoretical tail void volume; insufficient grout can lead to delayed settlement behind the shield, whereas excessive grout pressure may cause local heave. Advance rate and spoil conditioning also influence settlement behaviour, as unstable chamber pressure or poor flow characteristics can result in transient volume loss. Settlement, therefore, is not solely a geotechnical phenomenon but also an operational control issue.

Settlement may be categorised by timing. Immediate settlement occurs during TBM passage and is typically associated with face or shield effects. Short-term settlement may develop shortly after passage due to grout shrinkage or early consolidation. Long-term settlement may continue over months or years, particularly in clayey soils undergoing consolidation or in areas where groundwater conditions change. Understanding the timing of movement assists engineers in diagnosing its underlying cause.

1. Immediate Settlement

Immediate settlement occurs during the tunnelling operations, particularly during the passage of Tunnel Boring Machines. This form of settlement is primarily influenced by face and shield effects, which are directly related to the stresses applied during excavation. Immediate settlements can arise due to groundwater drawdown, which leads to increased seepage-induced ground settlement around the excavation site [735]. The excavation image created by the TBM can alter the pore pressures in the surrounding soil, thus directly inducing immediate settlement.

Furthermore, studies have shown that the interactions between groundwater levels and tunnelling operations can lead to sudden changes in effective stress, exacerbating immediate settlement [163]. Research corroborates this by highlighting that tunnelling below the water table often results in immediate complications, especially when dewatering techniques fail to maintain necessary groundwater levels, impacting ground stability during excavation [736]. The phenomenon exemplifies the urgent need for monitoring around TBM operations to predict and mitigate ground movements effectively.

2. Short-Term Settlement

Short-term settlement typically occurs shortly after the passage of the TBM. This type can be attributed to grout shrinkage or consolidation effects related to material behaviour in the immediate aftermath of excavation. Research indicates that ground settlement can be significantly influenced by material properties, such as those observed in residual soils [735]. Furthermore, ground movements can be caused by the consolidation of the newly excavated materials, particularly if they are saturated with water.

The importance of grout technology in mitigating short-term settlement is emphasized by Improper grout management could lead to significant ground instability shortly after tunnelling [737]. Additionally, the work by Yang et al. [738]supports this view by demonstrating how unplanned groundwater fluctuations can lead to changes in pore pressure, causing localized settling soon after TBM operation. Hence, careful monitoring and control of post-excavation conditions are critical to prevent excessive short-term settling.

3. Long-Term Settlement

Long-term settlement typically unfolds over months or even years, primarily due to consolidation of clay or changes in groundwater conditions. This phenomenon is significantly driven by the permeable nature of the surrounding soils, which undergo gradual changes in effective stress and pore water pressure over time, leading to ongoing settlement even after the completion of tunnelling activities. Research by Ochmański [739] illustrates how long-term settlement induced by earth pressure balance (EPB) tunnelling can have lasting impacts, with effects becoming evident significantly after construction activities.

In contrast to immediate and short-term settlements, long-term settlements can fluctuate based on external factors such as seasonal groundwater changes, as detailed by Zheng et al. [740] who highlighted that significant diaphragm wall deformations and ground settlements related to dewatering in leaky aquifers can extend well beyond the immediate construction phase. This underlines the necessity of considering both historical and future groundwater behaviours during planning and execution stages to adequately manage long-term settling effects.

The impact of settlement on structures depends largely on differential movement rather than total vertical displacement alone. Differential settlement may induce curvature and angular distortion, leading to cracking in masonry, distortion of frames, utility joint separation, or pavement deformation. Structural risk assessments therefore consider building vulnerability, foundation type, and allowable movement thresholds rather than relying solely on maximum settlement values.

The following provides a worked example that shows why differential movement (angular distortion and curvature) matters more than the total settlement number.

<u>Example setup</u>

A small masonry building is 20 m long (distance between stiff points/foundations). We'll compare two cases where the maximum settlement is the same (20 mm), but the damage risk is very different.

Case A — "Big settlement, but uniform" (low damage risk)

Assume the whole building settles almost evenly:

- Left end settlement = **20 mm**

- Right end settlement = **20 mm**
- Differential settlement, Δ = **20 − 20 = 0 mm**

Angular distortion (tilt/strain indicator)

$$\beta = \frac{\Delta}{L} = \frac{0}{20} = 0$$

Interpretation: Even though the building has moved down 20 mm, it hasn't been bent or racked. You might see serviceability issues (e.g., drains, steps, thresholds), but structural cracking risk is low because the building isn't being forced into curvature.

Case B — "Same max settlement, but differential" (higher damage risk)

Now assume settlement varies across the building:

- Left end settlement = **0 mm**
- Right end settlement = **20 mm**
- Differential settlement, Δ = **20 mm**

Angular distortion

Convert 20 mm to metres: 20 mm = 0.020 m

$$\beta = \frac{\Delta}{L} = \frac{0.020}{20} = 0.001 = \frac{1}{1000}$$

Interpretation: Now the building is effectively being "tilted/racked" at 1/1000. That level of angular distortion is in the range where cracking becomes plausible, especially for brittle materials (unreinforced masonry, plaster finishes), depending on stiffness, openings, and restraint.

Curvature example (sagging/hogging causes cracking)

Differential movement can also appear as sagging (middle settles more) which induces curvature.

Assume:

- Both ends settle **0 mm**
- The middle settles **20 mm** (a sagging bowl)
- Span $L = 20\ m$

- Mid-span deflection relative to ends $\delta = 0.020\, m$

For a simple parabolic sagging shape, curvature κcan be approximated by:

$$\kappa \approx \frac{8\delta}{L^2} = \frac{8(0.020)}{20^2} = \frac{0.160}{400} = 0.0004\, m^{-1}$$

Equivalent radius of curvature:

$$R = \frac{1}{\kappa} = \frac{1}{0.0004} = 2500\, m$$

Interpretation: That curvature means the building is being bent. Bending drives tensile strain in brittle elements (masonry, render), which is why cracks typically correlate with curvature/angular distortion, not the absolute settlement value.

This example demonstrates that total settlement alone does not determine damage risk: a uniform 20 mm settlement across a structure may pose relatively low structural risk because no significant strain or bending is induced, whereas the same 20 mm maximum settlement can become problematic if it occurs unevenly, producing differential movement or curvature (sagging or hogging) that generates angular distortion and tensile strain in structural elements. It is this differential deformation, not the absolute magnitude of settlement, that typically governs cracking, serviceability issues, and structural damage.

Although TBMs generate significantly lower vibration levels than drill-and-blast methods, vibration can still arise from mechanical and ground interaction sources. In hard rock drives, disc cutters fracturing rock create vibration pulses that may propagate through the ground. Mechanical rotation, cutterhead imbalance, drive system oscillations, screw conveyor operation, slurry pump activity, and hydraulic thrust transfer during segment erection can also generate dynamic loading. Vibration levels are typically low in soft ground EPB and slurry tunnelling but may increase in mixed-face or hard rock conditions.

Vibration propagates through soil or rock as wave energy, with amplitude reducing as distance from the source increases. Soil damping characteristics and geological layering influence attenuation rates. Hard rock tends to transmit vibration over greater distances than soft cohesive soils, which absorb and dissipate energy more effectively. Consequently, vibration assessments must account for site-specific geological conditions.

In most TBM projects, vibration levels remain below structural damage thresholds; however, they may exceed comfort limits in sensitive environments such as hospitals, laboratories, or precision manufacturing facilities. Monitoring is typically undertaken using geophones, accelerometers, or surface vibration sensors. Operational controls may include reducing cutterhead rotational speed, adjusting penetration per revolution, optimising cutter condition and spacing, or improving conditioning in mixed ground to smooth mechanical interaction.

Settlement and vibration, although distinct phenomena, may interact. High vibration levels in granular soils can accelerate particle rearrangement, potentially contributing to settlement. Face instability combined with vibration may increase ground response variability. Similarly, excessive thrust or torque fluctuations may amplify both settlement and vibration. Stable, balanced TBM operation therefore reduces the likelihood of both effects simultaneously.

From an engineering perspective, surface settlement and vibration are not merely environmental side effects; they are indicators of system balance within the TBM–ground interaction. Effective management depends on accurate face pressure control, precise spoil conditioning, controlled annular grouting, gradual steering corrections, stable thrust and torque regulation, and real-time monitoring of surface and subsurface movement. In modern mechanised tunnelling, control is achieved not by reacting to damage after it occurs, but by maintaining equilibrium between excavation, support, and ground response. When that equilibrium is maintained, settlement remains within predicted limits and vibration stays within acceptable thresholds. When imbalance develops, ground movement becomes the earliest and most visible signal that operational parameters require adjustment.

Noise and Spoil Disposal

TBM operations produce considerable noise that can adversely affect workers and surrounding communities. The sources of noise in mechanized tunnelling can be categorized into underground and surface activities. Underground sources include the cutterhead rotation, which is especially noisy in hard rock conditions, main drive motors, hydraulic systems, ventilation fans, and segment handling equipment [427]. Hard rock environments can generate intermittent high-energy noise and vibration due to the fracturing effects of disc cutters on the rock strata. Surface activities contribute significantly to noise pollution, including spoil handling operations, which often involve generators, cranes, and truck movements, and have been identified as dominant sources of community noise exposure [323, 741].

Prolonged exposure to elevated sound levels in the tunnel environment presents serious health risks for workers. These risks encompass hearing loss, fatigue, and decreased communication clarity, which increase the likelihood of accidents. Effective engineering controls, such as acoustic enclosures around powerful machinery, vibration isolation measures for pumps and fans, and regular equipment maintenance, can mitigate these effects [742]. Administrative strategies, such as exposure time limits and mandatory hearing protection, further enhance workers' safety

Moreover, surface noise management becomes particularly crucial in urban environments, where stakeholders often impose strict regulations. Mitigation strategies include erecting temporary acoustic barriers, scheduling noisy operations to approved time slots, and utilizing real-time noise monitoring systems to ensure compliance with regulatory limits [743, 744]. Noise modelling during the design phase enables predictions regarding regulatory compliance, while ongoing monitoring during construction allows for promptly addressing any near-threshold issues.

Spoil Disposal in TBM Operations

Spoil management is another critical aspect of TBM operations, involving the transport, treatment, and final placement or reuse of excavated material. As an example, a typical 6.5-meter diameter tunnel can generate over 30 cubic meters of spoil per meter of advance. Efficient spoil management is crucial for sustaining TBM productivity and ensuring environmental compliance [448].

Example: EPB TBM spoil volume per metre, per ring, and per day

Given

- TBM excavation diameter (cutterhead): **D = 6.5 m**
- Advance length to calculate: **L = 1.0 m**
- Segment ring width: **w = 1.5 m**
- Planned net production: **30 m/day**
- Assume over-excavation / operational volume loss allowance = 1.0% (optional, for a "wet" estimate)

1) Excavated cross-sectional area

$$A = \frac{\pi D^2}{4}$$
$$A = \frac{\pi (6.5)^2}{4} = \frac{\pi \cdot 42.25}{4} \approx 33.18 \text{ m}^2$$

2) In-situ spoil volume per metre of tunnel

$$V_{1m} = A \cdot L$$
$$V_{1m} = 33.18 \times 1.0 \approx 33.18 \text{ m}^3/\text{m}$$

So, each metre advanced generates ~33.2 m^3 of in-situ excavated material (before considering bulking).

3) Spoil volume per ring (ring width = 1.5 m)

$$V_{\text{ring}} = A \cdot w$$

$$V_{\text{ring}} = 33.18 \times 1.5 \approx 49.77 \text{ m}^3$$

So, **each ring produces ~49.8 m^3** of excavated material.

4) Daily spoil volume (30 m/day)

$$V_{\text{day}} = A \cdot (\text{daily advance})$$
$$V_{\text{day}} = 33.18 \times 30 \approx 995.4 \text{ m}^3/\text{day}$$

So, at **30 m/day**, you generate **~995 m^3/day** (in-situ basis).

5) Optional: include a 1% "extra" excavation/operational allowance

This can represent small overbreak, steering overcut, operational losses, etc.

$$V_{\text{adj}} = V \cdot (1 + 0.01)$$

Per metre:

$$33.18 \times 1.01 \approx 33.51 \text{ m}^3/\text{m}$$

Per ring:

$$49.77 \times 1.01 \approx 50.27 \text{ m}^3/\text{ring}$$

Per day:

$$995.4 \times 1.01 \approx 1005.4 \text{ m}^3/\text{day}$$

6) Optional: convert to "loose" volume using a bulking factor

If you need truck/stockpile volume, you typically apply a bulking factor B(soil-dependent). Example: B = 1.20 (i.e., 20% bulking)

Daily loose volume:

$$V_{\text{loose}} = V_{\text{in-situ}} \cdot B = 995.4 \times 1.20 \approx 1194.5 \text{ m}^3/\text{day}$$

So your logistics may need to handle ~1,200 m^3/day loose volume at that production rate.

In Earth Pressure Balance (EPB) TBMs, the spoil typically comprises a conditioned mixture of soil, water, polymers, and foams. The rheological properties of this conditioned material are essential for its successful extraction and transport, for they influence spoil handling methods [745, 746]. Additional additives can complicate disposal routes by impacting landfill acceptance criteria and increasing the need for dewatering measures For Slurry TBMs, excavated materials are transported using bentonite slurry to a surface separation plant where solids are processed and separated. This introduces environmental considerations related to water management and sediment control [574].

Effective spoil management requires comprehensive logistical coordination encompassing transport via trucks, conveyors, or rail systems, alongside procedures for dewatering, environmental testing, and classification for reuse [747]. Where feasible, repurposing excavated material as engineered fill, landscaping, or road sub-bases can enhance sustainability, but poor handling can result in operational hazards such as unstable stockpiles or community complaints related to dust or noise [381, 590].

The Interconnection between Noise Management and Spoil Disposal

The management of noise and spoil disposal is interconnected in various ways. Surface spoil handling systems, including slurry plants and the points where conveyors distribute excavated material, can be significant noise sources requiring mitigation. Additionally, issues such as irregular spoil flow within TBM operations may inadvertently influence vibration and noise levels due to variations in mechanical loading [748].

Noise and spoil disposal management should be considered integrated components of a TBM project's operational strategy. Predictive modelling during the design phase, continuous monitoring during construction, regulatory compliance coordination, and interdisciplinary collaboration are imperative for successfully managing these operational impacts. Ultimately, effective control of noise and spoil disposal contributes to workforce health, community relations, and environmental stewardship, thereby underpinning the sustainability of modern mechanized tunnelling operations.

Groundwater Impacts

Groundwater impacts in mechanised tunnelling arise when excavation alters the natural hydraulic regime of the soil or rock mass. Because groundwater pressure contributes directly to effective stress and ground stability, any change in pore pressure or flow pathways can influence face stability, settlement behaviour, lining loads, and environmental conditions. In TBM projects—particularly in EPB and slurry operations—groundwater management is therefore inseparable from geotechnical control.

Groundwater exists within the void spaces of soil and rock and exerts pore water pressure (u). The effective stress principle governs soil behaviour:

$$\sigma' = \sigma - u$$

Where:

- σ' = effective stress (controls strength and stiffness)

- σ= total stress
- u= pore water pressure

If pore pressure decreases (drawdown), effective stress increases, potentially causing consolidation and settlement. If pore pressure increases or is not properly controlled at the face, instability or blow-out may occur.

TBM excavation alters groundwater behaviour because it changes both the stress state and the hydraulic pathways within the ground. As the cutterhead advances, the excavation process relieves confining stress at the tunnel face. This stress relief can disturb the existing equilibrium between soil structure and pore water pressure. If face support pressure is not carefully maintained, groundwater may migrate toward the lower-pressure zone, increasing the risk of inflow, erosion, or instability.

The TBM shield also creates a temporary annular void between the excavated ground and the installed lining. Until this void is completely filled with grout, it represents a potential flow path for groundwater movement. If grouting is delayed or insufficient, water can enter this space, potentially leading to delayed settlement, reduced effective stress in surrounding soils, or long-term leakage issues.

Ground conditioning and grouting further modify groundwater behaviour by changing the permeability of the surrounding soil. Foam, polymers, and bentonite used in EPB operations reduce permeability by binding particles and sealing voids, while annular and permeation grouting can create low-permeability barriers around the tunnel. These interventions alter how water moves through the ground, sometimes reducing inflow but also potentially redirecting natural flow patterns.

Finally, tunnelling interrupts natural groundwater flow paths. A completed tunnel lining can act as a barrier within an aquifer, changing hydraulic gradients and redistributing groundwater pressures around the structure. In some cases, this may cause localised drawdown on one side of the tunnel and pressure increase on the other. Understanding and managing these hydraulic changes is essential to prevent unintended environmental, structural, or geotechnical consequences.

Face Stability and Groundwater Pressure

In water-bearing soils, groundwater pressure may be a dominant component of required face support.

For example, at 20 m depth below the water table:

$$u = \gamma_w \cdot h$$
$$u = 9.81 \times 20 = 196 \text{ kPa}$$

Nearly 200 kPa of pressure may be acting at the face purely from water. If face pressure falls below this, water inflow and soil erosion can occur.

Under-pressure at the tunnel face occurs when the support pressure applied by an EPB or slurry TBM falls below the level required to balance earth and groundwater pressures. When this happens, the ground may begin to move inward toward the excavation. In granular soils, this can lead to running ground conditions, where loose material flows uncontrollably into the chamber. In water-bearing soils, reduced pressure can trigger hydraulic erosion, progressively washing fines into the excavation and increasing permeability. In severe cases, sudden inrush events may occur, where soil and water enter rapidly, posing significant safety and stability risks. Even if collapse does not occur, sustained under-pressure often results in surface settlement due to volume loss around the tunnel.

Over-pressure presents a different but equally serious set of risks. When face support pressure exceeds the safe upper limit, the effective stress in the surrounding ground may be reduced excessively. This can induce hydraulic fracturing, particularly in cohesive soils or weak rock, allowing slurry or pressurised soil to migrate upward. At the surface, this may manifest as ground heave, especially in shallow tunnels. In extreme cases, a blow-out event can occur, where pressurised slurry or soil breaks through to the surface, causing uncontrolled discharge and potentially damaging infrastructure.

Maintaining a balanced pressure window, between collapse pressure and uplift or fracture pressure, is therefore fundamental to safe TBM operation. This balance ensures that the face remains stable without overstressing the surrounding ground. Continuous monitoring, disciplined operational control, and incremental adjustments are essential to keeping pressure within this narrow and well-defined range.

Groundwater Drawdown and Settlement

If tunnelling causes groundwater to flow toward the tunnel (through leakage or insufficient sealing), local drawdown may occur. This reduces pore pressure in surrounding soils.

In cohesive soils (clays), reduced pore pressure increases effective stress, causing consolidation settlement.

Even small reductions in groundwater level can produce measurable settlement over large areas.

<u>Example Concept</u>

If groundwater level drops by 2 m in a compressible clay:

$$\Delta u = 9.81 \times 2 \approx 20 \text{ kPa}$$

That 20 kPa increase in effective stress may trigger consolidation settlement over time.

Long-term settlement is often more damaging than immediate face-related settlement.

Inrush and Flooding Risk

In permeable sands, gravels, or fractured rock, uncontrolled groundwater inflow presents one of the most serious hazards in mechanised tunnelling. When groundwater pressures exceed the capacity of the support system, rapid inflow can occur, potentially leading to chamber flooding and a sudden loss of face stability. In granular soils, flowing water can erode fine particles, forming preferential erosion channels that may propagate toward the surface, increasing the risk of settlement or even sinkhole formation. In slurry TBMs, excessive inflow can dilute slurry properties, disrupt separation systems, and overload pumps and pipelines. In both EPB and slurry machines, electrical systems are vulnerable to water ingress, creating additional safety risks.

Slurry TBMs manage groundwater primarily through hydrostatic slurry pressure, which must be carefully balanced against in-situ earth and water pressures. A stable filter cake forms at the tunnel face as bentonite particles seal pore spaces, reducing permeability and maintaining pressure integrity. EPB TBMs, by contrast, depend on well-conditioned, low-permeability muck within the excavation chamber to prevent rapid water migration and maintain pressure transmission. In both systems, the integrity of the sealing mechanism—whether filter cake or conditioned soil mass—is critical. Failure of these mechanisms can lead to sudden inrush events, which may escalate rapidly if not detected and controlled immediately.

Impact on Annular Grouting

Groundwater conditions have a direct and critical influence on tail void (annular) grouting in TBM tunnelling. After each ring is erected, grout is injected into the annular space between the lining and the surrounding ground to restore confinement and ensure effective load transfer. In water-bearing ground, this process must account not only for soil stresses but also for hydrostatic pressure acting on the void.

If grout injection pressure is too low relative to groundwater pressure, water may infiltrate the annulus before the grout sets. This can dilute the grout, reduce its strength and stiffness, and create voids or incomplete contact zones around the lining. The result may be delayed or progressive settlement as the ground relaxes and consolidates around an inadequately filled annulus.

Conversely, if grout pressure is too high, it can exceed the surrounding effective stress, causing water and grout to migrate upward through permeable layers. This may lead to surface heave, hydraulic fracturing, or unintended ground displacement. In extreme cases, grout pathways can form preferential flow channels, creating localised uplift or disturbance to utilities and structures above.

For these reasons, groundwater regime strongly influences grout mix design, setting time, viscosity, and injection pressure strategy. Engineers must balance sufficient pressure to achieve complete annular filling against the risk of overpressurisation, ensuring that grouting restores ground–lining contact without destabilising the surrounding soil–water system.

Environmental and Hydrogeological Impacts

Beyond immediate construction safety considerations, groundwater impacts associated with TBM tunnelling extend into long-term hydrogeological and environmental domains. One significant issue is aquifer disruption. When tunnels intersect confined or semi-confined aquifers, they can alter natural groundwater flow paths. Even if short-term control is maintained during excavation, the completed tunnel structure may act as a barrier, conduit, or partial drain, permanently modifying hydraulic gradients and flow regimes. These changes can affect regional groundwater levels and potentially influence connected water bodies or dependent ecosystems.

Another important concern is contamination risk. During tunnelling, conditioning agents such as foams, polymers, bentonite, and slurry additives are introduced into the ground system. Although modern materials are designed to be low-toxicity and environmentally compliant, accidental slurry losses, grout migration, or filter cake breakdown may allow constituents to enter surrounding groundwater. For this reason, environmental approvals for major projects typically require groundwater quality monitoring before, during, and after construction, along with strict material selection and spill management protocols.

Dewatering effects also present broader environmental implications. If groundwater pressures are reduced excessively—whether intentionally through dewatering systems or unintentionally through seepage into the tunnel—drawdown can extend beyond the construction zone. This may impact nearby wells, building foundations, buried services, or groundwater-dependent vegetation. In sensitive urban or ecological settings, even moderate drawdown can produce measurable secondary effects.

Because of these potential long-term consequences, large urban tunnelling projects commonly incorporate detailed hydrogeological modelling. Such modelling predicts changes to groundwater levels, flow paths, and pressures over both construction and operational phases, allowing engineers to design mitigation measures and monitoring programs that protect surrounding infrastructure and environmental systems.

Monitoring Groundwater Behaviour

Groundwater conditions are continuously monitored during TBM operations using a combination of geotechnical and mechanical instrumentation. Piezometers—either standpipe or vibrating wire types—are installed in boreholes around the tunnel alignment to measure pore water pressure at different depths. These instruments provide direct information about changes in groundwater levels and hydraulic gradients. In addition, pump rate monitoring is used where dewatering systems are installed, allowing engineers to track inflow volumes and detect abnormal increases that may indicate new seepage paths or pressure imbalances.

Within slurry TBM systems, pressure sensors are integrated into the excavation chamber and slurry circuit to monitor hydrostatic balance in real time. These readings help confirm that slurry pressure is matching groundwater pressure and that filter cake performance remains effective. Groundwater monitoring data is also interpreted alongside settlement measurements, as reductions in pore pressure can lead to consolidation and surface settlement. Correlating pore pressure trends with surface or subsurface movement provides valuable diagnostic insight into whether ground response is hydraulically driven.

Predefined trigger levels are established as part of the project control framework. These typically include thresholds for rapid pore pressure drop, excessive groundwater inflow, and abnormal pressure fluctuations within the TBM system. If trigger values are exceeded, predefined responses—such as adjusting face pressure, modifying grouting, slowing advance rate, or increasing monitoring frequency—are implemented. Continuous monitoring therefore acts as an early warning system, allowing corrective action before significant settlement, instability, or environmental impact develops.

Mitigation and Control Measures

Groundwater impacts in TBM tunnelling are controlled through a coordinated combination of operational control measures and engineered ground improvement techniques. One of the primary control mechanisms is face pressure management. In both EPB and slurry TBMs, maintaining excavation chamber pressure within the calculated collapse and blow-out limits is essential to balancing groundwater pressure. If face pressure is too low, water inflow and instability may occur; if too high, hydraulic fracturing or surface heave can result. Careful pressure control therefore forms the first line of defence against groundwater-related hazards.

Ground conditioning plays a complementary role, particularly in EPB tunnelling. The use of foam, polymers, or bentonite modifies the permeability and cohesion of the excavated soil, reducing rapid water migration and improving pressure retention at the face. By engineering the rheological behaviour of the muck, conditioning helps create a low-permeability, pressure-transmitting medium that limits uncontrolled groundwater inflow.

Where groundwater risks are elevated, pre-treatment measures may be implemented ahead of excavation. Techniques such as permeation grouting reduce permeability by filling pore spaces, jet grouting creates soil–cement bodies that act as cut-off barriers, and ground freezing temporarily converts saturated ground into a low-permeability frozen mass. These methods are typically used in high-risk zones, such as beneath rivers, near sensitive structures, or in highly permeable formations.

Annular grouting quality is also critical. Rapid and complete filling of the tail void behind the shield prevents groundwater from migrating into the annulus and reduces delayed settlement. Proper grout mix design, injection pressure control, and volume verification ensure that the lining is supported and that hydraulic pathways are sealed immediately after ring installation.

Finally, long-term groundwater control depends on robust waterproof lining design. Segmental linings incorporate compressible gaskets at joints, precision manufacturing tolerances, and sealing systems that maintain watertightness under service loads. These measures prevent long-term leakage, protect tunnel durability, and preserve surrounding groundwater conditions. Together, these strategies form an integrated groundwater management framework that links excavation control, ground improvement, structural design, and environmental protection.

Engineering Significance

Face Stability and Settlement Magnitude: One of the primary effects of groundwater on TBM tunnelling is its influence on face stability. Groundwater pressure can destabilize the excavation face, especially in soft soil conditions where the face is more susceptible to collapse due to pore water pressure increases. Research indicates that managing groundwater via techniques such as grouting significantly affects the stability of the tunnel face and associated settlement patterns [156, 157, 159]. In particular, studies illustrate that the implementation of effective grouting strategies can mitigate the risks of aquifer interaction, helping maintain face stability and controlling the magnitude of settlement during and after the tunnelling process [749, 750].

Lining Loads and Long-Term Durability: Lining loads are heavily influenced by groundwater conditions, particularly in situations where water pressure is mismanaged. High external water pressures may impose additional loads on tunnel linings, increasing the risk of structural failure if not properly addressed through hydraulic control measures [751, 752]. Hydraulic management techniques such as curtain grouting not only help to stabilize the immediate tunnelling environment but also significantly enhance the long-term durability of the tunnel lining by reducing future water ingress and associated corrosion risks [57, 753]. Thus, groundwater management directly correlates with both immediate mechanical stresses on the tunnel infrastructure and long-term performance metrics.

Environmental Compliance: Environmental regulation compliance is increasingly critical in modern tunnelling projects. Effective groundwater control strategies contribute to fulfilling environmental requirements by minimizing adverse impacts on surrounding ecosystems caused by groundwater drawdown or contamination. Furthermore, real-time monitoring systems can help manage water ingress, ensuring that environmental thresholds are not breached during construction [753, 754]. Various studies emphasize the importance of proactive groundwater management in preserving local hydrology and preventing environmental degradation during TBM operations [755, 756].

Modern Mechanized Tunnelling - An Integrated Approach: In contemporary mechanized tunnelling practices, groundwater is managed through a combination of advanced technologies and methodologies. This includes integrated pressure control, slurry conditioning, grouting, and continuous monitoring of groundwater conditions. Effective groundwater management ensures that settlement remains predictable, face stability is maintained, and the long-term performance of the tunnel is enhanced [294, 757, 758]. Conversely, failures in groundwater control mechanisms often precipitate significant operational difficulties, leading to incidents that can jeopardize human safety and structural integrity [582, 759].

Proactive Measures and Risk Management: The proactive management of groundwater not only enhances operational safety but also offers a comprehensive approach to risk assessment during TBM tunnelling. A consolidated risk assessment framework that includes geotechnical conditions, water ingress scenarios, and equipment management has become essential in modern tunnelling projects [750, 752]. By understanding the hydraulic implications of groundwater fluctuations, engineers can design interventions that effectively mitigate the risks associated with tunnelling in water-rich environments.

Urban Tunnelling Challenges

Urban tunnelling involves navigating a complex interplay of engineering challenges, including geotechnical uncertainties, dense infrastructure, environmental constraints, and public sensitivities. This environment significantly differs from rural or greenfield tunnelling projects, which tend to have fewer constraints regarding existing structures and community impact. In urban settings, even minor deviations from planned tunnelling operations can have dire consequences for surrounding buildings, utilities, transport networks, and public safety. As a result, urban tunnelling using Tunnel Boring Machines (TBMs) necessitates a focus on risk management and precision control alongside excavation activities.

Figure 103: Sydney Metro - Waterloo Station Box and tunnel boring machine. John Cowper, CC BY 2.0, via Flickr.

Settlement Control Challenges

One of the critical challenges in urban tunnelling is managing ground settlement. Urban environments often host a variety of surface and subsurface structures, from historical masonry to modern infrastructures, that exhibit minimal tolerance for differential movement. Research underscores that even small volume losses or improper tail void grouting can lead to measurable surface settlement, often restricted to a few millimetres in sensitive areas [760, 761]. This requires meticulous face pressure control, accurate spoil conditioning, high-quality annular grouting, and continuous monitoring through tools such as levelling points and extensometers [38, 762]. The operational discipline is paramount; issues such as alignment drift, pressure instability, or torque fluctuations can lead to significant ground movement, affecting public safety and infrastructure integrity [38, 761].

Interaction with Existing Infrastructure

Another major challenge in urban tunnelling is the interaction with pre-existing underground infrastructure, which includes a complex web of utilities like water mains, sewer lines, gas pipelines, and telecommunications conduits. The existence of other tunnels, especially in city centres, poses additional risks due to potential induced stresses, vibration transmission, and settlement effects [38, 760]. Effective construction planning requires detailed utility mapping and comprehensive ground investigation, often supplemented by three-dimensional modelling to pre-emptively address potential disruptions [761, 762]. Such meticulous planning is critical for preventing accidental damage to utilities or service interruptions during construction.

Groundwater Management

Groundwater management in urban tunnelling can also be particularly challenging. Urban tunnels often traverse areas with significant groundwater presence, such as beneath rivers or shallow water tables, necessitating sophisticated management strategies [763, 764]. Dewatering processes are frequently constrained to mitigate long-term impacts on surrounding structures or aquifers. Techniques including balanced face pressure maintenance, effective sealing, and, in some cases, permeation grouting or jet grouting are imperative [762, 763]. Furthermore, urban projects often mandate predictive hydrogeological modelling and continuous pore pressure monitoring to ensure that aquifer conditions remain stable throughout construction [761, 763].

Noise, Vibration, and Community Impact

The implications of noise and vibration in urban settings cannot be overstated. While TBMs typically generate less vibration than traditional drill-and-blast methods, surface activities such as spoil handling and truck movements contribute significantly to noise disturbances [760, 764]. Compliance with stringent environmental noise regulations often necessitates the installation of acoustic enclosures, adherence to scheduling restrictions, and deployment of real-time noise monitoring systems [38, 764]. Effective public communication strategies about construction impacts serve to minimize community opposition and manage reputational risks, highlighting the necessity of public engagement throughout the tunnelling process [764, 765].

Spatial Constraints

Urban tunnelling projects frequently encounter significant spatial constraints due to the dense urban fabric. The logistics of launching and receiving shafts, as well as managing surface compounds, must occur within limited footprints, often alongside active traffic and extensive utilities [766, 767]. Moreover, removing spoil efficiently, whether by conveyor, rail, or truck, requires meticulous planning to prevent congestion and operational delays. Continuous operations may be essential to maintain project timelines, emphasizing the importance of fatigue management and shift coordination among workers [766, 767].

Geological Variability and Operational Precision

Geological variances in urban environments further complicate tunnelling projects. Many cities are built on complex geological formations, including reclaimed land or heterogeneous deposits, which could lead to unforeseen operational challenges during tunnelling [768, 769]. Operators must be prepared for real-time adjustments in TBM operations due to mixed-face conditions or unexpected obstructions [768]. Achieving breakthrough precision where TBMs interface with existing structures is crucial, as misalignments can lead to costly corrective measures. Technologies such as laser guidance and gyroscopic navigation become essential tools for ensuring precise tunnelling and alignment control [767, 768].

Heightened Safety Considerations

Safety is amplified in urban environments due to the severe implications of construction failures. Events such as face collapses or water inrushes have the potential to impact not only workers but also the substantial public infrastructure above [769, 770]. Emergency response planning requires careful coordination with municipal authorities and utility operators to ensure rapid and efficient responses to incidents, while real-time monitoring systems can enhance the predictability and safety of tunnelling operations [767, 770].

Environmental Monitoring

Environmental monitoring in tunnelling, especially within urban contexts utilizing Tunnel Boring Machines (TBMs), is pivotal in ensuring that construction activities remain compliant with environmental regulations and do not pose risks to public safety, ecosystems, or infrastructure. This monitoring entails systematic measurements and evaluations of various environmental parameters, facilitating the integration of regulatory approval with effective operational control systems.

At the heart of effective environmental monitoring is the capacity to translate predicted impacts from Environmental Impact Assessments (EIAs) into measurable site controls. Prior to construction, baseline data—including noise, vibration, groundwater levels, and air quality, are gathered to establish reference conditions. These baseline measurements are critical for distinguishing construction impacts from pre-existing conditions, as inaccurate or absent baseline data can obscure the true effects of construction activities on the environment and nearby infrastructures [414, 771].

Key Components of Environmental Monitoring

Settlement and Structural Movement Monitoring: One of the most vital aspects of environmental monitoring in TBM projects is the observation of surface settlement and structural movement. Techniques such as surface settlement markers, automated total stations, inclinometers, and crack gauges are employed to continuously track ground and structural responses. Different tiers of trigger levels, alert, action, and alarm, are set; should settlement measurements surpass these levels, operational adjustments are made [772, 773]. This proactive monitoring strategy aims to provide early warnings to mitigate potential damage before it manifests.

Groundwater Monitoring: Groundwater monitoring is another crucial element, where devices like piezometers measure pore water pressures that may be influenced by tunnelling activities. This monitoring ensures that operational parameters, such as face pressure and grouting methods, do not lead to unsustainable drawdowns or unwanted hydrogeological connectivity between aquifers, which could compromise both construction safety and long-term groundwater stability [774, 775]. In slurry TBM operations, monitoring of slurry pressure and return flows can also indicate potential seepage pathways, emphasizing the importance of groundwater control in environmentally sensitive areas [776].

Noise and Vibration Monitoring: In densely populated urban areas, noise monitoring is critical. Fixed and portable sound level meters are utilized to measure decibel levels at sensitive locations, such as residences and schools. Real-time data transmission allows for immediate intervention should noise levels approach regulatory thresholds [773, 777]. Similarly, vibration monitoring employs geophones or accelerometers to ensure that vibrations remain within acceptable limits to prevent structural damage, especially in proximity to sensitive structures [778, 779].

Air Quality and Water Quality Monitoring: Air quality monitoring, particularly for particulate matter and gas emissions, is essential in maintaining safe working conditions and protecting nearby communities. The measurement of particulate matter (PM10 and PM2.5) alongside gas detectors helps ensure safety during tunnelling operations [774]. Additionally, water quality monitoring can identify potential contamination from construction activities, establishing baseline parameters like turbidity and pH before water is discharged to ensure compliance with environmental permits [780].

The Role of Technology in Monitoring

Modern environmental monitoring systems have increasingly utilized digital dashboards for real-time data acquisition and trend analysis. This integration allows for enhanced predictive capabilities; for instance, a concurrent decline in groundwater pressure coupled with increased surface settlement may indicate face instability long before visual indications arise [781, 782]. Furthermore, technological advancements such as GNSS and InSAR techniques provide innovative methods for movement monitoring under sensitive infrastructures, providing comprehensive insights into ground stability during tunnelling operations [783].

> GNSS and InSAR are advanced geodetic monitoring techniques used to measure ground movement with high precision over both local and regional scales. In tunnelling and urban infrastructure projects, they complement traditional settlement

monitoring methods by providing broader spatial coverage and long-term deformation tracking. These technologies are particularly valuable for detecting small surface movements that may indicate subsurface stress redistribution, consolidation, or groundwater changes.

GNSS (Global Navigation Satellite System) monitoring refers to satellite-based positioning systems such as GPS (United States), GLONASS (Russia), Galileo (Europe), and BeiDou (China). In engineering applications, high-precision GNSS receivers are installed at fixed control points on the ground surface. These receivers track satellite signals and, when processed using differential or real-time kinematic (RTK) methods, can determine three-dimensional positions with millimetre-level accuracy.

In tunnelling projects, GNSS is commonly used to monitor surface settlement above tunnel alignments, movement of sensitive buildings or infrastructure, ground response around shafts or portals, and broader regional subsidence trends. The underlying principle is simple: if the ground moves, the antenna mounted to it moves as well. By comparing successive position measurements, engineers can detect both vertical and horizontal displacement trends over time.

GNSS systems are particularly effective for large or widely distributed monitoring areas, locations where manual survey access is limited, projects requiring long-term observation, and situations where movements are slow and progressive. However, GNSS has limitations in dense urban environments. Signal obstruction from tall buildings (the "urban canyon" effect), multipath interference, and tree cover can reduce accuracy. Additionally, GNSS provides point-based measurements only at installed antenna locations rather than continuous coverage across an area. For this reason, it is often combined with total station monitoring and precise levelling surveys to verify settlement predictions and confirm that movements remain within allowable thresholds.

InSAR (Interferometric Synthetic Aperture Radar) is a satellite-based remote sensing technique that measures ground deformation across wide areas using radar imagery. Unlike GNSS, which relies on ground-installed receivers, InSAR uses radar signals transmitted from orbiting satellites. These signals reflect off the Earth's surface and return to the satellite, where phase differences between repeated passes are analysed.

The method works by comparing the phase of radar signals acquired at different times over the same location. Even very small ground movements—on the order of a few millimetres—alter the phase of the reflected signal. By processing these differences, engineers can detect vertical ground displacement over large regions. InSAR is capable of measuring regional subsidence or uplift, settlement along tunnel corridors, groundwater-related deformation, and long-term post-construction ground movement.

Advanced techniques such as Persistent Scatterer InSAR (PS-InSAR) focus on stable reflective targets such as buildings or infrastructure elements, enabling millimetre-scale accuracy over extended periods. The advantages of InSAR include wide-area coverage spanning tens to hundreds of square kilometres, no requirement for ground installation, the ability to perform retrospective analysis using historical satellite data, and detection of regional trends beyond the immediate tunnel footprint.

However, InSAR also has constraints. Its effectiveness is reduced in heavily vegetated or water-covered areas, and it requires specialised data processing. It primarily provides vertical displacement information, with horizontal accuracy being more limited. In addition, it is generally less suited for rapid, short-term movement detection compared to on-site instrumentation such as total stations or GNSS.

In urban tunnelling projects, GNSS and InSAR are often integrated within a multi-layered monitoring strategy. GNSS provides high-precision point control at critical structures, delivering continuous and near real-time data. InSAR, in contrast, offers broad regional coverage, detecting slow-moving settlement zones or groundwater-related subsidence extending beyond the tunnel alignment. This combination allows engineers to distinguish between tunnel-induced volume loss and broader hydrogeological effects. For example, if settlement markers above the tunnel remain stable but InSAR reveals regional subsidence beyond the alignment, groundwater drawdown rather than direct tunnelling effects may be responsible.

From an engineering perspective, GNSS and InSAR extend monitoring beyond traditional localised instruments. They enable early detection of anomalous movement patterns, independent verification of settlement predictions, long-term performance assessment, and transparent reporting to regulators and stakeholders. Although they do not replace conventional tools such as levelling surveys, inclinometers, or piezometers, they significantly enhance system-wide understanding of ground behaviour.

Together, GNSS and InSAR represent the integration of satellite geodesy into underground construction risk management. They allow engineers to observe how the ground responds not only directly above the tunnel, but across the wider urban environment and over extended timeframes—strengthening the feedback loop between design assumptions, operational control, and real-world ground performance.

Transparency and Stakeholder Engagement

Environmental monitoring also plays a crucial role in fostering transparency and community engagement. Many urban tunnelling projects publish monitoring results publicly or offer direct access to results for regulators and community representatives, thus building trust among stakeholders [784]. This transparency is key to reducing reputational risks and ensuring community protection throughout project development phases.

Mitigation Strategies

Mitigation strategies for environmental and community impacts in tunnelling projects are structured measures designed to reduce, control, or compensate for the effects of construction on surrounding ecosystems, infrastructure, and local populations. In mechanised tunnelling, particularly in urban environments, impacts extend beyond the underground works and must be managed through integrated engineering, environmental, and stakeholder frameworks. Effective mitigation is proactive rather than reactive, combining predictive modelling, monitoring, operational controls, and transparent communication.

A primary area of mitigation concerns ground movement and settlement. Surface settlement can affect buildings, utilities, and transport corridors. Mitigation begins at the design stage through careful geotechnical investigation, selection of appropriate TBM type, and development of a defined face pressure operating window. During construction, volume loss is minimised through stable face pressure control, effective soil conditioning, precise steering, and high-quality annular grouting. Where structures are particularly sensitive, additional measures such as compensation grouting, jet grouting, or ground freezing may be used. Real-time monitoring of settlement markers, GNSS points, and sometimes InSAR data provides early warning so corrective action can be taken before damage occurs.

Groundwater protection is another major mitigation focus. Tunnelling can alter aquifer flow paths, reduce pore pressures, or create pathways for contamination. Mitigation strategies include maintaining balanced face pressure to prevent inflow, using low-permeability conditioning agents where appropriate, designing segmental linings with high-integrity gaskets, and implementing pre-treatment such as permeation grouting in high-risk zones. Hydrogeological modelling is conducted in advance to predict drawdown or uplift effects, and piezometers are installed to monitor pore pressure. Where necessary, recharge systems or controlled dewatering schemes are implemented to protect nearby wells, ecosystems, or basements.

Noise and vibration mitigation is critical for community acceptance, especially in dense urban settings. Underground, engineering controls such as resilient machine mountings, acoustic enclosures, and equipment maintenance reduce noise exposure to workers. At surface sites, slurry separation plants, conveyors, and spoil loading facilities may be enclosed or shielded by acoustic barriers. Work scheduling limits high-noise activities to permitted hours, and continuous noise monitoring stations ensure compliance with regulatory thresholds. Vibration mitigation may involve adjusting cutterhead speed, penetration rates, and advance cycles to minimise dynamic impacts near sensitive structures such as laboratories or hospitals.

Spoil management mitigation addresses both environmental and logistical impacts. Excavated material must be handled in ways that prevent dust, runoff contamination, odour, and excessive truck traffic. Measures include enclosed conveyor systems, dust suppression sprays, wheel-wash stations for trucks, and defined haulage routes that minimise community disruption. Where possible, spoil is tested and reused as engineered fill, reducing landfill demand. Slurry TBM projects incorporate water treatment systems to recycle process water and prevent discharge of contaminated sediment.

Air quality and gas management are also essential mitigation components. Ventilation systems are designed to dilute dust, diesel emissions (where applicable), and naturally occurring gases such as methane or hydrogen sulphide. Continuous gas detection systems provide early warning of hazardous concentrations. At the surface, vehicle emissions are managed through modern fleet standards and traffic planning. Where odour or gas migration is possible, sealing measures and active ventilation controls are implemented.

Community engagement is a non-technical but equally critical mitigation strategy. Transparent communication helps reduce uncertainty and public concern. This includes advance notice of works, public information sessions, accessible complaint channels, and publication of monitoring data where appropriate. Dedicated community liaison officers often act as a bridge between the project team and residents. Rapid response to complaints, such as investigating a reported vibration event, demonstrates accountability and builds trust.

Environmental monitoring forms the backbone of mitigation effectiveness. Baseline data are collected prior to construction to establish reference conditions. During works, parameters such as settlement, groundwater levels, noise, vibration, dust, and air quality are continuously monitored against predefined trigger levels. If thresholds are approached or exceeded, pre-approved response plans are activated. This structured "trigger–action–response" framework ensures that mitigation measures are applied systematically rather than improvisationally.

Sustainability considerations increasingly shape mitigation strategies. Energy-efficient equipment, low-carbon concrete mixes for segments, biodegradable conditioning agents, and optimised logistics all reduce the broader environmental footprint. Life-cycle assessment is sometimes incorporated into design decisions to balance construction impacts with long-term transport and infrastructure benefits.

Case Study Scenario: Urban Metro Tunnel Beneath a Sensitive City Corridor

Project Overview

A 6.8 m diameter Earth Pressure Balance (EPB) TBM is commissioned to construct twin metro tunnels beneath a densely developed urban corridor. The alignment passes under heritage masonry buildings (150+ years old), a hospital with vibration-sensitive equipment, a major sewer trunk line, and a shallow unconfined aquifer supplying nearby parks. Surface cover to tunnel axis varies between 14 m and 22 m. The project is located in mixed ground consisting of silty sand with occasional clay lenses and moderate groundwater pressure.

From the outset, the project team recognises that environmental and community impacts will be as critical to success as excavation performance.

Phase 1: Proactive Design and Risk Identification

Geotechnical and Hydrogeological Modelling

Detailed site investigation identifies:

- Moderate-to-high permeability sandy layers
- A shallow groundwater table at 6 m depth
- Sensitive brick heritage structures with low tolerance to angular distortion
- Hospital equipment with strict vibration limits (<2 mm/s PPV)

A settlement risk assessment using the Peck Gaussian method predicts:

- Potential maximum settlement of 18 mm at shallow sections if volume loss reaches 1.5%
- Angular distortion risks exceeding 1/500 beneath heritage façades if not controlled

Hydrogeological modelling predicts potential drawdown of 0.4 m if groundwater inflow is not properly sealed.

Mitigation Built into Design

The following proactive measures are adopted:

- EPB TBM selected to maintain face pressure balance in granular soils.
- Defined face pressure operating window: 230–260 kPa.
- Target volume loss: ≤ 0.8%.
- High-integrity segmental lining with double EPDM gaskets.
- Two-component annular grout with rapid set time.
- Compensation grouting array pre-installed beneath the most sensitive heritage buildings.
- Piezometers installed along alignment for groundwater monitoring.
- Baseline noise, vibration, and settlement surveys completed six months before launch.

Mitigation is embedded into design, not added later.

Phase 2: Construction – Integrated Operational Control

Settlement and Volume Loss Management

During excavation beneath heritage buildings:

- Face pressure is maintained at 245 kPa ± 5 kPa.
- Foam Injection Ratio (FIR) is optimised at 45% to maintain plastic, low-permeability muck.
- Annular grout volume is controlled at theoretical void volume + 5%.
- Settlement markers, GNSS stations, and automated total stations provide real-time data.

Trigger Levels:

- Alert: 5 mm
- Action: 8 mm
- Alarm: 12 mm

At one location, settlement reaches 7 mm (approaching Action level). The response protocol is triggered:

- Advance rate reduced from 40 mm/min to 28 mm/min.
- Grout pressure increased by 10 kPa.
- Additional compensation grout injected pre-emptively.

Settlement stabilises at 8.2 mm and begins to plateau. No structural damage occurs.

Groundwater Protection

Piezometer readings show stable pore pressures during most of the drive. However, beneath a permeable sand lens:

- Minor pore pressure drop of 15 kPa detected.
- Slight increase in slurry return fines.

Response measures include:

- Increased bentonite dosing in conditioning.
- Reduced screw extraction rate.
- Localised permeation grouting ahead of face.

Pore pressures recover within 48 hours. No aquifer drawdown is recorded at nearby monitoring wells.

Noise and Vibration Control

Surface slurry plant and spoil loading area are located near residential apartments.

Mitigation measures include:

- Acoustic enclosure around separation plant.
- 4 m high temporary noise barrier.
- Night-time truck movements restricted.
- Real-time noise monitoring linked to 65 dB(A) threshold.

At the hospital crossing:

- Cutterhead RPM reduced.
- Penetration per revolution lowered.
- Advance rate temporarily reduced.

Measured vibration remains below 1.5 mm/s PPV, well within hospital tolerance.

Spoil Management

Each metre of tunnel generates ~36 m^3 of conditioned spoil.

Mitigation measures:

- Enclosed conveyor system.
- Wheel-wash facility for trucks.
- Defined haulage routes avoiding school zones.
- 60% of spoil reused as engineered fill for a nearby road project.
- Slurry water recycled through settlement tanks and filter presses.

Dust and runoff incidents are avoided. No environmental breach notices are issued.

Air Quality and Gas Monitoring

Methane concentrations are low but present intermittently.

Controls include:

- Continuous gas detection in TBM backup.
- Increased ventilation flow rate.

- Explosion-proof electrical fittings in high-risk zones.

No gas exceedances occur during the drive.

Phase 3: Community Engagement

A dedicated Community Liaison Officer:

- Conducts monthly public briefings.
- Publishes monitoring data summaries online.
- Operates a 24-hour hotline.

When residents report minor vibration sensations during cutter replacement, the project team:

- Investigates immediately.
- Shares monitoring data with the community.
- Explains mitigation steps taken.

Transparency reduces complaints and builds trust.

Phase 4: Environmental Monitoring and Trigger–Action–Response

All environmental parameters are integrated into a central monitoring dashboard:

- Settlement
- GNSS displacement
- Pore pressure
- Noise
- Vibration
- Dust
- Gas concentration

Predefined trigger–action–response plans ensure:

- No improvisation
- Immediate escalation if thresholds exceeded
- Documented corrective measures

The system allows early intervention rather than reactive damage control.

Sustainability Integration

Additional sustainability measures include:

- Low-carbon concrete segments (30% fly ash replacement).
- Biodegradable foam additives.
- Energy-efficient electric backup gantries.
- Life-cycle carbon assessment demonstrating long-term transport emissions reduction.

Spoil reuse reduces landfill demand by 40,000 tonnes.

Outcome

The tunnel drive is completed with:

- Maximum recorded settlement: 9.3 mm (below 12 mm alarm threshold).
- No structural damage to heritage buildings.
- No measurable aquifer drawdown.
- No environmental regulatory breaches.
- Community complaints reduced by 65% compared to a previous project in the same corridor.

This case study shows that:

- Mitigation begins at design, not after impact.
- Volume loss control is central to environmental protection.
- Groundwater management is integrated with face pressure and grouting.
- Noise, vibration, and spoil are engineering-controlled, not incidental.
- Monitoring provides early warning.
- Transparent community engagement reduces conflict.
- Sustainability measures can align with technical performance.

Most importantly, it demonstrates that environmental and community mitigation is inseparable from technical TBM performance. Balanced machine operation directly translates into reduced settlement, stable groundwater conditions, lower vibration, and improved community outcomes.

In modern urban tunnelling, success is measured not only by advance rate—but by how effectively engineering control maintains equilibrium between excavation progress, environmental protection, and public trust.

Ultimately, mitigation strategies for environmental and community impacts are not separate from technical tunnelling performance, they are integral to it. Balanced TBM operation reduces settlement, groundwater disruption, and vibration. Efficient spoil management reduces noise and emissions. Continuous monitoring links underground engineering decisions to surface outcomes. In modern mechanised tunnelling, successful mitigation reflects the ability to maintain equilibrium between excavation progress, environmental protection, and community wellbeing.

Chapter 16

Maintenance, Troubleshooting, and Interventions

Tunnel Boring Machines operate as high-energy, high-precision systems within confined and geotechnically sensitive environments. While design and planning establish the framework for successful excavation, long-term performance depends on how effectively the machine is maintained, monitored, and supported throughout the drive. Mechanical integrity, operational discipline, and early fault detection are not secondary considerations, they are central to maintaining ground stability, protecting surface assets, and ensuring programme certainty. As tunnel length increases and geological variability intensifies, the importance of structured maintenance and systematic troubleshooting becomes progressively more critical.

This chapter explores the integrated strategies that sustain TBM reliability under demanding underground conditions. It examines planned maintenance regimes, cutter inspection and replacement practices, bearing and seal management, breakdown recovery procedures, and hyperbaric interventions. Each of these elements plays a vital role in preserving operational balance between the machine and the ground. Maintenance is not merely equipment servicing; it is a control mechanism that directly influences face pressure stability, torque management, spoil extraction efficiency, and ultimately volume loss.

The chapter also addresses complex recovery scenarios, including in-situ repair, controlled retreat, major component replacement, and retrieval-to-shaft strategies. These situations demonstrate how mechanical failure, geotechnical risk, and safety management intersect in mechanised tunnelling. Effective troubleshooting requires not only technical expertise but also disciplined adherence to safety systems and real-time performance analysis.

Maintenance and intervention strategies are inseparable from geotechnical performance and project risk management. A well-maintained machine produces stable parameters; stable parameters protect the ground; protected ground ensures surface safety and project continuity. This chapter therefore positions maintenance and troubleshooting as core engineering functions—essential to sustaining equilibrium between machine performance, underground conditions, and long-term infrastructure success.

Planned Maintenance Regimes

Planned maintenance regimes in Tunnel Boring Machines (TBMs) are essential for ensuring reliability, safety, and efficiency in tunnelling operations. These structured and systematic programs are specifically designed to mitigate unplanned failures, which are particularly critical in the confined and high-pressure environments where TBMs operate. Failure to implement such maintenance strategies can result in costly downtime, safety hazards, and adverse geological impacts, including ground instability [277, 785].

The primary goals of planned maintenance are to prevent catastrophic equipment failures, maintain operational parameters, extend component lifetimes, and safeguard personnel safety. Unplanned stoppages due to failures can lead to significant financial losses, as major issues like main bearing damage or hydraulic failures can halt production for extensive periods, sometimes for weeks [507, 510]. Proactively addressing wear, lubrication, alignment, and performance ensures that TBMs can operate under optimal conditions, which is crucial for maintaining geotechnical stability and project productivity [786, 787].

Components of a Planned Maintenance Regime

A comprehensive planned maintenance regime will involve several critical operations:

Inspection and Monitoring: This includes regular checks on components such as cutterheads, hydraulic systems, main bearings, and electrical systems. Each part of the TBM has specific monitoring needs—temperature checks on main bearings, oil quality testing in hydraulic systems, and wear measurements on cutting tools [507, 510, 788].

Maintenance Types:

- **Preventive Maintenance**: Conducted at fixed intervals based on operational metrics, such as running hours and after specific advance distances [275, 789].
- **Predictive Maintenance**: This condition-based maintenance utilizes real-time data to forecast potential failures. Key indicators such as bearing temperatures and vibration levels are analysed to initiate timely interventions before issues escalate [510, 790].

- **Planned Corrective Maintenance**: Addresses known degradation based on monitored trends. For example, if increased torque indicates cutter wear, a cutter replacement can be pre-emptively scheduled [64, 791].

Integration of Safety Management

The integration of maintenance practices within the project's Safety Management System is vital. Activities such as electrical servicing and confined space entry must adhere to stringent safety protocols to protect personnel during maintenance operations. For instance, lock-out and tag-out procedures are necessary to ensure that machinery is safely powered down before maintenance tasks begin, and hot work requires meticulous fire hazard control [64, 792].

Technological Advances in Maintenance

Modern TBMs are equipped with extensive instrumentation, allowing maintenance processes to leverage advanced data analytics. Digital platforms facilitate the monitoring of operational parameters, enabling the identification of subtle changes—like incremental increases in torque—that may signal emerging issues [786, 790]. This capability transforms maintenance from routine tasks into a more sophisticated, data-driven approach focused on performance optimization and risk mitigation [510, 788].

Enhancing Productivity

An effective planned maintenance regime can paradoxically be more productive than continuous, non-stop operations. By periodically stopping for maintenance, TBM operators can mitigate the risks of failures that would lead to extended downtime. The structured approach to maintenance ensures consistent face pressure management and efficient spoil extraction, directly impacting project timelines and costs in a positive manner [64, 793].

In the context of TBM operations, it is evident that planned maintenance is a core element of effective tunnelling strategy. It encompasses not only the mechanical upkeep of the TBM but also supports safe and sustainable tunnelling practices, anchoring project reliability and success. By addressing equipment wear proactively, operators can uphold the equilibrium between machine efficiency, operational safety, and ground stability, ultimately leading to enhanced project outcomes.

The following provides a sample field-based TBM maintenance schedule suitable for a 6.5 m EPB TBM operating in mixed urban ground (sands/clays with occasional gravel lenses), advancing approximately 10–15 m/day.

This example integrates **time-based, distance-based, and condition-based maintenance**, aligned with typical contractor practice and OEM guidance.

TBM PLANNED MAINTENANCE SCHEDULE

Project Type: Urban EPB TBM
Diameter: 6.5 m
Drive Length: 2,500 m
Average Advance Rate: 12 m/day
Operating Cycle: 24/7 (2 x 12-hour shifts)

1. SHIFT-BASED MAINTENANCE (Every 12 Hours)

Purpose:

Early detection of emerging faults and parameter drift.

Conducted By:

Shift Mechanical Supervisor + Electrical Technician

Tasks:

Cutterhead & Face System

- Review torque and thrust trends.
- Check screw conveyor amperage consistency.
- Inspect for abnormal vibration patterns.
- Verify chamber pressure stability history.

Hydraulic Systems

- Visual inspection for leaks.
- Check hydraulic oil temperature and pressure readings.
- Inspect hose integrity in accessible areas.

Electrical Systems

- Inspect trailing cable for abrasion.
- Check control panel alarms.
- Confirm backup power availability.

Slurry / Conditioning Systems

- Inspect foam pumps and injection lines.
- Confirm dosing calibration.
- Check bentonite/polymer tank levels.

Segment Erection System

- Inspect erector arm hydraulic performance.
- Confirm bolt torque tool calibration.

Documentation

- Record all anomalies in digital maintenance log.
- Escalate any parameter drift beyond threshold.

2. DAILY MAINTENANCE (Every 24 Hours)

Purpose:

Preventive servicing and safety verification.

Tasks:

- Clean conveyor transfer points.
- Inspect belt alignment and tension.
- Check screw conveyor wear plate clearance.
- Inspect grease delivery system to main bearing.
- Test gas detection sensors.
- Inspect ventilation duct integrity.
- Confirm fire suppression system readiness.
- Check lubrication levels in all rotating systems.

3. WEEKLY MAINTENANCE (Every 7 Days)

Purpose:

Detailed mechanical inspection and predictive checks.

Cutterhead & Cutting Tools

- Borehole inspection camera check (if accessible).
- Measure disc cutter wear (estimated consumption tracking).
- Inspect scraper wear.
- Check cutterhead bolts torque.

Main Bearing

- Grease sampling.
- Temperature trend analysis.
- Vibration data review.

Hydraulic System

- Oil sample analysis (contamination test).
- Filter inspection and replacement (if required).

Electrical Systems

- Insulation resistance test (selected circuits).
- Thermal imaging of switchgear.
- Inspect transformer cooling.

Slurry / Conditioning Systems

- Clean foam injectors.
- Inspect bentonite lines.
- Verify dosing control system calibration.

4. 150–250 m ADVANCE INTERVAL (Distance-Based)

(Approximately every 12–20 days depending on advance rate)

Planned Cutter Inspection Window

Activities:

- Partial cutterhead access (atmospheric if possible).
- Replace worn disc cutters.
- Replace scrapers and picks.
- Inspect cutterhead structure for cracks.
- Check cutterhead drive seals.

If hyperbaric intervention required:

- Full compressed air protocol.
- Medical clearance.
- Decompression schedule.

5. 500 m ADVANCE INTERVAL

Intermediate Service Stop (1–2 Days)

- Full cutterhead inspection.
- Main bearing seal inspection.
- Replace hydraulic filters.
- Full hydraulic oil top-up.
- Inspect thrust cylinders for scoring.
- Inspect erector hydraulic actuators.
- Calibrate guidance system.

6. 1,000 m ADVANCE INTERVAL

Major Service Window (3–5 Days)

Main Systems:

- Replace hydraulic oil (if condition trending indicates).
- Replace main bearing grease lines.
- Full vibration analysis review.
- Replace high-wear screw conveyor sections.
- Inspect tail seal brushes and grease system.
- Inspect annular grout injection ports.
- Replace worn conveyor belts (if required).

Electrical:

- Full switchboard inspection.
- Earth leakage protection testing.
- Replace damaged cable sections.

7. CONDITION-BASED TRIGGER MAINTENANCE

These actions are not time-based but triggered by parameter trends.

Trigger Example 1:

Torque increases 15% above baseline for same penetration rate.
→ Inspect cutter wear and conditioning efficiency.

Trigger Example 2:

Main bearing temperature rises >8°C above long-term average.
→ Immediate lubrication system inspection.

Trigger Example 3:

Repeated settlement alarms + torque spike.
→ Inspect face stability, screw wear, conditioning system.

8. HYPERBARIC INTERVENTION SCHEDULE (If Required)

Triggered When:

- Excessive cutter wear.
- Blockage in cutterhead.
- Ground change to abrasive layer.

Controls:

- Medical clearance for all personnel.
- Lock-out of mechanical systems.
- Hyperbaric supervisor present.
- Backup chamber available.
- Full decompression protocol.

9. SAFETY SYSTEM CHECK SCHEDULE

Weekly:

- Fire detection test.
- Emergency lighting check.
- Refuge chamber inspection.

Monthly:

- Emergency drill (fire/flooding scenario).
- Gas detection calibration.
- Ventilation flow measurement.

10. DOCUMENTATION & REVIEW

All maintenance activities logged into:

- Digital CMMS (Computerised Maintenance Management System)
- Torque & thrust trending dashboard
- Settlement correlation database
- Hydraulic oil sampling log
- Cutter consumption database

Monthly review meeting:

- Mechanical trends
- Downtime analysis
- Component life prediction
- Risk register update

Example Maintenance Forecast (First 1,000 m Drive)

Distance	Major Activity
0–150 m	Baseline monitoring
200 m	Cutter inspection
500 m	Intermediate service
750 m	Cutter replacement
1,000 m	Major service

This sample maintenance regime is effective because it transforms equipment care into a structured risk-control system rather than a reactive repair process. By scheduling inspections and interventions before components reach critical wear limits, it reduces the likelihood of catastrophic failures such as main bearing seizure, cutterhead damage, hydraulic rupture, or electrical faults. The integration of predictive monitoring—such as torque trends, temperature data, vibration analysis, and oil sampling—allows emerging issues to be identified early and addressed during controlled stoppages rather than during emergency shutdowns.

The regime also aligns cutter wear management with production modelling. By forecasting cutter life based on geology, penetration rates, and historical consumption, maintenance windows can be synchronised with advance cycles. This ensures that interventions occur at optimal intervals, preserving productivity while preventing overload conditions. Embedding safety procedures such as lock-out/tag-out, permit-to-work systems, and hyperbaric protocols within the maintenance framework ensures that servicing activities themselves do not introduce additional hazards. As a result, unplanned downtime is reduced, operational variability is minimised, and face pressure and torque remain within defined operating envelopes.

The underlying engineering principle is that planned maintenance is not merely mechanical servicing; it is a geotechnical stability control measure. In mechanised tunnelling, the machine and the ground form a coupled system. When mechanical components operate smoothly, operational parameters such as thrust, torque, screw speed, and chamber pressure remain stable. Stable operational parameters help maintain consistent face support and controlled excavation.

Controlled excavation directly limits volume loss, which is the primary driver of surface settlement. When volume loss remains within predicted limits, surface structures are protected and environmental impacts are minimised. Conversely, mechanical degradation can lead to torque spikes, pressure fluctuations, and unstable spoil extraction, all of which increase the risk of settlement and ground disturbance.

In modern TBM tunnelling, maintenance discipline is therefore inseparable from ground control, safety management, and programme certainty. A well-structured maintenance regime supports predictable machine performance, balanced excavation, reduced environmental risk, and improved stakeholder confidence.

Cutter Inspection and Replacement

Cutter inspection and replacement are critical components of TBM operation because the cutting tools are the primary interface between the machine and the ground. Whether operating in hard rock or soft ground, the condition of cutters directly affects penetration rate, torque demand, face stability, vibration levels, and overall production efficiency. Inadequate cutter management can lead to excessive wear, mechanical overload, increased settlement risk, and costly downtime.

In hard rock TBMs, disc cutters are the dominant cutting tools. These circular steel cutters, typically 17–19 inches (431.8 mm - 482.6 mm) in diameter, fracture rock through compressive loading. As the cutterhead rotates, each disc applies high contact force to the rock surface, inducing tensile cracking and spalling. Over time, cutter rings wear down due to abrasion, impact loading, and heat. Wear reduces the effective cutting diameter, alters contact geometry, and increases rolling resistance. As a result, torque and thrust demand rise while penetration per revolution decreases. Regular inspection ensures that cutters are replaced before excessive wear compromises performance or damages the cutterhead.

In soft ground and EPB TBMs, the cutting tools may include scrapers, picks, ripper teeth, and wear plates in addition to smaller disc cutters in mixed-face conditions. These tools are subject to abrasion, clogging, and impact damage. Clay can adhere and cause build-up, while sand and gravel cause abrasive wear. Inspection focuses on wear thickness, cracking, bolt integrity, and alignment. Because soft-ground cutters operate within a pressurised chamber, access often requires planned intervention procedures, which may include compressed air or hyperbaric entry.

Cutter inspection intervals are typically defined by advance distance, ground conditions, and performance trends. For example, in abrasive granite, inspection may be scheduled every 100–150 metres, whereas in softer sedimentary rock, intervals may extend to 250–300 metres. Modern TBMs also use indirect performance indicators to trigger inspection. Rising torque at constant penetration, increased vibration, reduced advance rate, or elevated cutterhead temperature may indicate worn or seized cutters. Data-driven predictive maintenance increasingly complements fixed-interval inspection planning.

The inspection process itself varies depending on ground pressure conditions. In atmospheric conditions, access to the cutterhead may be achieved from within the excavation chamber once pressure is reduced. In pressurised environments, a manlock system is used, and workers enter under controlled decompression protocols. Safety controls include isolation of mechanical drives, confirmation of stable face pressure, ventilation checks, and communication systems. Hyperbaric interventions, when required, follow strict medical and procedural standards.

Replacement involves removing worn cutter rings or complete cutter assemblies and installing new units with correct torque specifications. Proper seating and alignment are essential to avoid uneven wear and imbalance. Cutterhead bolt integrity and mounting surfaces are also inspected to ensure structural integrity. In hard rock TBMs, cutter change time can significantly affect productivity; therefore, efficient logistics, pre-staged spares, and trained crews are essential.

Cutter consumption is a major cost driver in rock tunnelling projects. Engineers often track cutter wear rates in terms of metres excavated per cutter or cubic metres excavated per millimetre of wear. This data feeds back into geological assessment, cost forecasting, and advance rate modelling. Unexpectedly high cutter wear may indicate stronger rock, faulted zones, abrasive minerals such as quartz, or misaligned cutterhead loading.

From a geotechnical perspective, cutter condition affects ground response. Worn cutters increase required thrust and torque, which can induce higher vibration and potentially greater disturbance in sensitive ground. In soft ground TBMs, poorly performing cutters can affect soil conditioning efficiency and face stability. Therefore, cutter inspection is not only a mechanical maintenance activity but also a ground control measure.

Figure 104: Disc cutter from the tunnel boring machine (TBM) "Åsa" used to dig and line the Hallandsås tunnel. Erik Lundin, CC BY-SA 3.0, via Wikimedia Commons.

Changing a TBM Cutter

Changing a TBM cutter means safely accessing the cutterhead, removing a worn tool (disc cutter in rock or picks/scrapers in soft ground), and reinstalling a new one to the correct fit, torque, and alignment—then proving everything is secure before re-start. The exact method depends on TBM type (hard rock vs EPB/slurry), cutter type (disc vs pick), and whether the intervention is at atmospheric pressure or hyperbaric. Below is a detailed, practical step-by-step process used on most projects.

1) Pre-job planning and controls

Before anyone touches the cutterhead, the job is engineered and authorised.

Job planning

- Confirm which cutters need changing (from wear logs, torque/vibration trends, cutterhead inspections).

- Identify cutter positions (station number/radius), tool type, and spare parts required (ring only vs full cutter, seals, bearings, bolts).
- Plan access: cutterhead man-lock/hyperbaric entry, or open-face access at atmospheric.
- Confirm required tools: torque wrench, hydraulic bolt tensioner (if used), pullers, drift pins, measuring gauges, lifting aids, chain block, grease, thread-lock, cleaning tools.

Permits and isolations

- Permit-to-work for: confined space, stored energy, lifting, hot work (if any), and hyperbaric (if applicable).
- Lock-out / tag-out (LOTO) for main drive, hydraulics, and any systems that could move the cutterhead, conveyor, or articulation.
- Verify zero-energy state: attempt-start checks, pressure bleed-down, mechanical locks/pins engaged.

Roles

- Supervisor/foreman (controls permit and sequence)
- TBM operator (machine state control)
- Fitters/mechanics (tool change)
- Electrician (LOTO verification)
- Survey/engineer (records, QA)
- Hyperbaric team (if pressurised): supervisor, lock attendant, medic cover

2) Make the TBM safe and stable

A. Stabilise face and machine condition

- Bring TBM to a controlled stop with stable face support:
  - **EPB**: maintain chamber pressure, control screw conveyor to avoid pressure drop, keep chamber "full" and stable.
  - **Slurry**: maintain slurry pressure and circulation; ensure filter cake stability.
  - **Hard rock/open**: secure face and ensure no fall-of-ground risk at cutterhead.

B. Park the cutterhead in a safe position

- Rotate cutterhead to a planned "parking" orientation so the target cutter(s) are accessible.
- Engage rotation lock/brake. Many projects also use mechanical locking pins.

C. Verify ventilation and atmosphere

- Gas testing at the work area (O_2, CO, H_2S, CH_4, and others as required).

- Confirm ventilation flow and emergency lighting.

3) Decide: atmospheric vs hyperbaric intervention

Atmospheric access (lower risk, simpler)

Used when the cutterhead/work area is not pressurised and is physically accessible.

Typical steps:

- Open cutterhead access doors/hatches (if designed).
- Install barriers and fall protection (work is often awkward, wet, and slippery).
- Proceed to removal/installation steps (Section 5 onward).

Hyperbaric access (common for EPB/slurry)

Used when the excavation chamber must stay pressurised to hold the face.

Key additional steps:

- Medical clearance and briefing for the team.
- Entry via manlock, controlled compression, comms checks.
- Continuous monitoring of chamber pressure, oxygen %, temperature, and worker time under pressure.
- Follow decompression tables exactly after work is finished.

4) Identify the cutter and confirm the replacement scope

Confirm the worn condition

- Disc cutter: measure ring wear (diameter loss, edge rounding), check for cracks, flat spots, chipping, excessive heat discoloration.
- Check for secondary issues: seized bearing, damaged hub, missing seals, abnormal play.
- Decide whether you replace:
    - Ring only (if hub/bearing is OK and design allows ring replacement), or
    - Complete cutter assembly (common and faster in many systems).

Record baseline

- Cutter position (number/radius)
- Observed wear and any damage
- Adjacent cutters condition (outer cutters often wear faster)

5) Remove the worn cutter (disc cutter example)

Disc cutters are usually mounted in a cutterhead housing with a pin/axle and retention system.

Step-by-step removal

1. Clean the area

- Remove packed fines/muck around the housing so tools fit properly.
- Good cleaning prevents damaged threads and inaccurate seating.

2. Support the cutter weight

- Use a small chain block, hoist point, or mechanical support.
- Disc cutters are heavy; uncontrolled drop is a serious injury risk and can damage the housing.

3. Remove the retention system

- Depending on design, this may be:
    - Bolted clamp plates
    - Retaining rings
    - Keeper plates with locking tabs
    - A locking bolt and pin arrangement
- Mark and retain hardware if it's reused (many projects replace critical bolts each time).

4. Extract the axle/pin

- Drive out or pull the axle using:
    - Drift pins and a sledge (controlled blows),
    - A hydraulic puller, or
    - A threaded extraction tool.
- If seized, you may need penetrating oil, controlled heat (if permitted), or hydraulic force.

5. Remove the cutter assembly

- Slide the cutter out of the housing with the hoist support.
- Inspect the housing for damage: ovalised holes, cracking, deformation, wear steps.

6. Inspect and clean mating surfaces

- Clean the axle seat, clamp faces, and any shims/spacers surfaces.
- Remove burrs; do not "grind away" fit surfaces unless approved.

6) Inspect the housing and related components (this is where failures are prevented)

Before installing a new cutter, confirm:

- No cracks in the housing welds or base plate
- No excessive wear in axle bores
- No deformation that would misalign the cutter (misalignment accelerates wear and can overload bearings)
- Sealing surfaces intact (where applicable)
- Correct shim pack available (some systems use shims to control lateral play)

If a housing is damaged, installing a new cutter without repair often leads to rapid repeat failure.

7) Install the new cutter

Installation sequence

1. Confirm the correct cutter

- Correct diameter (e.g., 17", 18", 19")
- Correct ring type (standard vs heavy-duty)
- Correct hub type and seal configuration
- Correct location (outer gauge cutters may differ)

2. Pre-lubrication (if required)

- Some cutter types are sealed-for-life; others require grease or initial lubrication.
- Follow manufacturer procedure—over-greasing can damage seals.

3. Position the cutter in the housing

- Use hoist to align cleanly.
- Ensure washers/spacers/shims are correctly oriented.

4. Insert the axle/pin

- Insert fully; it should seat to the correct depth.
- Mis-seating is a common cause of later loosening.

5. Install retention/keeper plates

- Fit locking devices (tabs, cotter pins, locking wire, thread-lock where specified).
- Torque bolts to spec in the correct sequence.

- If a hydraulic tensioner is used, follow the tensioning procedure and verify final torque.

6. Check rotation and play

- Manually rotate the cutter—should spin smoothly.
- Check for:
    - Excessive side play,
    - Roughness/seizure,
    - Noise or binding.

7. Mark and record

- Paint-mark fasteners (witness marks) to detect later loosening.
- Record cutter ID/serial, position, date/time, and reason for change.

8) Repeat for multiple cutters with a smart sequence

When changing several cutters:

- Work from safest access positions outward.
- Avoid leaving housings open longer than necessary.
- Keep the chamber tidy—loose parts become hazards.
- Maintain time discipline in hyperbaric work.

9) Close-up, function checks, and controlled restart

Close-up checks

- Account for all tools, bolts, and parts (tool control is critical).
- Verify all access doors/hatches closed and secured.
- Confirm all personnel are out of the cutterhead/chamber.

Remove LOTO only after verification

- Supervisor confirms:
    - mechanical locks removed in correct order,
    - systems ready,
    - permits closed.

Restart proof

- Start cutterhead at low RPM initially.
- Monitor:
    - torque trend stability,

    - vibration levels,
    - temperature sensors (main bearing/drive),
    - chamber pressure stability (EPB/slurry).
- If anything abnormal appears (new vibration, sudden torque spikes), stop and re-check.

10) Special notes by TBM type

EPB TBMs

- Cutter changes often require maintaining face pressure—hyperbaric interventions are common.
- Conditioning quality and chamber behaviour matter: unstable muck increases risk during interventions.

Slurry TBMs

- Focus on maintaining slurry pressure and managing slurry displacement in the chamber.
- Strict control of slurry lines and valves to prevent sudden slurry movement.

Hard rock TBMs

- Often atmospheric access if the head is accessible, but hazards include rockfall from the face and water inflow.
- Disc cutters see high loads—housing and axle bore wear checks are especially important.

11) Common failure modes (and how good practice prevents them)

- **Loose retention bolts** → caused by poor torque/tensioning or missing locking devices.
- **Seized cutters shortly after install** → contamination, misalignment, damaged seals, incorrect lubrication.
- **Repeat failures in same position** → housing wear/deformation, abnormal ground abrasivity at that radius, wrong cutter type.
- **Sudden torque spikes after change** → tool not seated, adjacent cutters worn, muck packing, or installation error.

12) What you'd typically include in the cutter change record

- Chainage/ring number
- Date/time, shift
- Cutterhead RPM, thrust/torque trends leading up to change
- Cutter position (radius/number) and type/size
- Wear measurement (ring thickness or diameter loss)

- Any housing damage observed
- Parts used (cutter serial, bolts replaced)
- Torque/tension values used
- Post-restart observations

Hyperbaric Interventions

Hyperbaric interventions in Tunnel Boring Machine (TBM) operations are controlled maintenance or inspection activities carried out under elevated pressure inside the excavation chamber or cutterhead area. They are necessary when the chamber must remain pressurised to maintain face stability, typically in Earth Pressure Balance (EPB) or slurry TBMs operating below groundwater level or in unstable soils. Depressurising the chamber to atmospheric conditions could trigger face collapse, groundwater inrush, or excessive settlement. Instead, personnel enter the pressurised environment using compressed air or mixed-gas systems, allowing essential maintenance to occur without compromising geotechnical equilibrium. In essence, hyperbaric intervention enables mechanical servicing while preserving ground stability.

In pressurised TBM operations, the excavation chamber is maintained at a pressure equal to or slightly above surrounding earth and groundwater pressures. This pressure prevents inward ground movement, balances hydrostatic forces, and controls volume loss. If the chamber were depressurised during maintenance, water or slurry could rush in, running ground conditions could develop, settlement or sinkholes might occur, and overall face stability could be lost. For this reason, activities such as cutter inspection and replacement must sometimes be performed while maintaining full operational pressure.

Hyperbaric access is most commonly required for disc cutter replacement in EPB or slurry machines, removal of obstructions such as boulders or debris, clearing clogging in cohesive soils, inspection of cutterhead openings, and repair of damaged wear components. These interventions are particularly common in high groundwater pressure environments, river crossings, soft or unstable sands and silts, and deep urban tunnels where pressure balance is critical.

There are several types of hyperbaric intervention. The most common in EPB tunnelling is compressed air intervention, where the chamber is pressurised with compressed air and workers enter through a manlock (airlock) system. Pressures typically reach 3–4 bar, sometimes higher under strict procedural control. Soil cohesion must be sufficient to prevent blow-out when slurry is replaced by air. In slurry-supported interventions, used in slurry TBMs, the chamber contains bentonite slurry and may be partially displaced during access. Compressed air or mixed gas is introduced, and higher pressures may be encountered. Careful control is required to prevent destabilisation of the filter cake at the face. For deeper or higher-pressure environments, mixed-gas systems such as heliox or trimix may be used to reduce nitrogen narcosis and improve physiological safety. In extreme conditions, saturation systems may be employed, where personnel remain under pressure between shifts to reduce decompression cycles, although these are rare and highly specialised operations.

A hyperbaric access system typically includes a manlock chamber for entry and exit, a medical lock for emergencies, pressure control panels, gas supply and monitoring systems, communication equipment, redundant power systems, and on-site medical support trained in hyperbaric procedures. The manlock allows gradual pressurisation and decompression, preventing rapid pressure changes that could cause injury.

Hyperbaric interventions carry significant physiological risks. Decompression sickness (DCS) can occur if pressure is reduced too quickly, causing dissolved nitrogen to form bubbles in tissues and blood. Barotrauma may damage ears, sinuses, or lungs due to pressure changes. Nitrogen narcosis can impair cognitive function at elevated pressures, particularly in deeper interventions, while oxygen toxicity may occur if oxygen partial pressures are excessive. Emergency response is complex because evacuation from a pressurised environment must follow controlled decompression protocols.

A typical hyperbaric cutter intervention follows a structured sequence. Face pressure and machine parameters are stabilised, and mechanical and electrical systems are isolated. Gas composition and chamber conditions are verified. Workers enter the manlock, undergo controlled compression to match chamber pressure, and then access the excavation chamber to perform the required maintenance. After completing the task, they return to the manlock and undergo controlled decompression according to approved tables, followed by health monitoring. Every stage is supervised and documented under strict procedural control.

These activities are embedded within the project's Safety Management System (SMS). They require permit-to-work authorisation, medical clearance, approved decompression schedules, on-site medical capability, defined exposure time limits, redundant pressure and gas systems, and detailed emergency procedures. Deviation from decompression protocols is not permitted without medical authority.

From an engineering perspective, hyperbaric intervention is more than a maintenance technique—it is a geotechnical stability strategy. It allows cutter changes and inspections without depressurising the face, enables tunnelling in high groundwater environments, supports continued excavation in challenging ground, and protects surface structures by maintaining pressure balance. Many deep urban and underwater TBM projects would be infeasible without hyperbaric capability.

Operationally, hyperbaric interventions are high-risk and time-consuming, so minimising their frequency is a key objective. Accurate cutter wear prediction, careful production planning, and continuous monitoring of torque and performance trends help schedule interventions efficiently. Modern TBM design increasingly focuses on extending cutter life and improving condition monitoring to reduce the need for pressurised access.

Bearing and Seal Maintenance

Main bearings in TBMs are typically large-diameter roller bearings configured to withstand enormous loads. These loads include combined axial thrust produced by thrust cylinders, radial forces from the ground reaction, and bending moments due to varying ground conditions Especially in hard rock scenarios, the dynamic and fluctuating loads present unique challenges [794, 795]. The operational environment is characterized by continuous rotation and abrasive conditions, leading to a heightened risk of bearing failure. Once installed, these bearings are often difficult to access, making preventive maintenance critical to avoid catastrophic operational issues [794, 796].

The inherent risks associated with main bearing maintenance stem from several factors:

- Extreme loads that can cause mechanical fatigue.
- Continuous operational demands requiring robust lubrication management.
- Abrasive environments that can lead to rapid wear.
- Limited accessibility complicating maintenance interventions.

Failure modes typically begin with minor issues, such as seal degradation or improper lubrication, which, if left unchecked, can escalate to significant operational failures, including bearing seizure or catastrophic breakdowns [795, 797].

Main Bearing Maintenance Practices

Lubrication Management

Effective lubrication is essential in mitigating bearing wear and reducing the risk of overheating. Maintenance activities involve:

- Scheduled grease injections through automatic lubrication systems.
- Regular verification of grease flow to all bearing raceways.
- Monitoring trends in grease consumption to detect anomalies.
- Inspection of lubrication lines for blockages or pump failures [796, 798].

Inadequate lubrication can lead to severe consequences, including increased friction and accelerated wear rates.

Temperature Monitoring

Continuous temperature monitoring of the main bearing through embedded sensors is integral to predictive maintenance. An upward trend in temperature can indicate several potential issues:

- Insufficient lubrication.
- Seal wear or degradation.
- Increased mechanical loads due to cutter imbalance or internal bearing wear. Sudden spikes in temperature necessitate immediate intervention [539, 798].

Vibration Monitoring

Vibration analysis is particularly critical in hard rock TBMs where operational feedback can indicate bearing health. Changes in vibration amplitude or frequency may reveal issues such as:

- Damage to rollers.
- Pitting or wear on raceways.
- Misalignments or imbalances within the cutterhead [796, 799].

Grease Sampling and Analysis

Periodic sampling of lubricants serves as a diagnostic measure for bearing health, allowing for the detection of:

- Presence of metal particles indicating wear.
- Water contamination that can degrade lubrication properties.
- Ingress of fine soil particles. Laboratory analysis can provide insights into the effectiveness of the lubrication system without necessitating bearing disassembly [796, 800].

Seal Systems and Their Importance

Seal systems in TBMs are critical for maintaining the integrity of the main bearing. They prevent external contaminants from intruding and stop internal lubricants from leaking out, facilitating a stable operating environment. In Earth Pressure Balance (EPB) TBMs, for instance, seals must withstand high pressures from bentonite slurries, whereas in slurry TBMs, the focus is on preventing soil and water ingress [794, 795].

Seal Maintenance and Inspection

Regular maintenance and inspection of seals involve:

- **Pressure balance checks** to ensure the grease barrier pressure is maintained above external pressures.
- **Visual inspections** during maintenance cycles for any signs of damage or wear.

- **Monitoring for potential leaks** which can indicate seal failure [66, 539].

Early detection of seal failure often prevents subsequent bearing failure, making seal integrity a priority in TBM operations.

Operational Integration and Maintenance Scheduling

The maintenance of bearings and seals is closely linked to the operational parameters of TBMs. Excessive torque, uneven cutter wear, and misalignment can elevate loads on bearings, thus necessitating regular inspections and potential adjustments. The typical maintenance regime includes:

- Daily temperature and pressure monitoring.
- Weekly lubrication system verification.
- Monthly grease sampling.
- Major inspections at prescribed intervals, typically during scheduled cutterhead access [796, 798].

Utilizing predictive analytics can enhance fixed maintenance schedules by allowing for data-driven adjustments based on real-time monitoring results [799].

Safety Considerations

Maintenance of seals and bearings often requires compliance with strict safety protocols due to the dangers associated with confined space entry, hyperbaric access, and the handling of heavy components. Implementation of lock-out/tag-out (LOTO) systems and other safety measures is essential to ensure worker safety during these high-risk activities [795, 796].

Engineering Significance

Ultimately, the maintenance of bearings and seals is integral to risk containment in TBM operations. The performance of these components directly influences machine reliability, operational stability, and project continuity. By proactively managing the health of bearings and seals, operators can maintain the delicate balance between machine performance and ground control, ensuring both safety and efficiency in tunnelling operations [794, 798, 799].

Breakdown Recovery Strategies

Breakdown recovery strategies in Tunnel Boring Machine (TBM) operations are structured engineering and operational responses designed to restore safe and controlled operation following mechanical failure, geotechnical instability, or system malfunction. Because TBMs operate underground under high load, confined conditions, and often under groundwater pressure, breakdown recovery is not simply a repair task — it is a controlled risk management process that must protect ground stability, worker safety, and surface assets while restoring machine functionality.

In mechanised tunnelling, recovery strategy depends on three key factors:

- The nature of the failure (mechanical, electrical, geotechnical, hydraulic)
- The ground conditions and face stability risk
- The accessibility of the affected component

Breakdown response must always prioritise geotechnical equilibrium before mechanical repair.

Immediate Stabilisation Phase

The immediate stabilisation phase is the first and most critical step in responding to any TBM breakdown. Before mechanical repair begins, the priority is to secure the excavation environment and maintain geotechnical equilibrium. In mechanised tunnelling, ground stability and face support take precedence over equipment recovery, because uncontrolled ground movement can escalate rapidly into settlement, inrush, or structural damage at the surface.

Initial stabilisation measures focus on maintaining face pressure within the approved operational window. Advance is halted in a controlled manner while keeping thrust forces balanced to avoid uneven loading on the lining. The screw conveyor (in EPB machines) or slurry circuit (in slurry TBMs) must be secured to prevent uncontrolled extraction or pressure fluctuations. At the same time, engineers verify that annular grout remains stable and that groundwater control systems are functioning as intended.

If the breakdown directly affects face pressure control, emergency stabilisation actions may be required. These can include temporarily increasing chamber pressure to prevent collapse, reducing spoil extraction rates to maintain pressure balance, installing temporary bulkheads where the design allows, or injecting stabilising grout to reinforce the surrounding ground. Throughout this phase, monitoring of settlement, pore pressure, and machine parameters is intensified to detect any developing instability.

Only once the ground and pressure conditions are confirmed stable does mechanical diagnosis and repair begin. This disciplined sequence ensures that breakdown recovery does not compromise face support, groundwater balance, or surface protection. In TBM operations, stabilising the ground is always the first and non-negotiable step in safe recovery.

Failure Classification and Diagnosis

The second phase of breakdown recovery involves structured failure classification and accurate diagnosis. Once the excavation environment has been stabilised, the focus shifts to identifying the nature, location, and severity of the fault. Clear classification helps determine the appropriate response strategy, required resources, and expected downtime. In TBM operations, rapid but disciplined diagnosis is essential to avoid unnecessary dismantling and to prevent secondary damage.

Mechanical failures are among the most common categories. These include cutterhead drive failure, main bearing overheating, gearbox damage, screw conveyor seizure in EPB machines, or slurry pump failure in slurry TBMs. Such faults often manifest through abnormal torque trends, vibration increases, or temperature rise. Because these systems operate under extreme loads, early recognition of mechanical degradation can prevent catastrophic failure.

Hydraulic failures represent another critical class of breakdown. These may involve cylinder seal rupture, high-pressure hose bursts, or system-wide pressure imbalance. Hydraulic systems control thrust, articulation, segment erection, and conveyor functions; therefore, leakage or pressure loss can quickly affect both machine performance and safety. Hydraulic oil leaks also introduce fire risk, particularly in confined underground environments.

Electrical failures include transformer faults, trailing cable damage, drive motor malfunction, or sensor failure. Electrical systems are fundamental to TBM control and monitoring, and faults may lead to sudden stoppage or unreliable parameter readings. In some cases, what appears to be a mechanical issue may originate from sensor malfunction or signal loss, underscoring the importance of systematic diagnostics.

Geotechnical and operational failures are equally significant. These include face instability, excessive torque caused by clogging or abrasive conditions, settlement exceedance detected at the surface, or slurry loss in permeable ground. Such events may not originate from machine defects but from changes in ground behaviour or operational imbalance. Addressing the root cause often requires adjusting conditioning, face pressure, or grouting practices rather than replacing hardware.

Modern TBMs rely heavily on real-time monitoring data to support diagnosis. Trend analysis of temperature, torque, vibration, pressure, and flow rates enables engineers to distinguish between symptom and cause. By interpreting data patterns rather than reacting to isolated alarms, teams can pinpoint the underlying issue more accurately. This data-driven approach reduces unnecessary dismantling, shortens downtime, and prevents recurrence of the same failure mode.

Planned In-Situ Repair

The third phase of breakdown recovery involves planned in-situ repair, where corrective work is undertaken within the tunnel without removing or retracting the TBM. This approach is generally the most efficient and cost-effective recovery strategy because it avoids the extensive time, logistics, and risk associated with machine retrieval. However, its feasibility depends on component accessibility, available working space, and the stability of the excavation environment.

Typical in-situ repair activities include cutter replacement, which may be performed under atmospheric conditions in stable ground or under hyperbaric conditions where face pressure must be maintained. Other common repairs include hydraulic hose replacement, motor or gearbox swaps within accessible drive units, conveyor belt repairs, and replacement of faulty sensors or instrumentation components. These tasks are usually planned during controlled stoppages and executed under strict lock-out and permit-to-work procedures.

In-situ repair relies heavily on prior design considerations, such as modular component layout, lifting arrangements, spare parts availability, and maintenance platforms. TBMs are engineered with internal access routes and lifting systems specifically to facilitate underground maintenance. Even so, confined space constraints and restricted movement mean that repairs require careful coordination, risk assessment, and sequencing.

Where access to the excavation chamber or cutterhead is required in pressurised ground conditions, hyperbaric intervention planning becomes necessary. This introduces additional safety protocols, decompression scheduling, medical supervision, and extended downtime. As a result, while in-situ repair remains the preferred strategy, its execution must be carefully evaluated against geotechnical stability, worker safety, and operational constraints before proceeding.

Hyperbaric Recovery Strategy

The fourth phase of breakdown recovery involves a hyperbaric recovery strategy, which is required when cutterhead components or other front-end systems fail while the excavation chamber is under pressure. In these situations, depressurising the chamber is not an option because it could compromise face stability, trigger groundwater inflow, or cause surface settlement. Repairs must therefore be undertaken in a controlled pressurised environment.

The process begins with stabilising face pressure within the approved operating window to ensure geotechnical equilibrium. All mechanical and electrical systems in the affected area are then isolated, and a formal risk assessment is conducted. A permit-to-work is issued only after confirming medical clearance for personnel, verifying decompression schedules, and ensuring that emergency systems are operational. Controlled pressurisation of the manlock allows workers to enter the chamber safely.

Once inside, inspection, repair, or replacement of the failed component—such as disc cutters, scrapers, wear plates, or cutterhead bolts—is carried out under strict supervision. Following completion of the work, personnel exit through the manlock and undergo controlled decompression according to approved decompression tables. Post-intervention medical monitoring is mandatory to detect any signs of decompression-related illness.

Because hyperbaric recovery involves elevated physiological risk, specialised personnel, and extended stoppage time, it significantly increases downtime and operational complexity. For this reason, predictive maintenance, cutter wear monitoring, and early fault detection are critical strategies aimed at reducing the frequency and duration of hyperbaric interventions.

Controlled Shutdown and Partial Retreat

The fifth breakdown recovery approach involves a controlled shutdown and, where necessary, a partial retreat of the TBM to a safer position. This strategy is used when a failure cannot be repaired immediately at the excavation face, either due to access constraints, unsafe ground conditions, or the need for more extensive intervention. The primary objective is to maintain ground stability while repositioning the machine to a location where repairs can be carried out safely and efficiently.

Depending on the circumstances, the TBM may advance cautiously into more competent or stable ground before stopping, reducing the geotechnical risk associated with repair works. In other cases, the machine may be retracted slightly to relieve excessive thrust loads, reduce cutterhead contact forces, or stabilise face conditions. Additional segmental lining may be installed to provide enhanced structural support, particularly if the stoppage is expected to be prolonged.

In more challenging ground conditions, supplementary stabilisation measures may be required before any movement occurs. These can include permeation grouting to improve soil strength and reduce permeability, installation of temporary bulkheads, or even ground freezing in extreme cases where collapse risk is high. Controlled shutdown and retreat strategies are therefore not simply mechanical decisions—they are integrated geotechnical interventions designed to preserve excavation stability while enabling safe breakdown recovery.

Major Component Replacement Underground

Major component replacement underground represents one of the most complex and resource-intensive breakdown recovery scenarios in TBM operations. Failures involving large systems—such as the main drive motor, gearbox, or major hydraulic assemblies—cannot always be resolved with minor in-situ adjustments. Instead, substantial mechanical intervention must be carried out within the confined underground environment, often far from direct shaft access.

Such operations require the establishment of engineered lifting and handling systems within limited space. This may involve installing temporary gantries, chain blocks, hydraulic jacking systems, or rail-mounted lifting frames capable of safely supporting and manoeuvring heavy components. Backup gantries must be secured and stabilised to prevent unintended movement during lifting operations. Because TBM components are often extremely heavy and awkwardly positioned, detailed lift planning and load path analysis are essential to prevent secondary damage or safety incidents.

Electrical and mechanical isolation procedures form a critical part of the process. High-voltage systems must be locked out and verified safe before work begins, and hydraulic systems must be depressurised to eliminate stored energy hazards. At the same time, ventilation systems must be managed carefully to ensure adequate airflow, especially if hot work or welding is required. Fire risk increases significantly during major mechanical replacement, particularly in the presence of hydraulic oil and electrical cabling, so additional fire watch measures and suppression readiness are typically implemented.

Logistics planning becomes especially challenging during long tunnel drives where the TBM is several kilometres from the nearest shaft. Transporting heavy replacement parts underground requires coordinated use of rail haulage, conveyors, or specialised transport bogies. Advance preparation—including spare part staging, access clearance, and shift scheduling—is critical to minimise downtime. In modern TBM projects, contingency planning for major underground component replacement is integrated into the overall maintenance and risk management strategy, recognising that recovery efficiency directly affects both programme certainty and geotechnical stability.

Retrieval to Shaft Strategy

The retrieval-to-shaft strategy represents a worst-case recovery scenario in TBM operations and is typically considered only when catastrophic failure makes underground repair impractical. Examples include severe main bearing failure, structural damage to the cutterhead, or irreparable drive system breakdown. In such cases, continued forward advance may be impossible, and standard in-situ repair methods may not provide a viable solution.

One possible approach involves partial or full disassembly of the TBM underground. Major components may be removed in sections and transported back through the tunnel to an access shaft. This process requires detailed structural assessment, heavy lifting plans within confined space, and strict control of ground stability throughout the operation. The disassembly itself must be sequenced carefully to avoid compromising shield support or exposing the excavation to instability.

If retrieval through the completed tunnel is not feasible, a dedicated recovery shaft may need to be constructed above the disabled machine. This allows vertical extraction of large components or the cutterhead. Constructing such a shaft in urban or environmentally sensitive areas requires extensive geotechnical modelling, groundwater control planning, and environmental approvals. The shaft excavation must also ensure that the ground around the stranded TBM remains stable during exposure.

In some cases, a reverse drive may be attempted, retracting the TBM to a previously stabilised section of tunnel where safer access and repair can occur. However, reverse drives are mechanically complex and not always possible, particularly in segmentally lined tunnels where thrust and sealing systems are designed for forward motion. As a last resort, the machine head may be abandoned in place, and a secondary TBM drive launched from an alternative alignment or from the opposite direction to complete the tunnel.

These strategies are rare but extremely costly and time-consuming. They introduce significant programme delays, contractual implications, and additional geotechnical risk. For this reason, major failure contingency planning, predictive maintenance, and continuous condition monitoring are critical components of TBM risk management. Where retrieval is required, comprehensive ground stability analysis is essential to ensure that dismantling, shaft construction, or machine extraction does not trigger collapse, settlement, or groundwater instability during recovery operations.

Geotechnical Risk During Breakdown

Geotechnical risk during a TBM breakdown is often more critical than the mechanical failure itself. When excavation stops unexpectedly, the balance between face pressure, ground support, and lining performance must be maintained continuously. Breakdown recovery planning must therefore account for potential loss of face support, stability of the annular tail void, groundwater inflow, and progression of surface settlement. The excavation environment does not become stable simply because the machine has stopped; in many cases, the risk increases.

If face pressure cannot be maintained within the approved operating window, ground may begin to yield inward, increasing the risk of running ground or localised collapse. At the same time, the tail void behind the shield must remain adequately filled and supported. Any delay in grout injection or incomplete filling of the annulus can allow relaxation of the surrounding ground, which may lead to delayed settlement at the surface.

Groundwater conditions present additional complexity. A prolonged stoppage can alter pore pressure conditions around the face or lining. In slurry TBMs, filter cake integrity may degrade if slurry circulation is interrupted or improperly balanced, increasing the risk of inflow. In EPB machines, conditioned muck may stiffen, segregate, or lose its sealing properties, affecting pressure transmission and stability.

As downtime extends, secondary geotechnical effects may develop. Consolidation of cohesive soils may continue under changed stress conditions. Pressure loss in the chamber or grout shrinkage in the tail void can progressively increase volume loss. In slurry operations, filter cake degradation may reduce effective sealing at the face. These time-dependent processes mean that a breakdown lasting several hours may present manageable risk, whereas a breakdown lasting days can significantly alter ground behaviour.

For this reason, monitoring intensity is typically increased during breakdown events. Face pressure, groundwater levels, settlement markers, thrust distribution, and torque trends are reviewed more frequently. Additional survey readings may be taken, and piezometer or pressure sensor data closely examined for early signs of instability. Breakdown recovery is therefore not only a mechanical repair operation but also an active geotechnical risk management exercise, requiring continuous assessment of ground response while restoration works are undertaken.

Emergency Response Integration

Emergency response integration is a critical component of breakdown recovery planning in TBM operations. Certain failures can rapidly escalate beyond mechanical repair issues and become life-safety emergencies. These include fire within electrical or hydraulic systems, sudden water inrush due to face instability or seal failure, hazardous gas detection, and electrical short circuits affecting high-voltage systems. In such situations, the priority shifts immediately from equipment recovery to personnel safety and environmental control.

Fire underground presents a particularly severe hazard because smoke propagation can be rapid in confined tunnels with limited egress. A hydraulic leak onto hot components or an electrical fault can quickly escalate, requiring activation of fire detection and suppression systems and controlled evacuation procedures. Similarly, water inrush events may threaten both face stability and personnel safety, requiring immediate isolation of systems, potential bulkhead closure (if designed), and withdrawal of personnel to safe zones.

Gas detection—such as methane, hydrogen sulphide, or oxygen deficiency—can require immediate power isolation, ventilation adjustment, and evacuation depending on trigger levels. Electrical short circuits may introduce both fire and electrocution risk, necessitating rapid lock-out/tag-out procedures and confirmation of safe isolation before any inspection or repair begins.

Because of these potential escalation pathways, recovery planning must be fully integrated with the project's Emergency Response Plan (ERP). This includes clearly defined evacuation routes, operational refuge chambers with life-support capability, reliable communication systems, and trained emergency coordinators. Command hierarchy and decision authority must be established in advance, not improvised during an incident.

In modern TBM operations, breakdown recovery procedures are therefore developed alongside emergency response protocols. The ability to transition quickly from repair mode to emergency response mode is essential. Effective integration ensures that when mechanical failure intersects with fire, flooding, gas, or electrical hazards, the system responds in a structured, disciplined, and safety-first manner.

Production and Programme Impacts

Breakdown recovery strategies in TBM operations are inseparable from production modelling and programme certainty. Every significant stoppage affects advance rate, crew utilisation, logistics sequencing, and downstream activities such as segment supply, spoil handling, and fit-out works. For this reason, recovery planning is not treated as an isolated maintenance function, but as a core element of overall project planning.

Effective recovery planning begins with a structured spare parts strategy. Critical components—such as hydraulic seals, sensors, motors, gearboxes, cutter tools, and electronic modules—must be identified in advance, with appropriate stock levels maintained either on site or within rapid supply reach. Long-lead items, including main drive components or specialised bearings, are often pre-procured or reserved to prevent extended downtime in the event of failure.

Redundancy is another key planning consideration. Redundant slurry pumps, backup power systems, secondary compressors, and parallel hydraulic circuits reduce the risk that a single component failure will halt production entirely. In slurry TBMs especially, duplicate pump systems and pipeline isolation arrangements can allow continued operation while repairs are undertaken.

Cutter stock levels and wear forecasting are directly tied to advance rate modelling. Projects with accurate cutter wear prediction and adequate on-site inventory can schedule replacements efficiently and avoid emergency hyperbaric interventions. Similarly, ensuring the availability of specialist contractors—such as hyperbaric supervisors, electrical technicians, or heavy lifting crews—prevents delays when complex interventions are required.

Well-prepared projects that integrate maintenance planning, spare strategy, redundancy design, and contractor readiness into their production model consistently experience shorter recovery times and more stable advance rates. By contrast, reactive response—where parts are ordered after failure or expertise must be sourced urgently—often leads to extended downtime and significant programme impact. In modern mechanised tunnelling, proactive breakdown planning is therefore a productivity control measure as much as a maintenance strategy.

Preventive Strategy as Recovery Mitigation

The most effective breakdown recovery strategy in TBM operations is prevention. Rather than relying solely on response planning after failure occurs, modern tunnelling projects focus on mitigating recovery events through predictive and preventive measures. This approach reduces the likelihood of catastrophic failure, shortens intervention timeframes, and protects both geotechnical stability and programme certainty.

Predictive monitoring forms the foundation of this strategy. Continuous tracking of torque, thrust, bearing temperature, vibration levels, hydraulic pressure, and electrical load allows engineers to identify subtle parameter drift before performance deteriorates. Trend-based alarms are configured to trigger early investigation when values deviate from normal operating envelopes, enabling corrective action during planned stoppages rather than emergency shutdowns.

Planned maintenance further reduces risk by replacing or servicing components at controlled intervals based on operating hours, advance distance, or condition monitoring data. Conservative parameter control—such as maintaining face pressure comfortably within the approved window and avoiding aggressive penetration rates in abrasive or unstable ground—also protects mechanical systems from overload. In parallel, accurate cutter wear modelling ensures timely tool replacement before excessive load transfers to the cutterhead structure or main bearing.

Importantly, most catastrophic TBM failures are not sudden, unpredictable events. They are typically preceded by measurable warning signs: rising temperatures, increasing vibration amplitude, abnormal torque spikes, fluctuating pressure trends, or oil contamination indicators. When these signals are recognised and acted upon early, recovery becomes minor intervention rather than major breakdown. In modern mechanised tunnelling, prevention is therefore the most powerful form of recovery mitigation, linking mechanical discipline directly to safety, ground control, and production reliability.

Stuck TBM Scenarios

A stuck Tunnel Boring Machine (TBM) scenario occurs when the machine becomes immobilized and unable to advance, retract, or rotate its cutterhead. This immobilization can arise from mechanical failures, geotechnical instability, or operational mishaps, differing significantly from routine stoppages that are less severe and usually manageable. When a TBM is stuck, it faces immense resistance from the surrounding ground or internal failures, creating a combination of mechanical stress, potential geotechnical instability, substantial production delays, and increased financial risks [801, 802].

Common Causes of a Stuck TBM

1. **Ground Convergence or Squeezing Ground**: One of the primary causes of a stuck TBM is excessive ground convergence or the occurrence of squeezing soils, particularly in weak clay or highly plastic soils. When the surrounding ground deforms inward around the shield, the TBM may become encased due to inadequate lubrication and slow advancement rates. This results in increased friction along the shield skin and can culminate in severe entrapment, with the tail void closing prematurely, thereby escalating drag forces that exceed the machine's thrust capacity [65, 683, 789].

2. **Insufficient Face Pressure**: In Earth Pressure Balance (EPB) or slurry TBMs, maintaining adequate face pressure is critical. A sudden loss of ground or under-pressure conditions can lead to local collapses, where loose material flows into the chamber and consolidates around the cutterhead, rendering it immobilized. When spoil extraction is compromised, it leads to torque spikes, which often culminate in

stalling rotation. The inability to restart movement in such conditions is exacerbated by uneven pressure equalization surrounding the machine [65, 789, 802].

3. **Obstruction-Related Entrapment**: Physical obstacles such as large boulders, buried steel, or unexpected concrete foundations can obstruct the cutterhead. In mixed-face conditions, where part of the cutter encounters hard rock while other parts remain in soft soil, asymmetric loading can twist or fully stall the machine. Issues with screw conveyors can also manifest when conditioned soil becomes too stiff, leading to potential clogs in cohesive clay [801, 803].

4. **Mechanical Failures**: Indirect mechanical failures—such as seized bearings, broken gearboxes, or malfunctioning thrust cylinders—can immobilize the machine even in stable ground conditions. Moreover, electrical system failures that disrupt pressure control can lead to unsafe conditions whereby restarting operations is rendered impossible [65, 801, 804].

5. **Hydraulic Entrapment**: In high groundwater pressure environments, inflows can destabilize the face or fill the chamber with excessive water or slurry, creating pressure imbalances that prevent cutterhead rotation. Grout over-pressurization behind the shield can similarly exacerbate movement resistance [4, 803].

Risks Associated with a Stuck TBM

The implications of a stuck TBM are significant, involving both mechanical and geotechnical risks. Prolonged stoppages can lead to increased ground consolidation, heightened shield skin friction, grout shrinkage, filter cake breakdown in slurry machines, and overall face instability. In cases where ground conditions are water-bearing, uncontrolled inflow can rapidly escalate the risks and consequences [789, 802, 804].

Recovery and Mitigation Strategies

The recovery strategies employed depend largely on the cause of the stuck condition. Strategies may include:

- Reducing shield friction through bentonite or polymer injections.
- Increasing thrust force within operational limits.
- Undertaking localized grouting ahead of the face.
- Implementing pressure stabilization techniques.
- Using hyperbaric removal of obstructions when necessary and, in extreme cases, initiating ground freezing to create temporary stability [65, 789, 804].

Preventive controls often provide the most effective means of mitigation. Monitoring rising thrust demand, torque spikes, and advance rate reductions provides early warning signs. Maintaining sufficient lubrication, conditioning, balanced face pressures, and addressing cutter wear are critical for reducing the likelihood of entrapment [4, 683, 802].

A stuck TBM scenario exemplifies a critical loss of equilibrium between the machine's capabilities and ground behaviour. Effective management of tunnelling operations necessitates the maintenance of this balance. Understanding the multifaceted causes of entrapment and implementing robust recovery and preventive mechanisms can significantly enhance the reliability of mechanized tunnelling operations.

Chapter 17

Special Tunnelling Conditions

Chapter 17 introduces some of the most demanding and technically complex environments encountered in modern mechanised tunnelling. While earlier chapters may have focused on standard ground conditions and baseline TBM operations, this chapter shifts attention to scenarios where geological, hydraulic, mechanical, thermal, and logistical forces intensify and interact in unpredictable ways. These are the conditions that test not only machine capability, but engineering judgement, operational discipline, and integrated project planning.

Special tunnelling conditions arise when the ground or environment departs from uniform, predictable behaviour. High-pressure groundwater, mixed face transitions, squeezing and swelling formations, abrasive and corrosive materials, long tunnel logistics, and deep stress regimes all introduce amplified risks. In these environments, the TBM is no longer simply cutting ground; it is operating within a tightly coupled system where soil mechanics, rock mechanics, hydraulics, materials engineering, structural performance, and human factors must remain in equilibrium.

The engineering challenge in special conditions is rarely confined to a single variable. Elevated groundwater pressure affects face stability, lining design, seal performance, and hyperbaric intervention planning. Mixed faces create asymmetric loading and mechanical imbalance. Deep tunnels amplify stress redistribution and thermal effects. Long drives transform excavation into a complex underground supply chain. Each condition multiplies the interaction between ground behaviour and machine response.

A defining characteristic of special tunnelling conditions is that margins for error narrow significantly. Pressure windows become tighter, deformation becomes time-dependent, wear accelerates, and downtime consequences increase. Reactive approaches are rarely sufficient. Instead, success depends on predictive modelling, conservative parameter control, redundancy in systems design, enhanced monitoring, and experienced crews capable of interpreting early warning signals.

This chapter examines each of these environments in detail, explaining not only the mechanisms involved but also the operational, structural, and programme implications. The objective is to develop a systems-level understanding of how extreme or complex conditions alter excavation behaviour and risk profiles.

Mastery of special tunnelling conditions represents advanced professional competence. It requires integration across geotechnical engineering, structural design, mechanical systems, hydrogeology, materials science, and project management. In modern mechanised tunnelling, the ability to anticipate, interpret, and control these environments often defines the difference between controlled production and critical incident.

High-pressure Groundwater Tunnels

High-pressure groundwater tunnelling represents one of the most demanding environments in mechanised excavation. These conditions arise when tunnels are constructed below the natural water table at significant depth, beneath rivers or seabeds, or through highly permeable soils such as sands, gravels, or fractured rock formations where hydrostatic pressures are substantial. In such environments, the Tunnel Boring Machine must resist not only in-situ earth stresses but also considerable water pressure, which may exceed several bar.

Unlike dry or low-permeability ground, groundwater-active conditions introduce hydraulic forces that directly influence face stability, lining performance, equipment configuration, and worker safety. Inadequate groundwater control can result in sudden inflow, face collapse, excessive settlement, or catastrophic flooding. Consequently, groundwater management becomes a central engineering consideration rather than a secondary environmental issue.

Hydrostatic pressure increases with depth according to the relationship $P = \rho g h$, where $\rho$is water density, $g$is gravitational acceleration, and $h$is the depth below the groundwater level. At approximately 30 metres depth, groundwater pressure reaches about 3 bar, while at 50 metres it can exceed 5 bar. In deep urban tunnels or subsea crossings, pressures may reach 7 to 8 bar or more. Groundwater behaviour is further influenced by soil permeability, fracture networks in rock, confined aquifer systems, and hydraulic connectivity to rivers or coastal zones. Where tunnels intersect confined aquifers, inflow potential can be both extremely high and sustained over long durations.

Below are worked engineering examples demonstrating how hydrostatic groundwater pressure is calculated and interpreted in tunnelling design.

1. Hydrostatic Pressure Fundamentals

Hydrostatic pressure is calculated using:

$$P = \rho g h$$

Where:

- ρ= water density (kg/m^3)
- g= gravitational acceleration (9.81 m/s^2)
- h= depth below groundwater level (m)
- P= pressure (Pa or kPa)

For freshwater:

$$\rho \approx 1000 \text{ kg/m}^3$$

2. Worked Example 1 – 30 m Depth (Typical Urban Tunnel)

Given:

- $\rho = 1000 \text{ kg/m}^3$
- $g = 9.81 \text{ m/s}^2$
- $h = 30 \text{ m}$

$$P = 1000 \times 9.81 \times 30$$
$$P = 294{,}300 \text{ Pa}$$

Convert to kPa:

$$P = 294.3 \text{ kPa}$$

Convert to bar:

$$1 \text{ bar} = 100 \text{ kPa}$$
$$P = 2.94 \text{ bar}$$

Result:

At 30 m depth, groundwater pressure ≈ **3 bar**

This means:

- TBM face pressure must be close to or slightly above 3 bar.
- Segment gaskets must be rated above this pressure.
- Hyperbaric interventions may require compressed air at similar pressure.

3. Worked Example 2 – 50 m Depth (Deep Metro Tunnel)

$$P = 1000 \times 9.81 \times 50$$
$$P = 490{,}500 \text{ Pa}$$
$$P = 490.5 \text{ kPa}$$
$$P = 4.9 \text{ bar}$$

Result:

At 50 m depth, groundwater pressure ≈ **5 bar**

Engineering implications:

- Face support pressure must remain within a narrow window.
- Seal systems and main bearing seals experience high loading.
- Hyperbaric intervention becomes physiologically demanding.
- Tail grout pressure must exceed 5 bar but remain below fracture limits.

4. Worked Example 3 – 80 m Depth (Subsea Crossing)

$$P = 1000 \times 9.81 \times 80$$
$$P = 784{,}800 \text{ Pa}$$
$$P = 784.8 \text{ kPa}$$
$$P = 7.85 \text{ bar}$$

Result:

At 80 m depth, groundwater pressure ≈ **7.8 bar**

This is typical for:

- Deep subsea tunnels
- Fjord crossings
- Deep shaft launch areas

At this pressure:

- Mixed-gas hyperbaric systems may be required.
- Slurry TBMs are strongly preferred.
- Lining design must resist significant hydrostatic loading.
- Small pressure errors can trigger blow-out or inrush.

5. Confined Aquifer Scenario – Why Pressure Can Be Higher Than Depth

In confined aquifers, pressure is not controlled by local depth alone but by the elevation of the recharge zone.

Example:

- Tunnel depth below surface = 40 m
- Water table at surface = 10 m above tunnel crown
- Aquifer recharge zone is at hill 20 m higher

Effective head:

$$h = 40 + 10 + 20 = 70 \text{ m}$$
$$P = 1000 \times 9.81 \times 70$$
$$P = 686{,}700 \text{ Pa}$$
$$P = 6.87 \text{ bar}$$

Result:

Although tunnel depth is only 40 m, effective pressure is nearly 7 bar.

This explains why confined aquifers can generate unexpectedly high inflow pressures.

6. Worked Example – Potential Inflow Through Permeable Gravel

Assume:

- Hydraulic head difference = 50 m
- Permeability $k = 10^{-3}\ m/s$(coarse gravel)
- Face area = 30 m^2
- Hydraulic gradient $i = h/L \approx 1$(simplified)

Using Darcy's Law:

$$Q = k \cdot i \cdot A$$
$$Q = 10^{-3} \times 1 \times 30$$
$$Q = 0.03\, m^3/s$$

Convert to litres per second:

$$0.03 \times 1000 = 30\, L/s$$

Result:

Potential inflow ≈ **30 litres per second**

If uncontrolled, this could:

- Flood chamber rapidly
- Destabilise face
- Cause erosion channels
- Lead to collapse

This demonstrates why permeability matters as much as pressure.

7. Engineering Interpretation Summary

Depth	Pressure (bar)	Risk Level
20 m	~2 bar	Moderate
30 m	~3 bar	High
50 m	~5 bar	Very High
80 m	~8 bar	Extreme

But pressure alone is not enough — risk depends on:

- Soil permeability
- Fracture connectivity
- Aquifer recharge potential
- Hydraulic continuity to rivers/sea

Confined aquifers present a particularly serious hazard in mechanised tunnelling because the groundwater within them is pressurised and hydraulically connected to distant recharge zones. Unlike unconfined groundwater, which is governed primarily by the local water table, a confined aquifer can exert sustained pressure even when the tunnel is located at moderate depth. If a TBM intersects such an aquifer, the pressure is not easily relieved, and inflow may continue for an extended period rather than dissipating naturally.

When a pressurised confined aquifer is breached, inflow can be both rapid and persistent. Because the aquifer is recharged from remote sources—such as elevated terrain, rivers, or regional groundwater systems—water supply may effectively be continuous. This means that simple pumping strategies are often insufficient to control the situation. Even aggressive dewatering may fail to reduce pressure adequately, particularly if the recharge area is extensive. In addition, groundwater drawdown may propagate far beyond the tunnel alignment, potentially affecting building foundations, adjacent infrastructure, or environmentally sensitive zones.

The sustained hydraulic pressure and inflow increase the risk of ground instability and settlement. As water migrates toward the excavation, soil structure can weaken, erosion pathways may develop, and consolidation may accelerate. In extreme cases, face instability, blow-out events, or significant surface settlement can occur if pressure balance is not carefully maintained.

For these reasons, tunnelling in confined aquifer conditions typically requires proactive engineering controls rather than reactive measures. Pre-grouting may be used to reduce permeability and limit inflow before excavation begins. In highly unstable zones, ground freezing can create a temporary impermeable and structurally stable block around the tunnel alignment. Slurry TBMs are often preferred in such environments because they provide precise hydrostatic pressure control at the face. In addition, contingency planning for hyperbaric interventions is essential, as cutter inspections and repairs may need to be carried out under sustained pressurised conditions.

Confined aquifers are dangerous not simply because they contain water, but because they contain pressurised, hydraulically connected water that can sustain inflow and destabilise the ground unless carefully controlled through advanced tunnelling and groundwater management strategies.

The principal engineering risks associated with high-pressure groundwater are primarily linked to imbalance. If face support pressure falls below hydrostatic pressure, water can rapidly enter the excavation chamber. In granular soils, this may trigger running ground conditions, erosion pathways, and localised collapse. Conversely, excessive support pressure may induce hydraulic fracturing or upward migration of slurry or conditioned soil, leading to surface heave, blow-out craters, or uncontrolled slurry discharge. Sudden inflow events may occur in high-permeability zones such as gravel lenses, faults, or karst cavities, potentially flooding the chamber and destabilising surrounding ground. Inadequate grout pressure behind the shield can allow groundwater infiltration into the annulus, resulting in delayed settlement or lining instability. Additionally, high-pressure conditions increase mechanical demands on seals, cutterhead drives, slurry pumps, and structural components of the shield.

TBM selection is therefore critical in high-pressure groundwater environments. Slurry TBMs are often preferred because hydrostatic slurry pressure can precisely balance external water pressure, while bentonite slurry forms a stabilising filter cake at the excavation face. Advanced pressure control systems enable fine adjustments to maintain equilibrium. These machines are particularly effective in permeable sands and gravels. EPB TBMs can also operate under groundwater pressure, but they require highly controlled soil conditioning, low-permeability conditioned muck, and careful pressure regulation. At very high pressures, EPB operation becomes increasingly complex and may necessitate hyperbaric interventions.

Maintaining the correct pressure window is essential. Under-pressure conditions can result in inflow, face instability, and settlement, while over-pressure may cause hydraulic fracturing, surface heave, or slurry migration. The allowable pressure window is governed by earth pressure at the face, hydrostatic pressure, soil shear strength, and the fracture gradient of the surrounding ground. Continuous monitoring of chamber pressure, slurry density, and extraction rates is mandatory to ensure stable excavation.

Groundwater control strategies extend beyond face pressure balancing. Precise regulation of slurry pressure or EPB chamber pressure, combined with controlled extraction rates, allows dynamic response to changes in depth and ground conditions. In highly permeable or unstable zones, pre-treatment methods such as permeation grouting, jet grouting, compensation grouting, ground freezing, or, in limited cases, dewatering wells may be implemented to reduce permeability and improve stability before excavation. Tail void grouting is equally critical. Grout injection pressure must exceed groundwater pressure but remain below the fracture threshold. Proper and rapid filling of the annular void prevents water migration, lining flotation, and delayed settlement. Two-component grouts are often used to achieve rapid setting under wet conditions.

Waterproof lining design also plays a vital role. High-pressure tunnels typically require double or triple gasketed segment joints, pressure-rated seals, robust bolting systems, and provisions for secondary grouting. The segmental lining must resist sustained external hydrostatic pressure, uplift forces, and long-term seepage. Durability design must also account for potential chemical attack from groundwater.

High groundwater pressures increase the likelihood of hyperbaric interventions. Cutter changes and chamber inspections may require compressed air or mixed-gas systems, supported by medical supervision and strict decompression protocols. At elevated pressures, risks such as nitrogen narcosis and decompression sickness become significant, necessitating specialised procedures and experienced personnel.

Enhanced monitoring and instrumentation are fundamental in these environments. Piezometers track groundwater levels, while slurry density and pressure sensors provide real-time operational data. Surface settlement markers, lining convergence measurements, and grout pressure sensors help detect instability early. Trigger levels are established for rapid pore pressure changes, excessive inflow, or abnormal pressure fluctuations, and monitoring frequency is typically increased during critical stages.

Environmental and community impacts must also be carefully managed. Groundwater drawdown or pressure changes can affect nearby building foundations, adjacent tunnels, aquifer systems, and local ecosystems. Regulatory approvals often require comprehensive hydrogeological modelling, long-term groundwater monitoring programmes, and contingency planning for potential inflow events.

Ultimately, high-pressure groundwater tunnelling is fundamentally a pressure management challenge. The TBM, the ground, and the groundwater form a coupled hydraulic-mechanical system in which stability depends on maintaining equilibrium. Stable pressure control results in controlled volume loss, reduced settlement, and reliable lining performance. Loss of control can escalate rapidly into flooding, collapse, surface damage, and significant project delays. For this reason, successful tunnelling in high-pressure groundwater conditions demands advanced machine selection, redundant pressure systems, highly trained crews, conservative operational control, and rigorous geotechnical modelling. Mastery of these conditions represents one of the highest levels of technical and operational discipline in modern mechanised tunnelling.

Mixed Face Conditions

Mixed face conditions occur when the excavation face of a Tunnel Boring Machine (TBM) simultaneously encounters two or more distinctly different ground types. This may include combinations such as soft soil overlying hard rock, weathered rock transitioning into intact rock, sand lenses embedded within clay, boulders within granular deposits, or fractured zones within otherwise competent rock. These conditions are among the most technically demanding scenarios in mechanised tunnelling because the machine must respond to fundamentally different ground behaviours at the same time.

In uniform ground, operational parameters such as torque, thrust, penetration rate, and face pressure can be stabilised within predictable ranges. Under mixed face conditions, however, different portions of the cutterhead are subjected to different levels of resistance. One section may be cutting hard rock requiring high thrust and torque, while another may be excavating soft soil that demands careful pressure control to prevent collapse. This imbalance leads to asymmetric loading, variable cutter wear, instability risks, and complex operational control challenges.

Mixed face conditions commonly arise in geological transition zones. These include soil-to-rock interfaces, fault zones, weathered rock profiles, glacial deposits containing embedded boulders, karst formations, and river crossings with alternating sediment layers. For example, the upper part of the face may consist of loose sand while the lower part consists of sandstone, or one side of the cutterhead may engage stiff clay while the opposite side cuts fractured granite. Each material responds differently to excavation forces, spoil conditioning, and support pressure. In effect, the TBM must operate simultaneously as both a soft-ground machine and a rock-cutting machine.

Mechanically, mixed faces create significant challenges. Asymmetric loading occurs when the rock side of the cutterhead demands higher thrust and torque than the soil side. This imbalance can result in cutterhead tilting, uneven wear patterns, increased vibration, localised stress concentration in the main bearing, and long-term structural fatigue. Over time, such uneven loading may accelerate bearing degradation or damage cutterhead components. In addition, balancing torque and thrust becomes complex. Applying sufficient force to penetrate rock may overstress the softer soil zone, while reducing thrust to protect soft ground may cause stalling in harder sections.

Cutter wear disparity is another critical issue. Rock-facing cutters experience abrasive wear and impact damage, while soil-facing tools may experience clogging or comparatively minimal wear. This uneven wear alters cutterhead geometry, further increasing imbalance and potentially destabilising excavation performance.

Face stability becomes significantly more difficult to manage under mixed conditions, particularly for EPB and slurry TBMs. The soil portion of the face requires adequate support pressure to prevent collapse and groundwater inflow, while the rock portion may require little support and can be damaged by excessive pressure. Because pressure distribution across the face is not uniform, maintaining a balanced pressure window becomes technically demanding. Improper pressure control may lead to localised soil collapse, erosion pathways, surface settlement, or overbreak in fractured rock zones.

Spoil management and conditioning are also complicated by heterogeneous excavation material. Rock fragments, clay, sand, and groundwater may enter the chamber simultaneously. In EPB machines, conditioning must create a homogeneous, low-permeability plastic mass capable of transmitting pressure evenly. The presence of rock fragments disrupts this uniformity, reducing pressure transfer efficiency and increasing torque demand. In slurry TBMs, rock fragments increase wear on pumps and separation systems and can reduce separation efficiency.

Operational risks increase significantly in mixed face zones. These include settlement arising from unstable soil sections, torque spikes when encountering harder material, machine jamming if large blocks dislodge, cutterhead clogging in cohesive soils, and increased frequency of hyperbaric interventions for inspection and maintenance. Sudden transitions from soft to hard ground can generate abrupt torque increases, while unexpected soft pockets can cause rapid pressure loss.

Groundwater interaction adds further complexity. Mixed faces often occur in transitional hydrogeological zones where rock may be relatively dry while adjacent soil layers are water-bearing. This can create localised inflow zones, differential pore pressures, and erosion risk at soil–rock interfaces. Water may preferentially migrate through permeable soil sections, resulting in uneven hydraulic behaviour across the face.

These challenges influence TBM selection and cutterhead design. EPB machines may struggle in highly abrasive rock sections, while slurry machines provide more precise pressure control but may experience increased wear. Cutterhead designs may incorporate combination cutter arrangements, reinforced structural elements, adjustable opening ratios, and replaceable wear plates. Hybrid TBMs are sometimes selected for projects expected to encounter significant geological transitions.

Effective management of mixed face conditions requires intensive monitoring and adaptive control. Continuous torque and thrust trend analysis, vibration monitoring, settlement and pore pressure measurements, and penetration rate comparison across rotations help identify emerging imbalances. Operators may adjust rotation speed, thrust distribution, conditioning injection rates, and chamber pressure setpoints in response to changing conditions. Inspection intervals are often shortened to monitor cutter wear and structural integrity.

Mitigation techniques may include pre-grouting weak soil zones, localised ground improvement, reduced advance rates during transitions, increased soil conditioning, controlled thrust application, and targeted cutter replacement. In extreme cases, temporary stabilisation measures such as ground freezing may be employed to manage highly unstable transition zones.

Mixed face tunnelling represents a coupled mechanical–geotechnical challenge in which instability in one portion of the face can compromise the entire excavation. The engineering objective is to maintain equilibrium between mechanical loading, face pressure, ground response, and spoil flow behaviour. Successful management requires conservative operational control, adaptable machine configuration, and experienced crews capable of responding to rapidly changing ground conditions. In modern tunnelling practice, mixed face conditions are recognised as high-risk environments requiring detailed geological forecasting, flexible machine design, and proactive risk mitigation strategies.

Squeezing and Swelling Ground

Squeezing and swelling ground conditions represent two distinct but related geotechnical challenges in tunnelling. Both involve time-dependent ground deformation, but their mechanisms differ fundamentally. Squeezing ground develops when weak, overstressed rock or soil plastically deforms inward toward the excavation under high in-situ stresses. Swelling ground, by contrast, occurs when clay-rich soils or weak rocks expand after exposure to water or stress relief. In mechanised tunnelling, both conditions can impose significant loads on the TBM shield, segmental lining, and support systems, requiring careful design, monitoring, and operational control.

Squeezing Ground

Squeezing ground typically occurs in weak, ductile formations subjected to high overburden stress. Common materials associated with squeezing include weak shales, mudstones, claystones, phyllites, schists, highly weathered rock, and other soft rocks with low compressive strength. When a tunnel is excavated, the removal of confinement causes stress redistribution around the opening. If the ground strength is low relative to the in-situ stress, the material cannot respond elastically and instead yields plastically, deforming inward toward the excavation.

Unlike brittle failure, squeezing is gradual and time-dependent. Deformation may continue for days, weeks, or even months after excavation. The mechanism is governed by the interaction between in-situ stress magnitude, rock mass strength, overburden depth, and ground stiffness. Where the ratio of ground strength to applied stress is low, plastic flow develops around the tunnel perimeter. This leads to convergence of the tunnel walls, increased shield skin friction, loading on the segmental lining, and reduced tail void clearance. At depth, vertical stresses may exceed 5–10 MPa, and weak formations may be incapable of sustaining these stresses without significant deformation.

For TBM operations, squeezing ground introduces serious risks. Convergence increases shield drag, requiring higher thrust to maintain advance. The tail void may close before grouting is fully completed, potentially compromising lining performance. Segmental rings may distort or crack under excessive compressive stress. If inward deformation is rapid, the shield can become gripped by the surrounding ground, leading to a stuck TBM scenario. Mitigation measures include reducing advance rate, applying lubrication along the shield skin, providing sufficient overcut, using flexible or yielding lining systems, and strengthening segment design where necessary. Grout pressure must be carefully controlled to ensure rapid and complete annulus filling before significant convergence occurs. Continuous monitoring of thrust trends and tunnel convergence is essential in squeezing conditions.

Swelling Ground

Swelling ground is generally associated with clay-rich materials that expand when exposed to water. Typical examples include montmorillonite-rich clays, smectite-bearing mudstones, clay shales, and anhydrite-bearing formations where chemical reactions may occur. When these materials are excavated and exposed to moisture, they absorb water and increase in volume, generating swelling pressures that act against the tunnel lining.

The mechanism differs from squeezing because swelling is primarily water-driven rather than stress-driven. Swelling clays contain mineral structures capable of absorbing water into their lattice, increasing interlayer spacing and producing volumetric expansion. Swelling pressures in highly reactive formations can reach several MPa. Chemical swelling may also occur, such as when anhydrite converts to gypsum in the presence of water, producing significant volume increase. Swelling is typically time-dependent and may continue long after construction is complete.

In TBM tunnelling, swelling ground can cause progressive loading on the segmental lining, ring deformation, cracking of concrete segments, and increased tail void pressure. During excavation, swelling may increase shield friction and reduce annular clearance in a manner similar to squeezing, particularly if groundwater is present. Over time, swelling may generate long-term maintenance challenges if not properly accounted for in design.

Combined Squeezing and Swelling Behaviour

In some geological settings, both behaviours may occur simultaneously. Weak clay-rich rocks may deform plastically under high stress while also expanding upon water exposure. This creates complex, time-dependent loading conditions involving both stress-induced convergence and moisture-induced expansion. Such environments are particularly difficult to manage because deformation mechanisms are coupled and may continue well beyond initial excavation.

Design and Operational Considerations

Segmental lining systems in squeezing and swelling ground must be designed to resist radial compressive stress, ovalisation, long-term creep, and swelling pressure. In swelling formations, compressible layers or yielding elements may be incorporated between the ground and lining to accommodate expansion. Tail void grouting must fully fill the annulus rapidly, maintain appropriate pressure, and prevent groundwater ingress. In swelling clays, grout selection must consider chemical compatibility and permeability to avoid exacerbating expansion.

TBM shield design also plays a critical role. Smooth external surfaces, lubrication systems, adequate overcut, and reinforced structural components help manage ground interaction. In squeezing conditions, shield length and stiffness influence performance; longer shields may experience higher friction if convergence develops.

Enhanced monitoring is essential. Convergence measurements, thrust and torque trend analysis, segment stress instrumentation, piezometer readings where groundwater is present, and long-term lining deformation monitoring provide early warning of progressive loading. Trigger levels should be established for rapid convergence, rising thrust demand, and segment cracking.

Risk of Stuck TBM

Squeezing ground is a common cause of stuck TBM scenarios. Progressive inward deformation can trap the shield, particularly during prolonged stoppages for maintenance or breakdown. Swelling ground may similarly reduce clearance and increase drag, although typically over a longer time scale. In both cases, interruption of advance increases the risk of machine entrapment.

Engineering Significance

Squeezing and swelling ground conditions are fundamentally time-dependent deformation problems. Unlike instantaneous collapse mechanisms, they evolve progressively after excavation. The TBM, lining, and surrounding ground interact continuously under changing stress and moisture conditions. Successful management requires conservative advance control, accurate geological forecasting, flexible structural design, intensive monitoring, and rapid tail void stabilisation.

In mechanised tunnelling, squeezing and swelling ground are recognised as high-risk conditions demanding close integration between geotechnical modelling, structural design, and disciplined operational control. Effective management reflects advanced understanding of ground–structure interaction and long-term tunnel performance.

Abrasive and Corrosive Environments

Abrasive and corrosive environments represent two of the most demanding material-related challenges in mechanised tunnelling. Unlike instability-driven conditions such as squeezing ground or groundwater inrush, these environments primarily affect the durability and longevity of the TBM, its cutting tools, slurry or EPB systems, and the segmental lining. However, their indirect impacts on production rate, maintenance frequency, safety, and project cost can be substantial. Successful operation in such environments depends on careful machine specification, appropriate material selection, intensive monitoring, and disciplined preventive maintenance.

Abrasive Environments

Abrasive ground conditions occur when excavated material contains hard, angular, high-strength particles capable of wearing down steel surfaces. Typical abrasive formations include quartz-rich sands, high-silica sandstones, granite and other crystalline rocks, basalt, gravel deposits, and glacial tills containing hard rock fragments. Quartz is particularly problematic due to its high hardness (Mohs hardness of approximately 7). Even relatively small quartz content can significantly increase wear rates on cutting tools and structural components.

Abrasion occurs when hard particles slide, roll, or impact against metal surfaces under pressure. In TBM operations, this primarily affects the cutterhead, disc cutters, scrapers, buckets, wear plates, screw conveyors in EPB machines, and slurry pipelines and pumps in slurry systems. Three dominant wear mechanisms are typically observed: sliding abrasion, where particles scrape across surfaces under pressure; impact wear, where rock fragments strike components at high velocity; and erosive wear, particularly in slurry systems where high-speed fluid carrying fine particles gradually erodes pipe walls and pump components.

Wear rate depends on several factors, including rock hardness, abrasivity index (such as the Cerchar Abrasivity Index), particle angularity, contact pressure, relative velocity, and the hardness of the tool material. In highly abrasive formations, cutter consumption often becomes the primary production constraint. Rapid disc cutter wear increases cutter change frequency, raises torque demand, reduces penetration rate, and accelerates structural wear on the cutterhead. In slurry TBMs, pipelines and pump impellers may thin due to erosive wear, increasing failure risk. In EPB machines, screw conveyor flights and casings may require frequent refurbishment.

Managing abrasive conditions requires an integrated engineering approach. Hardened steel alloys, tungsten carbide inserts, wear-resistant liners, and replaceable wear plates are commonly used. Cutterhead design may incorporate optimised cutter spacing, reinforced bucket edges, and replaceable edge protection. Continuous monitoring of cutter wear, torque trends, penetration rates, and slurry pipeline thickness helps predict maintenance needs. Operational adjustments such as controlled rotation speed, optimised thrust application, and avoiding excessive penetration can reduce impact loading. Predictive modelling of cutter life is essential, particularly in pressurised environments where hyperbaric interventions are costly and risky.

Corrosive Environments

Corrosive environments arise when groundwater or soil contains chemical constituents capable of degrading metal or concrete over time. Common corrosive agents include chlorides in marine or coastal settings, sulphates, acidic groundwater with low pH, industrial contaminants, hydrogen sulphide gas, and dissolved carbon dioxide. Corrosion may affect both temporary TBM components and permanent segmental lining systems.

Corrosion is typically an electrochemical process in which metal reacts with its environment, leading to gradual material loss. Forms of corrosion encountered in tunnelling include uniform corrosion, where material loss occurs evenly across a surface; pitting corrosion, which produces localised deep cavities particularly in chloride-rich environments; stress corrosion cracking, resulting from combined tensile stress and chemical attack; and sulphate attack on concrete, where reactions with cement hydrates cause expansion and cracking. In slurry TBMs, corrosion may also affect pipelines and separation plant components exposed continuously to chemically active fluids.

The consequences of corrosion include degradation of bolts and structural steel, deterioration of seals, reduced service life of hydraulic systems, electrical system damage, and long-term deterioration of concrete segments. In permanent linings, sulphate-rich groundwater may compromise concrete durability, while chloride-rich environments can accelerate reinforcement corrosion if protection is inadequate. These effects may not immediately disrupt production but can significantly affect long-term performance and maintenance costs.

Combined Abrasive and Corrosive Conditions

In some projects, abrasive and corrosive conditions occur simultaneously. For example, subsea tunnels excavated through quartz-rich sand under saline groundwater are exposed to both intense mechanical wear and chemical attack. This combination accelerates component deterioration and increases maintenance frequency. In such environments, design must address both wear resistance and chemical resistance concurrently.

Design and Material Considerations

Effective management begins during design. TBM structural components may incorporate corrosion-resistant coatings, cathodic protection systems where appropriate, stainless or treated steel elements, and sealed bearing systems. Segmental lining design may include sulphate-resistant cement, low-permeability concrete mixes, increased reinforcement cover, high-quality gaskets, and waterproof membranes. In slurry and EPB systems, replaceable wear liners, hardfaced screw conveyors, abrasion-resistant pipelines, and corrosion-resistant pump materials are commonly specified. Careful material selection prevents premature deterioration and reduces lifecycle cost.

Monitoring and Maintenance

Enhanced inspection regimes are essential in abrasive and corrosive conditions. Regular cutterhead inspections, wear plate measurement, pipeline thickness gauging, groundwater chemical analysis, bolt and seal inspection, and concrete durability testing form part of a structured maintenance strategy. Predictive maintenance based on wear modelling and condition monitoring reduces unplanned stoppages and limits emergency interventions.

Production and Programme Implications

Abrasive conditions directly affect production by increasing cutter changes and reducing advance rates. Corrosive environments may not immediately reduce advance but influence long-term reliability and durability. High cutter consumption increases the frequency of hyperbaric interventions in pressurised conditions, raising safety risk and downtime. Spare parts logistics become critical, particularly in long drives far from shafts.

Engineering Significance

Abrasive and corrosive environments are fundamentally material degradation challenges. Although they may not trigger immediate ground instability, their cumulative effects can compromise mechanical reliability, elevate safety risk, and increase project cost. In abrasive formations, mechanical wear becomes the dominant operational constraint. In corrosive environments, long-term durability governs performance and lifecycle planning.

Successful management requires accurate geological and hydrochemical forecasting, appropriate machine and material specification, proactive maintenance planning, continuous monitoring, and conservative operational control. In modern mechanised tunnelling, these conditions underscore the importance of materials engineering alongside geotechnical design. Mastery of abrasive and corrosive environments ensures safe excavation, predictable production, and durable long-term tunnel performance.

Long Tunnels and Logistics Challenges

Personnel transport is another significant consideration. In long tunnels, workers may rely on dedicated trains or electric vehicles to reach the face. Travel times in a 10 km tunnel may exceed 30 to 45 minutes each way, reducing productive shift time and increasing fatigue. Fatigue management becomes essential, particularly when workers must perform physically demanding tasks after extended travel. Emergency response times are also extended due to the distance from shafts.

Emergency planning becomes more complex as tunnel length increases. In the event of fire, flooding, gas detection, or mechanical failure, evacuation distances are greater and response times longer. Refuge chambers must be positioned at strategic intervals, communication systems must function reliably over long distances, and rescue teams must have clear access routes. Fire risk is particularly critical in long tunnels because smoke dispersion and evacuation duration increase with length.

Maintenance and spare parts management become strategic functions in long drives. Critical components such as cutters, hydraulic hoses, electrical modules, conveyor rollers, and pump parts must be staged underground or readily accessible. If essential components are not immediately available, downtime increases significantly due to transport delays. Predictive maintenance becomes especially important to avoid breakdowns that are difficult and time-consuming to access.

Geological variability also compounds logistical challenges. Long tunnels often pass through multiple ground types, sometimes far from the nearest shaft. Transition zones may require hyperbaric interventions or specialised support systems deep within the drive. Groundwater inflow at distance may require higher pumping capacity due to increased hydraulic head, further complicating response efforts.

Distance amplifies downtime across all systems. A repair that might take two hours near the shaft may require six hours deep within a long tunnel due to travel, setup, and coordination. Conveyor repair time increases with system length, and emergency drills take longer to execute. Production modelling must account for these cumulative inefficiencies, as even minor breakdowns can significantly affect programme certainty.

Mitigation strategies for long tunnel logistics include redundant conveyor drives, booster pump stations, distributed electrical substations, staged spare parts stockpiling, multiple refuge chambers, advanced digital monitoring systems, and integrated scheduling between surface and underground teams. Digital twin modelling and production simulations are increasingly used to optimise logistics before excavation begins. In some projects, intermediate shafts or adits are constructed to reduce logistical distances and improve redundancy.

Long tunnel drives transform tunnelling into a systems engineering challenge. Mechanical, electrical, environmental, and logistical systems must function cohesively over many kilometres. In short tunnels, geological conditions may dominate risk; in long tunnels, logistics, reliability, and operational discipline often determine success. The primary engineering objective is maintaining continuous, balanced production despite increasing distance from surface support. Stability at the face must be matched by stability in supply chains, power systems, ventilation, and emergency preparedness.

Long tunnels do not simply extend excavation length—they multiply complexity. Mastery of long tunnel logistics reflects advanced project planning, redundancy design, predictive maintenance, and disciplined coordination across all engineering systems.

Deep Tunnelling Effects

Deep tunnelling introduces a complex set of geotechnical, structural, thermal, hydraulic, and operational challenges that differ significantly from shallow or intermediate-depth excavation. As depth increases, in-situ stresses, groundwater pressures, rock temperatures, and stress redistribution effects intensify. In mechanised tunnelling, particularly with Tunnel Boring Machines (TBMs), deep conditions transform excavation from primarily a stability and logistics challenge into a high-stress, high-pressure systems engineering problem. At depth, every governing force acting on the tunnel is amplified.

The most fundamental deep tunnelling effect is the increase in vertical stress due to overburden weight. Vertical stress is commonly approximated by the relationship $\sigma v = \gamma H$, where γ represents the unit weight of the overlying rock mass (typically 25–27 kN/m^3) and H is the depth below the ground surface. At 100 metres depth, vertical stress may be approximately 2.5 MPa. At 500 metres, it may exceed 12–13 MPa, and at 1000 metres it can surpass 25 MPa. In addition to vertical stress, horizontal stresses—particularly in tectonically active regions—may equal or exceed vertical stress. Consequently, tunnels at depth experience far greater stress concentrations than near-surface excavations.

When a tunnel is excavated, confinement is removed and stresses redistribute around the excavation boundary. At shallow depth, this redistribution may remain within elastic limits.

At greater depth, however, redistributed stresses may exceed the strength of the surrounding rock mass. This can result in plastic yielding, spalling, slabbing, squeezing behaviour, or even rock bursts in strong brittle formations. Stress concentrations are typically highest at the crown and sidewalls, where tangential stresses may reach several times the original in-situ stress. The deeper the tunnel, the greater the likelihood of brittle failure in strong rock and ductile yielding in weak rock.

In competent, brittle rock under high stress, deep tunnels may be susceptible to rock burst—an abrupt and violent release of stored strain energy. Rock bursts can involve sudden ejection of rock fragments, shock waves, damage to the TBM cutterhead, and serious safety risks. Spalling is a related but less violent phenomenon in which slabs detach from the tunnel boundary due to tensile stress development near the excavation surface. While rare in shallow tunnels, these behaviours become increasingly significant at depths exceeding several hundred metres.

In weaker rock masses at depth, high confining stress may exceed the rock mass strength, resulting in progressive inward deformation or squeezing. Deep squeezing is typically more severe than shallow squeezing because stress magnitudes are higher. Consequences may include significant convergence, increased shield skin friction, higher thrust demand, and compressive loading on the segmental lining. Time-dependent deformation or creep may continue well after excavation, complicating long-term lining performance.

Deep tunnels also encounter elevated hydrostatic pressures where groundwater is present. Groundwater pressure increases approximately 0.1 bar per metre below the water table. At 100 metres depth, pressure may reach 10 bar; at 300 metres, approximately 30 bar. Such pressures increase seal loading, intensify face pressure control requirements, complicate hyperbaric interventions, and elevate the risk of sudden inrush. Slurry TBMs are often preferred in deep, high-pressure groundwater conditions because of their precise pressure balancing capability.

Another significant effect of depth is increased geothermal temperature. Rock temperature typically rises according to the geothermal gradient, often between 25 and 35°C per kilometre. At depths approaching 1000 metres, rock temperatures may exceed 40–50°C depending on local conditions. Elevated temperatures affect worker safety, ventilation design, cooling requirements, hydraulic oil performance, and electrical system reliability. Heat management becomes a critical design and operational parameter in deep tunnels.

High confining stress at depth also affects TBM structural performance. Increased external pressure may require reinforced shield structures, higher-capacity main bearings, stronger thrust systems, and enhanced sealing systems. Cutter forces may increase due to stress confinement in rock, affecting penetration rate and torque demand. Elevated contact forces can accelerate cutter wear. Consequently, deep tunnelling demands robust machine specification and careful parameter control.

Segmental lining design must also account for depth-related effects. Linings must resist high external compressive stress, ovalisation, long-term creep, and elevated groundwater pressure. Structural design must consider both short-term construction loads and long-term ground–lining interaction. Deep tunnels often require thicker segments, higher-strength concrete, increased reinforcement, and enhanced gasket systems. In some cases, composite or secondary lining systems are adopted for additional structural resilience.

Stress and permeability interactions at depth add further complexity. High in-situ stress may close fractures and reduce rock permeability, but excavation-induced stress relief may reopen fractures locally, creating inflow pathways. This makes groundwater prediction more difficult. Intersecting a deep confined aquifer may result in sustained inflow under high pressure.

Given the elevated consequence of instability at depth, enhanced monitoring is essential. Instrumentation may include convergence monitoring, stress measurement within lining segments, microseismic monitoring for rock burst detection, piezometers for groundwater pressure, temperature sensors, and continuous torque and thrust analysis. Trigger–action–response frameworks must be conservative because instability can escalate rapidly under high stress conditions.

From a programme and risk perspective, deep tunnelling increases severity across all hazard categories. Rock burst may damage equipment, high groundwater pressure increases flood risk, elevated temperatures reduce productivity, and squeezing may slow advance rates. Maintenance and hyperbaric interventions at depth are more complex and higher risk. As a result, contingency allowances, insurance provisions, and safety planning are typically more stringent for deep projects.

Deep tunnelling fundamentally transforms excavation behaviour. Ground response becomes dominated by high stress rather than shallow gravitational effects. Mechanical systems must withstand elevated loads, and precise face pressure control becomes more critical. Successful deep tunnelling requires advanced geotechnical modelling, detailed stress analysis, conservative operational control, structural redundancy, and highly skilled crews. At depth, small errors in stress estimation, pressure balance, or support design can escalate rapidly.

Deep tunnels are not simply longer or deeper excavations; they operate within a fundamentally different stress regime where the earth's forces are magnified. Mastery of deep tunnelling conditions reflects integration of rock mechanics, machine capability, structural engineering, groundwater control, thermal management, and disciplined operational control.

Chapter 18

Project Management and Economics

Mechanised tunnelling is often viewed primarily through the lens of engineering performance, cutterhead torque, face pressure control, segmental lining behaviour, and advance rate. Yet beneath the mechanical and geotechnical systems lies an equally critical dimension: the economic and managerial architecture that determines whether a project succeeds or fails. Tunnel Boring Machines represent high-capital, high-risk production systems operating in uncertain geological environments. Their performance is not governed solely by engineering design, but by procurement strategy, contractual risk allocation, productivity modelling, cost control, and long-term financial planning.

This chapter moves from the mechanics of excavation to the management of uncertainty and capital. It examines how procurement decisions shape technical risk, how scheduling assumptions translate directly into financial exposure, and how contractual frameworks distribute responsibility among clients, contractors, designers, and manufacturers. In TBM projects, time is cost, and risk is capitalised. Small deviations in utilisation, cutter wear, or downtime accumulate into significant commercial consequences when multiplied over kilometres of tunnel and months of operation.

Project management in mechanised tunnelling therefore demands an integrated systems perspective. The TBM is not merely equipment; it is the central production asset, the primary risk-control mechanism, and one of the largest cost drivers in the project. Decisions made during procurement—whether contractor-supplied, client-supplied, or collaborative—define the technical envelope within which operations occur. Schedule modelling, productivity forecasting, and lifecycle costing determine whether the project is robust or vulnerable to geological variability.

Economic performance in tunnelling is fundamentally linked to engineering realism. Deterministic assumptions, optimistic advance rates, or under-specified machines frequently lead to programme extension and cost escalation. Conversely, conservative specification, probabilistic scheduling, structured benchmarking, and disciplined risk allocation can protect programme certainty and financial stability.

As the preceding sections of this book have demonstrated, TBM operations are governed by mechanical forces, ground response, and system reliability. This chapter demonstrates that they are equally governed by procurement philosophy, contractual clarity, productivity modelling, and economic foresight. Successful mechanised tunnelling requires not only technical competence, but strategic alignment between engineering design, commercial structure, and risk management.

TBM procurement strategies

Procurement of Tunnel Boring Machines is a pivotal strategic decision in mechanised tunnelling projects. Due to their highly specialised nature, TBMs are tailored to meet project-specific requirements, encompassing not only the anticipated ground conditions but also alignment geometry, groundwater pressures, and logistical constraints [805, 806]. The complexity inherent in TBM procurement necessitates a comprehensive approach that integrates engineering judgement with commercial and contractual considerations, profoundly impacting capital costs, operational risk, and long-term asset performance [455, 807].

Procurement Strategies and Models

Several procurement models exist for TBM acquisition, each with distinct risk allocation profiles among clients, contractors, and manufacturers. In general contractor-supplied arrangements, typical of Design & Construct (D&C) or Engineering, Procurement, and Construction (EPC) contracts, contractors assume the responsibility of selecting and procuring the TBM, which can yield benefits like alignment of machine selection with construction methodology. This model, however, diminishes client control over detailed specifications and might lead to a focus on minimum compliance rather than lifecycle durability [806, 808]. Conversely, in client-supplied models, project owners have direct responsibility for TBM procurement, allowing for stringent control over specifications and greater potential for standardisation across multiple drives [805]. This approach, while advantageous in some respects, may limit contractor flexibility and impose additional logistics burdens on the client [807].

Hybrid or collaborative models, commonly found in Public-Private Partnerships (PPP) and Early Contractor Involvement (ECI) frameworks, aim to combine the strengths of both approaches by allowing contractor input during the final configuration phase while maintaining a client-led baseline specification. This siding encourages a balanced risk-sharing environment, effectively marrying technical optimization with risk management [455, 806].

1.1 Contractor-Supplied TBM (Design–Build or EPC Contracts)

Under this model, the contractor is responsible for selecting, specifying, and procuring the TBM. The client defines performance requirements, but detailed machine design rests with the contractor and manufacturer.

Advantages:

- Risk largely transferred to contractor
- Contractor selects machine aligned with methodology
- Streamlined design–construction integration

Disadvantages:

- Reduced client control over machine specification
- Potential for minimum-compliance design
- Limited transparency in lifecycle durability considerations

This model is common in Design & Construct (D&C) or EPC contracts.

1.2 Client-Supplied TBM

In this strategy, the project owner procures the TBM directly and supplies it to the construction contractor.

Advantages:

- Full control over machine specification
- Standardisation across multiple contracts
- Potential cost savings in large programmes
- Asset reuse across projects

Disadvantages:

- Client assumes technical specification risk
- Storage and logistics responsibility
- Reduced contractor flexibility

This model is common in large metro programmes where multiple drives are planned.

1.3 Hybrid or Collaborative Procurement

A shared approach where the client leads baseline specification while contractors contribute to final machine configuration.

Often used in:

- Alliance contracts
- PPP (Public–Private Partnership) projects
- Early Contractor Involvement (ECI) models

This approach aims to balance risk, technical optimisation, and commercial certainty.

Selection of TBM Type and Ground Condition Considerations

One of the crucial aspects of TBM procurement is selecting the appropriate type of machine to match the geological conditions anticipated during tunnelling [807]. Variants of TBMs include Earth Pressure Balance (EPB) machines, slurry TBMs, and hard rock machines. The decision is influenced by factors such as ground type, abrasivity, permeability, groundwater pressure, and tunnel specifics like depth and diameter [805, 809]. Poorly chosen TBM types may lead to substantial financial losses, highlighting the importance of detailed geotechnical investigations and conservative geological interpretations prior to selection [807, 809].

Modern procurement practices increasingly adopt a risk-informed specification approach. In uncertain geological environments, over-specification of TBMs can enhance structural redundancy and operational flexibility. This strategy entails increased upfront costs but reduces the likelihood of catastrophic equipment failures and project delays due to unforeseen geological challenges [455, 805].

Timing and Logistics in TBM Procurement

Timing for TBM procurement is critical. The manufacturing lead times for TBMs can span from 12 to 24 months, contingent on the complexity and specifications of the machine. To safeguard against project delays, it is common for clients to pre-procure TBMs even before the final design is concluded [455, 807]. This pre-emptive action helps in maintaining schedule certainty while accommodating the necessary time for detailed design, fabrication, testing, and logistics planning.

In projects involving extensive metro or rail networks, procuring multiple identical TBMs can present advantages like streamlined spare parts management and training. Nonetheless, it is crucial to evaluate the trade-off between standardisation benefits versus adaptability to varying geological conditions along the tunnelling route [805, 807].

Financial Considerations and Lifecycle Costs

The financial implications surrounding TBM procurement extend beyond the initial purchase price. TBM acquisition can amount to tens of millions of dollars; thus, a comprehensive analysis of total lifecycle costs is imperative. Factors such as operational expenses, maintenance intervals, cutter consumption rates, and potential downtime risks should be critically assessed. Moreover, contractual arrangements may include performance guarantees, warranties, and refurbishment clauses that further complicate cost assessments [805, 810].

Ultimately, TBM procurement is far more than a mere acquisition of equipment; it represents a strategic decision-making process that significantly influences excavation safety, efficiency, and project delivery timelines. The interplay between procurement strategy, machine type selection, and logistical considerations must be appropriately navigated to optimise project outcomes. As the field of mechanised tunnelling evolves, modern procurement strategies incorporating risk management, sustainability, and technological integration will be crucial in addressing the complex demands of future tunnelling projects.

Case Study Scenario: Metro River Crossing – Strategic TBM Procurement Decision

Project Overview

A metropolitan transport authority plans a 7.2 km twin-bore metro tunnel beneath a dense urban corridor, including a 1.4 km river crossing section. The tunnel diameter is 6.8 m internal, with depths ranging from 18 m to 42 m below surface. Ground investigations indicate variable conditions:

- Alluvial sands and gravels near the river
- Interbedded clay and silt layers
- Occasional sandstone lenses
- Groundwater pressures up to 4.5–5.0 bar at maximum depth
- Moderate abrasivity due to quartz-rich sands

The programme is politically sensitive, with strict settlement limits due to heritage buildings above alignment. The project duration is 5 years, and the TBM drives must be completed within 24 months.

The authority must determine the optimal TBM procurement strategy.

Step 1: Procurement Model Decision

<u>Option A – Contractor-Supplied TBM (D&C Model)</u>

Under a Design & Construct contract, the contractor would select and procure the TBM based on performance requirements.

Advantages considered:

- Risk transfer for machine performance
- Contractor alignment between method and equipment
- Simplified contractual interface

Concerns identified:

- Risk of under-specification to reduce bid price
- Limited transparency on seal rating and pressure redundancy
- Potential minimum compliance design

Given the high groundwater pressures and settlement sensitivity, the client determines that insufficient control over machine specification presents unacceptable risk.

<u>Option B – Client-Supplied TBM</u>

The authority evaluates directly procuring two TBMs to ensure full specification control and potential reuse on future metro extensions.

Advantages:

- Full control over pressure rating and seal design
- Standardisation across twin drives
- Lifecycle cost optimisation
- Potential future asset reuse

Risks:

- Client assumes specification error risk
- Responsibility for storage and assembly logistics
- Potential misalignment with contractor's preferred method

After technical and commercial review, the authority selects a hybrid procurement model.

Option C – Hybrid / Collaborative Procurement (Selected Strategy)

The authority develops a baseline technical specification with:

- Slurry TBM configuration
- Main bearing seal rated to 6 bar (above maximum predicted 5 bar)
- Reinforced bulkhead
- Provision for mixed-face cutterhead tooling
- Hyperbaric intervention system rated for 6 bar
- Abrasion-resistant slurry pipeline lining

Shortlisted contractors participate in Early Contractor Involvement (ECI), refining cutterhead design, spoil handling layout, and maintenance access configuration.

This model balances client control with contractor operational optimisation.

Step 2: TBM Type Selection

Ground conditions indicate:

- High permeability near river
- Hydrostatic pressure up to 5 bar
- Mixed soil and occasional rock lenses
- Settlement sensitivity

Engineering analysis concludes:

- EPB machine may struggle with high permeability gravels
- Slurry TBM provides superior pressure control and filter cake formation
- Hybrid mixshield not necessary due to limited rock percentage

A slurry TBM is selected due to groundwater dominance and settlement control requirements.

Step 3: Risk-Based Specification

Because borehole spacing is limited beneath the river, geological uncertainty remains.

To mitigate risk, the specification includes:

- Cutterhead reinforcement for occasional sandstone lenses
- Replaceable wear plates
- Redundant slurry pumps
- Dual seal system with pressure monitoring
- Capability to operate at 20% above predicted maximum pressure
- Provision for compressed air hyperbaric intervention

This increases capital cost by approximately 8%, but modelling shows it reduces catastrophic failure risk by more than 40%.

Step 4: Custom vs Standard Machine

The manufacturer offers a catalogue slurry TBM rated to 4 bar.

However, predicted groundwater may reach 5 bar with confined aquifer potential.

The project elects for a custom-engineered machine with:

- 6 bar pressure rating
- Enhanced gasket design
- Reinforced cutterhead torque capacity
- Abrasion-resistant pump internals

Delivery time increases by 4 months, but schedule modelling confirms that avoiding unplanned intervention outweighs delay risk.

Step 5: Financial Evaluation

Capital cost per TBM: USD $22.5 million /AUD $32 million

Custom upgrade cost: +USD $1.76 million/+ AUD $2.5 million per machine

Lifecycle analysis estimates:

- Potential downtime cost per week: USD $2.12 million /AUD $3 million
- Hyperbaric intervention cost per event: USD $282,000/AUD $400,000
- Estimated intervention reduction with higher spec: 6–8 fewer events

Net present value modelling demonstrates that the higher-spec machine is economically justified.

Step 6: Contractual Safeguards

Procurement contract includes:

- Minimum availability guarantee of 85%
- Seal certification at 6 bar
- 24-month manufacturer on-site technical support
- Performance-based bonus for reduced cutter consumption
- Spare parts supply agreement for 3 years
- Buy-back option for future project reuse

This reduces commercial dispute risk.

Step 7: Factory Acceptance Testing (FAT)

The authority requires:

- Full pressure bulkhead testing to 6 bar
- Main bearing seal test at 110% operating pressure
- Cutterhead rotation under simulated torque load
- Control system redundancy verification
- Slurry pump performance testing

This reduces commissioning uncertainty once underground.

Step 8: Logistics and Assembly Planning

Due to urban constraints:

- TBMs delivered in modular sections
- Night-time heavy transport permits required
- Underground assembly hall pre-constructed
- 1,200-tonne crane scheduled months in advance

Procurement includes transport engineering review.

Step 9: Digital Integration

The TBM package includes:

- Real-time torque and thrust monitoring
- Slurry density and pressure telemetry
- Cutter wear prediction software
- Integration with project digital twin
- Remote diagnostics from manufacturer headquarters

This supports predictive maintenance and reduces breakdown risk.

Outcome

During excavation:

- Groundwater pressures peak at 5.2 bar in river section
- No major inrush events occur
- Hyperbaric interventions reduced by 35% compared to similar past projects
- Advance rate remains stable at 12–14 m/day
- Settlement remains within allowable limits

Although capital cost was higher than a minimal specification machine, programme certainty was maintained, and no catastrophic shutdown occurred.

This case study demonstrates that TBM procurement is fundamentally a risk engineering decision rather than a simple equipment purchase. The machine becomes the central production and risk control system for the project, and its specification directly influences operational stability, safety, and programme certainty. Procurement strategy must therefore be approached with the same level of rigour as geotechnical design or structural analysis.

The selection of the procurement model significantly affects control, accountability, and risk allocation between client, contractor, and manufacturer. Whether the TBM is contractor-supplied, client-supplied, or collaboratively specified determines who carries specification risk, how transparently design decisions are made, and how closely the machine aligns with project objectives. Strategic alignment between technical requirements and contractual structure is essential.

The case further highlights that TBM type selection must be closely matched to groundwater pressure, permeability, and overall ground behaviour. In environments with high hydrostatic pressure or permeable granular soils, machine capability in pressure control and face stability becomes critical. Incorrect type selection can introduce instability, increase intervention frequency, and elevate settlement risk.

A risk-based specification philosophy can justify higher upfront capital expenditure when geological uncertainty or extreme conditions are present. Designing for worst-case pressure, incorporating redundancy, and strengthening structural components may increase procurement cost but significantly reduce the probability of catastrophic failure, prolonged downtime, and costly remediation.

The use of custom-engineered machines, rather than standard catalogue models, may reduce lifecycle risk in complex or high-pressure environments. Although customisation extends manufacturing lead time and increases initial cost, it allows optimisation for specific geological and hydraulic conditions, improving reliability and performance over the duration of the drive.

Clear contractual provisions and comprehensive factory acceptance testing (FAT) play a protective role in project delivery. Performance guarantees, seal certification, structured testing regimes, and defined support arrangements reduce commissioning risk and minimise the likelihood of disputes during construction.

Digital integration, including real-time monitoring, predictive maintenance tools, and remote diagnostics, enhances operational reliability and supports proactive intervention planning. These systems enable early identification of abnormal trends, improving decision-making and reducing unplanned stoppages.

Finally, the case illustrates that under-specification in high-risk environments can jeopardise both programme and safety. Machines designed only for average conditions may struggle under peak loads, leading to excessive hyperbaric interventions, mechanical failure, settlement issues, and significant delays. In complex tunnelling environments, conservative and well-integrated procurement strategies are essential to maintaining control, protecting workers, and delivering the project successfully.

Scheduling and Productivity Analysis

Scheduling and productivity analysis are fundamental to the successful delivery of mechanised tunnelling projects. Unlike many surface construction activities, TBM excavation is a continuous, system-driven operation in which mechanical performance, ground response, logistics, and maintenance are tightly interconnected. Programme certainty therefore depends not only on a theoretical advance rate, but on a realistic understanding of how production is generated, sustained, interrupted, and recovered over time. Scheduling defines when activities occur and how they are sequenced, while productivity analysis determines how fast those activities can realistically proceed under actual site conditions. In TBM tunnelling, the two disciplines are inseparable.

At its most basic level, TBM productivity is expressed as advance rate, calculated as the length excavated divided by time. However, real production is influenced by multiple interacting variables, including penetration rate per revolution, cutterhead rotation speed, utilisation percentage, downtime frequency and duration, cutter change intervals, spoil removal efficiency, segment installation cycle time, and overall logistics performance. Theoretical maximum output rarely reflects actual performance. Effective scheduling must therefore rely on realistic productivity assumptions rather than idealised machine specifications.

At its simplest, TBM productivity is expressed as advance rate:

$$\text{Advance Rate} = \frac{\text{Length Excavated}}{\text{Time}}$$

However, actual productivity depends on multiple interacting components:

- Penetration rate (mm/rev)
- Cutterhead rotation speed (rpm)
- Utilisation rate (%)
- Downtime frequency and duration
- Cutter change time
- Spoil removal efficiency
- Segment installation cycle time
- Logistics performance

Theoretical maximum production rarely matches realised production. Scheduling must therefore use realistic productivity assumptions rather than idealised machine performance.

Below are practical worked examples to illustrate how theoretical TBM advance rate differs from realistic, scheduled production.

Example 1 – Simple Advance Rate (Idealised)

A TBM excavates 240 metres in 20 days.

$$\text{Advance Rate} = \frac{240\ \text{m}}{20\ \text{days}} = 12\ \text{m/day}$$

On paper, the average advance rate is **12 m/day**.

However, this number does not explain how that production was achieved or what interruptions occurred.

Example 2 – Penetration Rate and Rotation Speed

Assume:

- Penetration per revolution = **8 mm/rev**
- Cutterhead speed = **2 rpm**

Instantaneous Penetration Rate (IPR):

$$8 \text{ mm/rev} \times 2 \text{ rev/min} = 16 \text{ mm/min}$$

Convert to metres per hour:

$$16 \text{ mm/min} = 0.016 \text{ m/min}$$
$$0.016 \times 60 = 0.96 \text{ m/hour}$$

If boring continuously for 24 hours:

$$0.96 \times 24 = 23.0 \text{ m/day}$$

Theoretical maximum production = **23 m/day**

But this assumes:

- 100% utilisation
- No ring building time
- No maintenance
- No delays

This never occurs in reality.

Example 3 – Incorporating Utilisation

Assume realistic utilisation = **50%**

$$23 \text{ m/day} \times 0.50 = 11.5 \text{ m/day}$$

Now the realistic production becomes **11.5 m/day**, close to the 12 m/day average seen earlier.

This demonstrates how utilisation drastically affects scheduling.

Example 4 – Effect of Ring Build Time

Assume:

- Excavation stroke = 1.5 m
- Excavation time per stroke = 1 hour

- Ring build time = 45 minutes
- Auxiliary checks = 15 minutes

Total cycle time per 1.5 m:

$$1 + 0.75 + 0.25 = 2 \text{ hours}$$

Production per 24-hour period:

$$\frac{24}{2} = 12 \text{ strokes/day}$$
$$12 \times 1.5 = 18 \text{ m/day}$$

Now assume:

- 6 hours/day lost to minor stoppages

Effective working time = 18 hours:

$$\frac{18}{2} = 9 \text{ strokes/day}$$
$$9 \times 1.5 = 13.5 \text{ m/day}$$

Small delays significantly affect output.

Example 5 – Cutter Change Impact

Assume:

- Cutter change every 250 m
- Cutter change duration = 10 hours
- Average production = 12 m/day

Time to reach 250 m:

$$\frac{250}{12} \approx 21 \text{ days}$$

Lost time due to cutter change:

$$10 \text{ hours} \approx 0.42 \text{ days}$$

Adjusted average rate over 250 m section:

Total time = 21.42 days

$$\frac{250}{21.42} = 11.7 \text{ m/day}$$

One cutter intervention slightly reduces average rate.

If abrasive ground increases frequency to every 150 m:

Time to 150 m:

$$\frac{150}{12} = 12.5 \text{ days}$$

Add 10-hour intervention:

12.92 days total

$$\frac{150}{12.92} = 11.6 \text{ m/day}$$

Higher abrasivity reduces average production further.

Example 6 – Spoil Removal Constraint

Assume TBM theoretical cutting = 15 m/day.

But slurry plant maximum throughput limits removal to 600 m^3/day.

If tunnel cross-section generates 50 m^3 per metre:

Maximum excavation possible:

$$\frac{600}{50} = 12 \text{ m/day}$$

Even if cutting rate allows 15 m/day, logistics cap production at 12 m/day.

The bottleneck is no longer cutting performance—it is surface plant capacity.

Example 7 – Logistics Delay in Long Tunnel

Assume:

- Repair near shaft = 2 hours
- Same repair 8 km from shaft requires:
  - 1 hour travel each way
  - 2 hours repair

Total downtime:

$$2 + 2 = 4 \text{ hours additional}$$

Total = 6 hours

If daily boring time was 12 hours:

Utilisation drops from 50% to 37.5%.

Production decreases accordingly.

Distance amplifies downtime.

Example 8 – Hyperbaric Intervention Impact

Assume:

- 1 hyperbaric intervention every 1 km
- Duration = 24 hours
- Drive length = 6 km

Total intervention time:

$$6 \times 24 = 144 \text{ hours}$$

= 6 days lost

If planned production = 12 m/day:

$$6 \times 12 = 72 \text{ m lost equivalent}$$

Programme must absorb this loss.

Example 9 – Combined Realistic Scenario

Theoretical cutting capacity: 23 m/day
Realistic utilisation: 50%
Ring build constraints: -2 m/day
Logistics losses: -1 m/day
Cutter change effects: -0.5 m/day

Realistic scheduled rate:

$$23 \times 0.50 = 11.5$$
$$11.5 - 2 - 1 - 0.5 = 8 \text{ m/day}$$

Final realistic production = **8 m/day**

This is often the gap between optimistic planning and real-world delivery.

Several productivity metrics are used to analyse performance. Instantaneous Penetration Rate (IPR) measures the rate of advance while the cutterhead is actively cutting and is calculated as penetration per revolution multiplied by rotational speed. While useful for assessing cutting efficiency, IPR excludes stoppages and auxiliary activities. Shift production measures the total metres excavated per working shift and incorporates ring building, inspections, and minor delays. Daily Advance Rate (DAR) evaluates performance over a 24-hour period and is often a key programme indicator. Utilisation rate, defined as boring time divided by total available time, provides insight into operational efficiency. High-performing projects may achieve utilisation levels between 40 and 60 percent depending on geology and operational discipline.

Types of Productivity Metrics

Instantaneous Penetration Rate (IPR)

This measures the rate of advance while the cutterhead is actively boring:

$$\text{IPR} = \text{Penetration per revolution} \times \text{RPM}$$

This reflects cutting efficiency only and excludes stoppages.

Shift Production

Total metres excavated per shift (typically 8–12 hours).

This includes:

- Boring time
- Ring building
- Cutter inspections
- Minor delays

Daily Advance Rate (DAR)

Metres per 24-hour period. This is a key programme metric.

Utilisation Rate

$$\text{Utilisation} = \frac{\text{Boring Time}}{\text{Total Available Time}}$$

High-performing TBM projects may achieve 40–60% utilisation depending on ground conditions.

Below are worked numerical examples to clearly illustrate each productivity metric and how they relate to one another.

1. Instantaneous Penetration Rate (IPR)

Definition

$$\text{IPR} = \text{Penetration per revolution} \times \text{RPM}$$

This measures cutting performance only.
It assumes the cutterhead is actively boring and excludes all stoppages.

Example 1 – Soft Ground

Penetration per revolution = **10 mm/rev**
Rotation speed = **2.5 rpm**

$$IPR = 10 \times 2.5 = 25 \text{ mm/min}$$

Convert to metres per hour:

$$25 \text{ mm/min} = 0.025 \text{ m/min}$$
$$0.025 \times 60 = 1.5 \text{ m/hour}$$

If boring continuously for 24 hours:

$$1.5 \times 24 = 36 \text{ m/day (theoretical)}$$

This is purely cutting efficiency. It does not include ring build time or downtime.

Example 2 – Hard Rock

Penetration per revolution = **4 mm/rev**
Rotation speed = **3 rpm**

$$IPR = 4 \times 3 = 12 \text{ mm/min}$$
$$12 \text{ mm/min} = 0.012 \text{ m/min}$$
$$0.012 \times 60 = 0.72 \text{ m/hour}$$
$$0.72 \times 24 = 17.3 \text{ m/day (continuous boring)}$$

Even though RPM increased, lower penetration reduces IPR significantly.

2. Shift Production

Shift production measures total metres excavated during an operational shift and includes:

- Boring
- Ring build
- Inspections
- Minor delays

Example 3 – 12-Hour Shift

Assume:

- Boring time = 6 hours
- Ring building = 4 hours
- Inspections and minor delays = 2 hours
- IPR during boring = 1.2 m/hour

Metres bored during shift:

$$6 \times 1.2 = 7.2 \text{ m}$$

Shift Production = **7.2 metres**

Even though the machine is on shift for 12 hours, it only bores for 6 hours.

Example 4 – Improved Efficiency

If ring build time reduces by 1 hour (through crew learning):

- Boring time increases to 7 hours

$$7 \times 1.2 = 8.4 \text{ m}$$

Shift production increases by **1.2 m** without changing cutting performance.

This shows operational efficiency affects output.

3. Daily Advance Rate (DAR)

DAR measures metres excavated in a full 24-hour period.
It is the most commonly reported programme metric.

Example 5 – Two-Shift Operation

Assume:

- 2 shifts per day
- Each shift produces 7.2 m

$$DAR = 7.2 \times 2 = 14.4 \text{ m/day}$$

Example 6 – Unexpected Breakdown

Assume:

- One 4-hour breakdown during the day
- Boring time reduced by 4 hours

- IPR = 1.2 m/hour

Lost production:

$$4 \times 1.2 = 4.8 \text{ m}$$

New DAR:

$$14.4 - 4.8 = 9.6 \text{ m/day}$$

A short breakdown reduces daily advance by 33%.

This demonstrates why DAR fluctuates more than IPR.

4. Utilisation Rate

Definition

$$\text{Utilisation} = \frac{\text{Boring Time}}{\text{Total Available Time}}$$

It measures how much of the available time the TBM is actually cutting.

Example 7 – Moderate Performance

Total available time (24 hours)
Boring time = 11 hours

$$Utilisation = \frac{11}{24} = 0.46 = 46\%$$

This falls within typical high-performing ranges (40–60%).

Example 8 – High Utilisation

Boring time = 14 hours

$$\frac{14}{24} = 0.58 = 58\%$$

This would be considered very strong performance.

Example 9 – Low Utilisation Due to Maintenance

Boring time = 8 hours

$$\frac{8}{24} = 33\%$$

Even if IPR is high, low utilisation severely reduces overall production.

Comparing the Metrics Together

Assume:

IPR during boring = 1.5 m/hour
Utilisation = 50%

Maximum theoretical 24-hour cutting:

$$1.5 \times 24 = 36 \text{ m/day}$$

Realistic production:

$$36 \times 0.50 = 18 \text{ m/day}$$

If ring build and minor delays reduce effective boring time further to 10 hours:

$$1.5 \times 10 = 15 \text{ m/day}$$

The difference between theoretical and realised production is 21 metres per day.

TBM scheduling typically includes shaft construction, machine assembly and commissioning, launch, main drive excavation, breakthrough, and disassembly. The main drive phase dominates the overall programme and requires detailed modelling of cycle time and production variability. Each metre of tunnel involves a sequence of operations, including excavation stroke, TBM advance, segment installation, tail void grouting, spoil removal, survey verification, and routine maintenance checks. Cycle time is therefore the sum of excavation time, ring build time, and auxiliary activities. Even minor increases in cycle duration can significantly affect overall programme duration when multiplied over several kilometres.

The following are structured numerical examples to demonstrate how the TBM cycle works in practice and how small time increases affect programme duration.

1. Basic TBM Cycle Example

Assume:

- Stroke length = **1.5 m per ring**
- Penetration rate during cutting = **1.0 m/hour**

Step 1 – Excavation Stroke

Time to excavate 1.5 m:

$$\frac{1.5}{1.0} = 1.5 \text{ hours}$$

Step 2 – Ring Build

- Segment erection time = 40 minutes
- Bolting and alignment = 20 minutes

Ring build time = **1.0 hour**

Step 3 – Tail Void Grouting

Occurs during advance but final stabilisation takes
= 0.25 hours

Step 4 – Auxiliary Activities

- Spoil clearance checks = 0.10 hours

- Survey check = 0.15 hours
- Maintenance inspection = 0.20 hours

Auxiliary time = 0.45 hours

Total Cycle Time

$$CycleTime = 1.5 + 1.0 + 0.45$$
$$= 2.95 \text{ hours}$$

Production per 24 hours:

$$\frac{24}{2.95} = 8.1 \text{ cycles/day}$$
$$8.1 \times 1.5 = 12.2 \text{ m/day}$$

Daily advance rate ≈ **12 m/day**

2. Small Increase in Ring Build Time

Now assume ring build increases by 15 minutes due to crew inefficiency or segment handling delays.

New ring build time:

$$1.0 + 0.25 = 1.25 \text{ hours}$$

New cycle time:

$$1.5 + 1.25 + 0.45 = 3.2 \text{ hours}$$

Production:

$$\frac{24}{3.2} = 7.5 \text{ cycles/day}$$
$$7.5 \times 1.5 = 11.25 \text{ m/day}$$

A 15-minute delay reduces production by nearly 1 metre per day.

3. Impact Over Long Drive

Assume tunnel length = **6,000 m**

At 12.2 m/day:

$$\frac{6000}{12.2} \approx 492 \text{ days}$$

At 11.25 m/day:

$$\frac{6000}{11.25} \approx 533 \text{ days}$$

Difference:

$$533 - 492 = 41 \text{ days}$$

A 15-minute increase per cycle causes a 41-day programme extension.

4. Example – Excavation Rate Reduction

Assume harder ground reduces penetration rate from
1.0 m/hour → 0.8 m/hour

Excavation time:

$$\frac{1.5}{0.8} = 1.875 \text{ hours}$$

Cycle time:

$$1.875 + 1.0 + 0.45 = 3.325 \text{ hours}$$

Production:

$$\frac{24}{3.325} = 7.2 \text{ cycles/day}$$
$$7.2 \times 1.5 = 10.8 \text{ m/day}$$

Harder rock reduces daily advance by over 1.4 m/day.

5. Example – Spoil Removal Constraint

Assume spoil conveyor is temporarily restricted, adding 0.20 hours per cycle.

New cycle time:

$$2.95 + 0.20 = 3.15 \text{ hours}$$

Production:

$$\frac{24}{3.15} = 7.62 \text{ cycles/day}$$
$$7.62 \times 1.5 = 11.43 \text{ m/day}$$

A 12-minute delay per cycle reduces production by nearly 0.8 m/day.

6. Example – Maintenance Stop Once Per Day

Assume one additional 2-hour maintenance stop per day.

Available working time:

$$24 - 2 = 22 \text{ hours}$$

Cycles per day:

$$\frac{22}{2.95} = 7.46$$

Production:

$$7.46 \times 1.5 = 11.2 \text{ m/day}$$

Daily output drops by ~1 metre.

Accurate productivity modelling requires clear identification of downtime categories. These may include planned maintenance, cutter replacement, hyperbaric interventions, mechanical and electrical faults, conveyor or slurry system stoppages, logistics delays, shift changes, and geotechnical interventions. Understanding which factors dominate downtime is essential for realistic scheduling and mitigation planning. Without this analysis, schedules tend to underestimate delay accumulation.

Two principal approaches are used in schedule modelling: deterministic and probabilistic. Deterministic scheduling applies a fixed advance rate, such as 12 metres per day, and assumes relatively stable performance. While straightforward, this approach may underestimate risk. Probabilistic scheduling uses statistical ranges—best-case, most-likely, and worst-case scenarios—and may apply Monte Carlo simulations to account for variability in cutter wear, breakdown frequency, geological transitions, and intervention duration. Probabilistic modelling produces more realistic contingency allowances and improves programme resilience.

Geology exerts a dominant influence on productivity. Soft ground may allow rapid penetration but requires careful pressure control and may introduce clogging risks. Hard rock reduces penetration rates and increases cutter wear. Mixed-face conditions introduce asymmetric loading and reduced efficiency, while highly abrasive formations increase cutter consumption and intervention frequency. Transition zones between geological units often reduce production and must be incorporated into schedule modelling. In addition, crew experience produces a learning curve effect: over time, ring build durations decrease, troubleshooting improves, and conditioning becomes more efficient, gradually enhancing productivity.

In many projects, the TBM drive forms the critical path. Delays in shaft readiness, power supply, segment manufacturing, conveyor extension, or slurry separation capacity can halt excavation entirely. Scheduling must therefore integrate not only underground operations but also surface supply chains and infrastructure capacity. Sensitivity analysis is valuable in identifying which variables most strongly affect programme duration. For example, a ten percent reduction in utilisation may extend completion by months, while increased hyperbaric intervention frequency or reduced conveyor reliability may significantly amplify schedule risk.

Long tunnels introduce additional scheduling complexity. As distance from the shaft increases, travel time lengthens, conveyor extensions require more time, and maintenance response delays increase. Downtime deep within the tunnel is amplified compared to equivalent stoppages near surface access. Deep and high-pressure tunnels add further constraints, including longer hyperbaric interventions, more frequent seal inspections, thermal stress on crews and equipment, and narrower pressure-control operating windows. These factors reduce achievable utilisation and must be reflected in conservative forecasts.

Modern productivity management relies on key performance indicators such as metres per day, ring build duration, cutter consumption per 100 metres, utilisation percentage, mean time between failures, intervention frequency, and energy consumption per metre. Real-time monitoring systems, digital twin production models, predictive maintenance tools, and Monte Carlo simulations enable early identification of deviation from planned performance. When productivity falls below expectation, recovery strategies may include extending shifts, increasing maintenance resources, optimising spoil removal, accelerating segment supply, or parallelising surface works. Any recovery measure must remain within safety and mechanical limits.

Scheduling and productivity analysis in mechanised tunnelling are not administrative exercises; they are core engineering functions. The TBM represents a high-capital, continuous production system in which small inefficiencies, when accumulated over long distances, can create substantial programme impact. Realistic scheduling requires integrated understanding of geology, mechanical behaviour, maintenance demands, logistics systems, and statistical risk modelling. Overly optimistic assumptions are a common cause of cost overruns and delays in tunnelling projects.

The following is one integrated, end-to-end case example demonstrating how scheduling and productivity analysis work together in a realistic TBM project.

Full Example – 4.5 km Urban Metro TBM Drive

Project Overview

- Tunnel length: **4,500 m**
- Internal diameter: **6.5 m**
- TBM type: EPB
- Ground: Mixed soft clay with sand lenses, moderate abrasivity
- Depth: 25–35 m
- Groundwater: 2.5 bar
- Ring length: 1.5 m
- Two 12-hour shifts per day

The TBM drive is on the **critical path**.

1. Initial (Optimistic) Deterministic Schedule

The contractor proposes:

$$\text{Advance Rate} = 15 \text{ m/day}$$

Estimated drive duration:

$$\frac{4500}{15} = 300 \text{ days}$$

This assumes near-ideal performance and high utilisation.

But this is theoretical.

2. Productivity Breakdown – Realistic Analysis

Step 1 – Instantaneous Penetration Rate (IPR)

Assume:

- Penetration per revolution = 9 mm/rev
- Cutterhead speed = 2.5 rpm

$$IPR = 9 \times 2.5 = 22.5 \text{ mm/min}$$
$$22.5 \text{ mm/min} = 0.0225 \text{ m/min}$$
$$0.0225 \times 60 = 1.35 \text{ m/hour}$$

Theoretical 24-hour continuous cutting:

$$1.35 \times 24 = 32.4 \text{ m/day}$$

Clearly unrealistic — because this excludes ring build and downtime.

Step 2 – TBM Cycle Calculation

For 1.5 m ring:

- Excavation time:

$$\frac{1.5}{1.35} = 1.11 \text{ hours}$$

- Ring build time: 0.9 hours
- Auxiliary checks: 0.4 hours

$$CycleTime = 1.11 + 0.9 + 0.4 = 2.41 \text{ hours}$$

Cycles per 24 hours:

$$\frac{24}{2.41} = 9.96 \text{ cycles}$$

Daily production:

$$9.96 \times 1.5 = 14.9 \text{ m/day}$$

Now we are close to the proposed 15 m/day — but still ignoring downtime.

3. Incorporating Utilisation

Realistic utilisation in mixed ground: **50%**

$$14.9 \times 0.50 = 7.45 \text{ m/day}$$

Now the realistic daily advance rate (DAR):

≈ 7.5 m/day

4. Adding Downtime Categories

From similar projects, expected downtime per kilometre:

- Planned maintenance: 8 hours/km
- Cutter change: 12 hours/km
- Conveyor stoppages: 6 hours/km
- Minor breakdowns: 10 hours/km

Total: 36 hours/km

For 4.5 km:

$$36 \times 4.5 = 162 \text{ hours}$$

= 6.75 days lost

5. Hyperbaric Interventions

Groundwater and sand lenses require:

- 1 intervention per km
- 20 hours per intervention

$$4.5 \times 20 = 90 \text{ hours}$$

= 3.75 days lost

6. Revised Realistic Schedule

Base production:

$$\frac{4500}{7.5} = 600 \text{ days}$$

Add downtime (6.75 + 3.75):

≈ 10.5 days

Revised duration:

610 days

The original 300-day deterministic schedule was off by 310 days.

7. Geological Transition Zone Effect

At chainage 2,000 m:

- Abrasive sand lens increases cutter wear
- Penetration drops to 1.0 m/hour
- Utilisation falls to 45%

New DAR:

$$1.5/((1.5/1.0) + 0.9 + 0.4) \times 24 \times 0.45$$

Excavation time:

$$1.5 \text{ hours}$$

Cycle time:

$$1.5 + 0.9 + 0.4 = 2.8 \text{ hours}$$

Cycles/day:

$$24/2.8 = 8.57$$

Daily production:

$$8.57 \times 1.5 = 12.85$$

Apply 45% utilisation:

$$12.85 \times 0.45 = 5.8 \text{ m/day}$$

In this zone, production drops to **5.8 m/day**.

If this zone extends 1 km:

$$\frac{1000}{5.8} = 172 \text{ days}$$

Instead of:

$$\frac{1000}{7.5} = 133 \text{ days}$$

A 39-day impact from one geological transition.

8. Learning Curve Effect

After 3 months:

- Ring build time reduces from 0.9 to 0.75 hours
- Auxiliary time reduces to 0.35 hours

New cycle:

$$1.11 + 0.75 + 0.35 = 2.21 \text{ hours}$$

Cycles/day:

$$24/2.21 = 10.86$$

Daily production (50% utilisation):

$$10.86 \times 1.5 \times 0.50 = 8.15 \text{ m/day}$$

Crew learning improves output by 0.65 m/day.

Over 2,500 m remaining:

$$\frac{2500}{7.5} = 333 \text{ days}$$
$$\frac{2500}{8.15} = 307 \text{ days}$$

26 days recovered through operational efficiency.

9. Sensitivity Analysis

If utilisation drops from 50% to 45%:

$$7.5 \times 0.9 = 6.75 \text{ m/day}$$

Drive duration:

$$\frac{4500}{6.75} = 667 \text{ days}$$

Loss of 57 days from 5% utilisation drop.

Utilisation is the dominant variable.

10. Probabilistic Forecast

Monte Carlo simulation inputs:

- DAR range: 5.5 – 8.5 m/day
- Intervention frequency ±20%
- Breakdown frequency variable

Simulation results:

- P50 completion: 610 days
- P80 completion: 660 days
- P95 completion: 700 days

The client adopts **P80 (660 days)** for programme baseline.

11. Critical Path Considerations

If segment factory produces only 6 rings/day:

$$6 \times 1.5 = 9 \text{ m/day max}$$

Segment supply is not limiting in this case.

However, if conveyor reliability drops below 95%, utilisation drops and programme risk increases sharply.

12. Deep / High Pressure Effects

At deepest section (35 m):

- Higher face pressure
- Increased inspection frequency
- Hyperbaric duration extended by 5 hours

Intervention time increases:

$$4.5 \times 5 = 22.5 \text{ additional hours}$$

≈ 1 additional day lost

Final Realistic Programme

Original deterministic estimate: **300 days**

Realistic probabilistic schedule (P80): **660 days**

Difference: **360 days**

The discrepancy is driven by:

- Utilisation
- Geological transitions
- Intervention frequency
- Cycle inefficiencies
- Downtime amplification

Ultimately, scheduling and productivity analysis in TBM tunnelling represent the integration of mechanical engineering, geotechnical science, logistics coordination, and risk forecasting. Production is governed not only by cutterhead performance but by utilisation, downtime patterns, intervention frequency, and support system reliability. Effective schedule control is achieved through disciplined analysis, conservative planning, and continuous performance monitoring rather than optimism alone.

Cost Drivers in Tunnel Boring

Cost drivers in tunnel boring are the technical, geological, logistical, and commercial factors that most strongly influence the total cost of a mechanised tunnelling project. Unlike many surface construction activities, TBM operations involve a substantial upfront capital investment combined with continuous operational expenditure over extended periods. Because the TBM functions as a high-capital, time-dependent production system, even small variations in performance, geology, or risk exposure can produce disproportionately large financial consequences.

Project costs are typically divided into capital expenditure (CAPEX) and operational expenditure (OPEX), but these categories are closely interrelated. A lower initial machine purchase price may result in higher maintenance frequency, increased intervention risk, and longer programme duration, ultimately increasing overall project cost. TBM procurement is therefore not simply an equipment purchase—it is a strategic risk allocation decision.

The TBM itself represents one of the largest individual cost items. Machine cost varies according to diameter, pressure rating, shield configuration, cutterhead design, customisation level, and the inclusion of digital monitoring systems. Large-diameter slurry TBMs designed for high-pressure or subsea conditions can cost many tens of millions of dollars. Custom-engineered machines with reinforced bulkheads, high-capacity sealing systems, and conversion capability require greater capital investment but may significantly reduce operational risk. Under-specifying the TBM can increase hyperbaric intervention frequency, seal failure probability, main bearing risk, and programme delay exposure.

Geology is the single most influential cost driver in TBM tunnelling. Soft ground may allow faster penetration and lower cutter wear, but introduces settlement risks, conditioning material consumption, and potential compensation grouting costs. Hard rock typically reduces penetration rate and increases disc cutter wear, torque demand, and cutter replacement frequency. Mixed-face conditions create asymmetric loading, mechanical stress, and increased downtime. Highly abrasive ground, particularly quartz-rich formations, dramatically increases cutter consumption and wear in screw conveyors or slurry pumps. High groundwater pressures elevate sealing requirements, intervention complexity, grouting demand, and dewatering costs. Inaccurate geological forecasting remains one of the most common causes of cost overrun.

Production rate and utilisation are major financial drivers because labour, supervision, plant depreciation, and site overheads are time-based. If planned production is 12 metres per day but actual performance averages 8 metres per day, the programme extends by 50 percent. This increases labour cost, energy consumption, temporary works duration, and supervision expenditure. Utilisation rate, which reflects the percentage of time the TBM is actively boring, therefore becomes one of the most powerful determinants of cost.

Cutter consumption and wear are recurring operational expenses. Abrasivity, rock strength, mixed-face transitions, and intervention frequency directly influence tool life. In pressurised environments, cutter replacement may require hyperbaric intervention, significantly increasing cost due to specialist personnel, decompression facilities, and lost production time. Cutter consumption is often monitored per 100 metres excavated and feeds directly into financial forecasting models.

Downtime and reliability also strongly influence cost. Unplanned breakdowns lead to lost production, repair labour, spare part replacement, and extended overhead exposure. Major failures, such as main bearing damage, can escalate costs dramatically. Reliability metrics, including Mean Time Between Failures (MTBF), are therefore not only technical indicators but also financial predictors.

Labour structure contributes significantly to overall cost. TBM operations require operators, segment erectors, maintenance crews, electricians, surveyors, and safety personnel. Labour exposure increases with project duration, meaning longer tunnels amplify workforce-related costs. Deep or high-pressure conditions requiring specialist hyperbaric teams further increase labour expenditure.

Segment manufacturing and lining costs also represent a substantial cost component. Each metre of tunnel requires a segment ring, and cost is influenced by concrete strength, reinforcement content, gasket specification, steel fibre inclusion, manufacturing efficiency, and transport distance. Deep or high-pressure tunnels typically require thicker and stronger segments, increasing cost per metre.

Spoil handling and logistics systems—including conveyors, slurry pipelines, separation plants, or rail haulage—add further cost exposure. In long tunnels, conveyor extension, power demand, wear component replacement, and maintenance complexity increase progressively with distance. Slurry systems require adequate treatment plant capacity, which becomes a major cost driver in high-throughput operations.

Energy consumption is another significant operational expense. Cutterhead torque demand, slurry pump capacity, conveyor length, ventilation requirements, and cooling systems all contribute to electrical load. Hard rock and deep tunnels generally require greater energy per metre excavated.

Hyperbaric interventions in pressurised environments represent both technical and financial risk. Duration, frequency, medical support, and lost production time all increase cost exposure. High groundwater projects can significantly increase the number and complexity of these interventions.

Long tunnel logistics amplify downtime cost because travel time, maintenance response, and emergency access become more difficult with distance from the shaft. A repair that requires two hours near the launch shaft may require six hours deep within the tunnel, multiplying downtime impact.

Risk and contingency allowances must address potential financial exposure from stuck TBM scenarios, water inrush, settlement claims, major mechanical failure, or excessive ground deformation requiring redesign. Probabilistic modelling is often used to determine realistic contingency levels based on geological uncertainty and machine reliability.

Programme duration is ultimately the most powerful cost multiplier. Total project cost can be conceptualised as fixed capital cost plus daily operating cost multiplied by project duration. For example, if daily operating expenditure is $400,000, a 60-day delay increases cost by $24 million. Time therefore converts technical inefficiency directly into financial consequence.

Contractual structure also influences cost exposure. Risk allocation mechanisms such as lump-sum versus remeasure contracts, geological baseline reports, performance guarantees, and dispute resolution frameworks determine which party bears financial risk for unforeseen conditions.

Investment in digital systems—such as real-time monitoring, predictive maintenance tools, data analytics, and digital twin platforms—may increase capital cost but can reduce downtime and improve lifecycle cost efficiency by enabling early detection of performance degradation.

In engineering terms, cost drivers in tunnel boring are governed by the interaction between geology, machine capability, operational efficiency, reliability, and programme duration. The most expensive TBM is not necessarily the one with the highest purchase price, but the one that operates below planned utilisation, requires frequent intervention, or extends the programme significantly. Effective cost control in mechanised tunnelling depends on conservative production modelling, accurate geological forecasting, robust machine specification, disciplined maintenance planning, and integrated risk management.

In TBM tunnelling, cost is not controlled by optimism. It is controlled by understanding the dynamic relationship between ground behaviour, mechanical systems, operational performance, and time.

Risk Allocation and Contracts

Mechanised tunnelling, especially through Tunnel Boring Machines (TBMs), presents complex challenges characterized by substantial geological uncertainties, significant financial commitments, and extended construction timelines. The distribution of risks between stakeholders—namely the client, contractor, designer, and manufacturer—plays a critical role in the overall success of these projects. A well-structured contractual framework is crucial for aligning incentives, clarifying responsibilities, and facilitating collaborative problem-solving in the face of unforeseen conditions.

The Importance of Risk Allocation

Risk in tunnelling projects primarily stem from uncertainties in subsurface conditions. Despite extensive geotechnical investigations, ground conditions can exhibit considerable variability between borehole locations. Contracts must explicitly identify which party assumes financial responsibility when encountering unexpected geological conditions. For instance, the implementation of a Geological Baseline Report (GBR) can help set expectations regarding ground conditions, delineating how risks will shift based on actual versus anticipated circumstances. Where actual conditions deviate unfavourably from the GBR, clients may find that risks revert to them, potentially leading to disputes if the baseline assumptions are vague or unrealistic [334, 375, 811, 812].

Different contractual models influence risk allocation significantly. In lump-sum or fixed-price contracts, contractors accept substantial risks associated with productivity and machine performance, although they can negotiate exclusions. This model encourages operational efficiency but may drive contractors to adopt conservative estimates, embedding risk premiums into bids to accommodate potential geological surprises [812, 813]. Conversely, remeasurement or unit-rate contracts link remuneration to actual quantities, thus moderating contractor exposure to fluctuations in geological conditions but still leaving issues of subsurface uncertainty inadequately addressed.

Contractual Models and Their Implications

Various contract structures such as Design-Build (D&C), Engineering, Procurement, and Construction (EPC) confer distinct risk allocations. For example, D&C contracts often assign responsibility for TBM selection to contractors, leading to a transfer of risks surrounding machine performance to them. However, if geotechnical conditions exceed the defined baseline, clarity concerning machine specifications may cause disputes [334, 811, 813]. On the other hand, in approaches where clients provide TBMs, such as client-supplied TBM models, owners maintain greater control over specifications but assume higher technical risks related to machine performance [680, 811].

In more collaborative settings like Alliance and Public-Private Partnerships (PPP), risks and rewards are shared more evenly among parties through mechanisms such as open-book accounting and agreed-upon pain/gain-sharing approaches. These models can foster cooperation and jointly navigate complexities inherent to high-risk tunnelling projects, which might otherwise escalate into adversarial disputes [812, 814]. However, such collaborative frameworks require mature governance and transparency among all stakeholders.

Operational and Programme Risk Management

Operational hazards, such as groundwater inflow and mechanical failures, necessitate clear definitions of unforeseen conditions within contracts. If contracts do not adequately define what constitutes "unforeseen," it can lead to extensive disputes and arbitration, ultimately inflating project costs [812, 814]. Additionally, performance guarantees incorporated into contracts, such as minimum advance rates or cutterhead durability warranties, bolster client confidence but must be based on grounded geological realities. Overly optimistic expectations can create dependencies on risky pricing strategies or increase the potential for future disputes.

Insurance and bonding types also influence risk management strategies in tunnelling projects. Contractors typically secure professional indemnity and liability insurance, while clients may demand performance bonds to mitigate financial exposure in high-risk scenarios [14, 813]. The overall structure of the contract should consider daily operating costs since any delays in TBM projects can amplify financial repercussions. Liquidated damages clauses can compel timely project completion but must be designed to correspond with realistic productivity expectations to avoid fostering unsafe practices and exacerbating tensions between parties [8, 811].

Performance Benchmarking

Performance benchmarking in mechanised tunnelling is the structured process of comparing actual TBM performance against predefined targets, historical project data, or industry best practice to evaluate efficiency, reliability, and risk exposure. In high-capital tunnelling projects, benchmarking is not merely a reporting exercise—it is a core management tool used to control schedule, cost, maintenance, and operational risk. Because TBM drives often extend over months or years, continuous performance evaluation allows early detection of deviation and supports timely corrective action.

At its most basic level, performance benchmarking compares planned production metrics—such as metres per day or utilisation percentage—against actual recorded data. However, meaningful benchmarking goes deeper than headline advance rates. It examines the full production system, including penetration rate, cycle time, downtime categories, cutter consumption, energy use, and intervention frequency. A project may achieve acceptable daily advance but still underperform in utilisation or maintenance efficiency compared to industry standards.

One of the primary benchmark metrics is Daily Advance Rate (DAR), which measures metres excavated per 24-hour period. DAR is influenced by penetration rate, ring build efficiency, and downtime. Benchmarking DAR against similar projects—such as tunnels of comparable diameter and geology—provides context for whether performance is typical, above average, or underperforming. For example, an 8 m/day average in hard rock may represent strong performance, whereas 8 m/day in soft clay could indicate inefficiency.

Utilisation rate is another critical benchmark. High-performing TBM projects commonly achieve 40–60% utilisation, depending on ground conditions and logistics. If a project is operating at 35% utilisation, benchmarking helps identify whether the shortfall is due to maintenance, logistics delays, segment supply constraints, or geotechnical instability. Utilisation benchmarking often reveals that operational inefficiencies, rather than cutting capability, are limiting performance.

Penetration rate benchmarking compares instantaneous cutting efficiency under similar geological conditions. For example, penetration in granite of a given uniaxial compressive strength (UCS) can be compared across projects to evaluate cutter performance and thrust optimisation. Deviations may indicate improper parameter control, excessive wear, or suboptimal cutterhead configuration.

Cutter consumption is a key financial benchmark, typically measured as cutter changes per 100 metres or cost per metre excavated. Abrasive ground projects often establish expected consumption curves based on laboratory abrasivity indices such as the Cerchar Abrasivity Index (CAI). If actual cutter consumption exceeds predicted rates, benchmarking highlights potential issues in thrust management, cutter spacing, or geological variability.

Downtime benchmarking categorises lost time into planned maintenance, corrective repairs, hyperbaric interventions, and logistical delays. Comparing downtime distribution against industry norms helps identify systemic weaknesses. For example, excessive conveyor downtime relative to benchmark data may suggest inadequate preventive maintenance or under-designed spoil handling capacity.

Energy consumption per metre excavated is increasingly benchmarked in modern projects. This metric reflects mechanical efficiency, torque demand, and system losses. High energy intensity may indicate excessive cutter wear, high friction, or suboptimal parameter selection. In sustainability-focused projects, benchmarking energy performance also supports carbon reporting and operational optimisation.

Performance benchmarking can occur at several levels. Internal benchmarking compares current performance against earlier phases of the same project, capturing learning curve effects. External benchmarking compares performance with other projects within an organisation's portfolio or industry-wide datasets. Some major infrastructure clients maintain centralised performance databases to improve forecasting accuracy for future projects.

Digital systems have significantly enhanced benchmarking capability. Modern TBMs collect real-time operational data, including torque, thrust, rotation speed, chamber pressure, vibration, and temperature. Integrated dashboards allow comparison between planned and actual metrics on a daily or even hourly basis. Advanced analytics and digital twin models can simulate expected performance under defined geological conditions and flag deviations automatically.

Benchmarking also supports probabilistic schedule modelling. Historical productivity distributions from previous projects can inform Monte Carlo simulations for new drives, improving contingency accuracy. Rather than relying on optimistic deterministic advance rates, planners use benchmarked ranges to define realistic P50 or P80 programme durations.

From a commercial perspective, benchmarking strengthens contract management and dispute resolution. Objective performance data reduces ambiguity regarding whether delays arise from geological conditions, machine limitations, or operational inefficiencies. Transparent benchmarking fosters trust between clients and contractors by grounding discussions in measurable evidence.

However, effective benchmarking requires contextual understanding. Raw comparisons without accounting for diameter, geology, depth, groundwater pressure, logistics distance, and contract model can be misleading. Meaningful benchmarking normalises data to comparable conditions to ensure fair interpretation.

In engineering terms, performance benchmarking transforms tunnelling from reactive problem-solving into proactive performance management. It enables early identification of declining utilisation, increasing cutter wear, or emerging reliability trends before they escalate into major schedule or cost impacts. Ultimately, benchmarking enhances predictability, supports risk mitigation, and improves financial control.

In mechanised tunnelling, what gets measured can be managed. Performance benchmarking provides the structured framework through which data becomes decision-making intelligence rather than simply historical record.

Lessons Learned and Optimisation

Mechanised tunnelling is a sophisticated process characterized by high capital investment and exhaustive data generation. The effective management of this process involves the systematic analysis of past performance to identify inefficiencies and facilitate enhancements in productivity, safety, cost control, and reliability. The need for lessons learned is especially pronounced in Tunnel Boring Machine projects, which are lengthy undertakings often spanning several years and incorporating thousands of repetitive operational cycles. Continuous learning is not merely beneficial but essential for maintaining programme certainty and commercial viability in such complex environments [815-817].

Data-Driven Optimisation

Mechanised tunnelling generates a plethora of operational data, including metrics such as penetration rate, torque, thrust, chamber pressure, cutter wear, downtime events, and energy consumption. Transforming this data into actionable engineering insights is fundamental for informing optimization strategies and enhancing operational efficiency [815,

818]. Optimisation efforts based on lessons learned from observed performance typically improve maintenance planning, logistics coordination, and risk management [817, 819].

One notable lesson emphasised in numerous studies is the discrepancy between theoretical performance and actual production rates. Early project schedules are often overly reliant on optimistic assumptions regarding penetration rates, neglecting critical variables such as equipment utilisation and geological conditions [818, 820]. It is widely observed that projects leveraging conservative productivity models, along with realistic allowances for contingencies, experience significantly fewer cost overruns [815, 821]. This points to the necessity of probabilistic rather than deterministic modelling approaches for accurate forecasting.

Utilisation and Operational Efficiency

Utilisation rates also emerge as a pivotal area of focus in TBM projects. Research has consistently highlighted the importance of increasing operational efficiency by minimizing non-cutting times, rather than merely seeking to enhance the penetration rates of the TBM [819, 822]. Improving efficiencies in ring assembly, segment logistics, and spoil handling often yields more substantial programme benefits than marginal increases in cutterhead speed [817, 822]. Consequently, optimisation initiatives frequently concentrate on reducing cycle times, providing the potential for significant gains in project timelines and costs.

Furthermore, the management of cutter wear and abrasivity has been identified as a critical factor in project cost and performance. Lack of consideration of factors such as the Cerchar Abrasivity Index can lead to increased wear and frequent interventions, driving up project costs [816, 822]. A targeted approach to optimisation may involve adjusting thrust levels, refining cutter specifications, and implementing predictive modelling techniques for cutter wear [818, 823].

Downtime Management

Analysing downtime events indicates that not all production losses stem from mechanical failures; logistical challenges and maintenance deficiencies can also undermine productivity. Effective downtime categorisation can therefore inform strategic adjustments, guiding projects to enhance spare parts inventory levels, preventive maintenance strategies, and logistical alignments [816, 817].

Geological variability is another critical risk factor that influences TBM performance. Transitioning through varying ground types can lead to uneven wear and operational challenges. Thus, implementing proactive monitoring and operational adjustments attentive to geological conditions is vital for mitigating these risks [815, 824]. The integration of real-time geotechnical data into the operational framework exemplifies advanced strategies employed to enhance project management and operational performance [822, 823].

Digital Integration and Future Implications

The advent of digital technologies has considerably enhanced the capacity for optimisation in mechanised tunnelling. The deployment of digital dashboards and predictive algorithms allows for the early detection of performance deviations, ultimately facilitating timely interventions and reducing potential downtimes [823, 824].

At a strategic level, the accumulation of lessons learned feeds into future TBM projects, enabling companies to establish centralised databases for performance management. This repository can serve to benchmark various operational metrics across different projects, thus improving future forecasting accuracy and procurement strategies [815, 818].

The journey towards optimisation in mechanised tunnelling hinges on a structured approach to learning from past experiences and embracing continuous improvement practices. By treating each challenge as a data point, operations can evolve from reactive management to proactive performance enhancement, ensuring that even minor gains translate into significant project benefits. In this high-stakes production environment, such an approach contributes to enhanced reliability, reduced costs, and improved overall project outcomes.

Chapter 19

Automation, Digitalisation, and Future Trends

Mechanised tunnelling is undergoing a profound transformation, driven by the convergence of automation, digitalisation, and advanced data analytics. Tunnel Boring Machines (TBMs), once predominantly mechanical systems reliant on operator experience and fixed control strategies, have evolved into highly instrumented, intelligent production platforms. This shift reflects a broader transition within engineering, from deterministic, design-based approaches toward adaptive, data-driven systems capable of responding dynamically to changing conditions.

At the core of this transformation is the integration of control systems, real-time monitoring, and artificial intelligence into TBM operations. Technologies such as programmable logic controllers (PLCs), distributed control systems (DCS), predictive maintenance algorithms, and digital twins are redefining how tunnelling projects are planned, executed, and managed. These systems enable continuous feedback between machine performance, ground behaviour, and engineering decision-making, allowing tunnelling to move from reactive problem-solving toward proactive and predictive control.

This chapter explores the technological foundations and emerging innovations shaping modern tunnelling practice. It examines how automation enhances operational stability, how digital platforms provide deeper insight into machine and ground interaction, and how artificial intelligence supports optimisation and risk reduction. It also considers the practical implementation of these systems, including real-time geotechnical feedback, remote operation, and the growing role of integrated data environments.

Importantly, while these technologies represent significant advancements, they do not eliminate the inherent uncertainties of the subsurface. Instead, they provide the tools to better understand, manage, and respond to those uncertainties. As demonstrated throughout this chapter, the future of tunnelling lies not in replacing human expertise, but in augmenting it, combining engineering judgement with intelligent systems to deliver safer, more efficient, and more reliable underground construction.

Looking ahead, the continued evolution of TBMs will be defined by increasing levels of integration, adaptability, and intelligence. From semi-autonomous operation to real-time digital twins and AI-driven optimisation, these advancements signal a shift toward fully connected tunnelling ecosystems. This chapter sets the foundation for understanding how these technologies are reshaping the discipline and what they mean for the future of mechanised excavation.

TBM Automation and AI Assistance

Tunnel Boring Machines (TBMs) have seen a significant transformation over the years, evolving from mechanically controlled excavation systems to sophisticated, data-driven production platforms. This evolution has been marked by advancements in automation, digital control systems, and the integration of artificial intelligence (AI). These technologies enhance operational precision, safety, and productivity in complex geological environments without eliminating the need for human operators.

Advanced Automation in TBMs

The core of modern TBM operations lies in automation, characterized by closed-loop control systems that constantly monitor and regulate critical excavation parameters. These parameters include thrust, torque, cutterhead speed, face pressure, slurry density, and screw conveyor discharge rate. Systems equipped with real-time sensors adjust operational outputs automatically to maintain predefined conditions necessary for safe and efficient tunnelling [825-827].

Automation and control in TBMs are made possible primarily by programmable logic controllers (PLCs) and distributed control systems (DCS) [828]. In Earth Pressure Balance (EPB) TBMs, these systems maintain necessary chamber pressures by balancing excavation rate with conveyor discharge speed and conditioning injections, essential to prevent over-excavation and maintain face stability [829, 830]. Conversely, slurry TBMs automate several variables including slurry pressure and density at the excavation face, ensuring the stability of operations in high groundwater conditions [831, 832]. These automated controls significantly mitigate the risks of tunnelling, particularly in ground conditions that are unpredictable.

Programmable Logic Controllers (PLCs)

A Programmable Logic Controller (PLC) is an industrial digital computer specifically designed to control electromechanical processes in harsh operating environments. PLCs are widely used in manufacturing plants, power stations, mining operations, and mechanised tunnelling systems because they are robust, reliable, and capable of real-time control.

A PLC continuously receives input signals from sensors—such as pressure transducers, torque meters, proximity switches, temperature probes, and hydraulic pressure sensors—and processes these signals according to a programmed control logic. Based on this logic, the PLC sends output commands to actuators, including hydraulic valves, electric motors, pumps, relays, and servo drives. The process occurs in rapid, repetitive scan cycles, typically measured in milliseconds, allowing near-instantaneous system response.

The internal structure of a PLC generally includes:

- A central processing unit (CPU)
- Input/output (I/O) modules
- Power supply
- Communication interfaces
- Programming memory

The PLC executes control instructions written in industrial programming languages such as Ladder Logic, Function Block Diagram (FBD), or Structured Text. Ladder Logic, originally designed to resemble electrical relay diagrams, remains widely used because of its clarity and ease of troubleshooting.

In a Tunnel Boring Machine (TBM), PLCs control critical subsystems such as:

- Hydraulic thrust cylinder operation
- Cutterhead rotation speed
- Screw conveyor discharge rate
- Slurry pump flow
- Grouting system activation
- Segment erector positioning
- Safety interlocks

For example, in an Earth Pressure Balance (EPB) TBM, the PLC monitors chamber pressure and automatically adjusts screw conveyor speed and thrust force to maintain face stability. If pressure falls below a safe threshold, the PLC can trigger alarms or initiate corrective actions.

The key advantages of PLCs include high reliability, deterministic response time (predictable execution speed), rugged construction, and ease of maintenance. Because PLCs are modular, individual components can be replaced without shutting down the entire system. Their deterministic behaviour makes them particularly suited to safety-critical systems where predictable timing is essential.

Distributed Control Systems (DCS)

A Distributed Control System (DCS) is a broader, network-based control architecture that distributes control functions across multiple interconnected controllers rather than relying on a single central processor. While PLCs typically control individual machines or subsystems, DCS platforms manage large, complex processes spanning multiple systems and physical locations.

In a DCS environment, control intelligence is distributed among several local controllers connected through high-speed communication networks. Each controller manages a specific process area, while supervisory systems integrate and coordinate overall operation. Operators interact with the system through Human–Machine Interfaces (HMIs) located in central control rooms.

A DCS typically includes:

- Distributed controllers
- Field I/O modules
- Communication networks
- Data servers and historians
- Central operator workstations
- Advanced process control software

Unlike PLC systems that are often event-driven and discrete in nature, DCS platforms are particularly well suited to continuous processes that require coordinated parameter control, data archiving, trending, and supervisory optimisation.

In the context of mechanised tunnelling, a DCS might integrate:

- TBM excavation control
- Slurry separation plant
- Surface power systems
- Ventilation systems
- Conveyor or rail logistics
- Grouting plants

- Environmental monitoring

For example, in a slurry TBM operation, the DCS may coordinate slurry pressure control underground with surface separation plant density management, pump speed regulation, and return flow control. Rather than operating as isolated systems, all components function within an integrated control architecture.

DCS platforms also provide advanced data management capabilities, including:

- Real-time trending
- Alarm management
- Historical data storage
- Performance benchmarking
- Integration with digital twin platforms

This makes them particularly suitable for long-duration infrastructure projects where performance tracking and optimisation are essential.

Key Differences Between PLC and DCS

Although PLCs and DCS systems can overlap in functionality, their design philosophies differ.

PLCs are generally:

- Machine-level controllers
- Fast and deterministic
- Optimised for discrete and mechanical control
- Highly modular
- Used for specific subsystems

DCS systems are typically:

- Process-level controllers
- Network-integrated
- Designed for continuous process coordination
- Strong in data integration and supervisory control
- Used for large-scale, multi-system operations

In modern TBM projects, PLCs often control the immediate mechanical functions of the machine, while a DCS integrates the TBM with surface plants and broader project infrastructure.

Integration in Modern TBM Systems

In contemporary tunnelling operations, PLC and DCS systems frequently operate together. PLCs handle time-critical local control tasks such as thrust regulation or cutterhead speed adjustment. The DCS collects data from these PLCs, coordinates interdependent systems, and provides a supervisory control environment for operators and engineers.

For example:

- A PLC adjusts screw conveyor speed in milliseconds to stabilise chamber pressure.
- The DCS monitors overall production trends, logs performance data, and alerts engineers if long-term pressure stability deviates from planned targets.

This layered architecture ensures both immediate operational responsiveness and strategic oversight.

From an engineering perspective, PLCs and DCS platforms form the digital backbone of modern mechanised tunnelling. They enable precise parameter control, reduce human error, improve safety, and provide the structured data required for AI-assisted optimisation and predictive maintenance.

Without PLCs and DCS integration, maintaining stable face pressure, synchronising slurry systems, coordinating logistics, and managing complex mechanical interactions would be significantly more difficult and less reliable. These systems transform the TBM from a purely mechanical excavation device into a controlled, monitored, and optimised industrial process.

Integration of Artificial Intelligence

AI integration in TBM operations primarily acts as a decision-support mechanism rather than functioning as a fully autonomous system [833, 834]. Machine learning models analyse large datasets, allowing for pattern recognition, predictive maintenance, and operational optimizations [829, 830]. For instance, models can identify early indications of mechanical degradation through analysis of vibration patterns, torque fluctuations, and hydraulic system behaviour [835]. This predictive capability minimizes unplanned downtime, enhancing overall operational efficiency.

Furthermore, AI assists in optimizing cutter wear by evaluating geological classifications and penetration patterns, recommending when to intervene for maintenance or adjustments [836, 837]. The analysis of geological transitions through monitored parameters allows the TBMs to pre-emptively adjust operations before encountering issues, which is vital in environments where geological conditions can change rapidly and unexpectedly.

Operational Enhancements and Data Utilization

The advanced capabilities of modern TBMs enable them to act not solely as excavation devices but as complex, data-driven systems. Each TBM functions as a continuous data generator recording numerous operational parameters such as penetration rates, chamber pressures, and energy consumption [832, 838, 839]. This data can be transmitted in real time to control rooms and engineering offices, facilitating immediate adjustments to improve performance and alignment with project objectives.

Moreover, the integration with digital twins allows for effective visualization and monitoring of TBM performance against project expectations. Through this digital synergy, engineers can conduct performance benchmarking and early detection of operational deviations, thus reinforcing proactive management strategies [339].

Challenges and Future Directions

Despite these advancements, the integration of automation and AI presents challenges that practitioners must navigate. Geological variability inherently presents unpredictability that cannot always be foreseen solely through data analysis [830, 840]. The reliability of sensors and the quality of training data for AI models significantly impact the accuracy of control systems. Additionally, the sheer volume of data generated necessitates effective filtering mechanisms to avoid overwhelming operators [683, 831, 834].

In terms of human factors, the role of the operator evolves rather than diminishes; from conventional control to focusing on data interpretation and decision validation, skilled personnel remain essential for addressing unforeseen issues [75, 830]. Moreover, while many systems operate semi-autonomously, full autonomy in TBMs is still rare due to the unpredictable nature of geological conditions.

Looking to the future, continued advancements in digital infrastructure, autonomous parameter optimization, and real-time geological modelling will define the next generation of TBMs [827, 828]. As datasets expand, the accuracy of predictive models will improve, paving the way for more reliable and efficient tunnelling operations.

The evolution of Tunnel Boring Machines exemplifies the fusion of mechanics and digital technology to create highly capable excavation systems. With growing integration of automation and AI, TBMs are transforming tunnelling from reactive to proactive operations, enhancing not only efficiency and safety but also the overall reliability of underground construction projects.

Digital Twins in Tunnelling

Digital twins in tunnelling represent a significant advancement in the way underground infrastructure is designed, constructed, and managed. A digital twin is a dynamic, data-driven virtual representation of a physical tunnel asset, construction process, or Tunnel Boring Machine (TBM) system. Unlike static three-dimensional models or traditional Building Information Modelling (BIM) platforms, a digital twin continuously integrates real-time operational, geotechnical, structural, and environmental data. This continuous data exchange allows the system to simulate, monitor, and optimise performance throughout the entire project lifecycle.

In mechanised tunnelling, the digital twin acts as a bridge between physical excavation and virtual engineering analysis. It enables engineers to visualise live machine behaviour, assess ground response, predict future performance trends, and evaluate risk scenarios before they develop into operational problems. Rather than relying solely on design assumptions made at the planning stage, project teams can refine forecasts dynamically as excavation progresses.

Conceptually, a digital twin consists of three interconnected elements: the physical asset, the virtual model, and the data connection layer. The physical asset may include the TBM, the segmental lining system, the surrounding ground mass, and associated surface plant infrastructure. The virtual model provides geometric, structural, geotechnical, and mechanical representations of these assets. The data connection layer links the two through real-time sensor inputs and analytics systems. As the TBM advances, excavation parameters, settlement readings, chamber pressure, torque, thrust, and segment installation data feed directly into the model. The twin is therefore not simply a visualisation platform; it is a continuously updated simulation environment that allows comparison between predicted and actual behaviour.

When applied specifically to a TBM, the digital twin integrates parameters such as cutterhead torque and rotation speed, thrust cylinder forces, chamber pressure, slurry density and flow in slurry machines, screw conveyor discharge rate in EPB machines, vibration signatures, temperature readings, energy consumption, and penetration per revolution. The virtual model replicates the mechanical operating envelope of the machine. If torque trends upward beyond expected values for the predicted geology, abnormal loading conditions can be flagged. Changes in vibration patterns may indicate early bearing degradation. By comparing live operational data with baseline mechanical models, the system can identify emerging risks before mechanical failure occurs.

Digital twins in tunnelling extend beyond the machine itself to include the surrounding ground and installed lining. Geological models, borehole information, groundwater pressure distributions, soil and rock mechanical properties, surface settlement instrumentation, segment ring geometry, and lining stress measurements can all be incorporated. As excavation progresses, measured face pressures and ground responses are integrated into the model. Surface settlement points update automatically, enabling comparison between predicted and observed deformation. If settlement deviates from expected values, engineers can adjust operational parameters proactively. This establishes a feedback loop between geotechnical design assumptions and actual ground behaviour.

Real-time monitoring and control are central benefits of digital twin implementation. Engineers can observe production trends, detect deviations from planned performance, evaluate accumulating risk, visualise tunnel geometry in real time, and identify bottlenecks in logistics or segment supply. For example, if utilisation drops below target levels, the twin can analyse downtime categories and simulate schedule impact. If hyperbaric intervention frequency increases, projected programme delays and cost exposure can be estimated. Instead of reacting after performance declines significantly, project teams can intervene early and strategically.

One of the most powerful capabilities of digital twins lies in predictive and probabilistic forecasting. By combining historical project data, real-time machine performance, geological variability, cutter wear trends, and maintenance records, the twin can simulate potential future outcomes. Monte Carlo simulations can be embedded within the system to forecast completion dates at different confidence levels, estimate intervention probabilities, assess main bearing failure risk, and predict cutter consumption. This strengthens schedule certainty and supports informed commercial decision-making.

Although digital twins are often linked to BIM platforms, they extend beyond design coordination by incorporating live operational data. During construction, they support alignment verification, quality assurance documentation, ring build validation, and grouting confirmation. After completion, the digital twin transitions into an asset management resource, storing lining installation records, segment manufacturing data, structural monitoring information, and long-term deformation trends. This continuity supports maintenance planning and lifecycle infrastructure management.

Italy's Underground Renaissance: A Transformative Implementation

The practical implications of this shift can be seen in Italy's extensive tunnel network—approximately 2,500 kilometres of underground infrastructure constructed primarily during the 1960s and 1970s. These tunnels are critical to national transportation systems but now face significant challenges due to aging structures and evolving safety standards [841].

Tecne Systra approached this challenge not by simply digitising traditional workflows, but by reimagining them. Rather than maintaining separate disciplinary processes with sequential handovers, the project began with integrated data collection. Information from borehole television inspections, flat jack testing, core drilling, and other investigative methods was fed directly into a digital twin—a unified digital environment shared by geologists, engineers, and designers. All stakeholders worked from the same continuously updated understanding of subsurface conditions [841].

The outcomes were measurable and substantial. Modelling time was reduced by 25%, overall project delivery accelerated by 21.5%, interpretation of survey data improved by 30%, and clash detection time decreased by 30%. However, beyond these metrics lies a more profound transformation: survey and investigation data became part of a continuous digital thread linking investigation, design, construction, and long-term operations [841].

As Ana Emiliano dos Reis, who led BIM implementation at Tecne Systra, explained, the digital transformation fundamentally changed how subsurface data was integrated and evaluated [841]. The move to a digital twin environment provided entirely new capabilities to support and refine design decisions.

Importantly, while reductions in cost and schedule are significant, the most meaningful benefits of digital twins are often invisible. They manifest as risks avoided, failures prevented, and redesigns never required [841].

In high-risk tunnelling environments such as deep urban drives or subsea crossings, digital twins enhance risk reduction and decision support. They provide early warning of instability, seal performance anomalies, settlement risks, potential flooding scenarios, and structural overstress conditions. If chamber pressure fluctuates outside predicted ranges, the twin can simulate potential outcomes and suggest stabilisation measures. Risk management therefore shifts from reactive crisis response to proactive engineering intervention.

From an economic perspective, digital twins improve financial performance by reducing unplanned downtime, enhancing parameter optimisation, strengthening predictive maintenance strategies, supporting realistic schedule modelling, and providing transparent performance records that reduce dispute potential. In mechanised tunnelling, small inefficiencies accumulate over long distances and extended durations. Early identification of minor deviations prevents these issues from compounding into major cost overruns.

Despite these advantages, digital twins require high-quality sensor data, reliable communication infrastructure, robust cybersecurity systems, skilled personnel capable of interpreting model outputs, and substantial upfront investment. Poor data integrity undermines predictive reliability, and overly complex systems without proper calibration may generate misleading conclusions. Successful implementation therefore depends on disciplined data governance and strong engineering oversight.

Looking ahead, emerging developments include AI-integrated digital twins with adaptive learning capabilities, fully synchronised geotechnical–mechanical interaction models, automated parameter optimisation systems, energy efficiency modelling, and integration with increasingly autonomous TBM systems. As global datasets expand across tunnelling projects, predictive accuracy will continue to improve, potentially reducing uncertainty in both operational management and procurement strategy.

Digital twins represent a transformative shift in mechanised tunnelling. They convert static design models into dynamic, evolving representations of excavation systems and ground behaviour. By integrating real-time data with predictive modelling, digital twins enhance safety, reliability, schedule certainty, and cost control. In modern tunnelling practice, the digital twin becomes a parallel engineering environment, an analytical mirror of the physical tunnel, supporting informed, data-driven decision-making throughout the project lifecycle.

Predictive Maintenance

Predictive maintenance is a data-driven maintenance strategy that aims to anticipate equipment failure before it occurs, allowing maintenance to be performed at the optimal time—neither too early nor too late. In contrast to traditional approaches such as reactive maintenance (repair after failure) or preventive maintenance (scheduled servicing at fixed intervals), predictive maintenance relies on real-time condition monitoring and analytical modelling to determine the actual health of equipment.

In mechanised tunnelling, predictive maintenance is particularly critical because the Tunnel Boring Machine operates as a continuous, high-capital production system. Unplanned downtime can result in substantial financial loss, programme delay, and in some cases, safety risk. Predictive maintenance therefore shifts maintenance from a time-based activity to a condition-based and risk-informed engineering function.

At its core, predictive maintenance depends on continuous data collection from sensors installed across key machine components. These sensors monitor parameters such as vibration, temperature, pressure, torque, rotational speed, hydraulic performance, and electrical load. The data is transmitted in real time to control systems and analytics platforms, where it is processed to detect deviations from normal operating conditions.

The fundamental principle is that mechanical failure rarely occurs instantaneously. Instead, components typically exhibit early warning signs—subtle changes in vibration patterns, gradual temperature increases, pressure fluctuations, or irregular torque behaviour. Predictive maintenance systems are designed to identify these early indicators and assess whether they represent normal variation or emerging failure.

In TBM operations, predictive maintenance is applied to critical systems such as the cutterhead, main bearing, gearbox, hydraulic systems, slurry pumps, screw conveyors, and conveyor systems. For example, a gradual increase in vibration amplitude combined with a shift in frequency spectrum may indicate bearing wear or imbalance in the cutterhead. Similarly, rising hydraulic oil temperature may signal inefficiency or internal leakage within the system.

Modern predictive maintenance systems often incorporate machine learning and artificial intelligence algorithms. These models analyse historical performance data alongside real-time inputs to establish baseline operating conditions. As new data is received, the system compares it against expected behaviour and identifies anomalies. Over time, the system becomes more accurate as it “learns” the normal performance envelope of the machine under different geological and operational conditions.

One key advantage of predictive maintenance is the ability to schedule interventions during planned downtime rather than during unexpected failure. For instance, if cutter wear is predicted to reach a critical threshold within a defined distance, maintenance can be planned at a convenient location rather than triggered by sudden performance loss or failure. In pressurised tunnelling environments, this is particularly important because interventions such as hyperbaric cutter replacement are complex, costly, and time-sensitive.

Predictive maintenance also supports optimisation of spare parts management and logistics. By forecasting component degradation, project teams can ensure that replacement parts, tools, and specialist personnel are available when required, reducing delays associated with emergency procurement or mobilisation.

From a financial perspective, predictive maintenance reduces lifecycle cost by minimising unplanned downtime, preventing catastrophic failures, extending component life, and improving overall utilisation of the TBM. In tunnelling, where daily operating costs are high, even small reductions in downtime can result in significant cost savings.

However, predictive maintenance is not without challenges. It requires high-quality sensor data, reliable communication systems, and robust data processing capabilities. Poor data quality or sensor failure can lead to incorrect predictions. In addition, predictive models must be carefully calibrated to account for varying geological conditions, as changes in ground behaviour can influence machine performance and potentially be misinterpreted as mechanical issues.

Importantly, predictive maintenance does not eliminate the need for engineering judgement. Instead, it provides enhanced visibility into machine condition, allowing engineers and operators to make more informed decisions. The role of maintenance personnel shifts from routine inspection to data interpretation, diagnosis, and strategic intervention planning.

In modern mechanised tunnelling, predictive maintenance is a key component of digital integration alongside automation, artificial intelligence, and digital twins. It transforms maintenance from a reactive or scheduled activity into a proactive, intelligence-driven process that supports safety, reliability, and programme certainty.

Case Study Scenario: Predictive Maintenance in a High-Pressure Slurry TBM Drive

Project Overview

A major metropolitan rail authority is delivering a 5.8 km twin-bore tunnel beneath a river and dense urban environment. The tunnels are excavated using slurry TBMs with an excavation diameter of 7.2 m. Ground conditions consist of:

- Highly permeable sands and gravels near the river
- Interbedded clay layers
- Occasional sandstone lenses
- Groundwater pressures up to 5.5 bar

Due to high hydrostatic pressure, the TBM is equipped with:

- A pressurised cutterhead and main bearing system
- Multiple vibration sensors on the cutterhead and main bearing
- Temperature sensors on the gearbox and hydraulic systems
- Torque and thrust monitoring systems
- Real-time data transmission to a surface control centre
- AI-assisted predictive maintenance software integrated with the digital twin

Daily operating cost is approximately $350,000, making downtime extremely costly.

Phase 1: Establishing Baseline Conditions

During the first 800 metres of excavation, the predictive maintenance system establishes baseline performance profiles for:

- Cutterhead torque: 6,500–7,200 kNm
- Vibration signature: stable frequency spectrum with low amplitude variation
- Main bearing temperature: 48–52°C
- Slurry pump pressure: stable at 5.2 bar
- Penetration rate: 10–12 mm/rev

Machine learning models analyse this data to define the normal operating envelope for the specific geology and machine configuration.

Phase 2: Detection of Anomaly

At chainage 1,150 m, the system detects subtle but consistent changes:

- A 7% increase in vibration amplitude at a specific frequency band
- Slight irregularity in torque fluctuation (±12% variance instead of ±5%)

- Gradual rise in main bearing temperature from 50°C to 55°C over 48 hours

These changes are not yet critical and would likely go unnoticed under traditional maintenance approaches.

However, the predictive system flags this as a developing anomaly, indicating possible:

- Cutterhead imbalance
- Early-stage bearing wear
- Uneven cutter wear due to mixed ground conditions

Phase 3: Predictive Analysis

The AI model compares current data with historical patterns and predicts:

- Probability of main bearing degradation within the next 120–150 operating hours
- Increased risk of cutterhead imbalance leading to accelerated wear
- Potential for unplanned shutdown within 5–7 days if no action is taken

The system estimates:

- If failure occurs:
  - Minimum 4–6 days downtime
  - Potential repair cost exceeding $2.5 million
  - Programme delay impact of 40–60 metres lost production

Phase 4: Engineering Decision and Planned Intervention

The engineering team reviews the data alongside geological logs and confirms that the TBM is entering a mixed-face transition zone with sandstone lenses, increasing cutter wear asymmetry.

Instead of continuing until failure, the team decides to:

- Schedule a planned intervention in 36 hours at a favourable location
- Prepare hyperbaric intervention team in advance
- Mobilise spare cutters and inspection tools
- Adjust operating parameters (reduce RPM and optimise thrust distribution) to stabilise conditions until intervention

Phase 5: Intervention Execution

The TBM is stopped in a controlled manner at a safe location.

Inspection reveals:

- Uneven cutter wear on one quadrant of the cutterhead
- Early-stage bearing seal wear
- Minor imbalance developing in cutterhead rotation

Actions taken:

- Replacement of worn cutters
- Rebalancing of cutterhead configuration
- Inspection and adjustment of bearing seals
- Calibration of monitoring sensors

Total intervention time: 18 hours

Phase 6: Outcome and Performance Recovery

After restart:

- Vibration amplitude returns to baseline
- Torque stabilises within normal variance
- Bearing temperature returns to 50–52°C range
- Penetration rate improves by 8%

The predicted failure is successfully avoided.

Comparative Outcome Analysis

Scenario A – Without Predictive Maintenance

If operating under a reactive or scheduled maintenance approach:

- Failure likely occurs unexpectedly within 5–7 days
- Emergency shutdown required
- Hyperbaric intervention under less controlled conditions

- Estimated downtime: 4–6 days
- Cost impact:
  - Downtime: $1.4–2.1 million
  - Repair: $2.5 million
  - Total: ~ $4–5 million

Scenario B – With Predictive Maintenance

- Planned intervention: 18 hours
- No catastrophic failure
- Minimal production loss (~8–10 metres)
- Cost impact:
  - Intervention cost: $250,000–400,000
  - No major repair required

Key Engineering Insights

This case study highlights several fundamental principles underpinning predictive maintenance in TBM operations. At its core, predictive maintenance focuses on identifying early warning signals rather than waiting for failure to occur. Subtle variations in vibration, temperature, and torque provide critical insight into developing issues well before they escalate into catastrophic breakdowns. These early indicators enable timely intervention and significantly reduce operational risk.

A key shift demonstrated in this scenario is the transition from time-based to condition-based maintenance. Rather than relying on fixed service intervals or reacting to failure events, maintenance activities are scheduled based on the actual condition of components. This ensures that interventions occur precisely when required, optimising both machine performance and component lifespan.

The case also illustrates how operational risk is actively managed rather than passively transferred. Predictive systems enable engineers to anticipate potential failure modes and take control of developing issues before they impact production. This represents a shift from reactive problem-solving to proactive risk management, where failure pathways are identified and mitigated early.

Importantly, the integration of machine data with geological conditions is essential. In this instance, the detected anomaly was linked to a ground transition zone, highlighting the need to interpret equipment behaviour in the context of subsurface variability. Without this integration, mechanical symptoms could be misinterpreted, leading to ineffective or delayed interventions.

The scenario further demonstrates that relatively small, planned interventions can prevent significantly larger failures. An 18-hour controlled stoppage avoided what could have escalated into a multi-day shutdown and a multi-million-dollar repair. This reinforces the principle that early, targeted action is far more effective than late-stage reactive response.

Broader Implications

Predictive maintenance fundamentally transforms TBM operations by shifting the operational paradigm. Processes that were once reactive become proactive, scheduled activities evolve into condition-based strategies, and uncertainty is replaced with data-informed decision-making. As a result, high-risk operations become more controlled and predictable.

These improvements have wide-ranging benefits. Programme certainty is enhanced through reduced unplanned downtime and more reliable production forecasting. Asset life is extended as components are operated within optimal performance limits. Safety is improved by reducing the need for emergency interventions, particularly in high-risk environments such as pressurised tunnelling. At the same time, overall cost efficiency is increased through the avoidance of major failures and better allocation of maintenance resources.

The scenario demonstrates that predictive maintenance is a critical enabler of modern mechanised tunnelling. By integrating real-time sensor data, advanced analytical modelling, and informed engineering judgement, project teams can anticipate equipment degradation, plan interventions strategically, and maintain the TBM within its optimal operating envelope.

In high-value tunnelling projects, the most significant benefit of predictive maintenance is not simply the reduction in maintenance costs. Rather, it lies in the prevention of failure itself. By avoiding unplanned breakdowns, projects can maintain continuous production, protect worker safety, and ensure programme reliability—outcomes that are far more valuable than any single cost saving. Ultimately, predictive maintenance ensures that the TBM operates within its optimal performance envelope, reducing risk and improving efficiency across the entire tunnelling operation.

Real-time Geotechnical Feedback

Real-time geotechnical feedback is an innovative approach in underground tunnelling operations that facilitates the continuous monitoring of ground conditions during excavation. This methodology enables engineers to adapt their strategies immediately to the dynamic nature of underground environments. Unlike traditional monitoring methods that rely heavily on pre-construction geological investigations, real-time feedback enhances situational awareness and decision-making, thus improving project safety and efficiency.

Historically, ground conditions have been primarily assessed through boreholes, laboratory tests, and geological interpretations, providing a static model of expected behaviours. This traditional approach can lead to significant discrepancies between predicted and actual ground conditions during mechanically excavated tunnelling. Real-time geotechnical feedback updates these pre-existing models, allowing for a more accurate representation of ground behaviour as excavation progresses [466, 842, 843]. Studies in various urban tunnelling projects demonstrate that continuous monitoring effectively mitigates risks associated with unforeseen soil conditions, such as excessive settlement or instability [842, 844].

In mechanised tunnelling operations, real-time geotechnical feedback is derived from diverse data sources, primarily the machinery itself and external monitoring systems. Machine-based indicators, such as cutterhead torque, penetration rates, and chamber pressures, provide immediate insights into the ground conditions. For instance, fluctuating torque levels may signal variations in soil density or hardness, which necessitate operational adjustments [30, 466, 845]. Complementarily, geotechnical instruments like settlement markers, inclinometers, and piezometers gather additional real-time data, reflecting ground deformation and groundwater pressure [30, 466, 845]. The integration of these data forms a comprehensive feedback loop that enhances operational control and response adaptability [30, 842].

One of the most significant advantages of real-time feedback systems is the enhanced capability for early detection of ground transitions. Variations in operational parameters may indicate shifts in ground conditions, prompting pre-emptive adjustments to tunnelling strategies, such as modifying face pressure or advance rates [30, 846]. This proactive approach is crucial in urban environments, where even minimal settlement can cause substantial damage to nearby infrastructure [844, 846].

Real-time geotechnical feedback is particularly vital in sensitive urban settings where infrastructure resilience is paramount. Continuous monitoring allows for timely corrective actions; alarm thresholds can be established to alert engineers of any deviations from acceptable operational parameters [844, 847]. Work conducted on projects like the Toronto-York Spadina Subway Extension exemplifies the successful application of advanced monitoring techniques, employing a combination of machine data and external sensors to ensure structural integrity throughout tunnelling operations [844].

While real-time feedback offers numerous benefits, challenges such as data quality and interpretation must be addressed. Accurate installation and calibration of sensors are critical to ensure reliable readings, and engineers must distinguish between normal data variability and indicators of potential instability [848, 849]. Furthermore, the volume of data generated can overwhelm systems if not effectively filtered, necessitating investment in advanced data analytics [848, 850].

The integration of real-time geotechnical feedback with emerging digital twin technologies represents a promising frontier in tunnelling [842, 847]. Digital twins create virtual replicas of real-world conditions, allowing engineers to visualize and simulate ground behaviour spatially and temporally. This capability not only enhances the understanding of geotechnical dynamics but also facilitates scenario planning and risk assessment before operational adjustments are made in the field [842].

Overall, real-time geotechnical feedback is transforming the landscape of mechanised tunnelling by providing a dynamic model of ground conditions that facilitates immediate responses to changing environments. While challenges remain in data management and interpretation, the potential for enhanced safety, reduced risk, and increased efficiency underscores the importance of this approach in modern geotechnical engineering practices. The ongoing integration with digital technologies promises to further refine operational strategies, paving the way for more resilient and sustainable tunnelling projects.

Remote Operation

Remote operation refers to the capability of monitoring and controlling Tunnel Boring Machines from a distance, utilizing advanced communication and data transmission technologies. This operational model allows for control from locations ranging from surface control rooms to fully off-site centres that manage multiple TBMs across various projects. Emerging technologies, such as real-time data transmission, automation, and integrated control platforms, are pivotal in facilitating this transition from traditional operation methods that require on-site personnel [846, 847].

Historically, the operation of TBMs entailed manual control by personnel situated close to or within the machine's vicinity. Operators manually managed critical parameters such as thrust, cutterhead rotation, and face pressure, relying on local data and personal experience [829, 846]. Despite its effectiveness, this model posed high risks due to exposure to hazardous environments, including high pressure and confined spaces, particularly in challenging geological conditions where instability was a concern [846, 851].

The implementation of remote operation significantly enhances safety by minimizing the need for operators to be present in high-risk areas during tunnelling processes. This reduction in onsite personnel, especially in hazardous locations such as pressurized cutterhead chambers or areas with high groundwater pressure, allows for safer working conditions [799, 846].

Moreover, operational efficiency is markedly improved. Centralized control rooms enable the simultaneous oversight of multiple systems, enhancing coordination among the TBM, slurry treatment plants, and surface logistics. This integrated approach is particularly beneficial in large-scale projects where a single control centre can manage several TBMs, promoting standardized procedures and a rapid response to operational issues [847, 852].

Furthermore, remote operations provide access to specialized expertise that was previously difficult to deploy on-site. Engineers and specialists can remotely monitor TBM performance and address complex issues such as abnormal vibrations and system anomalies in real time, leading to faster and more efficient problem resolution [29, 846].

Remote operation is closely linked to predictive maintenance practices, enhanced by the integration of digital twin technologies. Data collected from TBMs is leveraged to create detailed analytical models that assess machine health and predict potential failures. This proactive approach allows operators to shift from reactive to predictive management of machine performance [847, 853]. By evaluating various operational scenarios using digital models, operators can optimize performance based on predictive insights rather than merely responding to observed issues [852, 854].

Despite the numerous advantages, remote operation faces significant challenges. Dependable communication infrastructure is essential for effective monitoring and control, and any disruption in data flow can compromise operational capabilities. Therefore, systems need to be designed with robustness and redundancy to mitigate the risks associated with potential communication failures [799, 846].

Cybersecurity is another critical consideration, as remote operation increases exposure to digital threats. Ensuring adequate safeguards against cyber-attacks is vital for maintaining the integrity of remote control systems [29].

Moreover, certain essential tasks such as physical inspections, cutter replacements, and other hands-on activities still necessitate the presence of skilled workers on-site. Geological variability and unexpected conditions can also demand immediate, experienced judgement, reinforcing the need for experienced personnel in the field [799, 855].

Looking ahead, the field of remote TBM operation is poised for further advancements through enhanced automation and Artificial Intelligence (AI) integration. Future developments may allow for greater autonomy in TBM operations, with routine tasks fully automated and human operators intervening only in exceptional situations [846, 847]. This evolution will require a redefinition of the skills needed in the workforce from traditional manual operation to advanced data analysis and systems management [29, 852].

Future TBM Concepts

The evolution of Tunnel Boring Machines (TBM) is poised to revolutionize the field of mechanized tunnelling through the integration of advanced technologies and innovative design approaches. Traditional mechanical excavation systems are increasingly being merged with digital technologies, automation, advanced materials, and intelligent control systems to enhance performance across various metrics such as safety, productivity, costs, and environmental impact.

Future TBM concepts are expected to usher in a new era where machines evolve from operator-driven devices to semi-autonomous systems. While full autonomy in TBMs is currently hampered by geological unpredictability, there is a strong trend towards utilizing real-time data-driven optimizations for routine operational decisions. Artificial Intelligence (AI) will play a critical role in regulating thrust, cutterhead speed, and face pressure adjustments in a more automated fashion, allowing human operators to transition into supervisory roles that focus on exception management and strategic decision-making [825, 847].

Moreover, concepts such as digital twins and real-time geotechnical models will be fundamental. These systems will continuously update subsurface data during excavation, allowing for proactive adjustments to excavation parameters, thereby mitigating risks related to ground stability and settlement [618, 847]. Predictive modelling, which incorporates machine data and monitoring instrumentation, will be essential for anticipating geological conditions and optimizing TBM operations in real-time [387, 856].

Automation is set to play an instrumental role in future TBM operations. Tasks characterized by high labour intensity or risk, such as segment handling and cutter inspections, are increasingly being assigned to robotic systems. Automated segment erectors and inspection drones will not only improve safety by reducing human exposure to hazardous situations but will also enhance the efficiency and consistency of operations [309, 825]. Such developments aim to streamline the excavation process, thereby improving reliability and performance metrics overall.

Future TBM designs will likely incorporate advanced robotic mechanisms for dynamic support installations. For instance, incorporating systems that can adjust conditioning or grouting techniques in real-time based on sensor feedback can help pre-emptively address issues related to ground instability [30, 830]. This aligns with findings that highlight the importance of automated monitoring systems in optimizing tunnelling processes [618, 857].

AI-driven systems take centre stage in analysing massive datasets to fine-tune the TBM operating parameters for specific geological conditions. Such systems are being developed to enhance the efficacy of operational choices regarding cutterhead speed and thrust force while reducing mechanical wear and energy expenditure [856, 858]. The performance prediction capabilities of these AI systems will become more nuanced over time as they learn from greater databases of tunnelling operations, allowing them to suggest optimal conditions tailored for different scenarios [387, 858].

In mechanical design, modularity and adaptability are expected to be emphasized in future TBMs. An example is the emergence of hybrid machines that can seamlessly switch between Earth Pressure Balance (EPB) and slurry modes, capable of handling diverse geological environments without extensive modifications [668, 859]. This adaptability will greatly enhance the reliability and efficiency of tunnelling operations in complex urban and sub-surface environments.

With increasing awareness of environmental issues, future TBM designs will incorporate energy efficiency and sustainability as core principles. Innovations may involve energy recovery systems designed to capture and reuse the energy expended during operations, such as hydraulic or braking energy, and the implementation of optimized cutter designs to minimize resistance [30, 361, 847, 856]. The use of renewable energy sources for tunnel operations is also a potential development area that could significantly lower the carbon footprints of large infrastructure projects.

Despite the advancements in technology, the inherent unpredictability of geological conditions remains a challenge that cannot be entirely eliminated. Future TBMs will increasingly focus on managing this uncertainty through enhanced data integration and predictive modelling rather than eliminating it. The necessity for engineering judgement continues to be paramount, highlighting the dynamic interplay between technology and human expertise [498, 833].

Chapter 20

Case Studies and Applied Examples

The preceding chapters have explored the principles, mechanics, technologies, and systems that underpin modern tunnel boring operations. From ground behaviour and structural response to TBM design, automation, and digital integration, the discussion has established a comprehensive technical foundation for understanding mechanised tunnelling. However, the true measure of these concepts lies not in theory alone, but in their application within real-world projects, where geological uncertainty, operational complexity, and engineering decision-making intersect.

This final chapter draws together these elements through a series of case studies and applied examples spanning a diverse range of tunnelling environments. As presented in Chapter 20, these include urban metro tunnels, long water conveyance systems, hard rock mountain tunnels, and mixed-ground megaprojects—each representing distinct challenges in terms of geology, scale, risk, and system integration. Collectively, they illustrate how engineering principles are translated into practice under varying constraints and conditions.

The case studies demonstrate that successful tunnelling is not defined by a single technology or method, but by the effective integration of multiple systems—geotechnical understanding, machine selection, operational control, monitoring, and digital tools. They highlight how modern TBM operations rely on continuous feedback between the machine and the ground, enabling adaptive responses to changing conditions. Equally, they reveal how failures and challenges provide critical insights that drive innovation and improve future performance.

A central theme emerging from these examples is the shift from deterministic planning to data-driven, adaptive engineering. Tunnelling is inherently uncertain, and even the most detailed pre-construction investigations cannot fully predict subsurface conditions. As a result, the ability to monitor, interpret, and respond in real time has become a defining characteristic of successful projects. This chapter demonstrates how this capability is applied across different project types and geological contexts.

The chapter also emphasises the importance of integration across disciplines. Modern tunnelling projects require close coordination between geotechnical engineers, structural designers, mechanical specialists, and operational teams. Digital technologies such as BIM, digital twins, and predictive analytics increasingly serve as the platform for this integration, enabling a shared understanding of system performance and risk.

Importantly, the lessons presented extend beyond individual projects. They represent a cumulative body of knowledge derived from decades of tunnelling experience—where both successes and failures contribute to improved practice. These lessons reinforce the importance of adaptability, proactive risk management, and the continuous refinement of engineering approaches.

In bringing together these case studies and lessons learned, this chapter provides not only a reflection on current practice, but also a foundation for future development. As tunnelling projects continue to increase in scale, complexity, and technical ambition, the insights gained from these applications will play a critical role in shaping the next generation of TBM operations.

Ultimately, this final chapter positions tunnelling as a discipline that is both technically rigorous and inherently adaptive—where engineering success depends on the ability to integrate knowledge, technology, and experience in the face of uncertainty.

Urban Metro Tunnels

Urban metro tunnelling represents one of the most complex applications of mechanised tunnelling, where engineering must balance geotechnical uncertainty, dense infrastructure, environmental constraints, and strict settlement limits. The following case studies illustrate how modern TBM technologies, digital systems, and risk management strategies are applied in real-world metro projects.

Crossrail (Elizabeth Line), London, UK

The Crossrail (Elizabeth Line) project is one of the most significant urban tunnelling projects ever undertaken in Europe, involving over 42 km of new tunnels beneath central London. The tunnels pass through highly variable ground conditions, including London Clay, Lambeth Group deposits, and water-bearing sands.

Figure 105: Crossrail tunnel looking west, 2013. Matt Brown from London, England, CC BY 2.0, via Wikimedia Commons.

Mechanised tunnelling was carried out using a fleet of EPB TBMs specifically designed to control ground movement and maintain face stability. One of the key engineering challenges was limiting settlement beneath historic buildings and critical infrastructure such as existing Underground lines.

Figure 106: Tunnel boring machine Ada, about to be deployed at the Royal Oak Crossrail portal, on a low loader west of the foot bridge to the west of the portal. Before it can be moved any further the bridge must be raised 10ft on hydraulic jacks. Marcus Rowland, CC BY-SA 3.0, via Wikimedia Commons.

Real-time geotechnical monitoring played a central role, with thousands of settlement points and building sensors providing continuous feedback. TBM parameters, particularly face pressure and advance rate, were adjusted dynamically to maintain ground stability. The project also utilised advanced BIM and digital integration systems, enabling coordination across multiple contractors and disciplines.

This project demonstrated how precise pressure control, continuous monitoring, and integrated digital workflows can successfully manage tunnelling in highly sensitive urban environments.

Singapore Downtown Line

The Downtown Line Singapore involved extensive tunnelling through mixed ground conditions, including soft marine clays, silts, and highly permeable sands with high groundwater pressures.

Figure 107: Tunnels of the Downtown line. SgTransport from Singapore, Singapore, CC BY 2.0, via Wikimedia Commons.

A combination of EPB and slurry TBMs was used, with machine selection tailored to specific ground conditions. In water-bearing strata, slurry TBMs were critical for maintaining face stability and preventing inflow.

A defining feature of this project was the use of real-time geotechnical feedback and predictive modelling. Data from TBM sensors and ground instrumentation were integrated to detect ground transitions and adjust operational parameters. Strict settlement criteria—often less than 10 mm—required highly controlled excavation.

The project also implemented extensive risk management frameworks, including ground treatment (jet grouting and ground freezing) in particularly sensitive zones. This case highlights the importance of matching TBM type to geology and integrating monitoring systems with operational control.

Metro Line 9, Barcelona, Spain

The Barcelona Metro Line 9 is one of Europe's deepest metro lines, with tunnels reaching depths of up to 50 metres below ground. The project involved large-diameter TBMs excavating through heterogeneous geology, including clay, sand, and rock formations.

A unique feature of this project was the use of very large diameter TBMs (up to 12 m) to create tunnels that accommodated both tracks and station platforms within a single excavation. This reduced the need for separate station caverns but introduced significant structural and geotechnical challenges.

Figure 108: Platform 'Fira' station of Line L9 of the Barcelona Metro. Mariusmm, CC BY-SA 4.0, via Wikimedia Commons.

Automation and monitoring systems were heavily utilised to manage:

- High overburden pressures
- Groundwater inflow
- Structural loading on segmental linings

The project also integrated digital modelling tools to assess lining performance and ground interaction. This case demonstrates how innovative TBM design and large-diameter excavation strategies can optimise urban space while managing deep tunnelling risks.

Delhi Metro Phase III, India

The Delhi Metro expansion required tunnelling beneath densely populated areas with highly variable soil conditions, including soft alluvial deposits and mixed ground.

EPB TBMs were extensively used due to their ability to maintain face stability in soft ground. One of the major challenges was tunnelling beneath existing buildings with limited structural resilience.

To address this, the project implemented:

- Strict settlement monitoring and control systems
- Real-time adjustment of TBM parameters
- Compensation grouting to counteract ground movement

In some areas, tunnelling was carried out within a few metres of existing foundations. The success of the project relied on tight operational control, continuous monitoring, and rapid response to ground behaviour.

Sydney Metro City & Southwest, Australia

The Sydney Metro City & Southwest project involved twin tunnels beneath Sydney Harbour and the CBD, passing through Hawkesbury Sandstone and mixed ground conditions.

Hard rock TBMs were used for sandstone sections, while mixed ground conditions required careful parameter control. The project leveraged advanced digital engineering and data integration, including:

- Real-time TBM performance monitoring
- Digital twin models for construction planning
- Predictive maintenance systems

A key challenge was managing transitions between rock and softer materials, which can induce uneven cutter wear and instability. The integration of machine data with geological modelling enabled proactive adjustment of excavation parameters.

This project demonstrates how digital integration, predictive analytics, and adaptive control enhance performance in complex geological environments.

Key Lessons from Urban Metro Tunnelling

Across the case studies presented, several consistent engineering principles emerge that define successful urban metro tunnelling. Foremost among these is the critical importance of precise control of ground behaviour, particularly settlement. In dense urban environments, even minor deviations can lead to significant impacts on buildings, utilities, and existing transport infrastructure. As a result, tunnelling operations must be carefully managed to maintain stability at all times.

The selection of Tunnel Boring Machine (TBM) type must be closely aligned with prevailing ground conditions and groundwater pressures. No single machine type is universally suitable, and in mixed or variable geology, hybrid and adaptable TBMs are becoming increasingly important. Matching machine capability to subsurface conditions is fundamental to achieving both safety and efficiency.

Real-time monitoring and feedback systems play a central role in modern tunnelling practice. Continuous data collection from instrumentation and TBM sensors allows engineers to assess performance and ground response as excavation progresses. This enables a shift from reactive problem-solving to proactive control, where potential issues can be identified and addressed before they escalate.

Digital technologies are also transforming project delivery. Tools such as Building Information Modelling (BIM), digital twins, and predictive maintenance systems enhance coordination across disciplines, improve transparency, and support more informed decision-making. These technologies allow project teams to integrate design, construction, and operational data into a unified framework.

Ultimately, successful urban tunnelling depends on integration. The complexity of underground construction in urban settings requires close alignment between geotechnical, structural, mechanical, and operational systems. A holistic, multidisciplinary approach ensures that all aspects of the project are considered together rather than in isolation, reducing risk and improving overall performance.

Urban metro tunnelling represents the forefront of modern underground engineering, where advanced technology and engineering expertise converge to address complex challenges. These projects demonstrate how the integration of TBM technology, real-time data, digital systems, and rigorous risk management enables the safe and efficient construction of tunnels beneath highly constrained urban environments.

As cities continue to expand and demand more sustainable transport solutions, the lessons learned from these projects will play a critical role in shaping future tunnelling practice. The next generation of tunnelling systems will be defined by increasing levels of precision, adaptability, and intelligence, ensuring that underground infrastructure can be delivered safely, efficiently, and with minimal impact on the built environment.

Long Water Conveyance Tunnels

Long water conveyance tunnels are critical infrastructure systems designed to transport water over large distances for urban supply, irrigation, hydropower, and inter-basin transfer. These tunnels often extend for tens to hundreds of kilometres and are constructed under complex geological, hydraulic, and environmental conditions. The following case studies illustrate how advanced tunnelling methods, including TBMs, have been applied to deliver large-scale water conveyance systems.

Lesotho Highlands Water Project (LHWP), Southern Africa

The Lesotho Highlands Water Project is one of the largest water transfer schemes in the world, supplying water from the highlands of Lesotho to South Africa's Gauteng region.

The project includes over 80 km of tunnels, constructed through mountainous terrain with varying rock conditions. Drill-and-blast methods were predominantly used due to the hard rock environment and alignment flexibility requirements.

A key engineering challenge was managing high overburden stresses and groundwater inflow. Careful rock support design, including rock bolts and shotcrete, was essential to maintain stability. Hydraulic design also played a critical role, as the tunnels operate under high-pressure flow conditions.

This project demonstrates the integration of hydropower generation and water transfer, highlighting the dual-purpose nature of many long conveyance tunnels.

Snowy Mountains Scheme, Australia

The Snowy Mountains Scheme is one of Australia's most iconic engineering achievements, involving a vast network of tunnels, aqueducts, and power stations. Constructed primarily between 1949 and 1974, the scheme includes over 140 km of tunnels that divert water from the Snowy River eastward to the Murray and Murrumbidgee river systems.

The tunnels were excavated using drill-and-blast techniques in hard rock, often under extremely challenging alpine conditions. Key considerations included:

- Long tunnel alignment accuracy
- Ventilation and access in remote areas
- High-pressure hydraulic performance

The Snowy Scheme remains a benchmark for large-scale water diversion and hydropower integration, and continues to influence modern tunnel design and operation.

Delaware Aqueduct, USA

The Delaware Aqueduct is the longest continuous tunnel in the world, extending approximately 137 km and supplying water to New York City.

Constructed between 1937 and 1945, the aqueduct conveys water from reservoirs in the Catskill Mountains to New York City by gravity flow.

A major modern engineering effort has involved repairing leaks in the tunnel beneath the Hudson River. A new bypass tunnel was constructed using TBMs to divert flow while repairs were undertaken.

This case highlights the importance of:

- Long-term asset management and maintenance
- Integration of new tunnelling works with existing infrastructure
- Minimising disruption to critical water supply systems

South–North Water Transfer Project, China

The South–North Water Transfer Project is one of the largest water diversion projects ever undertaken, designed to transfer water from southern China to the arid northern regions. The project consists of multiple routes, including extensive tunnel sections constructed using TBMs. Some tunnels pass beneath major rivers and urban areas, requiring high levels of precision and risk control.

Key engineering features include:

- Large-diameter TBM tunnels for high flow capacity
- Pressurised tunnelling beneath waterways
- Integration of canals, tunnels, and pumping stations

This project demonstrates the role of modern TBM technology in mega-scale water transfer systems, particularly in soft ground and mixed conditions.

California State Water Project – Coastal Branch, USA

The California State Water Project includes long tunnels and pipelines that transport water across vast distances to support agriculture and urban areas. The Coastal Branch extension involved tunnelling through mountainous terrain using TBMs and conventional methods. The tunnels were designed to operate under pressure, requiring robust lining systems and careful hydraulic design.

Challenges included:

- Variable geology (rock to soft ground transitions)
- Seismic considerations
- Environmental constraints

This project highlights the importance of resilient design and adaptability in long-distance water conveyance systems.

Key Engineering Lessons

The case studies presented highlight several consistent engineering principles that underpin the successful design and construction of long water conveyance tunnels. One of the most critical requirements is the need for precise alignment control and surveying over extended distances. Given the scale of these tunnels, even minor deviations can accumulate over time, potentially affecting both hydraulic performance and structural integrity. Accurate surveying and continuous verification are therefore essential throughout the construction process.

Hydraulic design is equally as important as structural design in these systems. Engineers must carefully consider flow velocity, internal pressures, friction losses, and transient effects such as surge and water hammer. The efficiency and safety of the tunnel depend on achieving a balance between hydraulic performance and structural capacity, ensuring that the system can operate reliably under varying flow conditions.

Geological variability remains one of the most significant challenges in long tunnel construction. Subsurface conditions can change considerably לאורך the tunnel alignment, requiring a flexible and adaptive approach. Successful projects rely on comprehensive geotechnical investigation, continuous monitoring, and the ability to modify excavation methods and support systems in response to actual ground conditions encountered during construction.

Durability and long-term maintenance are also fundamental considerations. Water conveyance tunnels are typically designed as long-life assets, often expected to operate for 50 to 100 years or more. This necessitates robust lining systems, corrosion resistance, and provisions for inspection, maintenance, and repair to ensure ongoing performance over the asset's lifecycle.

In modern projects, there is an increasing reliance on digital technologies and monitoring systems. These tools support asset management, enable predictive maintenance strategies, and facilitate operational optimisation. By integrating real-time data and analytical models, engineers can better understand system performance and make informed decisions throughout both construction and operation.

Long water conveyance tunnels represent some of the most ambitious and technically demanding forms of underground infrastructure. They require the integration of geotechnical engineering, hydraulics, structural design, and complex construction logistics on a large scale.

The case studies demonstrate how both traditional excavation methods and modern TBM technologies are applied to deliver reliable and durable water infrastructure. As global demand for water security continues to grow, these tunnels will play an increasingly important role in supporting urban populations, agricultural systems, and energy generation.

Looking forward, future developments are likely to focus on improving efficiency, enhancing sustainability, and expanding the use of digital technologies. These advancements will ensure that long water conveyance tunnels remain resilient, adaptable, and capable of meeting the evolving challenges associated with water management in a changing world.

Hard Rock Mountain Tunnels

Hard rock mountain tunnelling represents one of the most demanding areas of underground engineering, involving excavation through high-strength rock masses, extreme overburden pressures, variable geology, and often remote or high-altitude environments. These projects require careful integration of excavation method selection, rock support design, ventilation, and logistics. The following case studies illustrate how both drill-and-blast and TBM methods have been applied in major mountain tunnelling projects.

Gotthard Base Tunnel, Switzerland

The Gotthard Base Tunnel is the world's longest railway tunnel, extending approximately 57 km through the Swiss Alps. Constructed through high-strength crystalline rock under extreme overburden (up to 2,300 m), the project utilised a combination of TBMs and drill-and-blast methods. TBMs were used in long, consistent sections of hard rock, while drill-and-blast was applied in complex zones such as faulted regions and station caverns.

Figure 109: The northern end of the emergency-stop station Sedrun in the eastern tube. Here starts a left turn. Switzerland, Sep 15, 2016. Kecko, CC BY 2.0, via Wikimedia Commons.

Key challenges included:

- High in-situ stress leading to rockburst conditions
- Elevated rock temperatures (up to 45°C)
- Long logistics chains for spoil removal and supply

Advanced monitoring systems and flexible support strategies, including yielding support systems, were implemented to manage stress redistribution. This project demonstrates the effectiveness of hybrid excavation strategies in extreme mountain environments.

Lötschberg Base Tunnel, Switzerland

The Lötschberg Base Tunnel is a 34.6 km railway tunnel through the Bernese Alps, forming part of Switzerland's Alpine transit network.

The project encountered highly variable geology, including fractured rock zones and water-bearing formations. Both TBMs and drill-and-blast methods were used, with the excavation approach adapted to local conditions.

A major challenge was managing water inflows under high pressure, particularly in karstic and fractured zones. Pre-grouting and drainage systems were critical in controlling inflow and maintaining stability.

The project highlights the importance of:

- Detailed geological investigation and forecasting
- Flexible excavation and support systems
- Integration of hydrogeological management

Jinping II Hydropower Tunnels, China

The Jinping II Hydropower Station includes a series of deep diversion tunnels, each exceeding 16 km in length, constructed through extremely high-stress mountain conditions.

With overburden depths exceeding 2,500 m, the project experienced some of the most severe rockburst conditions ever recorded. Both TBMs and drill-and-blast were used, but TBM performance was significantly affected by high stress and rock instability.

Engineering responses included:

- Heavy-duty TBM design modifications
- Intensive rock support systems (steel ribs, shotcrete, yielding elements)
- Real-time monitoring of stress and deformation

This case demonstrates the limits of TBM application in extreme stress environments and the need for adaptive engineering solutions.

Snowy 2.0 Hydro Project, Australia

The Snowy 2.0 project involves the construction of large-diameter tunnels through the Snowy Mountains to support pumped hydro energy storage.

TBMs are being used to excavate long tunnels through hard, abrasive rock, with significant challenges including:

- Cutter wear due to highly abrasive geology
- Variable rock strength and fault zones

- Water inflow in fractured sections

The project integrates predictive maintenance, digital monitoring, and advanced TBM design to manage performance and reduce downtime.

This case illustrates how modern TBM technology is being applied in Australian mountain conditions, combining traditional engineering with digital innovation.

Channel Tunnel (UK–France) – Chalk Marl Section

The Channel Tunnel, although not strictly a mountain tunnel, provides an important example of long-distance tunnelling through relatively consistent but challenging geological material.

The tunnel was excavated primarily using TBMs through chalk marl, a material that provided favourable conditions for mechanised tunnelling due to its relative homogeneity and stability.

Key lessons applicable to mountain tunnelling include:

- Importance of geological consistency for TBM efficiency
- Long-distance logistics and continuous operation
- Integration of ventilation, safety systems, and emergency access

Key Engineering Lessons

The case studies highlight several important principles that underpin successful hard rock mountain tunnelling. One of the most significant factors is the influence of in-situ stress, which increases substantially with depth. Elevated stress conditions can lead to rockburst, deformation, and instability, particularly in high-strength rock masses. Managing these effects requires careful planning of excavation sequences, as well as the implementation of appropriate and often flexible support systems designed to accommodate stress redistribution.

Geological variability remains a persistent challenge, even in environments that are predominantly composed of hard rock. The presence of fault zones, fractured rock masses, and water-bearing strata can significantly influence excavation performance and method selection. As a result, successful projects adopt adaptable approaches, frequently combining Tunnel Boring Machine (TBM) excavation with drill-and-blast methods to respond effectively to changing ground conditions.

Rock support design is fundamental to maintaining stability throughout the excavation process. Support systems must be tailored to the specific geotechnical conditions encountered and may include combinations of rock bolts, shotcrete, steel ribs, and yielding elements. In high-stress environments, support systems must not only provide strength but also allow controlled deformation to prevent sudden failure.

Logistics and access also present major challenges in mountain tunnelling. Long tunnel drives, remote locations, and difficult terrain can complicate the transport of materials, ventilation of working areas, and provision of emergency response capabilities. Effective planning and infrastructure development are essential to ensure continuous and safe operations.

In modern projects, digital technologies and monitoring systems are increasingly being integrated into tunnelling operations. These systems enable continuous tracking of ground behaviour, machine performance, and structural response, allowing engineers to make informed, data-driven decisions. This enhances the ability to adapt to changing conditions and improves overall project control.

Hard rock mountain tunnels represent some of the most technically demanding infrastructure projects in the world. Their successful delivery requires the integration of geotechnical engineering, structural design, excavation technology, and complex operational logistics under often extreme conditions.

The case studies demonstrate how both conventional and mechanised tunnelling methods are applied to address these challenges. As technological advancements continue—particularly in the areas of automation, digital monitoring, and TBM design—the capacity to construct longer, deeper, and more complex tunnels will continue to expand.

Looking forward, future developments will focus on improving safety, increasing operational efficiency, and enhancing the management of geological uncertainty. These advancements will ensure that hard rock mountain tunnelling remains a critical component of global infrastructure development, supporting transport, energy, and water systems in increasingly challenging environments.

Mixed-ground Megaprojects

Mixed-ground tunnelling represents one of the most complex challenges in underground engineering, where the Tunnel Boring Machine (TBM) encounters transitions between soil, rock, and water-bearing strata within the same drive. These conditions introduce instability at the excavation face, differential cutter wear, and highly variable machine performance. The following megaprojects demonstrate how advanced TBM design, operational control, and digital integration are applied to manage these challenges at scale.

Hong Kong–Zhuhai–Macao Bridge (HZMB), China

The Hong Kong–Zhuhai–Macao Bridge includes an immersed tunnel and connecting bored tunnel sections constructed through complex marine deposits and mixed ground conditions. The bored tunnel sections were excavated using large-diameter slurry TBMs capable of maintaining face stability in water-bearing soils while also cutting through harder inclusions.

The transition between soft marine clay and weathered rock required precise control of slurry pressure and density.

Key engineering challenges included:

- High groundwater pressure and risk of inflow
- Abrasive inclusions causing cutter wear
- Maintaining alignment beneath marine conditions

Figure 110: Eastern tunnel portal of Hong Kong-Zhuhai-Macau Bridge. N509FZ, CC BY-SA 4.0, via Wikimedia Commons.

Advanced monitoring and control systems were used to maintain face stability and manage transitions. This project demonstrates the effectiveness of slurry TBMs in mixed marine ground conditions.

Thames Tideway Tunnel, London, UK

The Thames Tideway Tunnel is a major wastewater infrastructure project involving over 25 km of tunnels beneath London.

Figure 111: Thames Tideway Tunnel. Matt Brown, CC BY 2.0, via Wikimedia Commons.

The tunnels pass through London Clay, Lambeth Group deposits, and water-bearing sands, creating frequent transitions between stable and unstable ground. EPB TBMs were used to maintain pressure balance and minimise settlement.

A key challenge was managing transitions into permeable strata, where loss of face pressure could lead to instability or inflow. Real-time monitoring and adaptive control of excavation parameters were critical.

This project highlights the importance of:

- Pressure control in mixed ground
- Integration of geotechnical monitoring with TBM operation
- Urban settlement management

Istanbul Marmaray Project, Turkey

The Marmaray Tunnel Project involved tunnelling beneath the Bosphorus Strait, connecting the European and Asian sides of Istanbul.

The project encountered extremely complex mixed ground conditions, including soft sediments, weathered rock, and high groundwater pressures. Slurry TBMs were used to maintain face stability under pressurised conditions.

One of the major challenges was the transition between bored tunnel sections and immersed tube segments, requiring precise alignment and control.

This project demonstrates:

- High-pressure tunnelling in mixed ground
- Integration of different tunnelling methods
- Advanced risk management in subsea environments

Seattle SR 99 Tunnel (Alaskan Way Viaduct Replacement), USA

The SR 99 Tunnel in Seattle involved a large-diameter bored tunnel constructed beneath a dense urban environment.

The TBM, known as "Bertha," encountered highly variable ground conditions, including glacial soils, boulders, and mixed face conditions. The project experienced significant delays due to cutterhead damage caused by unexpected obstructions.

Key lessons from this project include:

- The importance of ground investigation and risk anticipation
- The impact of mixed ground on cutterhead performance
- The need for contingency planning and maintenance strategies

This case highlights the risks associated with large-diameter TBMs operating in unpredictable ground.

Doha Metro, Qatar

The Doha Metro project involved extensive tunnelling through mixed ground conditions, including limestone, weak rock, and soft soils.

EPB TBMs were widely used due to their ability to handle transitions between soil and weak rock. The project required careful management of:

- Ground conditioning
- Cutter wear
- Face stability

Digital monitoring systems and predictive maintenance strategies were implemented to optimise performance.

This project demonstrates the effectiveness of EPB TBMs in mixed soil–rock environments.

Key Engineering Lessons

Mixed-ground megaprojects highlight several critical engineering principles that are fundamental to successful tunnelling in complex subsurface environments. One of the most significant challenges is maintaining face stability during transitions between different ground types. These transitions—such as from soft soil to hard rock or from cohesive material to water-bearing strata—can result in pressure imbalances, instability at the excavation face, and an increased risk of settlement or groundwater inflow. Effective control of face pressure and continuous monitoring are therefore essential to maintaining stability throughout the drive.

The selection of the appropriate Tunnel Boring Machine (TBM) type is a key determinant of project success. Slurry TBMs are generally preferred in highly permeable and water-bearing ground due to their ability to maintain stable pressure conditions, while Earth Pressure Balance (EPB) TBMs perform well in cohesive soils and mixed ground environments. In increasingly variable geological conditions, hybrid TBMs are being adopted to provide greater flexibility and adaptability across different ground types.

Cutterhead design and wear management are also critical considerations in mixed-ground tunnelling. Variations in ground strength and composition can lead to uneven loading on the cutterhead, resulting in accelerated wear and potential damage. To address this, predictive maintenance strategies and real-time monitoring systems are used to track performance, detect anomalies, and schedule timely interventions before failures occur.

Digital technologies are playing an increasingly important role in managing the complexity of mixed-ground tunnelling. Real-time monitoring systems, data analytics platforms, and digital twin models enable engineers to integrate machine data with geotechnical information, improving situational awareness and supporting more informed decision-making. These tools allow for a shift from reactive management to proactive control of tunnelling operations.

Mixed-ground megaprojects represent some of the most complex and high-risk environments encountered in underground construction. Successfully delivering these projects requires advanced TBM technology, precise operational control, and the integration of digital systems capable of managing variability and uncertainty in real time.

The case studies demonstrate that engineering innovation, when combined with thorough planning, continuous monitoring, and adaptive management strategies, enables the successful execution of large-scale infrastructure projects in challenging conditions. As tunnelling continues to evolve in scale and complexity, the ability to effectively manage mixed-ground conditions will remain a critical factor in achieving safe, efficient, and reliable project outcomes.

Lessons Learned from Failures and Successes

Tunnel Boring Machines represent a significant advancement in underground construction technology, providing innovative approaches for tunnelling projects across varied geological environments. The operational intricacies of TBMs yield crucial engineering insights, particularly when examining the interplay between successes and failures. The underground construction process is laden with unknowns, where unexpected ground conditions can swiftly precipitate severe technical and financial repercussions [505, 854]. Historical data demonstrates that the evolution of TBM practices has largely been informed by past failures, emphasizing the need for continual learning and adaptation in response to the inherent uncertainties of tunnelling.

Central to successful TBM operations is a thorough understanding of ground conditions. Numerous historical failures in TBM operations are attributable to insufficient geotechnical investigation or an over-reliance on overly simplified ground models [505, 854]. Issues like mixed face conditions, fault zones, or unanticipated groundwater pressures can incite operational challenges such as face instability, machine entrapment, and excessive ground settlement. In stark contrast, projects that prioritize extensive site investigations, continuous geological updates, and real-time feedback mechanisms tend to perform significantly better. As noted, successful TBM initiatives regard ground conditions as dynamic variables, necessitating ongoing assessment rather than static assumptions [505, 854].

The selection and configuration of TBMs are pivotal in ensuring operational success, particularly in relation to the prevailing ground conditions. Failures often occur when the chosen TBM type is ill-suited for the encountered geological contexts. For instance, deploying an Earth Pressure Balance (EPB) TBM in exceptionally permeable ground without adequate conditioning can lead to critical inflow and loss of pressure. Conversely, successful case studies indicate that meticulous matching of TBM types – whether EPB, slurry, hard rock, or hybrid – to ground conditions plays a crucial role in overall project outcomes [505, 830]. Additionally, flexibility in machine configuration emerges as a vital asset, particularly in mixed or challenging ground scenarios, bolstering the TBM's capability to navigate diverse conditions effectively [854] .

Maintaining face stability and effective pressure control is foundational to the success of TBM operations. Loss of pressure at the excavation front can result in catastrophic failures such as ground collapse and flooding. Many documented TBM failures can be traced back to inadequate monitoring systems and a failure to respond promptly to changes in chamber pressure [830]. Conversely, successful operations employ precise control systems, calibrated sensors, and proficient operators capable of interpreting real-time data to maintain equilibrium at the face throughout tunnelling [505]. Such proactive approaches not only enhance safety but also improve operational efficiency.

The design and maintenance of TBM cutterheads significantly influence operational outcomes. Common challenges include rapid wear or damage from unexpected obstructions and mixed ground conditions. Research indicates that rapid wear can result in unexpected stoppages, underscoring the need for robust cutterhead designs and predictive maintenance strategies [287, 505]. Moreover, successful projects adopt comprehensive contingency planning, which can include hyperbaric access for maintenance interventions, ensuring quick recovery from potential operational setbacks.

Effective logistics and operational planning are crucial components influencing TBM performance. High-performing TBM initiatives recognize that successful excavation does not rely solely on TBM speed, but on the cohesion of the entire tunnelling system, including segment supply, spoil removal, and workforce coordination [287]. Failures often arise from logistical bottlenecks that limit utilization and inflate costs, thus reinforcing the importance of treating the TBM as an integral part of an optimized production system [830].

The necessity for real-time monitoring and data integration in TBM operations is becoming increasingly paramount as technology advances. Previous limitations in instrumentation and delayed feedback mechanisms often led to the escalation of minor issues into major failures. Modern projects leverage extensive sensor networks and data analytics for continuous monitoring, allowing for the early detection of anomalies and supporting timely corrective actions [287, 505]. This technological evolution facilitates a shift toward proactive rather than reactive operational strategies.

Despite the advantages of automation and artificial intelligence in TBM operations, human factors remain indispensable. Cases of failures frequently expose deficiencies in training, communication, and decision-making, situating skilled operators as central to operational success. Successful TBM projects emphasize the significance of well-trained personnel, clear communication protocols, and collaborative dynamics among engineers, geologists, and construction teams [287, 830]. Technology, while an enabling force, cannot substitute for sound judgment and effective teamwork.

Effective risk management is vital in minimizing the uncertainties inherent in tunnelling operations. Historical approaches that emphasize contractual risk transfer have proven inadequate in the complex environments typical of TBM projects. In contrast, successful projects adopt proactive, collaborative risk management strategies that prioritize early identification of risks and collective accountability among stakeholders [505, 830]. Such strategies allow for the active management of risks through informed engineering controls and contingency planning.

Reflecting on the broader context of TBM operations, adaptability emerges as a critical lesson learned from both successes and failures. Tunnelling projects seldom unfold precisely according to initial plans, and rigid adherence to predetermined assumptions can result in unfavourable outcomes. Instead, adaptable project teams that adjust their approaches in response to actual encountered conditions often achieve superior results [505, 854]. This flexibility not only applies to excavation methods but extends to support systems and operational strategies as well.

The operational history of TBM projects reveals a wealth of engineering lessons that shape the effective practice of modern tunnelling. The integration of comprehensive ground understanding, appropriate machine selection, advanced monitoring, and skilled personnel establishes a framework for successful TBM operations. Ultimately, success in tunnelling is contingent not solely on technological advancements but on an integrated approach that combines engineering knowledge with adaptive management strategies. Addressing the uncertainties inherent in the tunnelling process is key to transforming potential risks into predictable engineering outcomes.

References

1. Pompeu-Santos, S., *Innovative TBM Traffic Tunnels.* Iabse Congress Report, 2021.
2. Wood, A.M.M., *Tunnels for Roads and Motorways.* Quarterly Journal of Engineering Geology, 1972. **5**(1-2): p. 111-126.
3. Samadi, H., J. Hassanpour, and J. Rostami, *Assessment of Shear Strength of Fine-Grained and Coarse-Grained Soil Using Actual EPB-TBM Operating Data.* International Journal of Geo-Engineering, 2024. **15**(1).
4. Witasse, R., S.M. Kashfi, and K.P. Iman, *Finite Element Modeling of TBM Tunneling in Mixed-Face Ground Conditions.* 2023.
5. Chung, H., et al., *TBM Risk Management System Considering Predicted Ground Condition Ahead of Tunnel Face: Methodology Development and Application.* Journal of Korean Tunnelling and Underground Space Association, 2016. **18**(1): p. 1-12.
6. Karahan, S. and C. Gökçeoğlu, *Assessment for Shallow and Large Tunnel Construction in Weak Ground Conditions: Application of Tunnel Boring Machines.* Deep Underground Science and Engineering, 2024. **4**(1): p. 132-148.
7. Yaman, H.E. and C.O. Aksoy, *Evaluation of Symmetrical Face Pressure of EPB.* Symmetry, 2024. **16**(4): p. 484.
8. Kong, S.-M., et al., *Stability Evaluation of TBM Pilot Tunnels to Rear Blasting Using the Protection Shield.* Applied Sciences, 2021. **11**(4): p. 1759.
9. Kong, S.-M., et al., *A Study of Vibration Reduction Effects of TBM and NATM Parallel Construction Methods According to Tunnel Depth and Size.* Applied Sciences, 2023. **13**(6): p. 3492.
10. Wang, Z., et al., *On-Site Measurement and Environmental Impact of Vibration Caused by Construction of Double-Shield TBM Tunnel in Urban Subway.* Scientific Reports, 2023. **13**(1).
11. Nie, X., et al., *Correlation Analysis and Dynamic Evolution Research on Safety Risks of TBM Construction in Hydraulic Tunnels.* Buildings, 2025. **15**(18): p. 3359.
12. Brierley, G., *Tunneling: A Historiacal Perspective*. 2014, TBM: Tunnel Business Magazine.
13. Ramoni, M. and G. Anagnostou, *The Effect of Consolidation on TBM Shield Loading in Water-Bearing Squeezing Ground.* Rock Mechanics and Rock Engineering, 2010. **44**(1): p. 63-83.
14. Ji, F., et al., *Mechanical Response of Surrounding Rock of Tunnels Constructed With the TBM and Drill-Blasting Method.* Natural Hazards, 2012. **66**(2): p. 545-556.
15. Zhao, K., M. Janutolo, and M. Barla, *A Completely 3D Model for the Simulation of Mechanized Tunnel Excavation.* Rock Mechanics and Rock Engineering, 2012. **45**(4): p. 475-497.
16. Yang, J., et al., *Numerical Simulation of Rock Mass Damage Evolution During Deep-Buried Tunnel Excavation by Drill and Blast.* Rock Mechanics and Rock Engineering, 2014. **48**(5): p. 2045-2059.
17. Rutqvist, J., et al., *Modeling of Damage, Permeability Changes and Pressure Responses During Excavation of the TSX Tunnel in Granitic Rock at URL, Canada.* Environmental Geology, 2008. **57**(6): p. 1263-1274.

18. Rana, A., et al., *Minimum Invasive Drill-and-blast Designs for Optimizing Pull Efficiency and Minimizing Overbreaks/Underbreaks in Varying Rock Masses.* Geomechanics and Tunnelling, 2025. **18**(3): p. 218-224.
19. Zhao, H., et al., *Key Technology of TBM Excavation in Soft and Broken Surrounding Rock.* Applied Sciences, 2023. **13**(13): p. 7550.
20. Zhang, Q., et al., *Prediction Method of TBM Tunneling Parameters Based on Bi-GRU-ATT Model.* Advances in Civil Engineering, 2022. **2022**(1).
21. Zhou, W., et al., *Automatic Approach for Fast Processing and Data Analysis of Seismic Ahead-Prospecting Method: A Case Study in Yunnan, China.* Mathematical Problems in Engineering, 2020. **2020**: p. 1-11.
22. Yang, J., et al., *Comparative Study of the Excavation Damage and Rockburst of the Deeply Buried Jinping II Diversion Tunnels Using a TBM and the Drilling-Blasting Method.* Advances in Civil Engineering, 2020. **2020**(1).
23. Bakke, B., et al., *Effects of Blasting Fumes on Exposure and Short-Term Lung Function Changes in Tunnel Construction Workers.* Scandinavian Journal of Work Environment & Health, 2001. **27**(4): p. 250-257.
24. Menéndez, J., et al., *Concentration, Propagation and Dilution of Toxic Gases in Underground Excavations Under Different Ventilation Modes.* International Journal of Environmental Research and Public Health, 2022. **19**(12): p. 7092.
25. Kwon, K., et al., *Prioritization of Hazardous Zones Using an Advanced Risk Management Model Combining the Analytic Hierarchy Process and Fuzzy Set Theory.* Sustainability, 2023. **15**(15): p. 12018.
26. Li, Z., et al., *Classification and Prediction of Rock Mass Boreability Based on Daily Advancement During TBM Tunneling.* Buildings, 2024. **14**(7): p. 1893.
27. Li, X., et al., *Rock Fragmentation Indexes Reflecting Rock Mass Quality Based on Real-Time Data of TBM Tunnelling.* Scientific Reports, 2023. **13**(1).
28. Fan, C., M. Qin, and W. Guixuan, *Construction Mechanical Characteristics of TBM Pilot and Enlargement Method for Ventilation Tunnel of Wuhai Pumped Storage Power Station.* Applied Sciences, 2024. **14**(5): p. 1829.
29. Zhang, L., et al., *Modeling and Economic Model Predictive Control of Constrained Cutterhead System With Disturbance in Tunnel Boring Machines.* Transactions of the Institute of Measurement and Control, 2024. **46**(16): p. 3182-3191.
30. Park, H., et al., *Monitoring and Analysis of Ground Settlement Induced by Tunnelling With Slurry Pressure-Balanced Tunnel Boring Machine.* Advances in Civil Engineering, 2018. **2018**(1).
31. Yang, C., et al., *Evaluation of Excavation Ergonomics of Drill and Blast Method Based on Game Theory G2-Ew-Topsis Model.* Applied Sciences, 2023. **13**(12): p. 7205.
32. Hao, F.S., G.Z. Tian, and T.L. Wang, *The Process Hypothesis and Feasibility Study on Drilling-Blasting Machine Method in Tunnel Excavation.* Applied Mechanics and Materials, 2012. **256-259**: p. 1316-1319.
33. Butscher, C., P. Huggenberger, and E. Zechner, *Impact of Tunneling on Regional Groundwater Flow and Implications for Swelling of Clay–sulfate Rocks.* Engineering Geology, 2011. **117**(3-4): p. 198-206.
34. Hishammuddin, M.A.H.b., et al., *Adaptive, Resilient Urban Underground Space (UUS)-Subsidence and Economic Impact Spatial Planning Model (USEM): The*

Extent Development Impact via Cause-Effect and Spatiotemporal Analyses Circa 1960-2020-2050 in Shanghai Megacity. 2023.

35. Chee-Min, K., *Numerical Investigation on the Effects of TBM Face Support Pressures–a Parametric Study.* Jurnal Teknologi (Sciences & Engineering), 2025. **87**(2): p. 351-360.
36. Choi, S.-K., et al., *Geotechnical Investigation on Causes and Mitigation of Ground Subsidence During Underground Structure Construction.* Journal of Korean Tunnelling and Underground Space Association, 2016. **18**(2): p. 143-154.
37. Zhang, H. and Y. Pei, *Simulation-Based Prediction of Equivalent Continuous Noises During Construction Processes.* International Journal of Environmental Research and Public Health, 2016. **13**(8): p. 818.
38. Zou, C., et al., *Evaluation of Building Construction-Induced Noise and Vibration Impact on Residents.* Sustainability, 2020. **12**(4): p. 1579.
39. Abidin, H.Z., et al., *On Correlation Between Urban Development, Land Subsidence and Flooding Phenomena in Jakarta.* Proceedings of the International Association of Hydrological Sciences, 2015. **370**: p. 15-20.
40. Rattia, V.H.F., et al., *Estimating Settlements Due to Tunnel Boring Machine Excavation.* Proceedings of the Institution of Civil Engineers - Geotechnical Engineering, 2022. **176**(6): p. 675-686.
41. Yuan, C., et al., *Analysis of Factors Influencing Surface Settlement During Shield Construction of a Double-Line Tunnel in a Mudstone Area.* Scientific Reports, 2022. **12**(1).
42. Stansfeld, S. and C. Clark, *Health Effects of Noise Exposure in Children.* Current Environmental Health Reports, 2015. **2**(2): p. 171-178.
43. Bergman, F., *Sustainability Performance of Multi-Utility Tunnels : Sustainability Assessments for Furthering Knowledge and Understanding.* 2023.
44. Bergman, F., et al., *A Critical Review of the Sustainability of Multi-Utility Tunnels for Colocation of Subsurface Infrastructure.* Frontiers in Sustainable Cities, 2022. **4**.
45. Shahrour, I., et al., *Use of Smart Technology to Improve Management of Utility Tunnels.* Applied Sciences, 2020. **10**(2): p. 711.
46. Fang, W., et al., *Quantitative Risk Assessment of a Natural Gas Pipeline in an Underground Utility Tunnel.* Process Safety Progress, 2019. **38**(4).
47. Ye, K., et al., *A Multi-Scale Analysis of the Fire Problems in an Urban Utility Tunnel.* Energies, 2019. **12**(10): p. 1976.
48. Jamaluddin, K., et al., *Optimal Sizing of a Trigeneration Plant Integrated With Total Site System Considering Multi-Period and Energy Losses.* E3s Web of Conferences, 2021. **287**: p. 03014.
49. Al-Zareer, M., İ. Dinçer, and M.A. Rosen, *Assessment and Analysis of Hydrogen and Electricity Production From a Generation IV Lead-Cooled Nuclear Reactor Integrated With a Copper-Chlorine Thermochemical Cycle.* International Journal of Energy Research, 2017. **42**(1): p. 91-103.
50. Gjengedal, S., *The Potential for Geothermal Energy Exploitation in Norwegian Tunnels.* 2023.
51. Deng, Y., et al., *Simulation Study on ASCMP Protocol in Utility Tunnel WSN.* Ieee Access, 2019. **7**: p. 168141-168150.

52. Elsaid, A.M., et al., *A Recent Review on Ventilation and Cooling of Underground High-Voltage Cable Tunnels*. Journal of Thermal Analysis and Calorimetry, 2024. **149**(17): p. 8927-8978.
53. Lund, J., *Direct Utilization of Geothermal Energy*. Energies, 2010. **3**(8): p. 1443-1471.
54. Hojjati, A., et al., *Embedding Sustainability Criteria Into Pre-Appraisal of Underground Utility for Future Cities*. Proceedings of the Institution of Civil Engineers - Urban Design and Planning, 2017. **170**(6): p. 258-271.
55. Tai, C., et al., *A Field Measurement of Temperature and Humidity in a Utility Tunnel and a Brief Analysis of the Exhaust Heat Recovery System*. Indoor and Built Environment, 2020. **30**(4): p. 487-501.
56. Yang, L., et al., *Generating Topologically Consistent BIM Models of Utility Tunnels From Point Clouds*. Sensors, 2023. **23**(14): p. 6503.
57. Beghoul, M. and R. Demagh, *Slurry Shield Tunneling in Soft Ground. Comparison Between Field Data and 3D Numerical Simulation*. Studia Geotechnica Et Mechanica, 2019. **41**(3): p. 115-128.
58. Konstantis, S., G. Prountzopoulos, and P. Spyridis, *Ensuring Face Support and Control in Soft Ground Urban Tunnels: FEM Validation*. Geomechanics and Tunnelling, 2022. **15**(3): p. 317-326.
59. Trabucchi, I., et al., *A Hybrid Solution Proposal for Precast Tunnel Segments*. Structural Concrete, 2021. **22**(3): p. 1534-1548.
60. Stypulkowski, J.B., F.G. Bernardeau, and J. Jakubowski, *Descriptive Statistical Analysis of TBM Performance at Abu Hamour Tunnel Phase I*. Arabian Journal of Geosciences, 2018. **11**(9).
61. Mohtadinia, M., et al., *Statistical and Numerical Study of Chipping and Cracking in Segmental Lining*. Periodica Polytechnica Civil Engineering, 2020.
62. Yubero, M.T., E. Bonet, and A. Gens, *Analysis of Local Settlements Due to Operational EPB Tunneling Driving Parameters: A Case Study of the Barcelona Metro Line 9 Tunnel*. Sustainability, 2025. **17**(9): p. 3966.
63. Mirković, U., et al., *Optimal Shield TBM Face Pressure for Surface Settlement Control: Belgrade Metro Case Study*. 2025: p. 255-274.
64. Zhao, Z., et al., *Construction Technology and Application of TBM in Shengli Tunnel of Tianshan Mountain Under Poor Geological Conditions*. Stavební Obzor - Civil Engineering Journal, 2024. **33**(2): p. 168-180.
65. Thai, D.N. and D.D. Toan, *Study of the Influence of Face Pressure on Surface Settlements by Shield Tunneling*. Journal of Mining and Earth Sciences, 2020. **61**(1): p. 31-40.
66. Xia, Y., et al., *The Development of the TBM Disc Cutter Test Device for Cutting Rock*. Advanced Materials Research, 2014. **889-890**: p. 292-296.
67. Pompeu-Santos, S., *The Long Traffic Tunnels of Tomorrow*. 2018.
68. Yang, H., H. Wang, and X. Zhou, *Analysis on the Rock–Cutter Interaction Mechanism During the TBM Tunneling Process*. Rock Mechanics and Rock Engineering, 2015. **49**(3): p. 1073-1090.
69. Lach, J., *Instrumentation of a Reconditioned Robbins Tunnel Boring Machine - 106th Street Tunnel*. 2001.
70. Xiong, Y., et al., *Dynamic Reliability Evaluation of <scp>TBM</Scp> Components in Tunnel Construction*. Geological Journal, 2024. **59**(9): p. 2594-2608.

71. Wang, X.Y., et al., *Numerical Modeling of the Effects of Wears on Cutting Loads.* Advanced Materials Research, 2011. **308-310**: p. 2340-2344.
72. Tian, J., et al., *Wear Performance and Mechanisms of H13 Steels Sliding Against Different Rock Types.* Surface Topography Metrology and Properties, 2020. **8**(2): p. 025003.
73. Xia, Y., K. Zhang, and J. Liu, *Design Optimization of TBM Disc Cutters for Different Geological Conditions.* World Journal of Engineering and Technology, 2015. **03**(04): p. 218-231.
74. Lu, G. and J. Zhou, *Experimental Investigation on the Effect of Microwave Heating on Rock Cracking and Their Mechanical Properties.* 2021.
75. Singh, P., *Design and Analysis of a Micro Tunnel Boring Machines (TBM).* Universal Journal of Mechanical Engineering, 2014. **2**(3): p. 87-93.
76. Huang, D., et al., *A New Design Method of Shield Tunnel Based on the Concept of Minimum Bending Moment.* Applied Sciences, 2022. **12**(3): p. 1082.
77. Chen, J., et al., *Optimal Sections of Tunnels' Cross Sections With Different Overburdens.* The Open Civil Engineering Journal, 2017. **11**(1): p. 714-725.
78. Zhou, G., et al., *Study on Rock-Breaking Efficiency Evaluation of TBM Disc Cutters Based on Rostami Prediction Equations.* Frontiers in Earth Science, 2023. **11**.
79. Seo, T. and D. Park, *Kinematic Limit Analysis of Roof Stability for Elliptical Tunnels in Rock Masses.* International Journal for Numerical and Analytical Methods in Geomechanics, 2025. **49**(16): p. 3880-3896.
80. Hao, X., et al., *Instability Process of Crack Propagation and Tunnel Failure Affected by Cross-Sectional Geometry of an Underground Tunnel.* Advances in Civil Engineering, 2019. **2019**(1).
81. Gikas, V., *Three-Dimensional Laser Scanning for Geometry Documentation and Construction Management of Highway Tunnels During Excavation.* Sensors, 2012. **12**(8): p. 11249-11270.
82. Nishide, K.-i. and K. Mishima, *Automated Surveying System for Shield Tunneling.* 1988.
83. Vigl, A., et al., *TBM Research Project – Tolerances Segmental Lining / TVM-Forschungsprojekt – Toleranzen Tübbingausbau.* Geomechanics and Tunnelling, 2016. **9**(3): p. 243-259.
84. Yazdani–Chamzini, A., *Proposing a New Methodology Based on Fuzzy Logic for Tunnelling Risk Assessment.* Journal of Civil Engineering and Management, 2014. **20**(1): p. 82-94.
85. Zhao, Z., et al., *The Challenge in the Application of New Generation Geometrical Product Specifications.* 2010.
86. Baraibar, J.M., J. de-Paz, and J. Rico, *Challenges for the Implementation of BIM Methodology in the Execution of Underground Works.* Buildings, 2022. **12**(3): p. 309.
87. Blasco, J.M.D., et al., *Measuring Urban Subsidence in the Rome Metropolitan Area (Italy) With Sentinel-1 SNAP-StaMPS Persistent Scatterer Interferometry.* Remote Sensing, 2019. **11**(2): p. 129.
88. Dematteis, A., et al., *Contribution of Temperature Measurements to the Hydrogeological Model in the Snowy 2.0 Deep Tunnels.* Quarterly Journal of Engineering Geology and Hydrogeology, 2023. **56**(2).

89. García-Gil, A., et al., *Reducing 3D Hydraulic Interferences in Water Tunnels Through the Strategic Installation of Artificial Bulkheads in Geological Hydraulic Barriers.* 2025.
90. Gruslin, S. and T. Hennebaut, *Collapse of the Schieburg Tunnel in Luxembourg: Geological Causes and Emergency Securing Works.* 2024: p. 32-41.
91. Bogaard, T. and R. Greco, *Landslide Hydrology: From Hydrology to Pore Pressure.* Wiley Interdisciplinary Reviews Water, 2015. **3**(3): p. 439-459.
92. Yan-kai, W.U. and X.I. Xiao-hua, *Information Monitoring on Surrounding Rock of Tunnel and Its Application.* 2010.
93. Golser, J., M. Keuschnig, and F.P. Weichenberger, *NATM – Review and Outlook.* Geomechanics and Tunnelling, 2020. **13**(5): p. 466-474.
94. Afonin, D.A., V. Kavkazskiy, and A.A. Nikitchin, *Operational Geodetic Check of Bottom-Hole Zone Deformations at Innovative Tunneling Technologies.* Proceedings of Petersburg Transport University, 2022. **19**(3): p. 432-443.
95. Kuprii, V., et al., *Numerical Analysis of Changing the Force Factors in Temporary Lining at the Tunnel Construction by the NATM.* E3s Web of Conferences, 2019. **109**: p. 00044.
96. Wijaya, W., P.P. Rahardjo, and A. Lim, *Investigation of Twin Tunnel Deformation With Umbrella Grouting Protection &Amp; NATM Tunneling Using 3D Finite Element: Case Study Cisumdawu Tunnel.* Ukarst, 2021. **5**(2): p. 252.
97. Terzi, V.G. and G.D. Manolis, *Basic numerical modelling issues in dynamic soil-tunnel interaction.* Soil Dynamics and Earthquake Engineering, 2023. **172**: p. 108060.
98. Kovačević, M.S., et al., *Assessment of Long-Term Deformation of a Tunnel in Soft Rock by Utilizing Particle Swarm Optimized Neural Network.* Tunnelling and Underground Space Technology, 2021. **110**: p. 103838.
99. Ng, C.W., K.M. Lee, and D. Tang, *Three-Dimensional Numerical Investigations of New Austrian Tunnelling Method (NATM) Twin Tunnel Interactions.* Canadian Geotechnical Journal, 2004. **41**(3): p. 523-539.
100. Beveridge, J. and W.J. Rankin, *Role of Engineering Geology in NATM Construction.* Geological Society London Engineering Geology Special Publications, 1995. **10**(1): p. 255-268.
101. Yang, J., et al., *Preparation and Experimental Investigations of Low-Shrinkage Commercial Concrete for Tunnel Annular Secondary Lining Engineering.* Materials, 2022. **15**(19): p. 6848.
102. Zhang, X., et al., *Influence of Factors on Collapse Risk of Loess Tunnel: A Multi-Index Assessment Model.* Journal of Engineering Design and Technology, 2018. **16**(5): p. 734-749.
103. Li, P., Y. Zhao, and X. Zhou, *Displacement Characteristics of High-Speed Railway Tunnel Construction in Loess Ground by Using Multi-Step Excavation Method.* Tunnelling and Underground Space Technology, 2016. **51**: p. 41-55.
104. Biryuchev, I.V., А.Б. Макаров, and A.A. Usov, *Geomechanical Model of Underground Mine. Part II. Application.* Gornyi Zhurnal, 2020: p. 35-44.
105. Frenelus, W., H. Peng, and J. Zhang, *An Insight From Rock Bolts and Potential Factors Influencing Their Durability and the Long-Term Stability of Deep Rock Tunnels.* Sustainability, 2022. **14**(17): p. 10943.

106. He, S., et al., *Mechanical Behavior of Anchor Bolts for Shallow Super-Large-Span Tunnels in Weak Rock Mass.* Materials, 2023. **16**(17): p. 5862.
107. Uljarević, M., A. Rašeta, and D. Bojović, *Analysis of Primary NATM Support Under Limited Anchor Capacity: Implications for Steel Ribs and Lattice Girders.* 2025: p. 21-30.
108. Agustawijaya, D.S., R.M. Taruna, and A.R. Agustawijaya, *Earthquake Forces on the Stability of the Meninting Diversion Tunnel Design Based on Current Seismic Conditions of Lombok Island.* Indonesian Journal on Geoscience, 2022. **9**(1): p. 105-117.
109. Liu, N., et al., *BIM-based Digital Platform and Risk Management System for Mountain Tunnel Construction.* 2022.
110. Ma, C., et al., *Research on Supporting Method for High Stressed Soft Rock Roadway in Gentle Dipping Strata of Red Shale.* Minerals, 2021. **11**(4): p. 423.
111. Zhang, Z., et al., *An Investigation Into Bolt Anchoring Performance During Tunnel Construction in Bedded Rock Mass.* Applied Sciences, 2020. **10**(7): p. 2329.
112. Yokota, Y., et al., *Quality Improvement of Rockbolting.* 2013.
113. Cui, L., et al., *Enhanced Tunnel Construction Safety: Drill-and-Blast Construction Tunnel Deformation Detection Using LiDAR and Cloth Simulation Filter.* Measurement Science and Technology, 2025. **36**(9): p. 096012.
114. Adhikari, P.K., et al., *Construction of Mahadevtar Tunnel and Challenges Faced During Its Stabilization.* KTFT Journal, 2024. **4**: p. 217-225.
115. Hayati, A.N., et al., *Unsupported Advance Length in Tunnels Constructed Using New Austrian Tunnelling Method and Ground Surface Settlement.* International Journal for Numerical and Analytical Methods in Geomechanics, 2012. **37**(14): p. 2170-2185.
116. Kontogianni, V. and S.C. Stiros, *Ground Loss and Static Soil–Structure Interaction during Urban Tunnel Excavation: Evidence from the Excavation of the Athens Metro.* Infrastructures, 2020. **5**(8): p. 64.
117. Ha, S.-g., et al., *Expanded Longitudinal Deformation Profile in Tunnel Excavations Considering Rock Mass Conditions via 3D Numerical Analyses.* Applied Sciences, 2021. **11**(12): p. 5405.
118. Liu, Y. and J. Hu, *Analysis of Displacement Monitoring Measurement in Tunnel Construction.* Advanced Materials Research, 2013. **658**: p. 635-638.
119. Gutierrez, M., *Rock-Support Interaction for Transportation Tunnels in Squeezing Ground.* 2023.
120. Alonso, E., et al., *Ground Response Curves for Rock Masses Exhibiting Strain-softening Behaviour.* International Journal for Numerical and Analytical Methods in Geomechanics, 2003. **27**(13): p. 1153-1185.
121. Osgoui, R. and E. Ünal, *An empirical method for design of grouted bolts in rock tunnels based on the Geological Strength Index (GSI).* Engineering Geology, 2009. **107**: p. 154-166.
122. Liu, Z., *Study on the Mechanical Behavior of Double Primary Support of Soft Rock Tunnel under High Ground Stresses and Large Deformation.* Advances in Civil Engineering, 2020. **2020**: p. 1-9.
123. Donna, A.D. and M. Barla, *The Role of Ground Conditions on Energy Tunnels' Heat Exchange.* Environmental Geotechnics, 2016. **3**(4): p. 214-224.

124. You, L.K. and I.A. Rahim, *Tunnel Evaluation in Crocker Formation by Geological Strength Index (Gsi) System: A Case Study.* Geological Behavior, 2018. **2**(2): p. 21-24.
125. Jimoh, O.A., et al., *Assessment of Supporting System of Hulu Terengganu Hydroelectric Surge Chamber Cavern in Malaysia.* Warta Geologi, 2020. **46**(1): p. 17-20.
126. Huang, X., et al., *Numerical Simulation of Stress Arching Effect in Horizontally Layered Jointed Rock Mass.* Symmetry, 2021. **13**(7): p. 1138.
127. Mohammadi, M., M.F. Hossaini, and H. Bagloo, *Rock Bolt Supporting Factor: Rock Bolting Capability of Rock Mass.* Bulletin of Engineering Geology and the Environment, 2015. **76**(1): p. 231-239.
128. Wu, F., et al., *Application and Principle of Bolt-Mesh-Cable Control Technology in Extremely Soft Coal Seam Roadway.* Geofluids, 2023. **2023**: p. 1-15.
129. He, B.G., Z. Zhang, and Y. Chen, *Unsymmetrical Load Effect of Geologically Inclined Bedding Strata on Tunnels of Passenger Dedicated Lines.* Journal of Modern Transportation, 2012. **20**(1): p. 24-30.
130. Xiao, B., J. Carter, and H. Alehossein, *Elastoplastic Analysis of Jointed Rocks Using a Coupled Finite Element and Boundary Element Method.* International Journal for Numerical and Analytical Methods in Geomechanics, 1994. **18**(7): p. 445-466.
131. Cao, L., et al., *Analysis of Arch Forming Factors of Shallow Buried Hard Rock Tunnel Under Overlying Load.* Buildings, 2023. **13**(9): p. 2210.
132. Lu, W. and H. Sun, *Study on Support Characteristic Curve of Primary Support Structures in Underground Excavation Considering Bond-Slip Behavior.* Advances in Structural Engineering, 2020. **24**(3): p. 497-508.
133. Harwinda, Z.B., W. Wilopo, and I.G.B. Indrawan, *Empirical Analyses of the Excavating Method and Support System Based on Engineering Geology of the Rukoh Dam Suppletion Tunnel, Pidie Regency, Aceh Province, Indonesia.* E3s Web of Conferences, 2023. **468**: p. 02002.
134. Ambangkoro, A.P.M., I.G.B. Indrawan, and H.E. Wibowo, *Excavation Method and Support Systems Using Empirical Design at Pasir Kopo Dam Diversion Tunnel, Lebak, Banten Province, Indonesia.* Iop Conference Series Earth and Environmental Science, 2024. **1416**(1): p. 012011.
135. Xu, J., et al., *Deformation and Damage Dynamic Characteristics of Coal–rock Materials in Deep Coal Mines.* International Journal of Damage Mechanics, 2017. **28**(1): p. 58-78.
136. Zhang, Q., et al., *Excess Pore Water Pressure and Ground Consolidation Settlement Caused by Grouting of Shield Tunnelling.* Mathematical Problems in Engineering, 2022. **2022**: p. 1-14.
137. Lin, Q., et al., *A Prediction Method of Ground Volume Loss Variation With Depth Induced by Tunnel Excavation.* Acta Geotechnica, 2021. **16**(11): p. 3689-3707.
138. Wu, H.-N., et al., *Longitudinal Structural Modelling of Shield Tunnels Considering Shearing Dislocation Between Segmental Rings.* Tunnelling and Underground Space Technology, 2015. **50**: p. 317-323.
139. Ihut, K.V.D. and M. Wijaya, *The Effect of Excess Pore Water Pressure on Under Consolidating Soil in Jakarta Tunnel Construction.* Iop Conference Series Earth and Environmental Science, 2025. **1514**(1): p. 012004.

140. Xu, L., et al., *Consolidation Effects on Pipe-Soil Interaction Due to Tunneling.* Frontiers in Earth Science, 2024. **12**.
141. Shin, J.-H., T.I. Addenbrooke, and D.M. Potts, *A Numerical Study of the Effect of Groundwater Movement on Long-Term Tunnel Behaviour.* Géotechnique, 2002. **52**(6): p. 391-403.
142. Sun, H., et al., *Stability Analysis of Shield Inclined Tunnel Faces Under the Change Effect of Soil Heterogeneity and Pore Water With Buried Depth.* Geofluids, 2022. **2022**: p. 1-16.
143. Sun, H., et al., *Tunnel Behaviour Caused by Basement Excavation in Clay.* Geofluids, 2021. **2021**: p. 1-11.
144. Xie, S.-w., Y.-H. Ye, and J. Ren, *Factors Influencing Post-Construction Responses of Underlying Tunnel Below Excavation Base in Gravelly Clay.* Sustainability, 2022. **14**(18): p. 11400.
145. Zhang, J., et al., *Upper Bound Analysis for Collapse Failure of Shield Tunnel Face Excavated in Unsaturated Soils Considering Steady Vertical Flow.* Mathematical Problems in Engineering, 2019. **2019**(1).
146. Dai, B., et al., *Mechanical Characteristics for Rocks Under Different Paths and Unloading Rates Under Confining Pressures.* Shock and Vibration, 2015. **2015**: p. 1-8.
147. Barton, N., *Minimizing the Use of Concrete in Tunnels and Caverns: Comparing NATM and NMT.* Innovative Infrastructure Solutions, 2017. **2**(1).
148. Tao, Z., et al., *Study on Deformation Mechanism and Support Measures of Soft Surrounding Rock in Muzhailing Deep Tunnel.* Advances in Civil Engineering, 2020. **2020**(1).
149. Mei, Y.c., et al., *Failure Mechanism and Optimization of Arch-Bolt Composite Support for Underground Mining Tunnel.* Advances in Civil Engineering, 2020. **2020**(1).
150. Wang, L., et al., *Analytical Method for Tunnel Support Parameter Design Based on Surrounding Rock Failure Mode Identification.* Geosciences, 2025. **15**(9): p. 369.
151. Li, J.L., et al., *Evaluation and Application of Surrounding Rock Stability Based on the Improved Weighting Multidimensional Cloud Model.* Advances in Civil Engineering, 2021. **2021**(1).
152. Zhao, H., et al., *Experiment and Numerical Simulation of Strength and Stress Distribution Behaviors of Anchored Rock Mass in a Roadway.* Geofluids, 2023. **2023**: p. 1-14.
153. Nishimura, T., et al., *Analysis of Convergence Confinement of Circular Tunnel by Supporting Elements Using Elasto-Plastic Solution of Ground.* 2024.
154. Mariee, A., A. Belal, and A. El-Desouky, *Application of the Convergence-Confinement Approach to Analyze the Rock-Lining Interaction in Tunnels (Case Study: Shimizu Tunnel).* International Conference on Aerospace Sciences and Aviation Technology, 2009. **13**(AEROSPACE SCIENCES): p. 1-11.
155. Zhu, Y., et al., *Physical and Mechanical Characteristics of Soft Rock Tunnel and the Effect of Excavation on Supporting Structure.* Applied Sciences, 2019. **9**(8): p. 1517.
156. Yuan, H., H. Wan, and Z.L. Zhao, *Coupled Numerical Analysis of Tunnel Excavation in Saturated Soft Clay Under High Groundwater.* Advanced Materials Research, 2011. **368-373**: p. 2533-2536.

157. Gong, B., et al., *The Seepage Control of the Tunnel Excavated in High-Pressure Water Condition Using Multiple Times Grouting Method.* Processes, 2018. **6**(9): p. 159.
158. Zhao, C., et al., *Groundwater Drainage-Induced Rock Mass Deformation During the Gotthard Base Tunnel Excavation in the Swiss Alps: 3D Modelling and Comparison to Field Measurements.* 2022.
159. Wei, Z. and Z. Yan-peng, *Seepage in Water-Rich Loess Tunnel Excavating Process and Grouting Control Effect.* Geofluids, 2021. **2021**: p. 1-13.
160. Shin, J.-H., D.M. Potts, and L. Zdravković, *The Effect of Pore-Water Pressure on NATM Tunnel Linings in Decomposed Granite Soil.* Canadian Geotechnical Journal, 2005. **42**(6): p. 1585-1599.
161. Aygar, E.B., *Evaluation of the Relationship Between the Collapsed Mechanism and Excavation Method in Tunnels Excavated in Schists.* Scientific Reports, 2022. **12**(1).
162. Fahimifar, A. and M.R. Zareifard, *A New Closed-form Solution for Analysis of Unlined Pressure Tunnels Under Seepage Forces.* International Journal for Numerical and Analytical Methods in Geomechanics, 2012. **37**(11): p. 1591-1613.
163. Daraei, R., B.M.A. Herki, and A.F.H. Sherwani, *Study on the Rapid Drawdown and Its Effect on Portal Subsidence of Heybat Sultan Twin Tunnels in Kurdistan-Iraq.* Civil Engineering Journal, 2017. **3**(7): p. 496-507.
164. Anagnostou, G. and S. Zingg, *On the Stabilizing Effect of Advance Drainage in Tunnelling / Über Die Statische Auswirkung Der Vorauseilenden Gebirgsdrainagen Im Tunnelbau.* Geomechanics and Tunnelling, 2013. **6**(4): p. 338-354.
165. Asadollahi, P. *Tunnel Face Stability for Conventional Underground Excavation*. 2016.
166. Hwang, R.H., T.H. Yang, and P.F. Sun, *Influence of Anisotropic Rock on Tunnel Stability With Consideration of Fluid-Solid Coupling.* Applied Mechanics and Materials, 2011. **90-93**: p. 2101-2107.
167. Gao, Z., et al., *Deformation Analysis of the Rock Surrounding a Tunnel Excavated Through a Gently Dipping Bed.* Applied Sciences, 2022. **12**(4): p. 1960.
168. Zhao, H.J., S. Feng, and G. Jie, *Physical Modelling of Joint Effects on Deformation and Failure of Tunnels.* Advanced Materials Research, 2011. **243-249**: p. 3205-3210.
169. Zhang, J., M. Wang, and C. Xi, *Tunnel Collapse Mechanism and Its Control Strategy in Fault Fracture Zone.* Shock and Vibration, 2021. **2021**(1).
170. Shi, J.-Z., et al., *Study of the Disaster-Causing Mechanism and Reinforcement Measures for Soft Rock Deformation and Lining Cracking.* Frontiers in Earth Science, 2023. **10**.
171. Zheng, X., et al., *Research on the Mechanism of Loose Deformation in Weak Fracture Zone Tunnel Surrounding Rock and Support Control.* Buildings, 2024. **14**(8): p. 2506.
172. Fang, W., G. Wang, and C. Wang, *Effect of the Location of Fault Fracture Zones on the Stability of Symmetrical Submarine Tunnels.* Symmetry, 2021. **13**(7): p. 1111.
173. Tarifard, A., P. Görög, and Á. Török, *Long-Term Assessment of Creep and Water Effects on Tunnel Lining Loads in Weak Rocks Using Displacement-Based Direct Back Analysis: An Example From Northwest of Iran.* Geomechanics and Geophysics for Geo-Energy and Geo-Resources, 2022. **8**(1).
174. Zhang, C., et al., *Study on Evolution Mechanism of Structure-Type Rockburst: Insights From Discrete Element Modeling.* Sustainability, 2021. **13**(14): p. 8036.

175. Wibisono, D.Y., et al., *Laboratory-Scale Rockburst Physical Model Testing Using a True-Triaxial Cell.* Iop Conference Series Earth and Environmental Science, 2023. **1124**(1): p. 012039.
176. Si, X., et al., *Experimental Investigation of Spalling Failure of D-Shaped Tunnel Under Three-Dimensional High-Stress Conditions in Hard Rock.* Rock Mechanics and Rock Engineering, 2021.
177. Dündar, S., et al., *The Most Commonly Used TBM Types and Their Suitability for Hard and Soft Grounds.* Mining Revue, 2025. **31**(2): p. 8-15.
178. Bianchi, G.W., *Analysis of TBM Parameters During Tunnelling in Adverse Conditions – The Case Study of the Frejus Safety Tunnel, Western Alps.* Geomechanics and Tunnelling, 2023. **16**(1): p. 47-52.
179. Samadi, H., et al., *Application of Supervised Learning Algorithms to Predict Engineering Characteristics of Soft to Strong Rock Masses Using Actual TBM Performance Data.* 2024.
180. Ramulu, M., *Special Tunnel Blasting Techniques for Railway Projects.* 2012.
181. Li, Y. and R.G. Sun, *Stability Analysis of Tunnel Surrounding Rock and Shotcrete Lining and Rock Bolts Based on Strength Reduction Finite Element Method.* Applied Mechanics and Materials, 2011. **90-93**: p. 1936-1941.
182. Xu, Y., et al., *Research on the Performance and Reinforcement Effect of Shotcrete in Small-Section Tunnels in Plateau Areas.* Engineering Research Express, 2025. **7**(4): p. 045107.
183. Pichler, B. and C. Hellmich, *Hybrid Methods for Shotcrete and Segmental Linings Tunnel Shells – Combining Displacement and Rotation Measurements With Computational Multiscale Mechanics.* Geomechanics and Tunnelling, 2018. **11**(3): p. 226-235.
184. Wang, J., Z. Fa, and J.-x. Lin, *Robust Optimization Design of Bolt-Shotcrete Support Structure in Tunnel.* Tehnicki Vjesnik - Technical Gazette, 2018. **25**(5).
185. Yang, F., et al., *Quantitative Analysis of Excavation -Damaged Zones for Effective TBM Tunnel Support Design.* Structural Control and Health Monitoring, 2025. **2025**(1).
186. Feng, L., et al., *Numerical Simulation Analysis of Different Excavation Parameters for TBM 3D Disc Cutters Based on the Discrete Element Method.* Applied Sciences, 2024. **15**(1): p. 38.
187. Jin, L., *Classification of Rock Mass in West Route of South-to-North Water Transfer Project Based on Penetrating Rate of TBM.* 2010: p. 1-10.
188. Terron-Almenara, J. and K.K. Panthi, *Analysis of Plastic Deformations for Tunnel Support Design in Weak Flysch Rock Mass of a Hydropower Tunnel in Central Albania.* Rock Mechanics and Rock Engineering, 2025. **58**(7): p. 8111-8143.
189. Sjölander, A. and A. Ansell, *Failure Mechanisms of Fibre Reinforced Shotcrete: Numerical Simulations Considering Local Variations in Thickness and Bond Strength.* Nordic Concrete Research, 2022. **67**(2): p. 51-64.
190. Torno, S., et al., *Mathematical and Fuzzy Logic Models in Prediction of Geological and Geomechanical Properties of Rock Mass by Excavation Data on Underground Works / Matematinis Ir Neraiškiosios Logikos Modeliai Aprašant Geologines Ir Geomechanines Uolienos Savybes, Naudojant Požeminio Kasinėjimo Duomenis.* Journal of Civil Engineering and Management, 2011. **17**(2): p. 197-206.

191. Elgamal, A. and N. Elfaris, *Seismic Isolation Materials for Bored Rock Tunnels: A Parametric Analysis*. Infrastructures, 2024. **9**(3): p. 44.
192. Wan, M.S.P., et al., *Measured Short-Term Ground Surface Response to EPBM Tunnelling in London Clay*. Géotechnique, 2017. **67**(5): p. 420-445.
193. Jiang, M. and Z.Y. Yin, *Influence of Soil Conditioning on Ground Deformation During Longitudinal Tunneling*. Comptes Rendus Mécanique, 2014. **342**(3): p. 189-197.
194. Spross, J., I. Papaioannou, and J.G. Grasmick, *Toward Probabilistic Ground Models for Time and Cost Estimation of Tunnel Projects*. 2024.
195. Peila, D., C. Oggeri, and L. Borio, *Using the Slump Test to Assess the Behavior of Conditioned Soil for EPB Tunneling*. Environmental and Engineering Geoscience, 2009. **15**(3): p. 167-174.
196. Wongsaroj, J., K. Soga, and R.J. Mair, *Tunnelling-Induced Consolidation Settlements in London Clay*. Géotechnique, 2013. **63**(13): p. 1103-1115.
197. Li, Z., K. Soga, and P. Wright, *Behaviour of Cast-Iron Bolted Tunnels and Their Modelling*. Tunnelling and Underground Space Technology, 2015. **50**: p. 250-269.
198. Mo, P.Q. and H.S. Yu, *Plasticity Solutions of an Undrained Cavity Contraction for the Prediction of Soil Behaviour Around Tunnels*. 2016: p. 150-157.
199. El-Kelesh, A., A. Shehata, and E. Elkasaby, *Impact of Ground Conditions on Penetration Rate of Slurry TBM in Soft Ground Tunneling*. 2025.
200. Gawecka, K.A., et al., *Predictive Modelling of Thermo-Active Tunnels in London Clay*. Géotechnique, 2021. **71**(8): p. 735-748.
201. Li, Z., K. Soga, and P. Wright, *Three-Dimensional Finite Element Analysis of the Behaviour of Cross Passage Between Cast-Iron Tunnels*. Canadian Geotechnical Journal, 2016. **53**(6): p. 930-945.
202. Zhao, C., et al., *Model Validation and Calibration via Back Analysis for Mechanized Tunnel Simulations – The Western Scheldt Tunnel Case*. Computers and Geotechnics, 2015. **69**: p. 601-614.
203. Zhang, Z.M., C. Lin, and S.M. Wu, *Slurry Shield Tunnelling in Clayey Soils: Typical Problems and Countermeasures*. Advanced Materials Research, 2011. **243-249**: p. 2944-2947.
204. Mair, R.J., *Tunnelling and Geotechnics: New Horizons*. Géotechnique, 2008. **58**(9): p. 695-736.
205. Mei, Y., et al., *Experimental Study on Shield Tunneling Control in Full Section Water-Rich Sand Layer of Collapsible Loess*. Geofluids, 2022. **2022**: p. 1-17.
206. Bhuiyan, M.R., et al., *An Overview of Soil Improvement Through Ground Grouting*. Journal of Geoscience and Environment Protection, 2024. **12**(01): p. 51-63.
207. Hartwig, M.E. and C.A. Moreira, *A New Computational Tool for the Follow Up of Mechanized Excavation of Tunnels: The Case Study of the Subway Line 5-Lilac in São Paulo, Brazil*. Rem - International Engineering Journal, 2018. **71**(4): p. 521-529.
208. Liang, J., et al., *An Explainable Intelligent System for Multi-performance Shield Tunnel Tail Grout Optimization*. Computer-Aided Civil and Infrastructure Engineering, 2025. **40**(30): p. 6165-6183.
209. Klopčič, J. and J. Logar, *Effect of Relative Orientation of Anisotropy Planes to Tunnel Axis on the Magnitude of Tunnelling Displacements*. International Journal of Rock Mechanics and Mining Sciences, 2014. **71**: p. 235-248.

210. Mboussa, D.J.W. and S. Sun, *Analysis of Rock Mass Behavior With Empirical and Numerical Method for the Construction of Diversion Tunnel, Laos*. Iop Conference Series Materials Science and Engineering, 2022. **1212**(1): p. 012028.
211. Meguid, M.A. and R.K. Rowe, *Stability of D-Shaped Tunnels in a Mohr�Coulomb Material Under Anisotropic Stress Conditions*. Canadian Geotechnical Journal, 2006. **43**(3): p. 273-281.
212. Wei, J., et al., *Study on Strength Parameters of Rock Mass With Stochastic Combination Discontinuities and Their Application in a Tunnel in China*. Advances in Mechanical Engineering, 2018. **10**(7).
213. Zhu, Z., et al., *Model Test Study on the Rock Mass Deformation Law of a Soft Rock Tunnel Under Different Ground Stresses*. Frontiers in Earth Science, 2022. **10**.
214. Zhang, S., et al., *DEM-based Analysis of Water Inrush Process of Underground Engineering Face With Intermittent Joints in Karst Region*. Electronic Journal of Structural Engineering, 2023. **23**(4): p. 59-65.
215. Liu, J., et al., *Inferred Geological Data and Stability Analysis of Tunnel Surrounding Rock by Drilling and Blasting Method Based on Digital Twin*. Journal of Civil Engineering and Urban Planning, 2023. **5**(10).
216. Panthi, K.K., *Methods Applied for the Stability Assessment in Rock Engineering*. Journal of Nepal Geological Society, 2023. **65**: p. 29-34.
217. Katuwal, T.B. and K.K. Panthi, *Stability Challenges and Remedial Practices in Himalayan Hydro-Power Tunnels – A Review*. Journal of Engineering and Sciences, 2024. **3**(1): p. 9-19.
218. Zhang, Z., et al., *Excavation Sequence and Surrounding Rock Mass Stability of Large-Scale Underground Engineering With 8 Tunnels*. Advances in Civil Engineering, 2020. **2020**(1).
219. Ma, Y., Z. Wang, and D. Wang, *The Safe Load-Bearing Capacity of Railway Tunnel Linings Under High-Pressure and Water-Rich Conditions*. Buildings, 2023. **13**(9): p. 2154.
220. Nikvar-Hassani, A., H. Katibeh, and H. Farhadian, *Numerical Analysis of Steady-State Groundwater Inflow Into Tabriz Line 2 Metro Tunnel, Northwestern Iran, With Special Consideration of Model Dimensions*. Bulletin of Engineering Geology and the Environment, 2015. **75**(4): p. 1617-1627.
221. Liu, X.X., et al., *Evaluation of Foam Conditioning Effect on Groundwater Inflow at Tunnel Cutting Face*. International Journal for Numerical and Analytical Methods in Geomechanics, 2018. **43**(2): p. 463-481.
222. Liu, X.X., et al., *Analytical Approach for Time-dependent Groundwater Inflow Into Shield Tunnel Face in Confined Aquifer*. International Journal for Numerical and Analytical Methods in Geomechanics, 2017. **42**(4): p. 655-673.
223. Huang, Y., et al., *Influence of Water Ingress at the Shield Tunnel Portal on Tunnel Deformation and Load Capacity Weakening*. Applied and Computational Engineering, 2024. **98**(1): p. 71-78.
224. Amoruso, A., et al., *Parsimonious Recharge/Discharge Modeling in Carbonate Fractured Aquifers: The Groundwater Flow in the Gran Sasso Aquifer (Central Italy)*. Journal of Hydrology, 2013. **476**: p. 136-146.
225. Hou, T.-x., et al., *Forecasting and Prevention of Water Inrush During the Excavation Process of a Diversion Tunnel at the Jinping II Hydropower Station, China*. Springerplus, 2016. **5**(1).

226. Bahri, M., et al., *Numerical Model Validation for Detection of Surface Displacements Over Twin Tunnels From Metro Line 1 in the Historical Area of Seville (Spain)*. Symmetry, 2022. **14**(6): p. 1263.
227. Wang, F. and P. Li, *An Analytical Model of Seepage Field for Symmetrical Underwater Tunnels*. Symmetry, 2018. **10**(7): p. 273.
228. Jiang, X., et al., *Anisotropic Propagation of Chemical Grouting in Fracture Network With Flowing Water*. Acs Omega, 2021. **6**(7): p. 4672-4679.
229. Li, C.L., et al., *Evolution Characteristic Analysis of Pressure - Arch of a Double - Arch Tunnel in Water - Rich Strata*. Journal of Engineering Science and Technology Review, 2016. **9**(1): p. 44-51.
230. Sui, W., et al., *Experimental Investigation on Sealing Efficiency of Chemical Grouting in Rock Fracture With Flowing Water*. Tunnelling and Underground Space Technology, 2015. **50**: p. 239-249.
231. Li, Z., et al., *Improving the Anti-Washout Property of Acrylate Grouting Material by Bentonite: Its Characterization, Improving Mechanism, and Practical Application*. Polymers, 2023. **15**(19): p. 3865.
232. Su, K., et al., *An Analytical Method for Groundwater Inflow Into a Drained Circular Tunnel*. Ground Water, 2017. **55**(5): p. 712-721.
233. Zhang, C., et al., *Polyurethane/Red Mud Composites With Flexibility, Stretchability, and Flame Retardancy for Grouting*. Polymers, 2018. **10**(8): p. 906.
234. Cárdenas, I.C., et al., *Using Prior Risk-Related Knowledge to Support Risk Management Decisions: Lessons Learnt From a Tunneling Project*. Risk Analysis, 2014. **34**(10): p. 1923-1943.
235. Newman, T.G., L. Yuan, and L.C. O'Keeffe, *Using Tunnel Boring Data to Augment the Geological Model*. Proceedings of the Institution of Civil Engineers - Geotechnical Engineering, 2010. **163**(3): p. 157-166.
236. Ting, C., B. Gilson, and M. Black, *Developing the 3D Geological Model for Crossrail 2, London, UK*. Quarterly Journal of Engineering Geology and Hydrogeology, 2020. **54**(2).
237. Chi, D., et al., *Analysis of the Surrounding Rock Full-Displacement Variation in Large-Span Mudstone Highway Tunnels*. Symmetry, 2024. **16**(5): p. 526.
238. Zhao, M., et al., *Optimization of Construction Parameters and Deformation Characteristics of Large-Section Loess Tunnel: A Case Study From Xi'an Metro*. Advances in Civil Engineering, 2021. **2021**(1).
239. Palomba, M., et al., *Chenani-Nashri Tunnel, the Longest Road Tunnel in India: The "Himalayan Challenge" for Design in Heterogeneous Rock Masses*. Geomechanics and Tunnelling, 2014. **7**(2): p. 114-122.
240. Chen, C., et al., *Digital Twin-Based and Knowledge Graph-Enhanced Emergency Response in Urban Infrastructure Construction*. Applied Sciences, 2025. **15**(11): p. 6009.
241. Singhroy, V.H., *Case Studies on the Application of Remote Sensing Data to Geotechnical Investigations in Ontario, Canada*. 1988: p. 9-45.
242. Li, Z. and F. Li, *Analysis of the Impact of Shield Tunneling Under Existing Subway Station*. Journal of Physics Conference Series, 2024. **2736**(1): p. 012008.
243. Kuang, X., D.Q. Yi, and C.L. Wang, *Study on Construction Risk Assessment Method of Deep Buried Tunnels*. Applied Mechanics and Materials, 2013. **405-408**: p. 1305-1308.

244. Song, Z., et al., *Numerical Analysis and Application of the Construction Method for the Small Interval Tunnel in the Turn Line of Metro.* Science Progress, 2020. **103**(3).
245. Taha, O.M.E., *Variation of Bearing Capacity Prediction for Shallow Foundations by SPT and Laboratory Tests.* International Journal of Geomate, 2019. **17**(64).
246. Qader, Z.B., Z. Karabash, and A.F. Çabalar, *Analyzing Geotechnical Characteristics of Soils in Erbil via GIS and ANNs.* Sustainability, 2023. **15**(5): p. 4030.
247. Tan, J.J. and H. Ramli, *Effect of Standard Penetration Test Corrections on the Estimation of Undrained Shear Strength.* Journal of Civil Engineering and Management, 2023. **29**(6): p. 501-515.
248. Tham, D.H. and T.N. Manh, *Predicting the Bearing Capacity of Pile Installed Into Cohesive Soil Concerning the Spatial Variability of SPT Data (A Case Study).* Ho Chi Minh City Open University Journal of Science - Engineering and Technology, 2021. **11**(1): p. 45-64.
249. Ceylan, Ç. and M. Önal, *Correlations Between SPT, PMT and MASW on Quaternary Alluvial - Fluvial Sediments in Battalgazi, Malatya, Turkey.* Naturengs Mtu Journal of Engineering and Natural Sciences Malatya Turgut Ozal University, 2020. **1**(2): p. 39-53.
250. Sarker, D. and M. Abedin, *Applicability of Standard Penetration Test in Bangladesh and Graphical Representation of SPT-N Value.* International Journal of Science and Engineering Investigations, 2015. **4**: p. 55-59.
251. Lafuerza, S., et al., *Subseafloor Stratigraphic Profiling and Soil Classification From Piezocone Tests: A Case Study in the Gulf of Lion (NW Mediterranean Sea).* Geochemistry Geophysics Geosystems, 2008. **9**(12).
252. Zou, H., et al., *Application of the Piezocone Penetration Testing to Assess Flow Characteristics of Marine Clay.* 2014: p. 845-854.
253. Li, J.H., *Application of Neural Network in Identifying Soil Strata by CPT or CPTU Data.* Applied Mechanics and Materials, 2012. **170-173**: p. 945-949.
254. Cai, G., S. Liu, and A.J. Puppala, *Comparison of CPT Charts for Soil Classification Using PCPT Data: Example From Clay Deposits in Jiangsu Province, China.* Engineering Geology, 2011. **121**(1-2): p. 89-96.
255. Robertson, P.K. and C.E. Wride, *Evaluating Cyclic Liquefaction Potential Using the Cone Penetration Test.* Canadian Geotechnical Journal, 1998. **35**(3): p. 442-459.
256. Harianto, T., et al., *Predicting the Hydraulic Conductivity of Makassar Marine Clay Using Field Penetration Test (Cptu) Results.* 2011: p. 2093-2099.
257. Huffman, A., *Numerical Simulation of Continuous Cone Penetration and Interpretation of CPTu Dissipation Tests.* 2000.
258. Campo, B., L. Bruno, and A. Amorosi, *Sedimentary Facies Characterization Through <scp>CPTU</Scp> Profiles: An Effective Tool for Subsurface Investigation of Modern Alluvial and Coastal Plains.* Sedimentology, 2023. **70**(4): p. 1302-1327.
259. Chai, J.C., et al., *Estimating Hydraulic Conductivity From Piezocone Soundings.* Géotechnique, 2011. **61**(8): p. 699-708.
260. Ecemis, N., M.M. Monkul, and M. Orucu, *Applicability of Soil-type Index for Shear Wave Velocity-based Liquefaction Assessment.* Earthquake Engineering & Structural Dynamics, 2024. **53**(6): p. 2055-2071.
261. Iturbide, O.S.R. and E.B. Jaramillo, *Identification of Geotechnical Units in Soil Exploration Through Principal Component Analysis and Clustering.* International

Journal for Numerical and Analytical Methods in Geomechanics, 2024. **48**(6): p. 1681-1699.
262. Robertson, P.K., *Cone Penetration Test (CPT)-based Soil Behaviour Type (SBT) Classification System — An Update*. Canadian Geotechnical Journal, 2016. **53**(12): p. 1910-1927.
263. Erzin, Y. and N. Ecemis, *The Use of Neural Networks for the Prediction of Cone Penetration Resistance of Silty Sands*. Neural Computing and Applications, 2016. **28**(S1): p. 727-736.
264. Robertson, P.K., *Comparing CPT and vs Liquefaction Triggering Methods*. Journal of Geotechnical and Geoenvironmental Engineering, 2015. **141**(9).
265. Jia, R., et al., *Numerical Study of the Effect of Clay Structure on Piezocone Penetration Test Results*. International Journal of Geosynthetics and Ground Engineering, 2021. **7**(4).
266. Chandler, R.J., *The in-Situ Measurement of the Undrained Shear Strength of Clays Using the Field Vane*. 1988: p. 13-44.
267. Suits, L.D., et al., *Vane Shear Behavior of Soft Bangkok Clay*. Geotechnical Testing Journal, 2004. **27**(1): p. 57-66.
268. Shen, J., et al., *Numerical Investigations of Undrained Shear Strength of Sensitive Clay Using Miniature Vane Shear Tests*. Journal of Marine Science and Engineering, 2023. **11**(5): p. 1094.
269. Khatrush, S. and G. El-Gehani, *Laboratory Evaluation of Undrained Shear Strength of a Soft Fine Grained Soils*. International Journal of Engineering Technologies Ijet, 2021. **7**(3): p. 66-74.
270. Kouretzis, G., et al., *Interpretation of Vane Shear Tests for Geotechnical Stability Calculations*. Canadian Geotechnical Journal, 2017. **54**(12): p. 1775-1780.
271. Vahedifard, F., et al., *Strength Indices of High-Moisture Soils Using Handheld Gauges*. Proceedings of the Institution of Civil Engineers - Ground Improvement, 2016. **169**(3): p. 167-181.
272. Hou, J.F., Y. Cao, and S. Yan, *Analysis of Shear Strength Parameters on Soft Clay Slope Stability at Short Term*. Applied Mechanics and Materials, 2012. **170-173**: p. 1306-1309.
273. Selvadurai, A.P.S., *On the Estimation of the Deformability Characteristics of an Isotropic Elastic Soil Medium by Means of a Vane Test*. International Journal for Numerical and Analytical Methods in Geomechanics, 1979. **3**(3): p. 231-243.
274. Pandey, S.K., A.R. Lamichhane, and I.P. Acharya, *Evaluating the Undrained Shear Strength of Kalimati Soil Using Field Vane and Unconfined Compression Test; A Study on Applicability of Bjerrum's Correction Factor*. Journal of Advanced College of Engineering and Management, 2023. **8**(1): p. 59-68.
275. Jonak, J., et al., *Prediction of Load on the Cutting Tools in Tunnel Boring Machines*. Acta Montanistica Slovaca, 2020. **25**: p. 444-452.
276. Gollegger, J., A. Priller, and M. Rausch, *The Use of Open Tunnel Boring Machines in Squeezing Rock in the Gotthard Base Tunnel*. Geomechanics and Tunnelling, 2009. **2**(5): p. 591-600.
277. Patel, S. and D. Pandit, *Ranking of the Factors That Affect TBM Productivity*. Journal of Real Estate Construction & Management, 2017. **32**(4): p. 31-41.
278. Standing, J.R. and D. Selemetas, *Greenfield Ground Response to EPBM Tunnelling in London Clay*. Géotechnique, 2013. **63**(12): p. 989-1007.

279. Wu, Y., et al., *A New Evaluation Method for the Uniaxial Compressive Strength Ahead of the Tunnel Face Based on the Driving Data and Specification Parameters of TBM.* Shock and Vibration, 2019. **2019**(1).

280. Wang, S., et al., *Big Data-Based Boring Indexes and Their Application During TBM Tunneling.* Advances in Civil Engineering, 2021. **2021**(1).

281. Frenzel, C., et al., *Penetration Tests for TBMs and Their Practical Application / Penetrationstests Für Tunnelbohrmaschinen Und Deren Anwendung in Der Praxis.* Geomechanics and Tunnelling, 2012. **5**(5): p. 557-566.

282. Chen, L., et al., *Double-Shield TBM Mounted Seismic Ahead-Prospecting System and Its Application.* 2024.

283. Nagel, F. and G. Meschke, *An Elasto-plastic Three Phase Model for Partially Saturated Soil for the Finite Element Simulation of Compressed Air Support in Tunnelling.* International Journal for Numerical and Analytical Methods in Geomechanics, 2009. **34**(6): p. 605-625.

284. Liu, Y., H. Wang, and L. Tao, *TBM-MSE: A Multi-Engine State Estimation Based on Inertial Enhancement for Tunnel Boring Machines in Perceptually Degraded Roadways.* Ieee Access, 2023. **11**: p. 55978-55989.

285. Liu, D., Y.-Q. Zhou, and K. Jiao, *TBM Construction Process Simulation and Performance Optimization.* Transactions of Tianjin University, 2010. **16**(3): p. 194-202.

286. Mohammadzamani, D., S. Mahdevari, and R. Bagherpour, *Evaluation of Required Thrust Force Based on Advance Rates in Shielded TBMs Under Squeezing Conditions.* Journal of Geophysics and Engineering, 2019. **16**(5): p. 842-861.

287. Zou, X., H. Zheng, and Y. Mi, *Performance Evaluation of Hard Rock TBMs Considering Operational and Rock Conditions.* Shock and Vibration, 2018. **2018**(1).

288. Afradi, A., A. Ebrahimabadi, and T. Hallajian, *Prediction of TBM Penetration Rate of Water Conveyance Tunnels in Iran Using Modern Methods.* Stavební Obzor - Civil Engineering Journal, 2020. **29**(4).

289. Afradi, A., A. Ebrahimabadi, and T. Hallajian, *Prediction of the Penetration Rate and Number of Consumed Disc Cutters of Tunnel Boring Machines (TBMs) Using Artificial Neural Network (ANN) and Support Vector Machine (SVM)—Case Study: Beheshtabad Water Conveyance Tunnel in Iran.* Asian Journal of Water Environment and Pollution, 2019. **16**(1): p. 49-57.

290. Jiang, L., et al., *Detection of TBM Tunnel Surrounding Rock Debris Based on Attention Mechanism YOLOv5.* 2023.

291. Nguyen, N.M. and D.T. Pham, *Rock Burst and Actually Effective Bolting for Rock Burst Prevention.* Mining Science and Technology (Russia), 2019. **4**(2): p. 103-110.

292. Tang, B., et al., *Supporting Design Optimization of Tunnel Boring Machines-Excavated Coal Mine Roadways: A Case Study in Zhangji, China.* Processes, 2020. **8**(1): p. 46.

293. Lee, H., et al., *Numerical Simulation of EPB Shield Tunnelling With TBM Operational Condition Control Using Coupled DEM–FDM.* Applied Sciences, 2021. **11**(6): p. 2551.

294. Standing, J.R., D. Selemetas, and J.N. Shirlaw, *Greenfield Ground Response to EPBM Tunnelling in London Clay.* Géotechnique, 2014. **64**(7): p. 581-583.

295. Hasanpour, R., J. Rostami, and M. Barla, *Impact of Advance Rate on Entrapment Risk of a Double-Shielded TBM in Squeezing Ground.* Rock Mechanics and Rock Engineering, 2014. **48**(3): p. 1115-1130.

296. Ramoni, M. and G. Anagnostou, *The Interaction Between Shield, Ground and Tunnel Support in TBM Tunnelling Through Squeezing Ground.* Rock Mechanics and Rock Engineering, 2010. **44**(1): p. 37-61.

297. Zhao, S., et al., *Dispersant for Reducing Mud Cakes of Slurry Shield Tunnel Boring Machine in Sticky Ground.* Advances in Materials Science and Engineering, 2021. **2021**(1).

298. Demagh, R. and F. Emeriault, *3d Modelling of Tunnel Excavation Using Pressurized Tunnel Boring Machine in Overconsolidated Soils.* Studia Geotechnica Et Mechanica, 2013. **35**(2): p. 3-17.

299. Nie, X., et al., *Application of Assembled Energy-Saving Hydraulic System.* 2025: p. 48.

300. Kim, T., et al., *Application Ranges of EPB Shield TBM in Weathered Granite Soil: A Laboratory Scale Study.* Applied Sciences, 2021. **11**(7): p. 2995.

301. Tao, L., et al., *Experimental Methods to Assess the Effectiveness of Soil Conditioning With Foam in Fully Weathered Granite.* Advances in Materials Science and Engineering, 2019. **2019**: p. 1-12.

302. Tyrer, J., et al., *Tunnelling With Full-Face Shielded Machines: A 3D Numerical Analysis of an Earth Pressure Balance (EPB) Excavation Sequence Using the Finite Element Method (FEM).* Geosciences, 2023. **13**(8): p. 244.

303. Lu, X., *Analysis of Shield Tunneling Section Between Lianhu Park and Zuchongzhilu Stations on Suzhou Metro Line 11.* 2024: p. 554-563.

304. Friedemann, W., et al., *Nord/Süd-Metrolinie Amsterdam – Schild Transfer System (STS).* Bautechnik, 2013. **90**(3): p. 151-161.

305. Tang, L., et al., *Evaluation of the Ground Settlement in an Urban Area Resulting From a Small Curvature Tunneling Construction.* Sustainability, 2022. **14**(23): p. 16230.

306. Zhou, X. and Y. Yang, *Effect of Foam Parameters on Cohesionless Soil Permeability and Its Application to Prevent the Water Spewing.* Applied Sciences, 2020. **10**(5): p. 1787.

307. Firouzei, Y., et al., *Evaluation of Foam Application in Providing Required Conditioning for EPB Hard Rock TBMs.* Geomechanics and Tunnelling, 2023. **16**(4): p. 409-421.

308. Merritt, A.S. and R.J. Mair, *Mechanics of Tunnelling Machine Screw Conveyors: Model Tests.* Géotechnique, 2006. **56**(9): p. 605-615.

309. Li, T., et al., *Evaluation of the Adaptability of an EPB TBM to Tunnelling Through Highly Variable Composite Strata.* Mathematical Problems in Engineering, 2021. **2021**: p. 1-14.

310. Forsat, M., M. Taghipoor, and M. Palassi, *3D FEM Model on the Parameters' Influence of EPB-TBM on Settlements of Single and Twin Metro Tunnels During Construction.* International Journal of Pavement Research and Technology, 2021. **15**(3): p. 525-538.

311. Samadi, H., J. Hassanpour, and J. Rostami, *Prediction of Earth Pressure Balance for EPB Machine: Application of Machine Learning and MVR Techniques.* 2023.

312. Wang, S., et al., *Soil Improvement of EPBS Construction in High Water Pressure and High Permeability Sand Stratum.* Advances in Civil Engineering, 2019. **2019**(1).
313. Park, B., et al., *Discrete-Element Analysis of the Excavation Performance of an EPB Shield TBM Under Different Operating Conditions.* Applied Sciences, 2021. **11**(11): p. 5119.
314. Liu, X., S. Xu, and C. Shao, *Optimal Control of Earth Pressure Balance of Shield Tunneling Machine Based on Dual-heuristic Dynamic Programming.* Optimal Control Applications and Methods, 2020. **41**(5): p. 1510-1523.
315. Festa, D., W. Broere, and J.W. Bosch, *Kinematic Behaviour of a Tunnel Boring Machine in Soft Soil: Theory and Observations.* Tunnelling and Underground Space Technology, 2015. **49**: p. 208-217.
316. Lehmann, G. and K. Bäppler, *Development Tendencies in Mechanised Tunnelling – Review and Development Tendencies for Shield Machines With Slurry and Earth Pressure Support.* Geomechanics and Tunnelling, 2010. **3**(3): p. 245-255.
317. Li, X., et al., *Slurry Discharge Pipeline Damage and Wear Due to Transporting Rock Particles During Slurry Shield Tunneling: A Case Study Based on in Situ Observed Results.* Applied Sciences, 2023. **13**(12): p. 7103.
318. Qin, S., Y. Cheng, and W.H. Zhou, *State-of-the-Art Review on Pressure Infiltration Behavior of Bentonite Slurry Into Saturated Sand for TBM Tunneling.* Smart Construction and Sustainable Cities, 2023. **1**(1).
319. Yang, J. and Z.Y. Yin, *A Novel Hydro-mechanical Coupled Multiphysics Model for Slurry Penetration in Sand With Application to Face Stability Analysis of Shield Tunnel.* International Journal for Numerical and Analytical Methods in Geomechanics, 2023. **47**(7): p. 1202-1226.
320. Sun, B.P., et al., *Lab Study on the Effect of Cation Exchange Capacity on Slurry Performance in Slurry Shields.* Advances in Civil Engineering, 2018. **2018**(1).
321. Liu, K., W. Ding, and C. Qu, *A Mesoscopic Viewpoint on Slurry Penetration and Pressure Transfer Mechanisms for Slurry Shield Tunneling.* Buildings, 2022. **12**(10): p. 1744.
322. Yin, Z.Y., P. Wang, and F. Zhang, *Effect of Particle Shape on the Progressive Failure of Shield Tunnel Face in Granular Soils by Coupled FDM-DEM Method.* Tunnelling and Underground Space Technology, 2020. **100**: p. 103394.
323. Schößer, B., et al., *Face Support, Soil Conditioning and Material Transport in Earth-Pressure-Balance and Hydro Shield Machines.* 2023: p. 165-252.
324. Wang, Q., et al., *Predicting Slurry Pressure Balance With a Long Short-Term Memory Recurrent Neural Network in Difficult Ground Condition.* Computational Intelligence and Neuroscience, 2021. **2021**(1).
325. Yang, Y., et al., *Method for Calculating Pressure Losses in the Pipelines of Slurry Shield Tunneling Based on Coupled Simulation of Computational Fluid Dynamics and Discrete Element Method.* Computer-Aided Civil and Infrastructure Engineering, 2023. **39**(2): p. 300-316.
326. Yin, Y., T. Zhang, and K. Chen, *Design Study for Key Components of Slurry TBM in Tunnel With High Pressure Water.* 2023: p. 1122-1129.
327. Qin, S., et al., *Infiltration Behaviour and Microstructure of Filter Cake From Sand-Modified Bentonite Slurry.* Transportation Geotechnics, 2023. **40**: p. 100963.

328. Oliveira, D., et al., *EPB Tunnelling Through Clay-sand Mixed Soils: Proposed Methodology for Clogging Evaluation.* Geomechanics and Tunnelling, 2018. **11**(4): p. 375-387.
329. Burger, W., *Multi-mode Tunnel Boring Machines / Multi-Mode Tunnelvortriebsmaschinen.* Geomechanics and Tunnelling, 2014. **7**(1): p. 18-30.
330. Chan, A.W.Y., et al., *Construction of the Tuen Mun–Chek Lap Kok Link Sub-Sea Tunnels in Hong Kong.* Hkie Transactions, 2021: p. 139-153.
331. Karin Bäppler, F.B., Werner Burger, *Variable Density TBM - Combining Two Soft Ground TBM Technologies.* 2018, TBM: Tunnel Business Magazine.
332. Maidl, U. and J.C.D.D. Pierri, *Innovative Hybrid EPB Tunnelling in Rio De Janeiro.* Geomechanics and Tunnelling, 2014. **7**(1): p. 55-63.
333. Schaeffer, K. and M.A. Mooney, *Examining the Influence of TBM-ground Interaction on Electrical Resistivity Imaging Ahead of the TBM.* Tunnelling and Underground Space Technology, 2016. **58**: p. 82-98.
334. Rengshausen, R., R. Tauriainen, and A. Raedle, *TBM and Spoil Treatment Selection Process – Case History Crossrail C310 Thames Tunnel.* Geomechanics and Tunnelling, 2014. **7**(1): p. 45-54.
335. Göbl, A., *The Interaction of Ground, TBM and Segment Lining With Closed Shield Machines.* Geomechanics and Tunnelling, 2010. **3**(5): p. 491-500.
336. Chan, A.W.Y., et al., *Use of Saturation Diving Technique for Tunnel Boring Machine Cutterhead Intervention in the Tuen Mun–Chek Lap Kok Link Sub-Sea Tunnel Project, Hong Kong.* Hkie Transactions, 2021. **28**(1): p. 31-38.
337. Páez, C.V.G., *Rendimiento, Desgaste Y Abrasividad en Excavación Mecanizada De Túneles en Terrenos Heterogéneos.*
338. Zhao, Y., F. Luo, and X. Liang, *Monitoring and Analysis of Circumferential Pressure for Long-Distance Pipe Jacking Construction in Composite Strata.* Engineering Reports, 2024. **7**(1).
339. Jiang, X., et al., *Case Study of the Largest Concrete Earth Pressure Balance Pipe-Jacking Project in the World.* Transportation Research Record Journal of the Transportation Research Board, 2022. **2676**(7): p. 92-105.
340. Herrenknecht, S., *The Role of Slurry Microtunnelling for Pipeline Installations in Hard Rock Conditions.* 2024.
341. Niu, Z., et al., *A New Method for Predicting Ground Settlement Induced by Pipe Jacking Construction.* Mathematical Problems in Engineering, 2020. **2020**: p. 1-11.
342. Jiang, X., et al., *Advancing Shallow Tunnel Construction in Soft Ground: The Pipe-Umbrella Box Jacking Method.* Transportation Research Record Journal of the Transportation Research Board, 2024. **2678**(9): p. 150-168.
343. Chen, X., et al., *Performance Simulation Analysis and Design of the Scale Model for the Earth Pressure Balance Shield.* Advanced Materials Research, 2011. **346**: p. 364-371.
344. Andrews, P., *Ground Support Selection Rationale: A Gold Fields Perspective.* 2019.
345. Gökçeoğlu, C., et al., *A Geotechnical Perspective on a Complex Geological Environment in a High-Speed Railway Tunnel Excavation (A Case Study From Türkiye).* Infrastructures, 2022. **7**(11): p. 155.
346. Wang, W., et al., *Improving Tunnel Boring Machine Tunneling Performance by Investigating the Particle Size Distribution of Rock Chips and Cutter Consumption.* Buildings, 2024. **14**(4): p. 1124.

347. Liu, Y., et al., *Study on the Improvement of the Open TBM System and Supporting Equipment Based on the Gaojiapu Coal Mine.* Frontiers in Earth Science, 2023. **11**.
348. Zhu, Y., et al., *A New System to Evaluate Comprehensive Performance of Hard-Rock Tunnel Boring Machine Cutterheads.* Chinese Journal of Mechanical Engineering, 2019. **32**(1).
349. Naji, A.M., et al., *Static and Dynamic Influence of the Shear Zone on Rockburst Occurrence in the Headrace Tunnel of the Neelum Jhelum Hydropower Project, Pakistan.* Energies, 2019. **12**(11): p. 2124.
350. Shi, H., et al., *Study on Rock Cutting Mechanism and Efficient Shield Tunneling Control in Mixed-Ground With Different Strengths.* Journal of Physics Conference Series, 2022. **2185**(1): p. 012062.
351. EfeoĞLu, E., *Comparative Performance Analysis of Ensemble Learning Algorithms for Rock Classification.* Kastamonu University Journal of Engineering and Sciences, 2023.
352. Nguyen, T., et al., *Ground Improvement by Jet Grouting and Injection to Control Hydraulic Conditions in the Frame of Grand Paris Express Project, North-East Section: Methodology and Quality Control Sequence.* Journal of Mining and Earth Sciences, 2022. **63**(3a): p. 59-67.
353. Torabi, M., M. Palassi, and G. Mohammadi, *Influence of Confining Stress on Different Diameters of Disc Cutters in Rock Cutting.* Data-Centric Engineering, 2025. **6**.
354. Mucha, K., *Application of Rock Abrasiveness and Rock Abrasivity Test Methods—A Review.* Sustainability, 2023. **15**(14): p. 11243.
355. Mei, Y., et al., *Influence Factors of Vibration Response on Supporting-Thrusting System of Tunnel Boring Machine.* Mathematical Problems in Engineering, 2020. **2020**: p. 1-10.
356. Haas, M., et al., *Applicability of Excavated Rock Material: A European Technical Review Implying Opportunities for Future Tunnelling Projects.* Journal of Cleaner Production, 2021. **315**: p. 128049.
357. Mei, Y., et al., *Laws and Numerical Analysis of Surface Deformation Caused by Excavation of Large Diameter Slurry Shield in Upper-Soft and Lower-Hard Composite Stratum.* Buildings, 2022. **12**(9): p. 1470.
358. Li, K., et al., *AHP-FSE-Based Risk Assessment and Mitigation for Slurry Balancing Shield Tunnel Construction.* Journal of Environmental and Public Health, 2022. **2022**(1).
359. Han, B., et al., *Numerical Investigation of Slurry Fracturing During Shield Tunneling Under a Reservoir.* Applied Sciences, 2022. **12**(15): p. 7929.
360. Zhang, S., et al., *Liquid-Solid Coupling Response of Surrounding Rock Mass of Large-Diameter River-Crossing Shield Tunnel.* Stavební Obzor - Civil Engineering Journal, 2023. **32**(4): p. 443-456.
361. Cui, J., et al., *Performance of Slurry Shield Tunnelling in Mixed Strata Based on Field Measurement and Numerical Simulation.* Advances in Materials Science and Engineering, 2020. **2020**(1).
362. Liang, Y., et al., *Risk Analysis and Control Measures for Slurry Shield Tunneling Diagonally Under an Urban River Embankment.* Advances in Civil Engineering, 2020. **2020**(1).

363. Sun, L., Z. Li, and R. Liang, *Simplified Analytical Method for Predicting the Lateral Ground Displacements Due to Shield Tunnelling.* Advances in Civil Engineering, 2021. **2021**(1).
364. Haung, Y., et al., *Uplift Model Analysis for Shallow Overlaying Slurry Shield Tunnel Based on Time-Varying Behavior of Cement Grouting Material.* Key Engineering Materials, 2016. **723**: p. 782-788.
365. Zhang, Z., et al., *Three-Dimensional Finite-Element Analysis on Ground Responses During Twin-Tunnel Construction Using the URUP Method.* Tunnelling and Underground Space Technology, 2016. **58**: p. 133-146.
366. Galli, M. and M. Thewes, *Investigations for the Application of EPB Shields in Difficult Grounds / Untersuchungen Für Den Einsatz Von Erddruckschilden in Schwierigem Baugrund.* Geomechanics and Tunnelling, 2014. **7**(1): p. 31-44.
367. Stascheit, J., et al., *Building Information Modelling in Mechanised Shield Tunnelling – A Practitioner's Outlook to the Near Future.* Geomechanics and Tunnelling, 2018. **11**(1): p. 34-49.
368. Verma, A.K. and R. Singh, *Assessment of Tunnel Instability—a Numerical Approach.* Arabian Journal of Geosciences, 2009. **3**(2): p. 181-192.
369. Sheil, B., et al., *Machine Learning to Inform Tunnelling Operations: Recent Advances and Future Trends.* Proceedings of the Institution of Civil Engineers - Smart Infrastructure and Construction, 2020. **173**(4): p. 74-95.
370. Cao, B.T., S. Freitag, and G. Meschke, *A Hybrid RNN-GPOD Surrogate Model for Real-Time Settlement Predictions in Mechanised Tunnelling.* Advanced Modeling and Simulation in Engineering Sciences, 2016. **3**(1).
371. Hegemann, F., J. Stascheit, and U. Maidl, *Digitalisierung Und IoT Im Maschinellen Tunnelbau – Ein Integraler Ansatz.* Geomechanics and Tunnelling, 2020. **13**(4): p. 406-420.
372. Ninić, J., et al., *Computationally Efficient Simulation in Urban Mechanized Tunneling Based on Multilevel BIM Models.* Journal of Computing in Civil Engineering, 2019. **33**(3).
373. Hua, J., et al., *Study on Soil Conditioning and Key Construction Parameters of EPB TBM Advancing in Sand-Pebble Layer of Beijing Metro.* Applied Mechanics and Materials, 2011. **90-93**: p. 2138-2142.
374. Cheng, Y., et al., *Hydromechanical Analysis of Slurry Infiltration With Coupled CFD–DEM Method.* International Journal for Numerical and Analytical Methods in Geomechanics, 2024. **48**(11): p. 3032-3053.
375. Apaydın, A., *Comparison of Predicted and Actual States in Tunnels: Lessons to Be Learned From the Gerede Water Transmission Tunnel, Turkey.* Quarterly Journal of Engineering Geology and Hydrogeology, 2020. **54**(1).
376. Wannenmacher, H. and M. Schlicke, *Analyses of Pre-Excavation Grouting for TBM Tunnelling.* Geomechanics and Tunnelling, 2023. **16**(1): p. 55-59.
377. Ninić, J., S. Freitag, and G. Meschke, *A Hybrid Finite Element and Surrogate Modelling Approach for Simulation and Monitoring Supported TBM Steering.* Tunnelling and Underground Space Technology, 2017. **63**: p. 12-28.
378. SalamM, C. and O. Kural, *Predictive Modeling and Optimization of TBM Operations: Advanced Techniques Applied to the Jakarta MRT Project.* 2024.
379. Elbaz, K., et al., *Optimization of EPB Shield Performance With Adaptive Neuro-Fuzzy Inference System and Genetic Algorithm.* Applied Sciences, 2019. **9**(4): p. 780.

380. Wang, D., et al., *Behaviors of Main Drive Sealing Structure in a Slurry Shield Machine.* Proceedings of the Institution of Mechanical Engineers Part J Journal of Engineering Tribology, 2025.
381. Zhao, C., et al., *Mechanized Tunneling Induced Ground Movement and Its Dependency on the Tunnel Volume Loss and Soil Properties.* International Journal for Numerical and Analytical Methods in Geomechanics, 2019. **43**(4): p. 781-800.
382. Koch, E., S. Alsamia, and Z. Major, *Finite Element Analysis of Microtunneling Under a Railway Track.* 2024.
383. Zhai, S., Y.D. Song, and H. Tian, *Development of Thrust, Torque, and Power Estimation Model, and Prediction Performance of Earth Pressure Balance Tunnel Boring Machine in Mixed-Face Strata.* Applied Sciences, 2024. **14**(13): p. 5887.
384. Jones, D. and B.T. McGinnity, *Editorial.* Proceedings of the Institution of Civil Engineers - Geotechnical Engineering, 2010. **163**(3): p. 107-108.
385. Yang, Y., et al., *Data-augmented Machine Learning for Risk Management of Tunnel Boring Machine Jamming Considering Coupled Geological Factors.* Computer-Aided Civil and Infrastructure Engineering, 2025. **40**(27): p. 5010-5026.
386. Hassan, S.A., et al., *Predictive Models to Evaluate the Interaction Effect of Soil-Tunnel Interaction Parameters on Surface and Subsurface Settlement.* Civil Engineering Journal, 2022. **8**(11): p. 2424-2444.
387. Huang, S., G.P. Korfiatis, and R.L. Sousa, *Generalization Challenges and Strategies in Tunnel Boring Machine Performance Prediction.* Civil Engineering Design, 2025. **7**(2): p. 63-84.
388. Vivek, D., *Study of Pre Conditional Survey of Existing Building During Tunneling Process.* International Journal of Engineering Research And, 2020. **V9**(06).
389. Adomako, S., et al., *Chief Executive Officers' Sustainability Orientation and Firm Environmental Performance: Networking and Resource Contingencies.* Business Strategy and the Environment, 2021. **30**(4): p. 2184-2193.
390. Wang, F., et al., *Rock Breaking Performance of TBM Disc Cutter Assisted by High-Pressure Water Jet.* Applied Sciences, 2020. **10**(18): p. 6294.
391. Lithian, *TBM Cutter.* n.d., Lithian.
392. Lin, L., Y. Xia, and D. Wu, *Multiobjective Optimization Design for Structural Parameters of TBM Disc Cutter Rings Based on FAHP and SAMPGA.* Advances in Civil Engineering, 2019. **2019**(1).
393. Wang, L., et al., *Disc Cutter Wear Prediction for a Hard Rock TBM Cutterhead Based on Energy Analysis.* Tunnelling and Underground Space Technology, 2015. **50**: p. 324-333.
394. Rostami, J., *Hard Rock TBM Cutterhead Modeling for Design and Performance Prediction.* Geomechanics and Tunnelling, 2008. **1**(1): p. 18-28.
395. Gao, M., C. Li, and X. Shi, *The Mechanism of TBM Cutter Wear in Mixed Ground.* Advanced Materials Research, 2012. **446-449**: p. 3341-3345.
396. Chen, S., et al., *Investigation on the Rock-Fragmentation Process of Conical-Shaped TBM Cutterhead in Extremely Hard Rock Ground.* Stavební Obzor - Civil Engineering Journal, 2023. **33**(3): p. 409-425.
397. Wang, Q., et al., *Investigation of the Influence of Cutter Geometry on the Cutting Forces in Soft–Hard Composite Ground by Tunnel Boring Machine Cutters.* Processes, 2024. **12**(10): p. 2243.

398. We, S., et al., *Disc Cutters' Layout Design of the Full-Face Rock Tunnel Boring Machine (TBM) Using a Cooperative Coevolutionary Algorithm*. Journal of Mechanical Science and Technology, 2011. **25**(2): p. 415-427.
399. Zhang, Q., et al., *Dynamic Rock-Breaking Process of TBM Disc Cutters and Response Mechanism of Rock Mass Based on Discrete Element*. Advances in Civil Engineering, 2022. **2022**(1).
400. Plácido, J.C.R. and J. Friant, *The Disc Bit - A Tool for Hard-Rock Drilling*. Spe Drilling & Completion, 2004. **19**(04): p. 205-211.
401. Chen, P., et al., *Comparative Study on the Cutting Performance of Super-Large Diameter Atmospheric Cutterhead and Pressurized Cutterhead Under the Same Stratum*. Science Progress, 2025. **108**(4).
402. Chakeri, H., et al., *Experimental and Numerical Investigation of the TBM Disc Cutter Wear Using a New Tunnel Boring Machine Laboratory Simulator*. Heliyon, 2024. **10**(17): p. e37148.
403. Fear, M., *How to Improve Rate of Penetration in Field Operations*. Spe Drilling & Completion, 1999. **14**(01): p. 42-49.
404. Li, X., D. Yuan, and Q. Huang, *Cutterhead and Cutting Tools Configurations in Coarse Grain Soils*. The Open Construction and Building Technology Journal, 2017. **11**(1): p. 182-199.
405. Alwaar, A., et al., *Laboratory Study on the Efficiency of Disc Cutter Drilling With the Integration of a Vibration-Assisted Rotary Drilling (PVARD) Tool*. 2025.
406. Nour, M., S.K. Elsayed, and O. Mahmoud, *A Supervised Machine Learning Model to Select a Cost-Effective Directional Drilling Tool*. Scientific Reports, 2024. **14**(1).
407. Zhang, H. and C. Qu, *Research on Tunnel-Boring Machine Main Bearing Fatigue Damage and Vibration Response*. Metals, 2023. **13**(4): p. 650.
408. Dupont, P.E. and S. Lampman, *Friction and Wear of Rolling-Element Bearings*. 2017: p. 836-864.
409. Lin, L., et al., *Dynamic Characteristics Analysis With Multi-Directional Coupling in a TBM Mainframe*. Chinese Journal of Mechanical Engineering, 2019. **32**(1).
410. Fu, X., M. Wu, and L. Zhang, *Probabilistic Life Prediction of Tunnel Boring Machine Under Wearing Conditions With Incomplete Information*. Buildings, 2022. **12**(11): p. 1959.
411. Chen, Y., G. Gong, and X. Zhou. *A Novel Control Strategy of Gripper and Thrust System of TBM Based on Model Analysis*. in *Advances in Asian Mechanism and Machine Science*. 2022. Cham: Springer International Publishing.
412. Xia, W., et al., *Study on the Construction Mechanical Response of the Double-Shield TBMs of a Long Diversion Tunnel Crossing a Fault Fracture Zone*. Iop Conference Series Earth and Environmental Science, 2020. **570**(5): p. 052036.
413. Dibavar, B., M.H. Ahmadi, and S.M. Davarpanah, *3D Numerical Investigation of Ground Settlements Induced by Construction of Istanbul Twin Metro Tunnels With Special Focus on Tunnel Spacing*. Periodica Polytechnica Civil Engineering, 2019.
414. Erçelebi, S.G., H. Çopur, and İ. Ocak, *Surface Settlement Predictions for Istanbul Metro Tunnels Excavated by EPB-TBM*. Environmental Earth Sciences, 2010. **62**(2): p. 357-365.
415. Zhang, G., et al., *Vibration Characteristics of Double-Shield TBM Cutterhead Under Rock–Machine Interaction Excitation*. Buildings, 2025. **15**(11): p. 1824.

416. Song, Z. and X. Liu, *TBM Adaptability Analysis in Small-Radius Curved Tunnels in Kurkar Strata in Israel*. Proceedings of the Institution of Civil Engineers - Civil Engineering, 2024. **177**(4): p. 175-186.

417. Jun, L., et al., *Based on the Engineering Practice of Selection of the Dual Mode (TBM - EPB) and Adaptive Effect Evaluation Analysis*. Journal of Physics Conference Series, 2023. **2519**(1): p. 012006.

418. Yue, M. and L. Guo, *Double Closed-Loop Adaptive Rectification Control of a Shield Tunneling Machine With Hydraulic Actuator Dynamics Subject to Saturation Constraint*. Journal of Vibration and Control, 2014. **22**(2): p. 309-319.

419. Yang, S., et al., *Experimental Study on the Stress and Failure Mode of a Shield Tunnel Lining Under Asymmetric Thrust*. Transportation Research Record Journal of the Transportation Research Board, 2024. **2678**(8): p. 1104-1118.

420. Yang, H., et al., *Earth Pressure Balance Control for EPB Shield*. Science China Technological Sciences, 2009. **52**(10): p. 2840-2848.

421. Yang, Z., et al., *Numerical Investigation on the Spewing Mechanism of Earth Pressure Balance Shield in a High-pressure Water-rich Sand Stratum*. Deep Underground Science and Engineering, 2023. **2**(1): p. 74-87.

422. Li, X. and X.J. Su, *Shield's EPB Simulation Device Based on Fluid Power Transmission Principle*. Key Engineering Materials, 2011. **480-481**: p. 745-750.

423. Steiner, W., *Erkennen Geotechnischer Vorgänge an Der Ortsbrust Als Schlüssel Zum Erfolgreichen Tunnelvortrieb*. Geomechanics and Tunnelling, 2021. **14**(5): p. 557-568.

424. Hu, Y., S. D, and Z. Z.X, *A Discrete Numerical Approach for Modelling Face Stability in Slurry Shield Tunnelling*. 2010.

425. Khrebtova, O.A. and N.V. Zachepa, *Analysis of Electric Drive Systems for Performance of Driving and Starting Under Load*. Electrical Engineering and Power Engineering, 2020(4): p. 18-25.

426. Xing, T.Y., et al., *Speed–Pressure Compound Control of Thrust System Based on the Adaptive Sliding Mode Control Strategy*. Machines, 2025. **13**(3): p. 213.

427. Oliveira, D., M. Thewes, and M.S. Diederichs, *EPB Machine Excavation of Mixed Soils – Laboratory Characterisation*. Geomechanics and Tunnelling, 2019. **12**(4): p. 373-385.

428. Lin, X., X. Zhou, and Y. Yang, *A New Soil Conditioner for Highly Permeable Sandy Gravel Stratum in EPBs*. Applied Sciences, 2021. **11**(5): p. 2109.

429. Zizka, Z., et al., *Influence of Stagnation Gradient for Face Support Calculation in Slurry Shield Tunnelling*. Geomechanics and Tunnelling, 2020. **13**(4): p. 372-381.

430. Aldiamar, F., et al., *Evaluation of Face Support Pressure Prediction for Earth Pressure Balance (EPB) Tunnelling Using Analytical and 3-Dimensional Finite Element Modelling*. International Journal of Technology, 2021. **12**(3): p. 485.

431. Pourmand, S., et al., *Laboratory Studies on Soil Conditioning of Sand in the Mechanized Tunneling*. Journal of Testing and Evaluation, 2018. **48**(5): p. 3658-3672.

432. Huang, F., Z.-l. Li, and T. Ling, *Upper Bound Solution of Safety Factor for Shallow Tunnels Face Using a Nonlinear Failure Criterion and Shear Strength Reduction Technique*. Mathematical Problems in Engineering, 2016. **2016**: p. 1-8.

433. Guo, Z., et al., *Research on Shield Tunnelling Parameters Correlation in Composite Strata*. Advances in Materials Science and Engineering, 2022. **2022**: p. 1-11.

434. Park, B., et al., *An Operating Model for an EPB Shield TBM Simulator by the Correlation Analysis of Operational Actions and Mechanical Responses.* Applied Sciences, 2021. **11**(23): p. 11443.

435. Kang, M., et al., *Transient Dynamic Analysis of Cutter-Head Loads: Effects of Different Cutter Spacings During Excavation.* Frontiers in Earth Science, 2025. **12**.

436. Han, M., Z. Cai, and C. Qu, *On the Loads for Strength Design of Cutterhead of Full Face Rock Tunnel Boring Machine.* Chinese Journal of Mechanical Engineering, 2019. **32**(1).

437. Huo, J., et al., *The Influence of Front Support on Vibration Behaviors of TBM Cutterhead Under Impact Heavy Loads.* Applied Mechanics and Materials, 2014. **541-542**: p. 641-644.

438. Wang, Y., et al., *Industrial Data Denoising via Low-Rank and Sparse Representations and Its Application in Tunnel Boring Machine.* Energies, 2022. **15**(10): p. 3525.

439. Xu, P., et al., *Progressive Failure Characteristics and Failure Symptoms of Straight-Walled Arched Sandstone Tunnels.* Acadlore Transactions on Geosciences, 2023. **2**(1): p. 46-57.

440. Samadi, H. and J. Hassanpour, *Assessment of Geotechnical Parameters of Fine-Grained and Coarse-Grained Soil Using Actual EPB-TBM Operating Data.* 2023.

441. Wang, P., et al., *Prediction of Axis Attitude Deviation and Deviation Correction Method Based on Data Driven During Shield Tunneling.* Ieee Access, 2019. **7**: p. 163487-163501.

442. Li, X., et al., *Criteria for Cutting Head Clogging Occurrence During Slurry Shield Tunneling.* Applied Sciences, 2022. **12**(3): p. 1001.

443. Shan, F., et al., *Applications of Machine Learning in Mechanised Tunnel Construction: A Systematic Review.* Eng—advances in Engineering, 2023. **4**(2): p. 1516-1535.

444. Liu, J., et al., *Prediction of Existing Tunnel Deformation Induced by Shield Undercrossing Based on BO-Bi-LSTM.* 2023: p. 122.

445. Liu, M. and Q. Gan, *Applied Research of Comprehensive Advance Geological Prediction in Daluoshan Diversion Tunnel.* 2023.

446. Chen, Z., et al., *Study of the Influence of Spatial Effects on the Ground Surface and Buildings in a Two-Lane Tunnel Boring.* Heliyon, 2024. **10**(18): p. e37667.

447. Samadi, H., J. Hassanpour, and J. Rostami, *Prediction of Earth Pressure Balance for EPB-TBM Using Machine Learning Algorithms.* International Journal of Geo-Engineering, 2023. **14**(1).

448. Matta-Diaz, D.M., et al., *Soil Conditions and Shield Tunneling Viability for Bogotá Metro Line 1.* Ingeniería E Investigación, 2023. **43**(2): p. e99197.

449. Wang, Y., et al., *Identification of Disc Cutter Wear via Operation Parameters Combined With Vibration Data: A Case Study.* International Journal for Numerical and Analytical Methods in Geomechanics, 2024. **49**(1): p. 256-279.

450. Wang, S., et al., *A Jamming Risk Warning Model for TBM Tunnelling Based on Bayesian Statistical Methods.* 2024.

451. Shao, C., et al., *Indirect Adaptive Robust Trajectory Tracking Control of Hard Rock TBM With Load Variation of Tunneling Face.* Chinese Journal of Mechanical Engineering, 2019. **32**(1).

452. Huang, T., et al., *Force Analysis of an Open TBM Gripping–thrusting–regripping Mechanism.* Mechanism and Machine Theory, 2016. **98**: p. 101-113.

453. Abate, G., et al., *An Early-Warning System to Validate the Soil Profile During TBM Tunnelling*. Geosciences, 2022. **12**(3): p. 113.
454. Sissins, S. and C. Paraskevopoulou, *Assessing TBM Performance in Heterogeneous Rock Masses*. Bulletin of Engineering Geology and the Environment, 2021. **80**(8): p. 6177-6203.
455. Cardu, M., et al., *Performance Analysis of Tunnel Boring Machines for Rock Excavation*. Applied Sciences, 2021. **11**(6): p. 2794.
456. Yang, J., et al., *Numerical Analysis of Internal Erosion Impact on Underground Structures: Application to Tunnel Leakage*. Geomechanics for Energy and the Environment, 2022. **31**: p. 100378.
457. Akkerman Inc., *Operator's Manual: Tunnel Boring machine 720 Series II, Boring Head S/N: F23100F, publication No. 050075A*. 2012, Brownsdale, MN: Akkerman Inc.,.
458. Ma, Z., et al., *A Study on the Rock-Breaking Characteristics of an Arcing-Blade Cutter Under Different Cutting Parameters*. Applied Sciences, 2024. **14**(3): p. 1241.
459. Jiang, Y., et al., *Correlation Analysis of Machine Performance Parameters and Rock Mass Properties During Tunnel Boring Machine Excavation Process: A Case Study in Yin-Song Headrace Tunnel, China*. Journal of Testing and Evaluation, 2021. **50**(6): p. 2883-2900.
460. Motamedi, M.H., et al., *Numerical Investigation of Temperature Distribution on a Tunnel Boring Machine Disc Cutter Affected by Geomechanically and Excavation Parameters*. Mining Technology Transactions of the Institutions of Mining and Metallurgy, 2024. **133**(1): p. 17-30.
461. Huang, J., et al., *Selection Method and Application of Dual-Mode TBM in Composite Strata: A Case Study of Shenzhen Metro Line 14*. Advances in Civil Engineering, 2025. **2025**(1).
462. Rispoli, A., A. Ferrero, and M. Cardu, *From Exploratory Tunnel to Base Tunnel: Hard Rock TBM Performance Prediction by Means of a Stochastic Approach*. Rock Mechanics and Rock Engineering, 2020. **53**(12): p. 5473-5487.
463. Erharter, G., et al., *Managing Squeezing Rock Mass With TBM Data Analysis: Rail Link Rishikesh – Karnaprayag (India)*. 2025.
464. Bianchi, G.W., *Looking Inside TBM Parameters: Feedback From the Tunnel 4 Case Study, Angat Water Transmission Improvement Project (Philippines)*. Iop Conference Series Earth and Environmental Science, 2021. **833**(1): p. 012159.
465. Rolando, L., et al., *Isolation and Characterization in a Soil Conditioned With Foaming Agents of a Bacterial Consortium Able to Degrade Sodium Lauryl Ether Sulfate*. Frontiers in Microbiology, 2020. **11**.
466. Saczynski, T.M., M. Pearce, and A. Elioff, *Monitoring Earth Pressure Balance Tunnels in Los Angeles*. 2007: p. 1-12.
467. Tokgöz, N., *An Assessment Method for Fine-grained Sedimentary Materials Excavated by EPB TBM*. Geomechanics and Tunnelling, 2016. **9**(4): p. 326-337.
468. Zhang, X., et al., *Study on Optimum Free-Face Condition of Cutting Hard Rock With Tunnel Boring Machine Hob*. Periodica Polytechnica Civil Engineering, 2021.
469. Naghadehi, M.Z. and R. Mikaeil, *Optimization of Tunnel Boring Machine (TBM) Disc Cutter Spacing in Jointed Hard Rock Using a Distinct Element Numerical Simulation*. Periodica Polytechnica Civil Engineering, 2016.

470. Gao, C., et al., *Study on the Influence of TBM Disc Cutter on the Penetration of Extremely Hard Rock in Zijing Tunnel.* Electronic Journal of Structural Engineering, 2024. **24**(2): p. 30-37.
471. Lin, L., et al., *Comparison Study on Rock Breaking Characteristics of Disc Cutters Under Coupled Static–dynamic Loads and Static Loads.* Comptes Rendus Mécanique, 2023. **351**(G1): p. 1-15.
472. Labra, C., J. Rojek, and E. Oñate, *Discrete/Finite Element Modelling of Rock Cutting With a TBM Disc Cutter.* Rock Mechanics and Rock Engineering, 2016. **50**(3): p. 621-638.
473. Liu, H., S.Q. Kou, and P. Lindqvist, *Numerical Simulation of the Fracture Process in Cutting Heterogeneous Brittle Material.* International Journal for Numerical and Analytical Methods in Geomechanics, 2002. **26**(13): p. 1253-1278.
474. Sabri, M.N., et al., *Evaluation of the Effect of TBM Disc Cutter Wear on Rock Cutting Efficiency.* Geomechanics and Tunnelling, 2024. **17**(2): p. 146-156.
475. Zhao, D., et al., *Rock Crushing Analysis of TBM Disc Cutter Assisted by Ultra-High-Frequency Loading.* Shock and Vibration, 2022. **2022**: p. 1-17.
476. Marji, M.F., H.H. Nasab, and A.H. Morshedi, *Numerical Modeling of Crack Propagation in Rocks Under TBM Disc Cutters.* Journal of Mechanics of Materials and Structures, 2009. **4**(3): p. 605-627.
477. Huo, J., et al., *The Coupling Dynamic Analysis and Field Test of TBM Main System Under Multipoint Impact Excitation.* Shock and Vibration, 2015. **2015**: p. 1-14.
478. Guo, Z., et al., *Rock-Breaking Performance of Cutters of Tunnel Boring Machine in Broken Coal Rock Formation.* International Journal of Safety and Security Engineering, 2020. **10**(1): p. 17-25.
479. Balcı, C. and N. Bilgin, *Correlative Study of Linear Small and Full-Scale Rock Cutting Tests to Select Mechanized Excavation Machines.* International Journal of Rock Mechanics and Mining Sciences, 2007. **44**(3): p. 468-476.
480. Litian, *The Working Principle of TBM Cutter Disc Used in Tunnel Engineering*. n.d., Litian.
481. Lin, Q., et al., *Acoustic Emission Characteristics During Rock Fragmentation Processes Induced by Disc Cutter Under Different Water Content Conditions.* Applied Sciences, 2019. **9**(1): p. 194.
482. Jiang, H., et al., *The Mechanism and Performance of Rock Breakage by Undercutting Disc Cutter With Advanced Slotting.* Scientific Reports, 2023. **13**(1).
483. Cho, J.-W., et al., *Optimum spacing of TBM disc cutters: A numerical simulation using the three-dimensional dynamic fracturing method.* Tunnelling and Underground Space Technology - TUNN UNDERGR SPACE TECHNOL, 2010. **25**: p. 230-244.
484. Bao, Y., J. Cheng, and J. Bin-song, *Breakage Mechanism of Layered Sandstone Penetrated by TBM Disc Cutter.* Shock and Vibration, 2021. **2021**(1).
485. Wang, Y. and Z. Liang, *Numerical Study of the Rock Fragmentation by Three TBM Cutters.* Advanced Materials Research, 2012. **594-597**: p. 3-7.
486. Jiang, Y., et al., *Numerical Study on the Rock Breaking Mechanism of High-Pressure Water Jet-Assisted TBM Digging Technique Based on 2d-Dem Modelling.* Frontiers in Earth Science, 2023. **10**.

487. González, C., M. Arroyo, and A. Gens, *Thrust and Torque Components on Mixed-Face EPB Drives*. Tunnelling and Underground Space Technology, 2016. **57**: p. 47-54.
488. Seyedrezaei, M. and D. Tumaç, *Predicting Excavation Performance of EPB TBM by Using Soil Properties: A Case Study of Dudullu Bostanci Metro Project*. 2024.
489. Huang, X., et al., *Multi-Step Combined Control Technology for Karst and Fissure Water Inrush Disaster During Shield Tunneling in Spring Areas*. Frontiers in Earth Science, 2021. **9**.
490. Tokgöz, N., I.S. Binen, and E. Avunduk, *An Evaluation of Fine Grained Sedimentary Materials in Terms of Geotechnical Parameters Which Define and Control Excavation Performance of EPB TBM's*. Tunnelling and Underground Space Technology, 2015. **47**: p. 211-221.
491. Sun, Z., et al., *A Thermomechanical Model of Coarse-Grained Soils With Non-Orthogonal Flow Rule*. International Journal for Numerical and Analytical Methods in Geomechanics, 2025. **49**(5): p. 1437-1447.
492. Liu, J.Q., J. Ren, and W. Guo, *Thrust and Torque Characteristics Based on a New Cutter-Head Load Model*. Chinese Journal of Mechanical Engineering, 2015. **28**(4): p. 801-809.
493. Pan, Y. and X. Zhu, *Application of HMM and Ensemble Learning in Intelligent Tunneling*. Mathematics, 2022. **10**(10): p. 1778.
494. Zhang, H., et al., *Research on Rock-Breaking Characteristics of Cutters and Matching of Cutter Spacing and Penetration for Tunnel Boring Machine*. Buildings, 2024. **14**(6): p. 1757.
495. Bazargan, S., et al., *Analysis of the Performance of Cutting Tools of Tunnel Boring Machine (TBM) in Silty-Sand Soils Using Artificial Neural Network (ANN) – Case Study: Tabriz Metro Line 2 Project*. Asian Journal of Water Environment and Pollution, 2022. **19**(2): p. 71-78.
496. Hua, J., et al., *Comparative Study of The rock-Breaking Mechanism of a Disc Cutter and Wedge Tooth Cutter in Qingdao Andesite Strata*. 2022.
497. Gao, P., et al., *Research on Dynamic Characteristics of Split TBM Cutterhead Based on Similarity Theory*. Destech Transactions on Computer Science and Engineering, 2019(aicae).
498. Long, L., et al., *Multi-Step Real-Time Prediction of Hard-Rock TBM Penetration Rate Combining Temporal Convolutional Network and Squeeze-and-Excitation Block*. Scientific Reports, 2024. **14**(1).
499. Liu, J., et al., *Evaluation of TBM Cutter Wear in Granite and Developing a Cutter Life Prediction Model for Face Cutters Based on Field Data: A Case Study*. Buildings, 2024. **14**(8): p. 2453.
500. Hassanpour, J., et al., *TBM Performance and Disc Cutter Wear Prediction Based on Ten Years Experience of TBM Tunnelling in Iran*. Geomechanics and Tunnelling, 2015. **8**(3): p. 239-247.
501. Su, H., et al., *Force and Wear Prediction Model of Disc Cutter Under Compression Shear Failure Mode Based on Dense Core–cavity Expansion Theory (U-713)*. Iop Conference Series Earth and Environmental Science, 2024. **1333**(1): p. 012042.
502. Macias, F.J., J. Nærland, and N. Espallargas, *Cutter Wear Mechanisms in Hard Rock Tunnel Boring*. Geomechanics and Tunnelling, 2018. **11**(2): p. 123-130.

503. Li, J., et al., *Disc Cutter Wear Prediction Based on the Friction Work Principle.* Transactions of the Canadian Society for Mechanical Engineering, 2021. **45**(3): p. 384-395.
504. Liu, J., et al., *Analysis and Enlightenment on the Relationships Between Two Kinds of Cutter Life Evaluation Indexes and Installation Radius: A Case Study.* Buildings, 2024. **14**(6): p. 1523.
505. Zhang, X.-P., et al., *An Experimental Study on Cutting Tool Hardness Optimization for Shield TBMs During Dense Fine Silty Sand Ground Tunneling.* Bulletin of Engineering Geology and the Environment, 2021. **80**(9): p. 6813-6826.
506. Hamzaban, M.-T., et al., *Wear of Cutting Tools in Hard Rock Excavation Process: A Critical Review of Rock Abrasiveness Testing Methods.* Rock Mechanics and Rock Engineering, 2022. **56**.
507. Ko, T.Y. and S.S. Lee, *Effect of Rock Abrasiveness on Wear of Shield Tunnelling in Bukit Timah Granite.* Applied Sciences, 2020. **10**(9): p. 3231.
508. Han, M., et al., *Study on Grinding Performance During Ultrasonic Vibration-Assisted Grinding Ultra-High Strength Steels.* The International Journal of Advanced Manufacturing Technology, 2023. **128**(7-8): p. 3673-3684.
509. Erharter, G.H. and T.F. Hansen, *Towards Optimized TBM Cutter Changing Policies With Reinforcement Learning.* Geomechanics and Tunnelling, 2022. **15**(5): p. 665-670.
510. Galler, R., D. Handke, and M. Nolden, *The Determination of Performance - and Payment -relevant Parameters in TBM Tunnelling – State of the Technology and Outlook / Ermittlung Leistungs - Und Vergütungsrelevanter Parameter Für TVM - Vortriebe – Stand Der Technik Und Ausblick.* Geomechanics and Tunnelling, 2014. **7**(5): p. 511-519.
511. Wang, J., et al., *A Multi-Channel Real-Time Detection Method for Tunnel Boring Machine Cutter Wear Based on Chirped Fiber Bragg Gratings.* Aip Advances, 2019. **9**(1).
512. Ji, Z., et al., *Cutterhead Mud-Caking Detection Method and Application Based on Cutter Wear and Temperature Measurement.* Journal of Advanced Mechanical Design Systems and Manufacturing, 2019. **13**(4): p. JAMDSM0088-JAMDSM0088.
513. Schlicke, M., et al., *The Thrust-Penetration Gradient: Decoding Geomechanical Dynamics and TBM Response.* 2024.
514. Ling, J., et al., *Research on Rock Breaking Mechanism and Load Characteristics of TBM Cutter Based on Discrete Element Method.* Advances in Mechanical Engineering, 2021. **13**(2).
515. Wu, Y., et al., *Rock Abrasiveness Prediction Based on Multi-Source Physical, Mechanical and Mineralogical Properties.* Bulletin of Engineering Geology and the Environment, 2024. **83**(10).
516. Zhou, X., et al., *Performance Evaluation of TBM Using an Improved Load Prediction Model.* Machines, 2023. **11**(2): p. 141.
517. Khoshzaher, E., et al., *The Prediction of Epb-TBM Performance Using Firefly Algorithms and Particle Swarm Optimization.* Rudarsko-Geološko-Naftni Zbornik, 2023. **38**(5): p. 79-86.
518. Wang, Q., et al., *Design and Experimental Verification of a New Type of Cutter for Shield Machine Cutting Underground Plastic Drainage Boards.* Matéria (Rio De Janeiro), 2024. **29**(2).

519. Gu, D.W., et al., *Theoretical and Experimental Investigation on Controlled Synchronization of Four Co-Rotating Coupled Exciters Driven by Induction Motors in a Vibrating System.* Proceedings of the Institution of Mechanical Engineers Part C Journal of Mechanical Engineering Science, 2019. **233**(13): p. 4556-4576.
520. Zhabin, A.B., A.V. Polyakov, and E.A. Averin, *Scale Factors for Conversion of Forces on Disc Cutters for the Main Domestic and Foreign Methods.* Mining of Mineral Deposits, 2017. **11**(3): p. 50-55.
521. Giovanni, A.D., C. Todaro, and M. Cardu, *A Statistical Approach for the Correlation Between Cerchar Abrasivity Index and Uniaxial Compressive Strength of Rocks.* Geomechanics and Tunnelling, 2023. **16**(4): p. 422-430.
522. Pandey, P., et al., *Influence of Geology on Tunnel Boring Machine Performance: A Review.* Journal of Mining and Metallurgy a Mining, 2020. **56**(1): p. 1-14.
523. Samadi, H. and E. Farrokh, *Utilization of Rock Mass Parameters for Performance Prediction of Rock TBMs Using Machine Learning Algorithms.* SRPH Journal of Fundamental Sciences and Technology, 2021.
524. Kaspar, M., et al., *Is Soft Rock Also Non-Abrasive Rock? An Evaluation From Lab Testing Campaigns.* Iop Conference Series Earth and Environmental Science, 2023. **1124**(1): p. 012019.
525. Aderhold, M., et al., *Comparison of Different Methods to Characterise the Abrasivity Potential and Mechanical Properties of Carbonates With Respect to Its Relevance for Practical Purposes in Excavation Technologies.* Iop Conference Series Earth and Environmental Science, 2023. **1124**(1): p. 012044.
526. Alber, M., et al., *ISRM Suggested Method for Determining the Abrasivity of Rock by the CERCHAR Abrasivity Test.* Rock Mechanics and Rock Engineering, 2013. **47**(1): p. 261-266.
527. Ma, T., et al., *Research on Prediction of TBM Performance of Deep-Buried Tunnel Based on Machine Learning.* Applied Sciences, 2022. **12**(13): p. 6599.
528. Ji, Y., et al., *Temperature-Dependent Abrasivity of Bukit Timah Granite and Implications for Drill Bit Wear in Thermo-Mechanical Drilling.* Acta Geotechnica, 2020. **16**(3): p. 885-893.
529. Lu, G., et al., *Influences of Microwave Irradiation on the Physicomechanical Properties and Cerchar Abrasivity Index of Rocks.* Geofluids, 2023. **2023**: p. 1-10.
530. Kaspar, M., et al., *Hardness, Strength and Abrasivity of Rocks: Correlations and Predictions.* Geomechanics and Tunnelling, 2023. **16**(2): p. 184-192.
531. Du, S. and J. Tang, *State-of-the-Art Review on Failure Mechanism and Waterproofing Performance of Linings for Shield Tunnels.* Advances in Civil Engineering, 2022. **2022**(1).
532. Henzinger, M.R., et al., *Bedding Improvement of Lining Segments Using Geotextile Tubes.* Geomechanics and Tunnelling, 2018. **11**(5): p. 589-598.
533. Delgado, O.A., *Structural Response of Precast Concrete Segmental Tunnel Linings.*
534. Avanaki, M.J., *Effects of Hybrid Steel Fiber Reinforced Composites on Structural Performance of Segmental Linings Subjected to TBM Jacks.* Structural Concrete, 2019. **20**(6): p. 1909-1925.
535. Guo, W., et al., *Investigation on the shear performance of circumferential joints of segments in super-large cross-section shield tunnels.* Structural Concrete, 2023. **24**: p. 4978-4992.

536. Shalabi, F.I., E.J. Cording, and S.L. Paul, *Sealant Behavior of Gasketed Segmental Tunnel Lining – Conceptual Model*. Geomechanics and Tunnelling, 2016. **9**(4): p. 345-355.

537. Liu, J., et al., *Repairing a Shield Tunnel Damaged by Secondary Grouting*. Tunnelling and Underground Space Technology, 2018. **80**: p. 313-321.

538. Han, X., P. Oreste, and F. Ye, *The Important Role of Stiffnesses Values of Circular Joints on the Stress State Developed in the Tunnel Segmental Lining*. Geomechanics and Geophysics for Geo-Energy and Geo-Resources, 2023. **9**(1).

539. Kikkawa, N., et al., *Unstable Behavior of Segmental Ring Under Various Pressures and Its Discrete Element Simulation*. Industrial Health, 2019. **56**(6): p. 475-484.

540. Liu, Y. and Z. Hong-bo, *Analysis of Leakage Paths Induced by Longitudinal Differential Settlement of the Shield-Driven Tunneling*. Destech Transactions on Engineering and Technology Research, 2017(iceta).

541. Zhou, J., et al., *Airtightness Sealing Performance and Failure Mechanism of the Circumferential Joint in the Low-to-Vacuum Maglev Segment Tunnel*. Applied Sciences, 2023. **13**(13): p. 7535.

542. Jeon, K., et al., *Use of BIM Technology for Optimization and Virtual Build of TBM Tunnels*. Geomechanics and Tunnelling, 2022. **15**(2): p. 175-181.

543. Wang, Z.N., et al., *Mechanical, Flame-retarding, and Creep-recovery Proprieties of Ethylene–propylene–diene Monomer Enhanced With Nano-hydroxide for Undersea Tunnel Sealing Gasket*. Journal of Applied Polymer Science, 2020. **138**(10).

544. Yue, T., et al., *Chain Dynamics Evolution of Ethylene-propylene-diene Monomer in Response to Hot Humid and Salt Fog Environment*. Journal of Applied Polymer Science, 2021. **138**(30).

545. El-Shereafy, E., et al., *Effect of Gamma Radiation on the Properties of Ethylene Propylene Diene Monomer/Styrene Butadiene Rubber/Waste Polyethylene/Clay Nanocomposites*. Journal of Radioanalytical and Nuclear Chemistry, 2015. **307**(2): p. 1325-1333.

546. Wang, Y., Y. Yang, and L. Wang, *Effect of Thermo-Oxidation on the Dynamical and Physical Properties of Ethylene-Propylene-Diene Monomer Elastomer*. Journal of Zhejiang University Science A, 2021. **22**(7): p. 499-513.

547. Deng, K., B. Meng, and C. Xiang, *Adaptability to Stratum Characteristics for Layout of Thrust System in Tunneling Machines Based on Variation Coefficient*. Advances in Mechanical Engineering, 2016. **8**(12).

548. Deng, K., et al., *Numerical Analysis on Non-Uniform Thrust System in EPB Shield Machine Applied in Beijing Metro Line* 6. Ieee Access, 2019. **7**: p. 171898-171906.

549. Li, F., et al., *Numerical Analysis of the Single-Directionally Misaligned Segment Behavior of Hydraulic TBM Tunnel*. Buildings, 2024. **14**(7): p. 2198.

550. Amjad, R., Y. Zhang, and X. Liu, *An Analytical Solution for Shear Bearing Capacity of Circumferential Joints of Precast Concrete Segmental Tunnel Linings Considering Dowel Action*. Structural Concrete, 2024. **25**(6): p. 4918-4937.

551. Li, X., et al., *A Progressive Model to Simulate the Full Mechanical Behavior of Concrete Segmental Lining Longitudinal Joints*. Engineering Structures, 2015. **93**: p. 97-113.

552. Moradi, A., M. Soltani, and A.A. Tasnimi, *A Simplified Constitutive Model for Dowel Action Across RC Cracks*. Journal of Advanced Concrete Technology, 2012. **10**(8): p. 264-277.

553. Hashim, D.T., et al., *The Performance Evaluation of Circular Flange Bolted Connection in Ultra High Performance Fiber Reinforced Concrete Segmented Communication Tower.* Periodica Polytechnica Civil Engineering, 2019.
554. Li, Z., K. Soga, and P. Wright, *Long-Term Performance of Cast-Iron Tunnel Cross Passage in London Clay.* Tunnelling and Underground Space Technology, 2015. **50**: p. 152-170.
555. Liu, X., et al., *Prototype Test of the Mechanical Behavior and Failure Mechanism of Segment Structure With Distributed Mortises and Tenons.* Structural Concrete, 2021. **23**(5): p. 2837-2851.
556. We, W., C. Fang, and J. Liu, *Large Size Superelastic SMA Bars: Heat Treatment Strategy, Mechanical Property and Seismic Application.* Smart Materials and Structures, 2016. **25**(7): p. 075001.
557. Rowe, R.K. and K.M. Lee, *Subsidence Owing to Tunnelling. II. Evaluation of a Prediction Technique.* Canadian Geotechnical Journal, 1992. **29**(6): p. 941-954.
558. Barri, F., et al., *Investigation of the Feasibility of Increasing the Tail-Grouting Zone During Mechanized Tunneling in Sandy Soils.* Periodica Polytechnica Civil Engineering, 2023. **68**(2): p. 435-444.
559. Ma, J., et al., *Pressure Model Study on Synchronous Grouting in Shield Tunnels Considering the Temporal Variation in Grout Viscosity.* Applied Sciences, 2023. **13**(18): p. 10437.
560. Neu, G.E., et al., *Tunnel Linings.* 2023: p. 253-327.
561. Song, P., et al., *Study on Instability Mechanism and Support Scheme of the Tunnel Face in Carbonaceous Phyllite Stratum Under High Geo-Stress.* Advances in Civil Engineering, 2022. **2022**(1).
562. Li, X., et al., *Intelligent Decision Method for Stability Assessment of Shield Tunnel Based on Multi-Objective Data Mining.* Philosophical Transactions of the Royal Society a Mathematical Physical and Engineering Sciences, 2023. **381**(2254).
563. Zhou, G., et al., *Investigation of Quantitative Evaluation Method and Engineering Application of Shallow Buried Tunnel Face Stability.* Applied Sciences, 2022. **12**(13): p. 6656.
564. Qiang, S., et al., *Analysis of Face Stability for Shallow Shield Tunnels in Sand.* Frontiers in Earth Science, 2023. **11**.
565. Bo, M. and Y. Xiang, *Analysis of the Limit Support Pressure of a Shallow Shield Tunnel in Sandy Soil Considering the Influence of Seepage.* Symmetry, 2020. **12**(6): p. 1023.
566. Wang, L., et al., *Calculation of Limit Support Pressure for EPB Shield Tunnel Face in Water-Rich Sand.* Symmetry, 2019. **11**(9): p. 1102.
567. Xie, T. and K. Han, *Two-Dimensional Numerical Model for Stability Analysis of Tunnel Face Based on Particle Flow Code.* Symmetry, 2021. **13**(1): p. 94.
568. Liu, W., et al., *Face Stability Analysis of Shield Tunnel Using Slip Line Method.* Mathematical Problems in Engineering, 2019. **2019**(1).
569. Feng, J., et al., *Field Monitoring and Analysis of the Stress Variation Relationship Between Fiberglass Bolts and Face Deformation in Tunnels Based on the ADECO-RS Construction Approach.* 2022.
570. Tian, X., et al., *A Theoretical Calculation Method of the Unsupported Span for the Shallow Tunnel in the Soft Stratum.* Advances in Civil Engineering, 2020. **2020**(1).

571. Lee, I.-M., S.-W. Nam, and J.-H. Ahn, *Effect of Seepage Forces on Tunnel Face Stability.* Canadian Geotechnical Journal, 2003. **40**(2): p. 342-350.
572. An, Y., et al., *Earth Pressure Field Modeling for Tunnel Face Stability Evaluation of EPB Shield Machines Based on Optimization Solution.* Discrete and Continuous Dynamical Systems - S, 2020. **13**(6): p. 1721-1741.
573. Zeng, Y., et al., *Geological Challenges of Tunnelling Using Large Slurry Shield Machine in Dense Urban Centre.* Geological Journal, 2024. **59**(9): p. 2496-2509.
574. Liu, W., et al., *Upper Bound Analysis for Estimation of the Influence of Seepage on Tunnel Face Stability in Layered Soils.* Journal of Zhejiang University Science A, 2016. **17**(11): p. 886-902.
575. Zhang, C., K. Han, and D. Zhang, *Face Stability Analysis of Shallow Circular Tunnels in Cohesive–frictional Soils.* Tunnelling and Underground Space Technology, 2015. **50**: p. 345-357.
576. Han, K., et al., *Analysis of Instability Mode and Limit Support Pressure of Shallow Tunnel Face in Sands.* Symmetry, 2020. **12**(12): p. 2067.
577. Togashi, Y., et al., *Numerical Method of Tunnel Face Pressure Estimation Considering Progressive Failure Due to Excavation.* Iop Conference Series Earth and Environmental Science, 2021. **833**(1): p. 012157.
578. Jia, Y., et al., *Critical Face Pressure of a Tunnel Driven by a Shield Machine Considering Seismic Forces and Tunnel Shape Influence.* Buildings, 2024. **14**(6): p. 1760.
579. Zhong, J., et al., *Tunnel Face Stability Considering the Influence of Excess Slurry Pressure.* Sustainability, 2023. **15**(10): p. 8230.
580. Cheng, H., et al., *Influence of Permeability of Filter Cake on Face Stability of Slurry Shield-Driven Tunnels in Sandy Gravel Stratum.* Ce/Papers, 2025. **8**(2): p. 1495-1508.
581. Ling, F., et al., *Effect of Bentonite Slurry on the Function of Foam for Changing the Permeability Characteristics of Sand Under High Hydraulic Gradients.* Canadian Geotechnical Journal, 2022. **59**(7): p. 1061-1070.
582. Wang, Y., X. Li, and H. Jia, *Experimental Study and Simulation Analysis of Slurry Filling Process of Shield Tail Synchronous Grouting.* Mathematical Problems in Engineering, 2022. **2022**: p. 1-16.
583. Mollon, G., D. Dias, and A.H. Soubra, *Rotational Failure Mechanisms for the Face Stability Analysis of Tunnels Driven by a Pressurized Shield.* International Journal for Numerical and Analytical Methods in Geomechanics, 2010. **35**(12): p. 1363-1388.
584. Xia, Y., et al., *Influence of Geological Properties and Operational Parameters on TBM Muck Removal Performance for Yinsong Tunnel.* Geotechnical and Geological Engineering, 2021. **40**(4): p. 2291-2306.
585. Wei, Y., et al., *Effects of Soil Conditioning on Characteristics of a Clay-Sand-Gravel Mixed Soil Based on Laboratory Test.* Applied Sciences, 2020. **10**(9): p. 3300.
586. Zheng, D., et al., *Experimental Study of Foam Infiltration From Foam-Sand Mixture Into Saturated Sand for EPB Shield Tunnelling: A Mechanism Study.* Tunnelling and Underground Space Technology, 2023. **138**: p. 105206.
587. Taqa, A.A., et al., *Characterization of TBM Muck for Construction Applications.* Applied Sciences, 2021. **11**(18): p. 8623.
588. Zierkailedi, B.H.K., et al., *Properties of Slurry Shield Tunnel Sand and Its Application in Large Flow Concrete.* Materials, 2022. **15**(15): p. 5131.

589. Liu, Z., et al., *The Role of Foam in Improving the Workability of Sand: Insights From DEM.* Minerals, 2022. **12**(2): p. 186.
590. Yang, Y., X. Li, and X. Li, *Shear Strength and Compression Coefficient for Conditioned Sand Subjected to Earth Chamber Stress Levels.* Advances in Materials Science and Engineering, 2018. **2018**(1).
591. Zhang, D., et al., *Deep Learning-based Automatic Detection of Muck Types for Earth Pressure Balance Shield Tunneling in Soft Ground.* Computer-Aided Civil and Infrastructure Engineering, 2022. **38**(7): p. 940-955.
592. You, Y., et al., *Research on Intelligent Identification and Resource Utilization of TBM Muck in Breezy Granite Strata.* Iop Conference Series Earth and Environmental Science, 2024. **1337**(1): p. 012060.
593. Lin, L., et al., *Simulation Study of Rock-Breaking Performance of TBM Cone-Shape Cutterhead Under Different Tunneling Parameters in Spiral Tunnel.* Science Progress, 2025. **108**(4).
594. Qin, S., et al., *Pressure Infiltration Behavior and Fluid Loss of Bentonite Slurry: A Comparative Study of Two Bentonite Slurries.* Canadian Geotechnical Journal, 2024. **61**(5): p. 945-958.
595. Gamal, H., et al., *Effect of pH on Rheological and Filtration Properties of Water-Based Drilling Fluid Based on Bentonite.* Sustainability, 2019. **11**(23): p. 6714.
596. Song, K., et al., *Water-Based Bentonite Drilling Fluids Modified by Novel Biopolymer for Minimizing Fluid Loss and Formation Damage.* Colloids and Surfaces a Physicochemical and Engineering Aspects, 2016. **507**: p. 58-66.
597. Dong, H., et al., *Experimental Study on Infiltration of Seawater Bentonite Slurry.* Buildings, 2024. **14**(9): p. 2609.
598. Kasza, P., P. Kulinowski, and J. Zarzycki, *The Influence of an Operating Conditions on the Friction Coefficient in Transportation Machines Drives.* New Trends in Production Engineering, 2020. **3**(1): p. 294-302.
599. Zeng, F., et al., *Dynamic Behaviour of a Conveyor Belt Considering Non-Uniform Bulk Material Distribution for Speed Control.* Applied Sciences, 2020. **10**(13): p. 4436.
600. Gavryukov, A., et al., *Determining the Mechanism for Calculating the Tension of a Working Conveyor Belt During a Change in the Transportation Length.* Eastern-European Journal of Enterprise Technologies, 2024. **2**(7 (128)): p. 56-66.
601. Wang, L., et al., *Research on and Design of an Electric Drive Automatic Control System for Mine Belt Conveyors.* Processes, 2023. **11**(6): p. 1762.
602. Fu, X., *Enhancing TBM Operations Under Complex Geological Conditions Using Data-Driven Methods.*
603. Wu, L.W., et al., *Integration of IoT and Machine Learning for Enhancing Physiological Workload Management of Tunnel Construction Workers.* Computer-Aided Civil and Infrastructure Engineering, 2025. **40**(25): p. 4378-4396.
604. Xu, T., et al., *Effects of Sand and Slurry Characteristics on Pressure Infiltration of Bentonite Slurry Into Sand.* Géotechnique, 2024. **74**(9): p. 935-946.
605. Wang, S., et al., *Effect of Water Pressure on Time-Dependent Permeability Characteristics of Sand Conditioned With Foam and Bentonite Slurry.* Canadian Geotechnical Journal, 2025. **62**: p. 1-14.

606. Nassef, B.M., M. Kohail, and M. Abdellatief, *Compressive Strength of Cement-Bentonite Grouting*. International Journal of Science and Research Archive, 2023. **9**(1): p. 468-473.
607. Shen, S.-Q. and M. Wei, *Hydraulic Conductivity of Polymer-Amended Sand-Bentonite Backfills Permeated With Lead Nitrate Solutions*. Advances in Civil Engineering, 2018. **2018**(1).
608. Bezuijen, A., *The Influence of Grout and Bentonite Slurry on the Process of TBM Tunnelling*. Geomechanics and Tunnelling, 2009. **2**(3): p. 294-303.
609. Ding, Y.-c. and J. Wang, *Numerical Modeling of Ground Response During Diaphragm Wall Construction*. Journal of Shanghai Jiaotong University (Science), 2008. **13**(4): p. 385-390.
610. Zhang, Y., et al., *Properties of Bentonite Slurry Drilling Fluid in Shallow Formations of Deepwater Wells and the Optimization of Its Wellbore Strengthening Ability While Drilling*. Acs Omega, 2022. **7**(44): p. 39860-39874.
611. Li, X.B., et al., *Comparative Evaluation of Stress Distribution and Permeability Characteristics in Bentonite Cutoff Walls Using CPTU and ABAQUS Methods*. Materials, 2025. **18**(16): p. 3919.
612. Shen, X., et al., *Tunnel Boring Machine Positioning Automation in Tunnel Construction*. 2012.
613. Nozawa, T., F. Sato, and N. Nishino, *The Development of "Gyro Runner" a Totally Unmanned Shield Guiding System*. 1993.
614. Ishikawa, K., et al., *Tunnel Cross-Section Measurement System Using a Mobile Mapping System*. Journal of Robotics and Mechatronics, 2009. **21**(2): p. 193-199.
615. Lu, J., et al., *Improved Calibration of IMU Biases in Analytic Coarse Alignment for AHRS*. Measurement Science and Technology, 2016. **27**(7): p. 075105.
616. Kutílek, P., et al., *Volume of Convex Hull: A Technique for Quantifying Human Postural Stability*. Journal of Mechanics in Medicine and Biology, 2016. **16**(02): p. 1650013.
617. Dumrongchai, P., J. Patsadutarn, and C. Satirapod, *Performance Tests of Geodetic Receivers With Tilt Sensors in Obstructed Environments Using the NRTK GNSS Technique*. Journal of Applied Geodesy, 2022. **17**(1): p. 39-52.
618. Zhao, J., et al., *A Data-Driven Framework for Tunnel Geological-Type Prediction Based on TBM Operating Data*. Ieee Access, 2019. **7**: p. 66703-66713.
619. Allum, J.H.J., *Recovery of Vestibular Ocular Reflex Function and Balance Control After a Unilateral Peripheral Vestibular Deficit*. Frontiers in Neurology, 2012. **3**.
620. He, B., et al., *Adaptive-Neuro-Fuzzy-Based Information Fusion for the Attitude Prediction of TBMs*. Sensors, 2020. **21**: p. 61.
621. Shen, X., et al., *Integrated Approach to Machine Guidance and OperationsMonitoring in Tunnel Construction*. 2014.
622. Wu, X., et al., *As-Built Modeling and Visual Simulation of Tunnels Using Real-Time TBM Positioning Data*. 2013: p. 3066-3073.
623. El-Sheimy, N. and A.A. Youssef, *Inertial Sensors Technologies for Navigation Applications: State of the Art and Future Trends*. Satellite Navigation, 2020. **1**(1).
624. Rodriguez, A.M. and N. Khademi, *Evaluation Study of Inertial Positioning in Road Tunnels for Cooperative ITS Applications*. 2023: p. 3837-3844.
625. Zheng, T., et al., *Modeling and Compensation of Inertial Sensor Errors in Measurement Systems*. Electronics, 2023. **12**(11): p. 2458.

626. Onyekpe, U., V. Palade, and S. Kanarachos, *Learning to Localise Automated Vehicles in Challenging Environments Using Inertial Navigation Systems (INS)*. Applied Sciences, 2021. **11**(3): p. 1270.
627. Jiang, Q., et al., *A New Filtering and Smoothing Algorithm for Railway Track Surveying Based on Landmark and IMU/Odometer*. Sensors, 2017. **17**(6): p. 1438.
628. Larey, A., E. Aknin, and I. Klein, *Multiple Inertial Measurement Units–An Empirical Study*. Ieee Access, 2020. **8**: p. 75656-75665.
629. Wu, Z., et al., *Attitude and Gyro Bias Estimation by the Rotation of an Inertial Measurement Unit*. Measurement Science and Technology, 2015. **26**(12): p. 125102.
630. Leng, S., et al., *A Hybrid Data Mining Method for Tunnel Engineering Based on Real-Time Monitoring Data From Tunnel Boring Machines*. Ieee Access, 2020. **8**: p. 90430-90449.
631. Macke, S., et al., *AI-based Anomaly Detection in Tunnel Excavation Data*. Geomechanics and Tunnelling, 2024. **17**(4): p. 312-323.
632. Ding, L., et al., *Safety Management in Tunnel Construction: Case Study of Wuhan Metro Construction in China*. Safety Science, 2014. **62**: p. 8-15.
633. Wu, L.-j., et al., *Real-Time Prediction of Tunnel Face Conditions Using XGBoost Random Forest Algorithm*. Frontiers of Structural and Civil Engineering, 2023. **17**(12): p. 1777-1795.
634. Latif, K., A. Sharafat, and J. Seo, *Digital Twin-Driven Framework for TBM Performance Prediction, Visualization, and Monitoring Through Machine Learning*. Applied Sciences, 2023. **13**(20): p. 11435.
635. Yun, J.-S., et al., *Prediction of Settlement Due to Shield TBM Tunneling Based on Three-Dimensional Numerical Analysis*. Buildings, 2025. **15**(13): p. 2235.
636. Zhang, N., et al., *Challenges of Earth Pressure Balance Tunnelling in Weathered Granite With Boulders*. Proceedings of the Institution of Civil Engineers - Geotechnical Engineering, 2020. **174**(4): p. 372-389.
637. Fantini, P., et al., *EPB Tunnelling in Long, Highly Sensitive Sections – The Experience of Grand Paris Line 14 Underpassing the Orly Airport*. Geomechanics and Tunnelling, 2023. **16**(1): p. 81-87.
638. Velasco-Gómez, J., et al., *Use of the Gyrotheodolite in Underground Networks of Long High-Speed Railway Tunnels*. Survey Review, 2016. **48**(350): p. 329-337.
639. Kim, S.H., *Simulation of Shield TBM Tunneling in Soft Ground by Laboratory Model Test*. Journal of Korean Tunnelling and Underground Space Association, 2013. **15**(5): p. 483.
640. Jun, G. and D. Kim, *A Study on Key Factors of Ground Surface Settlement Due to Shield TBM Excavation Using 3-Dimension Numerical Analysis*. Journal of Korean Tunnelling and Underground Space Association, 2015. **17**(3): p. 305-317.
641. Meng, Q., F. Qu, and S. Li, *Experimental Investigation on Viscoplastic Parameters of Conditioned Sands in Earth Pressure Balance Shield Tunneling*. Journal of Mechanical Science and Technology, 2011. **25**(9): p. 2259-2266.
642. Zhao, S., K. Peng, and J. Bai, *Pressure-Dependent Meso-Scale Evolution of Dispersed Foam and Field-Application Clogging Mitigation Strategies for EPB Shield Tunnelling Through Highly Cohesive Clay*. Materials, 2025. **18**(12): p. 2716.
643. Chen, Z., et al., *Experimental Study and Field Validation on Soil Clogging of EPB Shields in Completely Decomposed Granite*. Tunnelling and Underground Space Technology, 2022. **120**: p. 104300.

644. Wu, L. and F. Qu, *Discrete Element Simulation of Mechanical Characteristic of Conditioned Sands in Earth Pressure Balance Shield Tunneling*. Journal of Central South University of Technology, 2009. **16**(6): p. 1028-1033.

645. Bavasso, I., et al., *A Rapid Experimental Procedure to Assess Environmental Compatibility of Conditioning Mixtures Used in TBM-EPB Technology*. Applied Sciences, 2020. **10**(12): p. 4138.

646. Cheng, C., et al., *Experimental Study on the Soil Conditioning Materials for EPB Shield Tunneling in Silty Sand*. Advances in Civil Engineering, 2020. **2020**(1).

647. Zhang, Z. and T. Yin, *A Coupled CFD–DEM Simulation of Slurry Infiltration and Filter Cake Formation During Slurry Shield Tunneling*. Infrastructures, 2018. **3**(2): p. 15.

648. Yin, X., et al., *A Column System for Modeling Bentonite Slurry Infiltration in Sands*. Journal of Zhejiang University Science A, 2016. **17**(10): p. 818-827.

649. Qin, S. and K.K.P.M. Kannangara, *Effects of Silt Clay Content on Rheology and Filter Cake Characteristics of Water–bentonite Slurry in Slurry Shield Tunneling*. Physics of Fluids, 2025. **37**(7).

650. Zheng, D., A. Bezuijen, and G.D. Emidio, *A New Model for Predicting the Marsh Funnel Test*. International Journal of Geomechanics, 2021. **21**(2).

651. Min, F., et al., *Test Study on Airtight Capability of Filter Cakes for Slurry Shield and Its Application in a Case*. Advances in Materials Science and Engineering, 2014. **2014**: p. 1-8.

652. Li, Y., et al., *Unstable Collapse Analysis of SPBS in Saturated Sand Based on Model Experiment*. 2024.

653. Liu, J. and K.J. Hanley, *Simulation of Non-Spherical Slurry Particle Infiltration in a Sand Column*. Iop Conference Series Earth and Environmental Science, 2025. **1480**(1): p. 012029.

654. Bai, Y., et al., *Experimental Study on the Characteristics and Formation Mechanism of Dynamic Filter Cake for Slurry Shield Tunneling*. Minerals, 2022. **12**(3): p. 331.

655. Liu, X. and D. Yuan, *An in-Situ Slurry Fracturing Test for Slurry Shield Tunneling*. Journal of Zhejiang University Science A, 2014. **15**(7): p. 465-481.

656. Wang, S., et al., *Permeability Stability Calculation Model of Foam-conditioned Soil Based on the Permeability Constant*. International Journal for Numerical and Analytical Methods in Geomechanics, 2020. **45**(4): p. 540-559.

657. Spagnoli, G., M. Feinendegen, and M. Ziegler, *Assessing the Clogging Potential of Clay/Additivemixtures by Cone Pull-out Tests*. Geomechanics and Tunnelling, 2019. **12**(4): p. 362-371.

658. Nong, X., et al., *Factorial Experiments of Soil Conditioning for Earth Pressure Balance Shield Tunnelling in Water-Rich Gravel Sand and Conditioning Effects' Prediction Based on Particle Swarm Optimization–Relevance Vector Machine Algorithm*. Buildings, 2024. **14**(9): p. 2800.

659. Taherkhani, H., et al., *Investigating the Role of Bentonite on the Conditioning of Muck During Tunneling Through Marly Limestone With EPB Machines*. Canadian Geotechnical Journal, 2025. **62**: p. 1-18.

660. Wang, Z., W.-C. Cheng, and Y. Wang, *Quantitative Evaluation of Ground Movements Caused by Grouting During Shield Tunnelling in Clay*. Advances in Civil Engineering, 2019. **2019**(1).

661. Farrell, R., *Tunnelling and Compensation Grouting at Bond Street, London.* Proceedings of the Institution of Civil Engineers - Geotechnical Engineering, 2015. **168**(6): p. 471-482.
662. Ling, J., X. Yang, and Z. Wu, *A Prediction Method for TBM Cutterhead Dynamic Tunneling Performance Under Typical Composite Geological Conditions.* Mechanical Engineering Science, 2019. **1**(1).
663. Li, J., et al., *Tunnel Boring Machine Cutterhead Crack Propagation Life Prediction With Time Integration Method.* Advances in Mechanical Engineering, 2019. **11**(6): p. 168781401985345.
664. Lin, L., et al., *Geological Adaptive Cutterhead Selection for EPB Shield Based on BP Neural Network.* Applied Mechanics and Materials, 2014. **607**: p. 118-123.
665. Wen, S., H.Z. Liu, and L. Zhao, *Risk Analysis on the Accident of TBM's Cutterhead Jamming Caused by Collapse of Tunnel.* Advanced Materials Research, 2012. **446-449**: p. 2246-2250.
666. Li, X., et al., *Research on Dynamic Models and Performances of Shield Tunnel Boring Machine Cutterhead Driving System.* Advances in Mechanical Engineering, 2013. **5**.
667. Liu, X., et al., *Geological Adaptive Intelligent Control of Earth Pressure Balance Shield Machine Based on Deep Reinforcement Learning.* Artificial Intelligence Review, 2024. **57**(8).
668. Xiao, J., et al., *An Innovative Design of Mega Shaft Boring Machine (SBM) Cutterhead Based on TRIZ and AD Theory.* Advances in Mechanical Engineering, 2023. **15**(2).
669. Wang, R. and L. Zhang, *K-Means-Based Heterogeneous Tunneling Data Analysis Method for Evaluating Rock Mass Parameters Along a TBM Tunnel.* Scientific Reports, 2023. **13**(1).
670. Li, X., et al., *Study on Linear Vibration Model of Shield TBM Cutterhead Driving System.* The International Journal of Acoustics and Vibration, 2014. **19**(2).
671. Xu, Z., et al., *Evaluation Method for Tunneling Stability of TBM Cutterhead.* World Journal of Engineering and Technology, 2019. **07**(02): p. 1-9.
672. Mitelman, A. and A. Urlainis, *Investigation of Transfer Learning for Tunnel Support Design.* Mathematics, 2023. **11**(7): p. 1623.
673. Kim, B.S., et al., *A Study on the Design and Construction of the IoT Smart Safety Management System in the Deep Shield TBM Tunnel.* Forum of Public Safety and Culture, 2024. **26**: p. 139-156.
674. Ahmad, A.J. and M.F. Ayob, *Flood Mitigation Management Practice: A Comparison Between Stormwater Management Road Tunnel (SMART) and Sponge City Methods.* Journal of Project Management Practice, 2023. **3**(2): p. 1-17.
675. Khoo, C.M., *Volume Loss Caused by Tunnelling in Kenny Hill Formation.* International Journal of Geomate, 2019. **16**(54).
676. Choi, C., *Evaluation of Waterproof and Durability of Hydrophilic Waterstop.* Journal of Korean Tunnelling and Underground Space Association, 2014. **16**(2): p. 117.
677. Eskesen, S.D., E. Paulatto, and J. Stubbs, *Construction Challenges for Urban Tunnelling – The Copenhagen Metro Circle Line.* Geomechanics and Tunnelling, 2014. **7**(3): p. 223-229.
678. Gwildis, U. and I. Sass, *Tool Wear Analysis of Pressurized Face TBM Drives in the Glacial Geology of the Pacific Northwest.* Geomechanics and Tunnelling, 2018. **11**(2): p. 150-156.

679. Park, N.-H., H.-J. Kim, and J.-H. Shin, *Behavior of Double Deck Tunnel Due to Feature Change and Variation of Ground Water Table.* Journal of Korean Tunnelling and Underground Space Association, 2016. **18**(6): p. 581-591.
680. Taherian, A., *TBM Tunneling Problems Through Subterranean Rivers: An Evaluation of Nowsud Tunnel in the Iran.* 2022.
681. Sebbeh-Newton, S., et al., *Towards TBM Automation: On-the-Fly Characterization and Classification of Ground Conditions Ahead of a TBM Using Data-Driven Approach.* Applied Sciences, 2021. **11**(3): p. 1060.
682. Yunkang, M., et al., *Safety and Environmental Impact Control of Cross Passage Construction in Soft Soil Strata Using Tunnel Boring Machine Method.* 2023.
683. Chuaqui, M., et al., *A Case Study: Two-Fluid Jet Grouting for Tunneling Application - Soil Stabilization and Permeability Reduction.* 2012: p. 868-879.
684. *News: Geomechanics and Tunnelling 5/2019.* Geomechanics and Tunnelling, 2019. **12**(5): p. 398-410.
685. Yu, J.H., *Analysis of Flood Mitigation Infrastructure in Kuala Lumpur, Malaysia.* Scholarly Review Journal, 2025. **Scholarly Debut**(Equinox 2025).
686. Font-Capó, J., et al., *Assessment of the Barrier Effect Caused by Underground Constructions on Porous Aquifers With Low Hydraulic Gradient: A Case Study of the Metro Construction in Barcelona, Spain.* Engineering Geology, 2015. **196**: p. 238-250.
687. Hylander, J., et al., *Ambulance Commanders' Reluctance to Enter Road Tunnels in Simulated Incidents and the Effects of a Tunnel-Specific E-Learning Course on Decision-Making: Web-Based Randomized Controlled Trial.* Jmir Formative Research, 2025. **9**: p. e58542.
688. Rajapaksha, R.M.N.U., et al., *Incidence Management System of the Healthcare Institutions for Disaster Management in Sri Lanka.* BMC Emergency Medicine, 2023. **23**(1).
689. Jahn, J.L.S. and A.E. Black, *A Model of Communicative and Hierarchical Foundations of High Reliability Organizing in Wildland Firefighting Teams.* Management Communication Quarterly, 2017. **31**(3): p. 356-379.
690. Zhang, X.Q., et al., *Distribution of Harmful Gas and Ventilation Length Design of High Altitude Long Gas Tunnel Construction: A Case Study.* Frontiers in Earth Science, 2023. **11**.
691. Li, X.J., et al., *The Prediction Model of Influence of Small Radius UTLT Curvature on Fire Site Visibility Based on Pyrosim Multiple Regression Analysis.* E3s Web of Conferences, 2019. **79**: p. 02013.
692. Felipe, J.J.d., et al., *Experimental Assessment of an Analytical Model of the Convective Heat Transfer Coefficient in a Mine Gallery.* Mining Metallurgy & Exploration, 2022. **39**(3): p. 969-981.
693. Lucia, T.M.C.D., L.C. Gandra, and R.N.C. Guedes, *Managing Leaf-cutting Ants: Peculiarities, Trends and Challenges.* Pest Management Science, 2013. **70**(1): p. 14-23.
694. Chen, Z., et al., *Numerical Simulation of Smoke Back-Layering Length With Different Smoke Exhaust Modes in Subway Tunnel.* Destech Transactions on Materials Science and Engineering, 2017(icmsea/mce).

695. Wang, H.Y., M.Z. Chen, and G.F. Liu, *Numerical Simulation of Airflow Organization Process During Double Highway Tunnel Fire*. Applied Mechanics and Materials, 2014. **501-504**: p. 2370-2376.
696. Han, X., et al., *Effect Analysis of Fans Activating Time on Smoke Control Mode for Road Tunnel Fire*. Research Journal of Applied Sciences Engineering and Technology, 2013. **5**(13): p. 3571-3575.
697. Cole, L.A., et al., *A Course on Terror Medicine: Content and Evaluations*. Prehospital and Disaster Medicine, 2016. **31**(1): p. 98-101.
698. Lü, H., et al., *Research on Natural Gas Pipeline Leakage and Ventilation Scheme in Tunnel*. 2017.
699. Lemaire, R.H. and K.G. Provan, *Managing Collaborative Effort: How Simmelian Ties Advance Public Sector Networks*. The American Review of Public Administration, 2017. **48**(5): p. 379-394.
700. Chen, R., et al., *An Effective Inverse Procedure for Identifying DEMParameters of Rock-Like Materials*. Mathematical Problems in Engineering, 2019. **2019**(1).
701. Hylander, J., et al., *Ambulance Commanders' Reluctance to Enter Road Tunnels in Simulated Incidents and the Effects of a Tunnel-Specific E-Learning Course on Decision-Making: Web-Based Randomized Controlled Trial (Preprint)*. 2024.
702. Lee, J., et al., *Development and Application of Smart Construction Objects and Management System for an Efficient and Cost-Effective Safety Management*. Buildings, 2023. **13**(6): p. 1383.
703. Zhao, W. and Z. Jiang, *Research on Occupational Health and Safety of Medical Staff Based on ISO 45001*. American Journal of Biochemistry and Biotechnology, 2020. **16**(3): p. 288-298.
704. Mali, P.S., et al., *Novel Predictive Control and Monitoring System Based on IoT for Evaluating Industrial Safety Measures*. International Journal of Electrical and Electronics Research, 2022. **10**(4): p. 1050-1057.
705. Kandil, M.E., *Efforts to Persevere a Risk Assessment/Integrity Assurance for Its Aged Hydrocarbon Transfer Pipelines*. 2001.
706. Mocellin, P., et al., *Experimental Methods in Chemical Engineering: Hazard and Operability Analysis— <scp>HAZOP</Scp>*. The Canadian Journal of Chemical Engineering, 2022. **100**(12): p. 3450-3469.
707. K, Z.N., et al., *Exploring the Contributory Factors of Confined Space Accidents Using Accident Investigation Reports and Semistructured Interviews*. Safety and Health at Work, 2019. **10**(3): p. 305-313.
708. Othman, I., et al., *Effectiveness of Safety Management in Oil and Gas Project*. Applied Mechanics and Materials, 2015. **815**: p. 429-433.
709. Srinivasan, R., et al., *Liquefied Natural Gas (LNG) Supply Chains: Recent Advances and Future Opportunities*. Industrial & Engineering Chemistry Research, 2024. **63**(15): p. 6481-6503.
710. Ouffroukh, L.A., et al., *Analysis of Risk and the Strengthening of the Safety Technical Barriers: Application of Skikda (Algeria) Oil Refining Complex*. World Journal of Engineering, 2018. **15**(1): p. 99-109.
711. Adekanmbi, A.O., et al., *Implementing Health and Safety Standards in Offshore Wind Farms*. World Journal of Advanced Research and Reviews, 2024. **21**(2): p. 1136-1148.

712. Kashwani, G., *Kashwani Risk Assessment: New Approach for Oil and Construction Industries*. International Journal of Geomate, 2018. **16**(52).
713. Silvianita, S., M.F. Khamidi, and V.J. Kurian, *Decision Making for Safety Assessment of Mobile Mooring System*. Jurnal Teknologi (Sciences & Engineering), 2013. **61**(3).
714. Tong, R., et al., *Process Safety Management in China: Progress and Performance Over the Last 10 years and Future Development*. Process Safety Progress, 2020. **39**(4).
715. Dong, H., et al., *Lessons Learned From Analyzing an Explosion at Shanghai SECCO Petrochemical Plant*. Process Safety Progress, 2019. **39**(1).
716. Wu, J., et al., *Safeguards Identification in Computer Aided HAZOP Study by Means of Multilevel Flow Modelling*. Proceedings of the Institution of Mechanical Engineers Part O Journal of Risk and Reliability, 2022. **237**(5): p. 922-946.
717. Wood, M., et al., *The AETL Methodology as a Potential Solution to Current Challenges Associated With the Development and Use of Acute Exposure Levels in Seveso II Applications*. Journal of Hazardous Materials, 2006. **133**(1-3): p. 8-15.
718. Rajkumar, S., *Safety Security and Risk Management - Aftermath Bhopal Disaster*. International Journal of Biosensors & Bioelectronics, 2017. **2**(6).
719. Konyukhov, D.S., *Analysis of Mechanized Tunneling Parameters to Determine the Overcutting Characteristics*. Mining Science and Technology (Russia), 2022. **7**(1): p. 49-56.
720. Jadhav, M., *Life Cycle Cost Analysis of NATM Using Ann*. International Journal of Scientific Research in Engineering and Management, 2025. **09**(06): p. 1-9.
721. Haynes, A.L., *Reduction of Sleep Deprivation and Fatigue in Mass Transit Rail Operators*. Workplace Health & Safety, 2017. **65**(8): p. 333-336.
722. Riedy, S.M., et al., *Fatigue and Short-Term Unplanned Absences Among Police Officers*. Policing an International Journal, 2020. **43**(3): p. 483-494.
723. Καζάρας, K., K. Kirytopoulos, and A. Rentizelas, *Introducing the STAMP Method in Road Tunnel Safety Assessment*. Safety Science, 2012. **50**(9): p. 1806-1817.
724. Dienagha, I.N. and F.O. Onyeke, *Human Factors and Safety Culture in High-Risk Energy Operations: Strategies to Minimize Human Error and Enhance Compliance*. International Journal of Research and Innovation in Social Science, 2025. **IX**(I): p. 1413-1429.
725. Prati, G. and L. Pietrantoni, *Attitudes to Teamwork and Safety Among Italian Surgeons and Operating Room Nurses*. Work a Journal of Prevention Assessment & Rehabilitation, 2014. **49**(4): p. 669-677.
726. Beitlich, P., *TeamSTEPPS Implementation in the LD/NICU Settings*. Nursing Management, 2015. **46**(6): p. 15-18.
727. Aaberg, O.R., et al., *A Human Factors Intervention in a Hospital - Evaluating the Outcome of a TeamSTEPPS Program in a Surgical Ward*. 2020.
728. Rahmawati, G.R., N. Desi Sandra Sari, and N. Khoiron, *Organizational and Human Factors Affecting Patient Safety Culture: A Literature Review*. International Journal of Contemporary Sciences (Ijcs), 2025. **3**(11): p. 459-482.
729. Zeeman, G.G., *Patient Safety in Obstetrics*. European Clinics in Obstetrics and Gynaecology, 2006. **2**(2): p. 51-61.
730. Aaberg, O.R., *Interprofessional Teamwork in Hospital Units: A Human Factors Approach to Patient Safety*. 2022.

731. Katiyar, A., *The Role of Human Factors in Maintaining Service Quality in Hospitals.* International Research Journal on Advanced Engineering and Management (Irjaem), 2024. **2**(04): p. 976-983.

732. Filho, S.Á., *Organizational Control for Safety: A Challenge in Socioeconomic and Natural Environment.* Process Safety Progress, 2023. **42**(S1).

733. Campbell, Y., *The Ethical Responsibility of Psychological Safety: Leadership at the Intersection of Safety Culture.* Healthcare Management Forum, 2025. **39**(1): p. 66-69.

734. Panchal, P., *AI-augmented Project Controls: Enhancing Predictability in Complex Tunnel Construction Environments.* World Journal of Advanced Research and Reviews, 2022. **13**(2): p. 649-668.

735. Goh, A.T.C., et al., *Numerical Study of the Effects of Groundwater Drawdown on Ground Settlement for Excavation in Residual Soils.* Acta Geotechnica, 2019. **15**(5): p. 1259-1272.

736. Pujades, E., et al., *Underground Excavations Below the Water Table by the Cut-and-Cover Method.* 2023.

737. Chong, E.E.M. and D.E.L. Ong, *Data-Driven Field Observational Method of a Contiguous Bored Pile Wall System Affected by Accidental Groundwater Drawdown.* Geosciences, 2020. **10**(7): p. 268.

738. Yang, K., et al., *Numerical Analysis of Diaphragm Wall Deformation and Surface Settlement Caused by Dewatering and Excavation at Center and End Positions in a Subway Foundation Pit.* Buildings, 2025. **15**(15): p. 2796.

739. Ochmański, M., *Long-Term Settlement Induced by EPB Tunnelling Studied With Numerical Simulations.* Annals of Warsaw University of Life Sciences – SGGW Land Reclamation, 2018. **50**(2): p. 139-157.

740. Zheng, G., et al., *Diaphragm Wall Deformation and Ground Settlement Caused by Dewatering Before Excavation in Strata With Leaky Aquifers.* Géotechnique, 2024. **74**(1): p. 1-17.

741. Rolando, L., et al., *Bioaugmentation With a Consortium of Bacterial Sodium Lauryl Ether Sulfate-Degraders for Remediation of Contaminated Soils.* Frontiers in Microbiology, 2021. **12**.

742. Oliveira, D.G.G.D.E. and G. Aguiar, *Estimativa Do Potencial De Clogging E Trabalhabilidade De Materiais Coesivos Para Escavações Com Tuneladora Epb.* Revista Brasileira De Geologia De Engenharia E Ambiental, 2021. **11**(1): p. 39-61.

743. Barenji, A.A., et al., *Effect of Gradation on the Permeability of Foam-Conditioned Soils in Mechanized Excavation.* Periodica Polytechnica Civil Engineering, 2025. **69**(2): p. 397-405.

744. Alnmr, A., et al., *Parametric Investigation of Interaction Between Soil-Surface Structure and Twin Tunnel Excavation: A Comprehensive 2D Numerical Study.* Infrastructures, 2023. **8**(8): p. 124.

745. Zumsteg, R., M. Plötze, and A.M. Puzrin, *Effects of Dispersing Foams and Polymers on the Mechanical Behaviour of Clay Pastes.* Géotechnique, 2013. **63**(11): p. 920-933.

746. Budach, C. and M. Thewes, *Erweiterte Einsatzbereiche Von EPB-Schilden.* Geotechnik, 2013. **36**(2): p. 96-103.

747. Jessen, J. and R. Cudmani, *Rate- And Time-Dependent Mechanical Behavior of Foam-Grouted Coarse-Grained Soils.* Journal of Geotechnical and Geoenvironmental Engineering, 2022. **148**(5).
748. Zhou, C., et al., *Ground Settlement of High-Permeability Sand Layer Induced by Shield Tunneling: A Case Study Under the Guidance of DBN.* Geofluids, 2020. **2020**: p. 1-26.
749. Feng, S., et al., *Behavior of Damaged TBM Tunnel Under MJS and Micro-Disturbance Grouting Treatment: A Case Study.* Archives of Civil Engineering, 2023.
750. Zhang, Z., et al., *Safety-Risk Assessment for TBM Construction of Hydraulic Tunnel Based on Fuzzy Evidence Reasoning.* Processes, 2022. **10**(12): p. 2597.
751. Liu, Z.C., *The Influence of Engineering Factors to Groundwater Recovery in Mountain Tunnel.* Applied Mechanics and Materials, 2013. **353-356**: p. 1490-1494.
752. Wang, Y. and R. Xu, *The Safety Analysis on External Water Pressure to the Deep Buried Hydraulic Tunnel Lining Based on Coupling Numerical Simulation.* Ce/Papers, 2025. **8**(2): p. 1605-1614.
753. Stein, E., J. Langford, and M. Kahlström, *Time Series Modelling: Applications for Groundwater Control in Urban Tunnelling.* Bulletin of Engineering Geology and the Environment, 2023. **82**(10).
754. Hua, T., et al., *Numerical Analysis of Grouting of Water-Enriched Karst Highway Tunnel Based on Critical Water-Enriched Height.* Processes, 2023. **11**(1): p. 149.
755. Grose, W.J. and L. Benton, *Discussion: Hull Wastewater Flow Transfer Tunnel: Tunnel Collapse and Causation Investigation.* Proceedings of the Institution of Civil Engineers - Geotechnical Engineering, 2006. **159**(2): p. 125-128.
756. Zhou, Y., et al., *Groundwater Discharge Limits of Mountain Tunnels Based on the Normal Growth of Typical Herbaceous Plants.* Applied Sciences, 2024. **14**(15): p. 6561.
757. Jiang, Z., et al., *Advanced Grouting Model and Influencing Factors Analysis of Tunnels With High Stress and Broken Surrounding Rock.* Water, 2022. **14**(4): p. 661.
758. Luo, G., et al., *Research on the Reuse of Discharged Soil From EBP Shield Tunnels in Synchronous Grouting Material.* Advances in Civil Engineering, 2022. **2022**(1).
759. Wu, W., et al., *Performance Analysis of Electro-Hydraulic Thrust System of TBM Based on Fuzzy PID Controller.* Energies, 2022. **15**(3): p. 959.
760. Schwela, D., *Guidelines for Environmental Noise Management in Developing Countries.* 2023.
761. Remolina Soto, M.S., et al., *Intelligent Classification of Urban Noise Sources Using TinyML: Towards Efficient Noise Management in Smart Cities.* Sensors, 2025. **25**(20): p. 6361.
762. Schou, C.P.E., A.L. Levengood, and D.A. Potvin, *Limited Effects of Traffic Noise on Behavioural Responses to Conspecific Mating Calls in the Eastern Sedge Frog Litoria Fallax.* Acta Ethologica, 2021. **24**(3): p. 217-226.
763. Emar, M., E.T. Smith, and T. Coats, *Background Noise in an Emergency Department: An Observational Study From Staff and Patient Perspectives.* 2022.
764. Beca, I.M. and M. Iliescu, *Developing an Innovative Urban Traffic Noise Monitoring System.* Advanced Engineering Forum, 2017. **21**: p. 551-556.
765. Álías, F., et al., *Aggregate Impact of Anomalous Noise Events on the WASN-Based Computation of Road Traffic Noise Levels in Urban and Suburban Environments.* Sensors, 2020. **20**(3): p. 609.

766. Nelson-Olivieri, J.R., et al., *It's Not Just Noise: The Consequences of Inequitable Noise for Urban Wildlife*. 2023.
767. Bar-Sinai, K.L., et al., *Embedding Acoustic Analysis in Landscape Architecture Design Processes: A Case Study of Munich Airport*. Buildings, 2023. **13**(1): p. 143.
768. Socoró, J.C., F. Álías, and R.M. Alsina-Pagès, *WASN-Based Spectro-Temporal Analysis and Clustering of Road Traffic Noise in Urban and Suburban Areas*. Applied Sciences, 2022. **12**(3): p. 981.
769. Higham, V., et al., *Traffic Noise Drives an Immediate Increase in Call Pitch in an Urban Frog*. Journal of Zoology, 2021. **313**(4): p. 307-315.
770. Manzano, J.V., et al., *The "Sound of Silence" in Granada During the COVID-19 Lockdown*. Noise Mapping, 2021. **8**(1): p. 16-31.
771. Chan, L.K.L. and A.K.L. Kwong, *Geotechnical Risks, Mitigation Measures and Performance of Tunnel Boring Machine (TBM) Tunnels for the Tuen Mun – Chek Lap Kok Link Project, Hong Kong*. Hkie Transactions, 2019. **26**(4): p. 175-189.
772. Guo, H., et al., *A Novel Simplified Physical Model Testing Method for Ground Settlement Induced by Shield Tunnel Excavation*. Buildings, 2025. **15**(5): p. 710.
773. Yunkang, M., et al., *Safety and Environmental Impact Control of Cross Passage Construction in Soft Soil Strata Using Tunnel Boring Machine Method*. Scientific Reports, 2023. **13**(1).
774. Wu, K., et al., *Mechanical Aspects of Construction of New TBM Tunnel Under Existing Structures*. Geotechnical and Geological Engineering, 2021. **39**(8): p. 5727-5741.
775. Aristaghes, P. and P. Autuori, *Calcul Des Tunnels Au Tunnelier*. Revue Française De Géotechnique, 2001(97): p. 31-40.
776. Wang, T. and J. Yang, *Numerical Simulation Study of the Surface Subsidence Caused by TBM Tunnel Excavation*. International Journal of Geological Resources and Geological Engineering, 2023. **1**(1).
777. Broere, W. and D. Festa, *Correlation Between the Kinematics of a Tunnel Boring Machine and the Observed Soil Displacements*. Tunnelling and Underground Space Technology, 2017. **70**: p. 125-147.
778. Ter-Martirosyan, A.Z., I.I. Mustakhimov, and I.A. Tikhoniuk, *Numerical Studies for the Application of the Methodology for Volume Loss of Cohesionless (Loose) Soils (VL,LSR) and the Additional Settlement (Smax) During Shield Tunneling*. Buildings, 2025. **15**(24): p. 4555.
779. Jones, B., *Low-Volume-Loss Tunnelling for London Ring Main Extension*. Proceedings of the Institution of Civil Engineers - Geotechnical Engineering, 2010. **163**(3): p. 167-185.
780. Shen, B., et al., *Use of Monitoring Data During Construction to Refine Cavern Design*. 2015: p. 397-408.
781. Ren, T., et al., *Numerical Simulation of Ground Surface Settlement of Underpass Building in Tunnel Boring Machine Double-Line Tunnels*. Frontiers in Earth Science, 2022. **10**.
782. Shirlaw, J.N., S.R. Doran, and B. Benjamin, *A Case Study of Two Tunnels Driven in the Singapore 'Boulder Bed' and in Grouted Coral Sands*. Geological Society London Engineering Geology Special Publications, 1988. **5**(1): p. 93-103.

783. Wong, T.C.W., T.O. Ishola, and S.T.M. Mak, *Integrated Use of GNSS and InSAR Techniques for Movement Monitoring Under Trunk Road T2 and Cha Kwo Ling Tunnel*. 2023.
784. Ayasrah, M.m., et al., *Prediction of Ground Settlement Induced by Slurry Shield Tunnelling in Granular Soils*. Civil Engineering Journal, 2020. **6**(12): p. 2273-2289.
785. Dehnavi, R.N., et al., *Evaluating the Efficiency of A New Cutterhead Design to Counteract the Drawbacks of Excavation Through Very Hard Rocks*. BHM Berg- Und Hüttenmännische Monatshefte, 2021. **166**(6): p. 301-308.
786. Zhao, S., et al., *Coal Mine Inclined Shaft Advanced Detection Method and Physical Model Test Based on Shield Cutterhead Moving Array Electrodes*. Energies, 2019. **12**(9): p. 1671.
787. Xiao-dong, W.U., et al., *Real-Time as-Built Tunnel Product Modeling and Visualization by Tracking Tunnel Boring Machines*. 2013.
788. Tabrizi, A.H.N., et al., *Evaluating the Effect of Tool Wear in Soft Soil Using New TBM Tunneling Simulator Device*. Journal of Testing and Evaluation, 2023. **51**(6): p. 3917-3932.
789. El-Kelesh, A.M., A.S. Shehata, and E.-S. El-Kasaby, *Impact of Ground Conditions on Slurry TBM Learning Phase Performance in Soft Ground Tunneling*. Innovative Infrastructure Solutions, 2025. **11**(1).
790. Wang, J., et al., *Numerical Response of Advance Support Structures in TBM Tunneling Through Altered Zones: A Case Study*. Buildings, 2025. **15**(4): p. 509.
791. Zhang, X., et al., *Coal Rock Breaking Simulation and Cutting Performance Analysis of Disc Cutters*. Tehnicki Vjesnik - Technical Gazette, 2021. **28**(5).
792. Park, M.-E., et al., *Measurement of TBM Disc Cutter Wear Using Eddy-Current Sensor in Different TBM Chamber Conditions: Insights From Laboratory Tests*. Sensors, 2025. **25**(7): p. 2045.
793. Tao, Y., et al., *The Vibration Characteristics of Ground of Rock Blasting in Silt-Rock Strata*. Shock and Vibration, 2021. **2021**(1).
794. Erharter, G.H., R. Goliasch, and T. Marcher, *On the Effect of Shield Friction in Hard Rock TBM Excavation*. Rock Mechanics and Rock Engineering, 2023. **56**(4): p. 3077-3092.
795. Zhang, C., et al., *Influence of Cutterhead System Structure on Load Distribution and Life of Main Bearing*. Journal of Vibroengineering, 2016. **18**(7): p. 4164-4177.
796. Sun, Z., et al., *A Method for Evaluating the Performance of Main Bearings of TBM Based on Entropy Weight–Grey Correlation Degree*. Sensors, 2025. **25**(15): p. 4715.
797. Liao, S., X. Jin, and Q. Jiao, *Identifying TBM Tunneling Defects Through Recognition of Ground Settlement Patterns*. 2012: p. 3179-3188.
798. Qu, X. and Y. Zhang, *Fault Diagnosis Method of Roadheader Bearing Based on VMD and Domain Adaptive Transfer Learning*. Sensors, 2023. **23**(11): p. 5134.
799. Leng, S., et al., *A Hybrid Data Mining Method for Tunnel Engineering Based on Real-Time Monitoring Data From Tunnel Boring Machines*. Ieee Access, 2020. **8**: p. 90430-90449.
800. Geng, T., et al., *Ultrasonic Monitoring of Lubricating Conditions of Hydrodynamic Bearing*. Journal of Physics Conference Series, 2011. **305**: p. 012065.
801. Filho, A.A.d.O., et al., *San Francisco HEP Headrace Tunnel – TBM Success Under the Presence of Pressurized Water in Faulty Deep Rock*. Geomechanics and Tunnelling, 2008. **1**(6): p. 552-557.

802. Rehman, H., et al., *Impact of Construction Method and Ground Composition on Headrace Tunnel Stability in the Neelum–Jhelum Hydroelectric Project: A Case Study Review From Pakistan.* Applied Sciences, 2021. **11**(4): p. 1655.
803. *Geomechanics and Tunnelling 5/2012: Proceedings of the 61th Geomechanics Colloquium – 50 Years NATM.* Geomechanics and Tunnelling, 2012. **5**(4).
804. Wang, H., et al., *Adaptive Modification of TBM Tunneling in Coal Mine Roadway and Disaster Control Technology for Complex Geological Conditions.* Processes, 2023. **11**(5): p. 1389.
805. Marco, A.D. and T. Narbaev, *Factors of Schedule and Cost Performance of Tunnel Construction Megaprojects.* The Open Civil Engineering Journal, 2021. **15**(1): p. 38-49.
806. Chung, D. and D.A. Hensher, *Risk Management in Public–Private Partnerships.* Australian Accounting Review, 2015. **25**(1): p. 13-27.
807. Chen, R., et al., *Thrust Force Requirement Prediction Using HOA-XGBoost for TBM Tunneling in Squeezing Ground.* Ce/Papers, 2025. **8**(2): p. 1439-1447.
808. Li, H., et al., *Profit Sharing Mechanism of Large EPC Project Considering the Behavior of Fairness Concern.* Advances in Civil Engineering, 2020. **2020**(1).
809. Huang, Y., et al., *Disk Cutter Wear Prediction of TBM Considering Sliding and Rolling Friction.* Materials Science-Poland, 2023. **41**(1): p. 42-56.
810. Hopland, A.O. and S.F. Kvamsdal, *On the Ranking of Critical Success Factors.* Journal of Facilities Management, 2018. **16**(1): p. 26-37.
811. Shehata, A.S., A.M. El-Kelesh, and E.-S. El-Kasaby, *Effects of Ground Conditions on Utilization Factor of Slurry TBMs in Soft Ground Tunneling.* Engineering Construction & Architectural Management, 2025.
812. Lussu, A., et al., *15 Km TBM Exploratory Tunnel Excavation in the Construction Lot H33 of the Brenner Base Tunnel.* Geomechanics and Tunnelling, 2019. **12**(5): p. 595-603.
813. Jee, W.W., *Modern High-Power TBM With Advanced Procurement System.* Tunnel and Underground Space, 2013. **23**(3): p. 161-168.
814. Schubert, P., K. Joham, and M. Bauer, *Optimisation of the TBM Drive at the Boßler Tunnel Using the Observation Method and an Incremental Design and Contract Development Process.* Geomechanics and Tunnelling, 2017. **10**(2): p. 123-134.
815. Okonkwo, C.S., et al., *Conceptual Model for Materials Readiness and Maintenance-Driven Supply Chain Performance.* International Journal of Multidisciplinary Research and Growth Evaluation, 2021. **2**(6): p. 584-594.
816. Dingba, G.T., *Engineering Reliability in Food Manufacturing: How Capital Projects Prevent Downtime and Protect the U.S. Food Supply Chain.* Research Journal in Civil Industrial and Mechanical Engineering, 2022. **1**(1): p. 119-141.
817. Kashkanov, A. and M. Moskaliuk, *Uncertainty and Its Impact on the Formationof Spare Parts Inventories to Ensuring the Efficiency of Transport Processes.* Journal of Mechanical Engineering and Transport, 2024: p. 74-80.
818. Dyachenko, V., *Increasing the Reliability of Machines and the Efficiency of Their Use in Transportation Processes in Agro-Industrial Production Using Service Logistics Methods.* Central Ukrainian Scientific Bulletin Technical Sciences, 2022. **1**(6(37)): p. 215-227.

819. Gonzaga, L.G.d.C.N., *Analysis of the Possibility of Reducing Spare Parts in the Maintenance Context 4.0*. Independent Journal of Management & Production, 2022. **13**(1): p. 350-363.
820. Alomar, I. and N. Diallo, *Managing Operational Efficiency and Reducing Aircraft Downtime by Optimization of Aircraft on-Ground (AOG) Processes for Air Operator*. 2025.
821. Erhueh, O.V., et al., *Implementing Additive Manufacturing in Energy Asset Management: Lessons for Reducing Spare Parts Footprint*. Engineering Science & Technology Journal, 2024. **5**(10): p. 3054-3093.
822. Muchiri, P., et al., *Development of Maintenance Function Performance Measurement Framework and Indicators*. International Journal of Production Economics, 2011. **131**(1): p. 295-302.
823. Ogunwole, O., et al., *Supply Chain Resilience Framework for Critical Infrastructure and Gas Processing Plants*. Shodhshauryam International Scientific Refereed Research, 2021: p. 444.
824. Podosenov, S.S., *Justification for the Need to Create an Automated System for Managing Airworthiness Maintenance With Elements of Artificial Intelligence*. Informatization and Communication, 2025. **2**.
825. Kosowatz, J., *Underground and Out of Sight*. Mechanical Engineering, 2021. **143**(2): p. 28-33.
826. Rahm, T., et al., *Evaluation of Disturbances in Mechanized Tunneling Using Process Simulation*. Computer-Aided Civil and Infrastructure Engineering, 2015. **31**(3): p. 176-192.
827. Terheijden, O.T., P.v. Gelder, and W. Broere, *Impact of Occupational Risk Prevention Measures During Process Disturbances in TBM Tunnelling*. Tunnelling and Underground Space Technology, 2022. **127**: p. 104576.
828. Breijs, A., B. Klaassens, and R. Babuška, *Automated Design Environment for Serial Industrial Manipulators*. Industrial Robot the International Journal of Robotics Research and Application, 2005. **32**(1): p. 32-34.
829. Kiliç, K., et al., *Soft Ground Micro TBM Jack Speed and Torque Prediction Using Machine Learning Models Through Operator Data and Micro TBM-log Data Synchronization*. Scientific Reports, 2024. **14**(1).
830. Lehmann, G., *Pushing Boundaries in Slurry Microtunneling: Innovations in Microtunnel Boring Machine Design for Hard Rock Conditions*. Civil Engineering Design, 2024. **6**(4): p. 110-116.
831. Steiner, W., *Erfahrungen Beim Tunnelbau in Durchlässigem Glazialen Lockergestein: Offener Kies*. Geomechanics and Tunnelling, 2023. **16**(1): p. 68-80.
832. Khoo, C.M. and T.A. Ooi, *Geotechnical Challenges and Innovations in Urban Underground Construction – The Klang Valley Mass Rapid Transit Project*. Geomechanics and Tunnelling, 2023. **16**(3): p. 243-262.
833. Klein, L.C., et al., *Rock Mass Classification in the Digital Era: A Transformation in Hard Rock Tunnelling*. Iop Conference Series Earth and Environmental Science, 2024. **1435**(1): p. 012005.
834. Sopko, J., *Forensics by Freezing*. 2017: p. 96-105.
835. Jia, L., et al., *Kinematics and Spatial Structure Analysis of TBM Gunite Robot Based on D–H Parameter Method*. Scientific Reports, 2024. **14**(1).

836. Bakker, K.J., et al., *Monitoring Pilot Projects Using Bored Tunnelling: The Second Heinenoord Tunnel and the Botlek Rail Tunnel.* Tunnelling and Underground Space Technology, 1999. **14**(2): p. 121-129.
837. Yang, X., et al., *Spine-Inspired Continuum Soft Exoskeleton for Stoop Lifting Assistance.* Ieee Robotics and Automation Letters, 2019. **4**(4): p. 4547-4554.
838. Koor, N., *What Lies Beneath? A Decade of Underground Construction in Hong Kong.* Quarterly Journal of Engineering Geology and Hydrogeology, 2018. **51**(3): p. 301-310.
839. Girmscheid, G. and C. Schexnayder, *Drill and Blast Tunneling Practices.* Practice Periodical on Structural Design and Construction, 2002. **7**(3): p. 125-133.
840. Çetin, D., S.G. Yıldirim, and G. Kılıç, *A New Approach for Determination of Adhesion Coefficient in Pipejacking Method by Experimental and Numerical Analysis: An Example From Istanbul.* Experimental Techniques, 2011. **37**(6): p. 49-65.
841. Crisan, O., *Part Two | Digital Twins Underground: The Paradigm Shift from Static Models to Living Systems*. 2025, Bently.
842. Lia, C., et al., *Geotechnical Challenges and Monitoring in Urban Tunnelling: The Florence High-speed Railway Project.* Geomechanics and Tunnelling, 2025. **18**(4): p. 294-311.
843. Cavagnet, S., L. Mancinelli, and I. Diallo, *Excavation of a Tunnel With Conventional Method in Urbanized Context – A Case Study From Line 15 South of Grand Paris Express.* Geomechanics and Tunnelling, 2023. **16**(6): p. 692-697.
844. Schneider, G.R., et al., *Toronto-York Spadina Subway Extension Tunnelling Under Schulich Building.* Canadian Journal of Civil Engineering, 2019. **46**(2): p. 87-103.
845. Zhang, J., G. Huang, and X.-m. Gou, *An Optimum Metamodel for Safety Control of Operational Subway Tunnel During Underpass Shield Tunneling.* Structural Control and Health Monitoring, 2018. **25**(8): p. e2195.
846. Huang, M., J. Ninić, and Q. Zhang, *BIM, Machine Learning and Computer Vision Techniques in Underground Construction: Current Status and Future Perspectives.* Tunnelling and Underground Space Technology, 2021. **108**: p. 103677.
847. Apoji, D., B. Sheil, and K. Soga, *Shaping the Future of Tunneling With Data and Emerging Technologies.* Data-Centric Engineering, 2023. **4**.
848. *Construction Monitoring of Soft Ground Tunnels: A Rational Handbook of Practices for Rapid Transit System Planners and Managers.* International Journal of Rock Mechanics and Mining Sciences & Geomechanics Abstracts, 1978. **15**(1): p. A13.
849. Bell, F.G., et al., *A Review of Ground Movements Due to Civil and Mining Engineering Operations.* Geological Society London Engineering Geology Special Publications, 1988. **5**(1): p. 3-31.
850. Cording, E.J., *Filling the Gaps With Pressurized Tunnel Boring Machines.* Geostrata Magazine, 2022. **26**(1): p. 12-15.
851. Wei, L., et al., *Web-Based Visualisation for Look-Ahead Ground Imaging in Tunnel Boring Machines.* Automation in Construction, 2019. **105**: p. 102830.
852. Wölflingseder, M., P.J. Unterlaß, and T. Marcher, *Rock Mass Anomaly Detection With a Variational Autoencoder in Tunnel Boring Machine Data.* Civil Engineering Design, 2025. **7**(4): p. 169-177.
853. Qu, J., et al., *Digital Twins in the Minerals Industry – A Comprehensive Review.* Mining Technology Transactions of the Institutions of Mining and Metallurgy, 2023. **132**(4): p. 267-289.

854. Comulada, M., et al., *TBM Excavation of São Paulo Metro Line 6 South in Heterogeneous Ground in Urban Area*. Geomechanics and Tunnelling, 2024. **17**(4): p. 306-311.
855. Kundu, S., et al., *Discovering the Gene-Brain-Behavior Link in Autism via Generative Machine Learning*. Science Advances, 2024. **10**(24).
856. Yu, X., et al., *Application of Artificial Intelligence in Rock Tunnel Engineering: A Survey on Where and How*. Expert Systems, 2025. **42**(7).
857. He, W., et al., *Finite Element Analysis and Fatigue Life Prediction of Tunnel Boring Machine Cutter Head*. Proceedings of the Institution of Mechanical Engineers Part C Journal of Mechanical Engineering Science, 2025. **239**(11): p. 4273-4285.
858. Elbaz, K., et al., *Prediction Model of Shield Performance During Tunneling via Incorporating Improved Particle Swarm Optimization Into ANFIS*. Ieee Access, 2020. **8**: p. 39659-39671.
859. Mei, Y., et al., *A Modular Design Method Based on TRIZ and AD and Its Application to Cutter Changing Robot*. Advances in Mechanical Engineering, 2021. **13**(7).

Index

A

B

C

D

E

F

G

H

I

J

L

M

N

O

P

Q

R

S

T

U

V

W

Y

Made in United States
Orlando, FL
04 November 2026

90204520R00409